ELECTRICAL WIRING
Commercial

BASED ON THE 2011
NATIONAL ELECTRICAL CODE®

ELECTRICAL WIRING
Commercial

14TH EDITION

RAY C. MULLIN AND
PHIL SIMMONS

DELMAR
CENGAGE Learning™   Australia Canada Mexico Singapore Spain United Kingdom United States

DELMAR
CENGAGE Learning™

Electrical Wiring, Commercial, 14th Edition
Ray C. Mullin, Phil Simmons

Vice President, Career and Professional Editorial: Dave Garza

Director of Learning Solutions: Sandy Clark

Acquisitions Editor: Stacy Masucci

Managing Editor: Larry Main

Senior Product Manager: John Fisher

Editorial Assistant: Andrea Timpano

Vice President, Career and Professional Marketing: Jennifer Baker

Marketing Director: Deborah Yarnell

Marketing Manager: Kathryn Hall

Associate Marketing Manager: Scott A. Chrysler

Production Director: Wendy Troeger

Production Manager: Mark Bernard

Content Project Manager: Barbara LeFleur

Senior Art Director: David Arsenault

Technology Project Manager: Christopher Catalina

Cover Images:
Electronic schematic: © 2011 iStockphoto/ Shane White
Lightbulb illustration: © 2011 Joseph Villanova
Light wall: © Paul Prescott/Shutterstock

For product information and technology assistance, contact us at
**Professional Group Cengage Learning Customer & Sales Support,
1-800-354-9706**
For permission to use material from this text or product,
submit all requests online at
cengage.com/permissions.
Further permissions questions can be e-mailed to
permissionrequest@cengage.com.

Library of Congress Control Number: 2010930585

ISBN-13: 978-1-4354-9829-7

ISBN-10: 1-4354-9829-1

Delmar
5 Maxwell Drive
Clifton Park, NY 12065-2919
USA

Cengage Learning is a leading provider of customized learning solutions with office locations around the globe, including Singapore, the United Kingdom, Australia, Mexico, Brazil and Japan. Locate your local office at: **international. cengage.com/region**

Cengage Learning products are represented in Canada by Nelson Education, Ltd.

For your lifelong learning solutions, visit **delmar.cengage.com**

Visit our corporate website at **cengage.com.**

Notice to the Reader

Publisher does not warrant or guarantee any of the products described herein or perform any independent analysis in connection with any of the product information contained herein. Publisher does not assume, and expressly disclaims, any obligation to obtain and include information other than that provided to it by the manufacturer. The reader is expressly warned to consider and adopt all safety precautions that might be indicated by the activities described herein and to avoid all potential hazards. By following the instructions contained herein, the reader willingly assumes all risks in connection with such instructions. The publisher makes no representations or warranties of any kind, including but not limited to, the warranties of fitness for particular purpose or merchantability, nor are any such representations implied with respect to the material set forth herein, and the publisher takes no responsibility with respect to such material. The publisher shall not be liable for any special, consequential, or exemplary damages resulting, in whole or part, from the readers' use of, or reliance upon, this material.

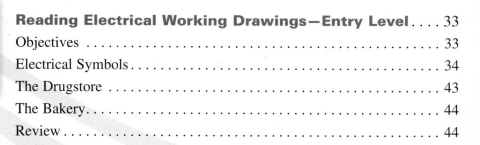

Contents

CHAPTER 1

CHAPTER 2

CHAPTER 3

CHAPTER 4

CHAPTER 5

CHAPTER 6

CHAPTER 7

CHAPTER 8

CHAPTER 9

CHAPTER 10

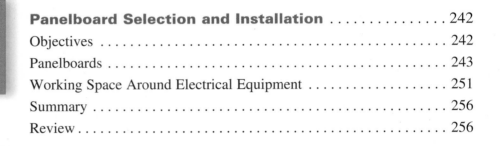

CHAPTER 15

CHAPTER 16

CHAPTER 17

CHAPTER 18

Short-Circuit Calculations and Coordination of Overcurrent Protective Devices

CHAPTER 19

CHAPTER 20

CHAPTER **21**

CHAPTER **22**

**Plans for a Commercial Building
(Attached to the Inside Back Cover)**

Preface

INTENDED USE AND LEVEL

Electrical Wiring—Commercial is intended for use in commercial wiring courses at two-year and four-year colleges, as well as in apprenticeship training programs. The text provides the basics of commercial wiring by offering insight into the planning of a typical commercial installation, carefully demonstrating how the load requirements are converted into branch circuits, then to feeders, and finally into the building's main electrical service. An accompanying set of plans at the back of the book allows the reader to step through the wiring process by applying concepts learned in each chapter to an actual commercial building, in order to understand and meet *Code* requirements set forth by the *National Electrical Code*.

SUBJECT AND APPROACH

The fourteenth edition of *Electrical Wiring—Commercial* is based on the 2011 *National Electrical Code*.* This new edition thoroughly and clearly explains the *NEC* changes that relate to typical commercial wiring.

The *National Electrical Code* is used as the basic standard for the layout and construction of electrical systems. To gain the greatest benefit from this text, the learner must use the *National Electrical Code* on a continuing basis.

State and local codes may contain modifications of the *National Electrical Code* to meet local requirements. The instructor is encouraged to furnish students with any variations from the *NEC* as they affect this commercial installation in a specific area.

This book takes the learner through the essential minimum requirements as set forth in the *National Electrical Code* for commercial installations. In addition to *Code* minimums, the reader will find such information above and beyond the minimum requirements.

The commercial electrician is required to work in three common situations: where the work is planned in advance, where there is no advance planning, and where repairs are needed. The first situation exists when the work is designed by a consulting engineer or by the electrical contractor as part of a design/build project. In this case, the electrician must know the installation procedures, be able to read and follow the plans for the project, be able to understand and interpret specifications, and must know the applicable *Code* requirements. The second situation occurs either during or after construction when changes or remodeling are required. The third situation arises any time after a system is installed. Whenever a problem occurs with an installation, the electrician must understand the operation of all equipment included in the installation in order to solve the problem. And as previously stated, all electrical work must be done in accordance with the *National Electrical Code* and any local electrical codes.

The electrician must understand that he or she is a part of a construction team with the goal of getting the project completed on time and within the budget. Cooperation and "pulling your load" are the keys to success. The general contractor and owner count on every trade and specialist to get the components on the job when they are needed and install them so as to keep the project moving ahead smoothly.

When the electrician is working on the initial installation or is modifying an existing installation, the circuit loads must be determined. Thorough explanations and numerous examples of calculating these loads help prepare the reader for similar problems on the job. The text and assignments make frequent reference to the Commercial Building drawings at the back of the book.

The electrical loads (lighting, outlets, equipment, appliances, etc.) were selected to provide the reader with experiences that he or she would encounter when wiring a typical commercial building. The authors also carry many calculations to a higher level of accuracy as compared to the accuracy required in many actual job situations. This is done to demonstrate the correct method according to the *National Electrical Code*. Then, if the reader and/or the instructor wish to back off from this level, based upon installation requirements, it can be done intelligently.

FEATURES

- **Safety** is emphasized throughout the book and fully covered in the first chapter. Special considerations in working with electricity, such as how to avoid arc flash, as well as guidelines for safe practices, provide readers with an overview of what dangers are to be expected on the job.

- **Commercial Building Drawings** are included in the back of the book, offering readers the opportunity to apply the concepts that they have learned in each chapter as they step through the wiring process. A description of working drawings and an explanation of symbols can be found in the first chapter.

- *National Electrical Code* references are integrated throughout the chapters, familiarizing readers with the requirements of the *Code* and

including explanations of the wiring applications. Revisions to the *NEC* between the 2008 and 2011 editions are carefully identified.

- **Review Questions** at the end of each chapter allow readers to test what they have learned in each chapter and to target any sections that require further review.

NEW TO THIS EDITION

Every *Code* reference in the fourteenth edition of *Electrical Wiring—Commercial* is the result of comparing each and every past *Code* reference with the 2011 *NEC*. As always, the authors review all comments submitted by instructors from across the country, making corrections and additions to the text as suggested. The input from current users of the text ensures that what is covered is what electricians need to know.

- Emphasis is given to making the wiring of the Commercial Building conform to energy saving Standards. In other words, the wiring and connected loads in *Electrical Wiring—Commercial* are "Green."

- One of the most far-reaching new requirements in the *2011 National Electrical Code* is that the grounded circuit conductor must be either brought to every switch or provisions made to easily get it there if needed. This new requirement has been addressed in *Electrical Wiring—Commercial*, with all wiring diagrams revised accordingly. This means that more 3-wire and 4-wire cable, and possibly larger boxes, will be required.

- The lighting layout for the Commercial Building has been totally redesigned to conform to energy savings requirements and today's desired lighting in typical commercial lighting applications. Many LED luminaires have been added. Exit lighting has been added. Battery-backup luminaires provide illumination in the event of a power outage. Thanks go to the lighting engineers at Cooper Lighting for their valuable contributions in bringing the lighting design up to current trends.

- A major change in the emergency lighting feature via the use of battery backup luminaires is found in all of the Commercial Building occupancies.

- An elevator has been added to make the Commercial Building ADA compliant.

- Replacing receptacles with proper devices is required in existing facilities where weather-resistant receptacles are now required. The replacement must be of the weather-resistant type.

- The electrical design and the architectural design have been revised to meet the current trend to "GO GREEN." Each tenant now has individual heating and air conditioning.

- All of the wiring diagrams have been updated to show the latest system of electrical symbols. This is based on the NECA/NEIS Standards. Thanks are extended to the National Electrical Contractors Association for permission to use the NECA/NEIS electrical symbols in this text.

- Major revisions of many diagrams and figures have been made to improve the clarity and ease of understanding the *Code* requirements.

- Many new full-color illustrations have been added.

- A fire alarm system has been added to the building to comply with applicable building and electrical codes.

- Exit lighting has been added to meet egress requirements.

- The service equipment and metering equipment have been relocated to the outdoors to save valuable indoor space.

- Because of concern and confusion over how to cope with the heat generated in confined areas such as circular raceways like EMT, RMC, and IMC, the *2011 National Electrical Code* calls attention to the difference between circular raceways and other wireways such as surface metal raceways, auxiliary gutters, and the like installed on rooftops. The new term *circular raceways* has been addressed in this text.

- A new chapter was added for Commercial Utility Interactive Photoelectric System.

- All *National Electrical Code* references have been updated to the 2011 *NEC*. Changes between the 2008 and 2011 editions of the *NEC* are marked with these symbols: ▶◀

- A major rewrite and formatting of the more difficult text, tables, and calculations was done, making them easier to understand.

- Expanded the list of the NECA/ANSI installation standards. These standards are not *Code* requirements, but rather are installation standards an electrician should follow in order to make an installation in a workmanlike manner.

- Greatly expanded list of construction terms to the Glossary to help the student better understand and interpret plans and specifications.

- Updated the electrical symbols to NECA 100-2006, *Electrical Symbols for Electrical Construction Drawings*. These are reprinted with permission of National Electrical Contractors Association.

SUPPLEMENTS

The following supplemental materials are available with the text.

- **Instructor's Manual**—contains the answers to all review questions included in the book. (Order #1-4354-9827-5)

- **Instructor Resource**—components include a PowerPoint presentation, a computerized test bank, an image library database of images from the text, and an electronic copy of the Instructor's Manual. (Order #1-4354-9828-3)

To access additional course materials including CourseMate, please visit www.cengagebrain.com. At the CengageBrain.com home page, search for the ISBN of your title (from the back cover of your book) using the search box at the top of the page. This will take you to the product page where these resources can be found.

ABOUT THE AUTHORS

This text was prepared by Ray C. Mullin and Phil Simmons.

Mr. Mullin is a former electrical circuit instructor for the Electrical Trades, Wisconsin Schools of Vocational, Technical and Adult Education. A former member of the International Brotherhood of Electrical Workers, Mr. Mullin is presently an honorary member of the International Association of Electrical Inspectors, an honorary member of the Institute of Electrical and Electronic Engineers, and an honorary member of the National Fire Protection Association, Electrical Section. He served on Code Making Panel 4 for the *National Electrical Code*, NFPA-70 for the National Fire Protection Association.

Mr. Mullin completed his apprenticeship training and has worked as a journeyman and supervisor. He has taught both day and night electrical apprentice and journeyman courses and has conducted engineering seminars. Mr. Mullin has contributed to and assisted other authors in their writing of texts and articles relating to overcurrent protection and conductor withstand ratings. He has had many articles relating to overcurrent protection published in various trade magazines.

Mr. Mullin attended the University of Wisconsin, Colorado State University, and Milwaukee School of Engineering.

He served on the Executive Board of the Western Section, International Association of Electrical Inspectors. He also served on their National Electrical Code Committee and on their Code Clearing Committee. He is past chairman of the Electrical Commission in his hometown.

Mr. Mullin has conducted many technical Code workshops and seminars at state chapter and section meetings of the International Association of Electrical Inspectors and served on their Code panels.

Mr. Mullin is past Director, Technical Liaison, and Code Coordinator for a large electrical manufacturer and contributed to their technical publications.

Phil Simmons is self-employed as Simmons Electrical Services. Services provided include consulting on the *National Electrical Code* and other Codes, writing, editing, illustrating, and producing technical publications and inspection of complex electrical installations. He develops training programs related to electrical codes and safety and has been a presenter on these subjects at numerous seminars and conferences for Universities, the NFPA, IAEI, Department of Defense, and private clients. Phil also provides plan review of electrical construction documents. He has consulted on several lawsuits concerning electrical shocks, burn injuries, and electrocutions.

Mr. Simmons is the co-author and illustrator of *Electrical Wiring—Residential* (17th edition) and *Electrical Wiring—Commercial* (14th edition) and author and illustrator of *Electrical Grounding and Bonding* (3rd edition), all published by Delmar, Cengage Learning. While at IAEI, Phil was author and illustrator of several books, including the *Soares Book on Grounding of Electrical Systems* (five editions), *Analysis of the NEC* (three editions), and *Electrical Systems in One- and Two-Family Dwellings* (three editions). Phil wrote and illustrated the National Electrical Installation Standard (NEIS) on *Standard on Types AC and MC Cables* for the National Electrical Contractors Association.

Phil presently serves NFPA on Code Making Panel-5 of the *National Electrical Code* Committee (grounding and bonding). He previously served on the *NEC* CMP-1 (*Articles 90, 100,* and *110*), as Chair of CMP-19 (articles on agricultural buildings and mobile and manufactured buildings), and member of CMP-17 (health care facilities). He served six years on the NFPA Standards Council, as NFPA Electrical Section President and on the NEC Technical Correlating Committee.

Phil began his electrical career in a light-industrial plant. He is a master electrician and was owner and manager of Simmons Electric Inc., an electrical contracting company. He is also a licensed journeyman electrician in Montana and Alaska. Phil passed the certification examinations for Electrical Inspector General, Electrical Plan Review, and Electrical Inspector One- and Two-Family.

He previously served as Chief Electrical Inspector for the State of Washington from 1984 to 1990 as well as an Electrical Inspector Supervisor, Electrical Plans Examiner and field Electrical Inspector. While employed with the State, Phil performed plan review and inspection of health care facilities including hospitals, nursing homes, and boarding homes.

Phil served the International Association of Electrical Inspectors as Executive Director from 1990 to 1995 and as Education, Codes, and Standards Coordinator from 1995 through June 1999. He was International President in 1987 and has served on local and regional committees.

He served Underwriters Laboratories as a Corporate Member and on the Electrical Council from 1985 to 2000 and served on the UL Board of Directors from 1991 to 1995. Phil is a retired member of the International Brotherhood of Electrical Workers.

IMPORTANT NOTE

Most all of the work in updating this edition of *Electrical Wiring—Commercial* was completed after all normal steps of revising the *National Electrical Code NFPA 70* were taken but before the actual issuance and publication of the 2011 edition of the *NEC*.

Every effort has been made to be technically correct, but there is always the possibility of typo-graphical errors or appeals made to the NFPA Board of Directors after the normal review process that could result in reversal of previous actions taken in processing the *NEC*.

If changes in the *NEC* do occur after the printing of this text, these changes will be incorporated in the next printing.

The National Fire Protection Association has a standard procedure to introduce changes between *NEC Code* cycles after the actual *NEC* is printed. These are called "Tentative Interim Amendments," or TIAs. TIAs and a list of errata items can be downloaded from the NFPA Web site, http://www.nfpa.org, to make your copy of the *Code* current.

Acknowledgments

The authors and Publisher wish to thank the following reviewers of this and past editions for their contributions:

Warren DeJardin
Northeast Wisconsin Technical College
Green Bay, WI

Charlie Eldridge
Indianapolis Power and Light, Retired
Indianapolis, IN

Greg Fletcher
Kennebec Valley Technical College
Fairfield, ME

David Gehlauf
Tri-County Vocational School
Glouster, OH

Wesley Gubitz
Cape Fear Community College
Castle Hayne, NC

Fred Johnson
Champlain Valley Tech
Plattsburgh, NY

Thomas Lockett
Vatterott College
Quincy, IL

Gary Reiman
Dunwoody Institute
Minneapolis, MN

Lester Wiggins
Savannah Technical College,
Savannah, GA

Ray Mullin and Phil Simmons want to join in thanking our friends and colleagues who over the years have provided us with many helpful comments and suggestions. These individuals are in the electrical industry, members of Code Making Panels, electrical inspectors, instructors, training directors, electricians, and electrical contractors. To name but a few . . . Madeline Borthick, David Dini, John Dyer, Paul Dobrowsky, Joe Ellwanger, Ken Haden, David Hittinger, Michael Johnston, Robert Kosky, Richard Loyd, Neil Matthes, Bill Neitzel, Don Offerdahl, Cliff Redinger, Jeff Sargent, Gordon Stewart, Clarence Tibbs, Charlie Trout, Ray Weber, J.D. White, Lester Wiggins, David Williams, and the electrical staff at NFPA headquarters. We apologize for any names we might have missed.

The authors gratefully acknowledge the contribution of the chapter on Commercial Utility Interactive Photovoltaic Systems by Pete Jackson, electrical inspector for the City of Bakersfield, CA.

Applicable tables and section references are reprinted with permission from NFPA 70-2011, *National Electrical Code*, copyright © 2011, National Fire Protection Association, Quincy, MA 02169. This reprinted material is not the complete and official position of the NFPA on the referenced subject, which is represented only by the standard in its entirety.

RECOGNITION:

Past revisions of *Electrical Wiring—Commercial* were done by the combined efforts of Robert R. Smith and Ray C. Mullin. Each was responsible for specific chapters.

Sincere thanks go to Bob, who contributed so much to past revisions of this text. Unfortunately, Mr. Smith has passed away after an extended illness.

Without skipping a beat, Ray welcomes Phil Simmons to this edition and future editions of *Electrical Wiring—Commercial*. Phil has an outstanding background in the *National Electrical Code*. He is recognized as one of the country's top *Code* instructors. Read more about Phil in the "About the Authors" section in the front material of this text.

Commercial Building Plans and Specifications

OBJECTIVES

After studying this chapter, you should be able to

- understand how the *NEC*® is organized and how the articles relate.

- understand the process for updating the *NEC*.

- understand the basic safety rules for working on electrical systems.

- define the project requirements from the contract documents.

- demonstrate the application of building plans and specifications.

- locate specific information on the building plans.

- obtain information from industry-related organizations.

- apply and interchange International System of Units (SI) and English measurements.

INTRODUCTION TO *ELECTRICAL WIRING—COMMERCIAL*

You are about to explore the electrical systems of a typical small commercial building. You may find this text to be challenging depending on your experience and understanding in installing electrical equipment and wiring, along with the many requirements in the *National Electrical Code®* (*NEC*). This text and the *NEC* may seem easy at times and difficult at other times. As you study, you may want to have both this text and the *NEC* open, as well as to spread out the drawings located in the back of this text.

As you study this text, you will learn about safety, wiring methods, electrical equipment, luminaires, and *NEC* requirements. You will be using the text, the set of Plans, and the *NEC*.

The set of Plans and Specifications in the back of this text will be used and referred to continually. The objective is to correlate what you are learning to a typical commercial installation. Tying the text, the Plans, and the *NEC* together is much preferred over merely presenting a stand-alone *NEC* rule without associating the rule to a real situation. The Plans are those of an actual building, not just a convenient drawing to illustrate a specific *Code* rule. For all intents and purposes, upon completing this text you will have wired a commercial building.

Throughout this text, red triangles ▶◀ indicate a change in the 2011 edition of the *NEC* from the previous 2008 edition.

Let us begin with probably the most important part of learning the electrical trade: *safety*.

SAFETY IN THE WORKPLACE

Before we get started on our venture into the wiring of a typical commercial building, let us talk about safety.

Electricity can be dangerous! Occupational Safety and Health Act (OSHA) regulations and *National Fire Protection Association (NFPA) 70E*, the standard on *Electrical Safety in the Workplace*, consider working on energized equipment over 50 volts to represent a shock hazard. Working on electrical equipment with the power turned on can result in death or serious injury, either as a direct result of electricity flowing through a person or from an indirect secondary reaction, such as falling off a ladder or falling into the moving parts of equipment. Dropping a metal tool onto live parts or allowing metal shavings from a drilling operation to fall onto live parts of electrical equipment generally results in an arc flash and arc blast, which can cause deadly burns and other physical trauma. The heat of an electrical arc flash has been determined to be as much as 35,000°F (19,427°C), or about four times hotter than the sun. Pressures developed during an arc blast can blow a person across the room and inflict serious injuries. Dirt, debris, and moisture can also set the stage for catastrophic equipment failures and personal injury. Neatness and cleanliness as well as wearing appropriate personal protective equipment and following all safety procedures in the workplace are a must.

The *OSHA Code of Federal Regulations (CFR) Number 29, Subpart S,* in paragraph *1910.332,* discusses the training needed for those who face the risk of electrical injury. Proper training means *trained in and familiar with the safety-related work practices required by paragraphs 1910.331 through 1910.335.* Numerous texts are available that cover the OSHA requirements in great detail.

NFPA 70E, the *Standard for Electrical Safety in the Workplace*, should be used in conjunction with the OSHA regulations to develop and implement an effective electrical safety program for the workplace. The OSHA rules state what is required. *NFPA 70E* provides information on how to comply with the OSHA rules and achieve a safe workplace. The *NEC* defines a *qualified* person as *One who has skills and knowledge related to the construction and operation of the electrical equipment and installations and has received safety training to recognize and avoid the hazards involved.** Merely telling someone or being told to be careful does not meet the definition of proper training and does not make the person qualified. This definition emphasizes not only recognizing hazards but also avoiding them. Avoiding an electrical accident is usually worth much more than "an ounce of prevention" and certainly much more than "a pound of cure." Shock and burn injuries usually happen so fast that it is difficult

---

*Reprinted with permission from NFPA 70-2011.

to react quickly enough to get out of harm's way. Yet these injuries can almost instantly change your life in a very negative manner. Most often, victims are never the same as before the incident.

Important requirements for training are found in *NFPA 70E Article 110*. The training required is specifically related to the tasks to be performed. The rule includes a statement: *A person can be considered qualified with respect to certain equipment and methods but still be unqualified for others.*** If you have not been trained to do a specific task, you are considered unqualified in that area. The training given and received is required to be documented. If you are ever in an electrical accident that is reportable to OSHA, one of the first things they will ask for is a copy of your personnel record to prove you were trained for the task you were performing. Employers are required to provide appropriate training and safety procedures. Employees are required to comply with the safety training they have received.

Only qualified persons are permitted to work on or near exposed energized equipment. To become qualified, a person must

- have the skill and training necessary to distinguish exposed live parts from other parts of electrical equipment;
- be able to determine the voltage of exposed live parts; and
- be trained in the use of special precautionary techniques, such as personal protective equipment, insulations, shielding material, and insulated tools.

An unqualified person is defined in *Article 100* of *NFPA 70E* as *A person who is not a qualified person.* Although this seems simplistic, a person can be considered *qualified* for performing some tasks and yet be *unqualified* for other tasks. Training and experience make the difference.

Subpart S, paragraph 1910.333, of the OSHA regulations, requires that safety-related work practices be employed to prevent electrical shock or other injuries resulting from either direct or indirect electrical contact. Live parts to which an employee may be exposed are required to be de-energized

before the employee works on or near them, unless the employer can demonstrate that de-energizing introduces additional or increased hazards.

Working on "live" equipment is acceptable only if there would be a greater hazard if the system were de-energized. Examples of this would be life-support systems, some alarm systems, certain ventilation systems in hazardous locations, and the power for critical illumination circuits. Working on energized equipment requires properly insulated tools, proper flame-resistant clothing, rubber gloves, protective shields and goggles, and in some cases insulating blankets. As previously stated, OSHA regulations allow only qualified personnel to work on or near electrical circuits or equipment that has not been de-energized. The OSHA regulations provide rules regarding lockout and tagout (LOTO) to make sure that the electrical equipment being worked on will not inadvertently be turned on while someone is working on the supposedly dead equipment. As the OSHA regulations state, *A lock and a tag shall be placed on each disconnecting means used to de-energize circuits and equipment. . . .*

Some electricians' contractual agreements require that, as a safety measure, two or more qualified electricians must work together when working on energized circuits. They do not allow untrained apprentices to work on live equipment but do allow apprentices to stand back and observe.

According to *NFPA 70E, Standard for Electrical Safety in the Workplace,* circuits and conductors are not considered to be in an electrically safe work condition until all sources of energy are removed, the disconnecting means is under lockout/tagout, and the absence of voltage is verified by an approved voltage tester. Proper personal protective equipment (PPE) is required to be worn while testing equipment for absence of voltage during the lockout/tagout procedure. Equipment is considered to be energized until proven otherwise.

Safety cannot be compromised. Accidents do not always happen to the other person.

Follow this rule: *Turn off* and *lock off* the power, and then properly tag the disconnect with a description as to exactly what that particular disconnect serves.

**Reprinted with permission from NFPA 70E-2009.

Arc Flash and Arc Blast

An electrician should not get too complacent when working on electrical equipment. A major short circuit or ground fault at the main service panel, or at the meter cabinet or base, can deliver a lot of energy. On large electrical installations, an arc flash can generate temperatures of 35,000°F (19,427°C). This is hotter than the surface of the sun. This amount of heat will instantly melt copper, aluminum, and steel. For example, copper expands 64,000 times its original volume when it changes state from a solid to a vapor. The resulting violent blast will blow hot particles of metal and hot gases all over, often resulting in personal injury, fatality, or fire. An arc blast, Figure 1-1, also creates a tremendous air-pressure wave that can cause serious ear damage or memory loss due to the concussion. Damage to internal organs such as collapsed lungs is common in these events, Figure 1-2. The blast might blow the victim away from the arc source, causing additional injuries from falls.

A series of tests were performed to determine the temperatures and pressures an arc flash and blast event would produce. The results of test No. 4 are shown in Figure 1-2. For this test, the voltage was 480, with approximately 22,600 amperes short-circuit current available. The overcurrent device on the supply side of the fault was an electronic power circuit breaker set to open in 12 cycles.

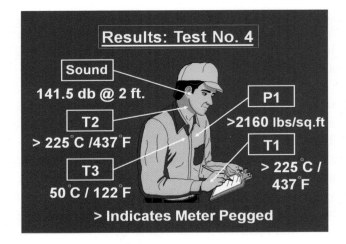

FIGURE 1-2   Results of arc-flash and arc blast event. (*Courtesy Cooper Bussmann*)

The significance of the test results are as follows:

- Sound: hearing protection is required for sound levels above 85 db.

- T1: the temperature on exposed skin exceeded 437°F (225°C). No doubt third- or fourth-degree burns will occur almost instantly at that temperature.

- T2: Same comment as for T1.

- T3: The temperature probe was on the skin under the clothing. A significant reduction in temperature resulted in no injury to the skin.

- P1: The pressure on the chest exceeded 2160 lbs per square ft. At these pressures, damage to internal organs is very likely.

An electrician should not be fooled by the size of the service. Commercial installations often have very large services, providing a potential for a significant arc flash and arc blast hazard. The Commercial Building discussed in this text is served by three 350-kcmil (thousand circular mils) copper Type XHHW-2 conductors that total 930 amperes in the 75°C column of *NEC Table 310.15(B)(16)*.

It is important that an arc flash hazard analysis is performed to determine the arc flash protection boundary as well as the level of personal protective equipment that people are required to wear within the arc flash boundary. New requirements are contained within *NFPA 70E* for posting the level of incident energy that is available or the rating of

FIGURE 1-1   Arc-flash and arc blast event. (*Courtesy Cooper Bussmann*)

flame-resistant personal protective equipment that must be worn. This posting is so important because the incident energy can vary from one piece of equipment to another. With this information, electricians can select the personal protective equipment that is needed so they are protected in the hazardous area. In some cases, the arc flash study may dictate that an arc flash suit with a beekeeper-type hood be used. The best approach continues to be that work on the equipment only be done while it is de-energized.

Electricians seem to feel out of harm's way when working on small electrical systems and seem to be more cautious when working on commercial and industrial electrical systems. Do not allow yourself to get complacent. Nearly half of the electrocutions each year are from 120-volt systems. A very small current is all that is needed when flowing through our nervous system to cause paralysis so the electrician is "hung up." This occurs when the external voltage flowing through the electrician's nervous system prevents him or her from releasing contact with the energized part.

Now consider the effects of 60-hertz (60-cycle) ac currents on humans in the study by Charles F. Dalziel ("Dangerous Electric Currents," reported in *AIEE Transactions*, Vol. 65 [1946], p. 579; discussion, p. 1123), presented in Table 1-1. (The effects vary depending on whether the current is dc or ac and on the frequency if it is ac.)

Mr. Dalziel is credited with inventing the ground-fault circuit interrupter (GFCI), which, for the Class A personnel protection version, is required to open between 4 and 6 mA of current flow. This device has saved countless lives and reduced the electric shock injuries.

A fault at a small main service panel, however, can be just as dangerous as a fault on a large service. The available fault current at the main service disconnect, for all practical purposes, is determined by the kilovolt-ampere (kVA) rating and impedance of the transformer. Other major limiting factors for fault current are the size, type, and length of the service-entrance conductors. If you want to learn more, visit Bussmann's Web site, http://www.bussmann.com. There you will find an easy-to-use computer program for making arc-flash and fault-current calculations. An Excel spreadsheet designed to simplify fault-current calculations is available for free download at http://www.mikeholt.com.

Short-circuit calculations are discussed in Chapter 18 of this text.

Electricians should not be fooled into thinking that if they cause a fault on the load side of the main disconnect, the main breaker will trip off and protect them from an arc flash. An arc flash will release the energy that the system is capable of delivering, for as long as it takes the main circuit breaker or fuse to open. How much current (energy) the main breaker will let through depends on the available fault current and the breaker or fuse opening time. A joke in the electrical trade is that a power company will sell power to you a little at a time—or all in one huge arc blast.

Although not required for dwelling units, *NEC 110.16* specifies that electrical equipment, such as switchboards, panelboards, industrial control panels, meter socket enclosures, and motor control

TABLE 1-1		
Current in milliampere (mA), 60 hertz.		
Effect(s)	**Men**	**Women**
Slight sensation on hand	0.4	0.3
Perception of "let go" threshold, median	1.1	0.7
Shock, not painful, and no loss of muscular control	1.8	1.2
Painful shock—muscular control lost by half of participants	9	6
Painful shock—"let go" threshold, median	16	10.5
Painful and severe shock—breathing difficult, muscular control lost	23	15

⚠ DANGER

Arc flash and shock hazard.

Follow ALL requirements in NFPA 70E for safe work practices and for Personal Protective Equipment.

FIGURE 1-3   Typical pressure-sensitive arc-flash and shock-hazard label to be affixed to electrical equipment as required by *NEC 110.16*. (*Delmar/Cengage Learning*)

centers in commercial and industrial installations that are likely to be worked on while energized, shall be field-marked to warn qualified personnel of potential arc-flash hazard. This marking must be clearly visible to any qualified persons who might have to work on the equipment.

Figure 1-3 is an example of a commercially available label.

Electrical Power Tools on the Job

On the job, you will be using portable electric power tools. Although many of these tools are battery powered, several larger tools like threaders, benders, bandsaws, and pullers are powered by 120 or 240 volts. The electrical supply on construction sites is often in the form of temporary power, covered by *Article 590* of the *NEC*.

NEC 590.6(A) and *(B)* require that ground-fault circuit-interrupter protection for personnel be provided for all 125-volt, single-phase, 15-, 20-, and 30-ampere receptacle outlets irrespective of whether they are a part of the permanent wiring of the building or structure, or are supplied from a portable generator. The issue is whether these power sources supply receptacle outlets that are in use by the worker. An exception is provided for receptacle outlets of other ratings that have protection by the testing protocols of an assured equipment grounding conductor–testing program.

Because the GFCI requirement is sometimes ignored or defeated on job sites, as part of your tool collection you should carry and use a portable ground-fault circuit interrupter (GFCI) of the type shown in Figure 1-4—an inexpensive investment that will protect you against possible electrocution. Remember, "The future is not in the hands of fate, but in ourselves."

Refer to Chapter 5 for details on how GFCIs operate and where they must be installed.

FIGURE 1-4   Two types of portable plug-in cord sets that have built-in GFCI protection. (*Courtesy Hubbell Inc.*)

Stand to One Side!

A good suggestion is that when turning a standard disconnect switch on, *do not* stand in front of the switch. Instead, stand to one side. For example, if the handle of the switch is on the right, then stand to the right of the switch, using your left hand to operate the handle of the switch, and turn your head away from the switch. That way, if an arc flash occurs when you turn the disconnect switch on, you will not be standing in front of the switch. You will not have the switch's door fly into your face. There is a good chance that the molten metal particles resulting from an arc flash will fly past you.

More Information

You will find more information about the hazards of an arc flash and when conditions call for personal protective equipment (PPE) in *Electrical Safety in the Workplace NFPA 70E* and in Chapter 13 of this text.

Information on the content of warning signs can be found in the ANSI Standard Z535.4, *Product Safety Signs and Labels.*

Just about every major manufacturer of electrical equipment has arc-flash information on its Web site.

Where Do We Go Now?

With safety the utmost concern in our minds, let us begin our venture on the wiring of a typical commercial building.

COMMERCIAL BUILDING SPECIFICATIONS

When a building project contract is awarded, the electrical contractor is given the plans and specifications for the building. These two contract documents govern the construction of the building. It is very important that the electrical contractor and the electricians employed by the contractor to perform the electrical construction follow the specifications exactly. The electrical contractor will be held responsible for any deviations from the specifications and may be required to correct such deviations or variations at personal expense. Thus, it is important that any changes or deviations be verified—in writing. Avoid verbal change orders.

It is suggested that the electrician assigned to a new project first read the specifications carefully. These documents provide the detailed information that will simplify the task of studying the plans. The specifications are usually prepared in book form and may consist of a few pages to as many as several hundred pages covering all phases of the construction. This text presents in detail only that portion of the specifications that directly involves the electrician; however, summaries of the other specification sections are presented to acquaint the electrician with the full scope of the document.

The specification is a book of rules governing all of the material to be used and the work to be performed on a construction project. The specification is usually divided into several sections.

General Clauses and Conditions

The first section of the specification, *General Clauses and Conditions*, deals with the legal requirements of the project. The index to this section may include the following headings:

Notice to Bidders
Schedule of Drawings
Instructions to Bidders
Proposal
Agreement
General Conditions

Some of these items will impact the electrician on the job, and others will be of primary concern to the electrical contractor. The following paragraphs give a brief, general description of each item.

Notice to Bidders. This item is of value to the contractor and his estimator only. The notice describes the project, its location, the time and place of the bid opening, and where and how the plans and specifications can be obtained.

Schedule of Drawings. The schedule is a list, by number and title, of all of the drawings related to the project. The contractor, estimator, and electrician will each use this schedule prior to preparing the bid for the job: the contractor to determine whether all the drawings required are at hand, the estimator to do a takeoff and to formulate a bid, and the electrician to determine whether all of the drawings necessary to do the installation are available.

Instructions to Bidders. This section provides the contractor with a brief description of the project, its location, and how the job is to be bid (lump sum, one contract, or separate contracts for the various construction trades, such as plumbing, heating, electrical, and general). In addition, bidders are told where and how the plans and specifications can be obtained prior to the preparation of the bid, how to make out the proposal form, where and when to deliver the proposal, the amount of any bid deposits required, any performance bonds required, and bidders' qualifications. Other specific instructions may be given, depending on the particular job.

Proposal. The proposal is a form that is filled out by the contractor and submitted at the proper time and place. The proposal is the contractor's bid on a project. The form is the legal instrument that binds the contractor to the owner if (1) the contractor completes the proposal properly, (2) the contractor does not forfeit the bid bond, (3) the owner accepts the proposal, and (4) the owner signs the agreement. Generally, only the contractor will be using this section.

The proposal may show that alternate bids were requested by the owner. In this case, the electrician on the job should study the proposal and consult with the contractor to learn which of the alternate bids has been accepted in order to determine the extent of the work to be completed.

On occasion, the proposal may include a specified time for the completion of the project. This information is important to the electrician on the job because the work must be scheduled to meet the completion date.

Agreement. The agreement is the legal binding portion of the proposal. The contractor and the owner sign the agreement, and the result is a legal contract. After the agreement is signed, both parties are bound by the terms and conditions given in the specification.

General Conditions. The following items are normally included under the *General Conditions* heading of the *General Clauses and Conditions*. A brief description is presented for each item:

- General Note: Includes the general conditions as part of the contract documents.
- Definition: As used in the contract documents, defines the owner, contractor, architect, engineer, and other people and objects involved in the project.
- Contract Documents: Lists the documents involved in the contract, including plans, specifications, and agreement.
- Insurance: Specifies the insurance a contractor must carry on all employees and on the materials involved in the project.
- Workmanship and Materials: Specifies that the work must be done by skilled workers and that the materials must be new and of good quality.
- Substitutions: Specifies that materials used must be as indicated or that equivalent materials must be shown to have the required properties.
- Shop Drawings: Identifies the drawings that must be submitted by the contractor to show how the specific pieces of equipment are to be installed.
- Payments: Specifies the method of paying the contractor during the construction.
- Coordination of Work: Specifies that each contractor on the job must cooperate with every other contractor to ensure that the final product is complete and functional.
- Correction Work: Describes how work must be corrected, at no cost to the owner, if any part of the job is installed improperly by the contractor.
- Guarantee: Guarantees the work for a certain length of time, usually one year.

- Compliance with All Laws and Regulations: Specifies that the contractor will perform all work in accordance with all required laws, ordinances, and codes, such as the *NEC* and city codes.

- Others: Sections added as necessary by the owner, architect, and engineer when the complexity of the job and other circumstances require them. None of the items listed in the General Conditions has precedence over another item in terms of its effect on the contractor or the electrician on the job. The electrician must study each of the items before taking a position and assuming responsibilities with respect to the job.

Supplementary General Conditions

The second main section of the specifications is titled *Supplementary General Conditions*. These conditions usually are more specific than the *General Conditions*. Although the *General Conditions* can be applied to any job or project in almost any location with little change, the *Supplementary General Conditions* are rewritten for each project. The following list covers the items normally specified by the *Supplementary General Conditions*:

- The contractor must instruct all crews to exercise caution while digging as any utilities damaged during the digging must be replaced or repaired by the contractor responsible. Most communities have a Call Before You Dig program. Services are available to locate and mark all underground utilities in the area such as power, water, sewer, telephone, and cable systems.

- The contractor must verify the existing conditions and measurements.

- The contractor must employ qualified individuals to lay out the work site accurately. A registered land surveyor or engineer may be part of the crew responsible for the layout work.

- Job offices are to be maintained as specified on the site by the contractor; this office space may include space for owner representatives.

- The contractor may be required to provide telephones at the project site for use by the architect, engineer, subcontractor, or owner.

- Temporary toilet facilities and water are to be provided by the contractor for the construction personnel.

- The contractor must supply an electrical service of a specified capacity to provide temporary light and power at the site.

- The contractor may have to supply a specified type of temporary heating to keep the temperature at the level specified for the structure.

- According to the terms of the guarantee, the contractor agrees to replace faulty equipment and correct construction errors for a period of one year.

The previous listing is by no means a complete catalog of all of the items that can be included in the section *Supplementary General Conditions*.

Other names may be applied to the *Supplementary General Conditions* section, including *Special Conditions* and *Special Requirements*. Regardless of the name used, these sections contain the same types of information. All sections of the specifications must be read and studied by all of the construction trades involved. In other words, the electrician must study the heating, plumbing, ventilating, air-conditioning, and general construction specifications to determine whether there is any equipment furnished by the other trades and where the contract specifies that such equipment is to be installed and wired by the electrical contractor. The electrician must also study the general construction specifications because the roughing in of the electrical system will depend on the types of construction that will be encountered in the building.

This overview of the *General Conditions* and *Supplementary General Conditions* of a specification is intended to show the student that construction workers on the job are affected by parts of the specification other than the part designated for their particular trade.

Contractor Specification

In addition to the sections of the specification that apply to all contractors, separate sections exist for each of the contractors, such as the general contractor who constructs the building proper, the plumbing contractor who installs the water and sewage systems, the heating and air-conditioning contractor, and the electrical contractor. The contract documents usually do not make one contractor responsible for work specified in another section of the specifications. However, it is always considered good practice for each contractor to be aware of how he or she is involved in each of the other contracts in the total job.

⎓⎓⎓⎓ WORKING DRAWINGS

The construction plans for a building are often called *blueprints*. This term is a carryover from the days when the plans were blue with white lines. Today, a majority of the plans used have black lines on white because this combination is considered easier to read and more economical to produce. The terms *plans* and *working drawings* will be commonly used in this text.

A set of 10 plan sheets is included at the back of the text, showing the general and electrical portions of the work specified:

- *Sheet A1—Architectural Floor Plan; Basement*

- *Sheet A2—Architectural Floor Plan; First Floor*

- *Sheet A3—Architectural Floor Plan; Second Floor:* The architectural floor plans give the wall and partition details for the building. These sheets are drawn to scale (dimensioned); the electrician can find exact locations by referring to these sheets. The electrician should also check the plans for the materials used in the general construction, as these will affect when and how the system will be installed.

- *Sheet A4—Site Plan, East and West Elevations:* The plot plan shows the location of the commercial building and gives needed elevations. The east elevation is the street view of the building, and the west elevation is the back of the

building. The index lists the content of all the plan sheets.

- *Sheet A5—Elevations; North and South:* The electrician must study the elevation dimensions, which are given in feet and hundredths of a foot above sea level. For example, the finished second floor, which is shown at 218.33 ft, is 218 ft 4 in. above sea level.

- *Sheet A6—Building Cross-Sections*

- *Sheet E1—Basement Electrical Plan*

- *Sheet E2—First Floor Electrical Plan*

- *Sheet E3—Second Floor Electrical Plan*

- *Sheet E4—Panelboard & Service Schedules, One-Line Diagram*

These sheets show the detailed electrical work on an outline of the building. Because dimensions usually are not shown on the electrical plans, the electrician must consult the other sheets for this information. It is recommended that the electrician refer frequently to the other plan sheets to ensure that the electrical installation does not conflict with the work of the other construction trades.

To assist the electrician in recognizing components used by other construction trades, the following illustrations are included: Figure 1-5A and Figure 1-5B, Architectural drafting symbols; Figure 1-6, Standard symbols for plumbing, piping, and valves; and Figure 1-7, Sheet metal ductwork symbols. A comprehensive list of electrical symbols typically used for commercial building wiring is included in Chapter 2 of this text. Electrical symbols that are important for reference are included in Appendix H of this text. However, the electrician should be aware that variations of these symbols may be used, and the specification and/or plans for a specific project must always be consulted.

Submitting Plans

In most communities, building plans and specifications must be submitted to a building department for review prior to the issuance of a construction permit. *NEC 215.5* states, *If required by the authority having jurisdiction, a diagram showing feeder*

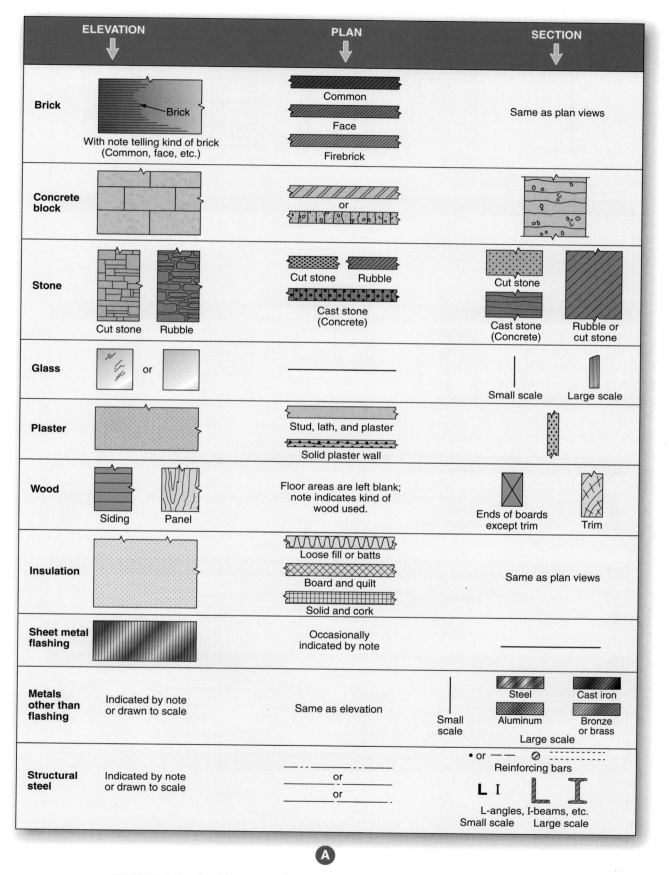

FIGURE 1-5 Architectural drafting symbols. (*Delmar/Cengage Learning*)

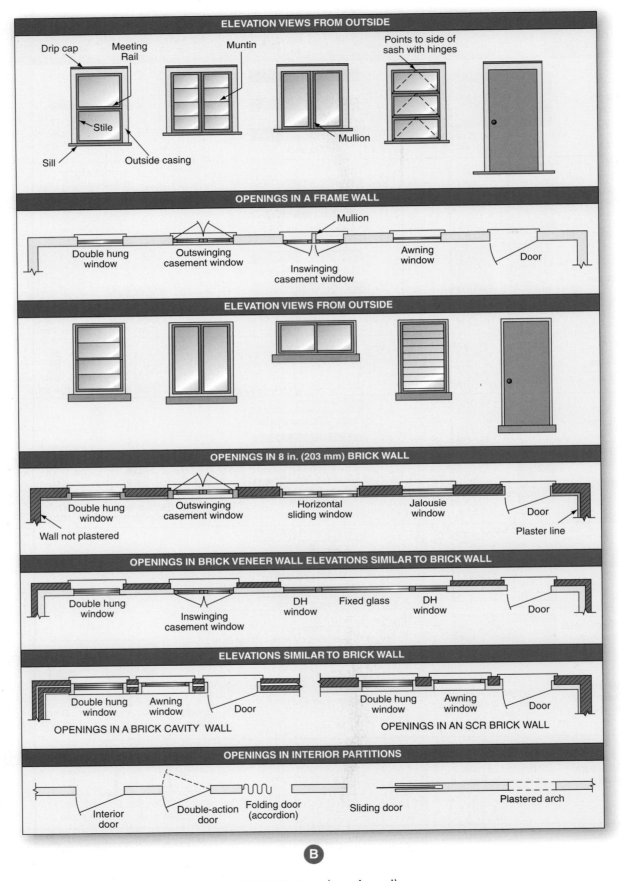

B

FIGURE 1-5 (continued)

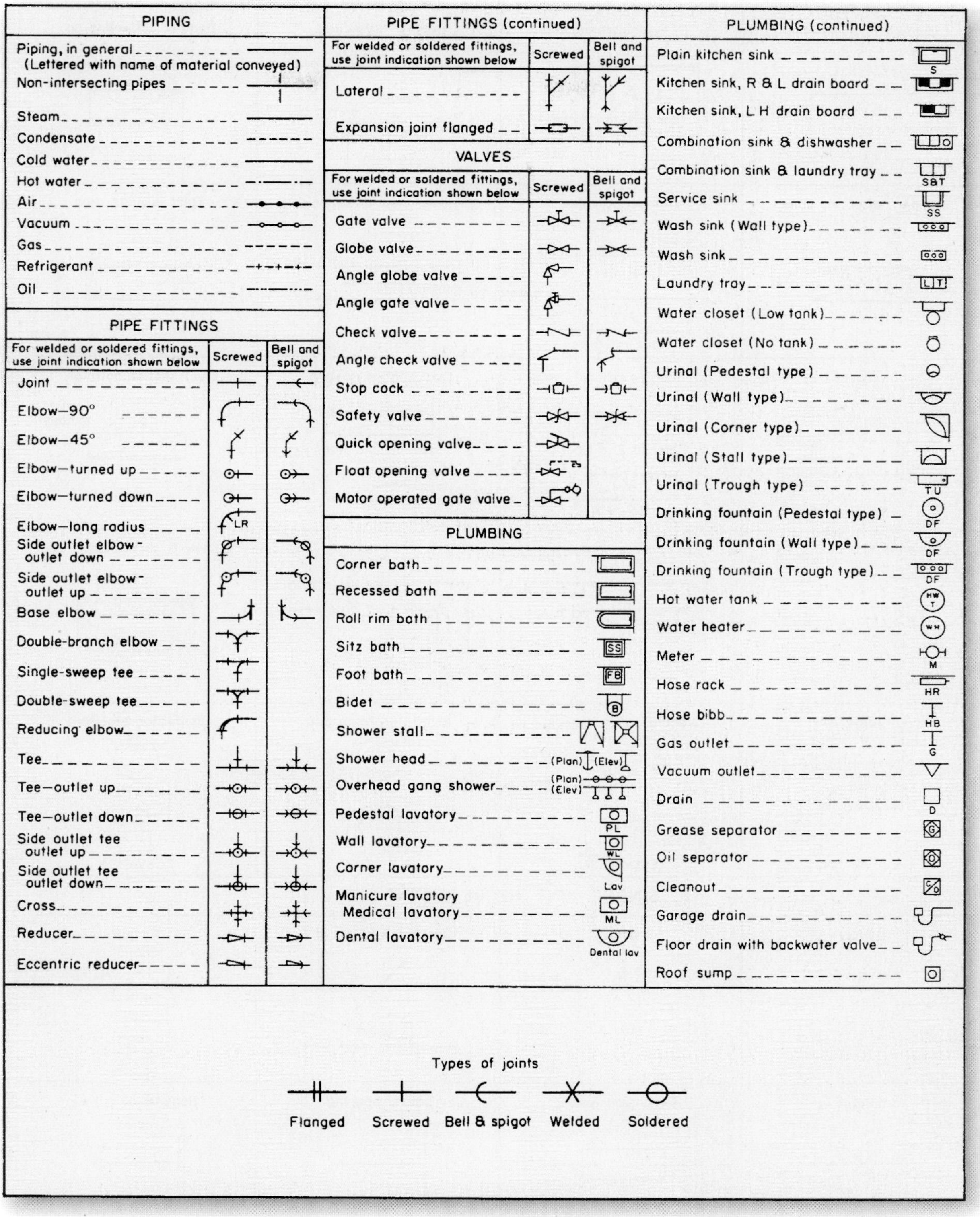

PIPING		
Piping, in general _ _ _ _ _ _ _ _ _ _		
(Lettered with name of material conveyed)		
Non-intersecting pipes _ _ _ _ _ _		
Steam _ _ _ _ _ _ _ _ _ _ _ _ _		
Condensate _ _ _ _ _ _ _ _ _ _		
Cold water _ _ _ _ _ _ _ _ _ _		
Hot water _ _ _ _ _ _ _ _ _ _		
Air _ _ _ _ _ _ _ _ _ _ _ _ _		
Vacuum _ _ _ _ _ _ _ _ _ _ _		
Gas _ _ _ _ _ _ _ _ _ _ _ _ _		
Refrigerant _ _ _ _ _ _ _ _ _ _		
Oil _ _ _ _ _ _ _ _ _ _ _ _ _		

PIPE FITTINGS		
For welded or soldered fittings, use joint indication shown below	Screwed	Bell and spigot
Joint _ _ _ _ _ _ _ _ _ _ _		
Elbow—90° _ _ _ _ _ _		
Elbow—45° _ _ _ _ _		
Elbow—turned up _ _ _ _ _		
Elbow—turned down _ _ _ _		
Elbow—long radius _ _ _ _		
Side outlet elbow- outlet down		
Side outlet elbow- outlet up _ _ _ _ _ _ _		
Base elbow _ _ _ _ _ _ _		
Double-branch elbow _ _ _		
Single-sweep tee _ _ _ _ _		
Double-sweep tee _ _ _ _ _		
Reducing elbow _ _ _ _ _ _		
Tee _ _ _ _ _ _ _ _ _ _ _		
Tee—outlet up _ _ _ _ _ _		
Tee—outlet down _ _ _ _ _		
Side outlet tee outlet up _ _ _ _ _ _ _		
Side outlet tee outlet down _ _ _ _ _ _		
Cross _ _ _ _ _ _ _ _ _ _		
Reducer _ _ _ _ _ _ _ _ _		
Eccentric reducer _ _ _ _ _		

PIPE FITTINGS (continued)		
For welded or soldered fittings, use joint indication shown below	Screwed	Bell and spigot
Lateral _ _ _ _ _ _ _ _ _		
Expansion joint flanged _ _		

VALVES		
For welded or soldered fittings, use joint indication shown below	Screwed	Bell and spigot
Gate valve _ _ _ _ _ _ _		
Globe valve _ _ _ _ _ _ _		
Angle globe valve _ _ _ _ _		
Angle gate valve _ _ _ _ _		
Check valve _ _ _ _ _ _ _		
Angle check valve _ _ _ _		
Stop cock _ _ _ _ _ _ _ _		
Safety valve _ _ _ _ _ _ _		
Quick opening valve _ _ _		
Float opening valve _ _ _ _		
Motor operated gate valve _		

PLUMBING	
Corner bath _ _ _ _ _ _ _ _ _ _	
Recessed bath _ _ _ _ _ _ _ _ _	
Roll rim bath _ _ _ _ _ _ _ _ _	
Sitz bath _ _ _ _ _ _ _ _ _ _ _	SS
Foot bath _ _ _ _ _ _ _ _ _ _ _	FB
Bidet _ _ _ _ _ _ _ _ _ _ _ _ _	B
Shower stall _ _ _ _ _ _ _ _ _ _	
Shower head _ _ _ _ _ _ _ _ _	(Plan) (Elev)
Overhead gang shower _ _ _ _ _	(Plan) (Elev)
Pedestal lavatory _ _ _ _ _ _ _ _	PL
Wall lavatory _ _ _ _ _ _ _ _ _	WL
Corner lavatory _ _ _ _ _ _ _ _ _	Lav
Manicure lavatory Medical lavatory _ _ _ _ _ _ _	ML
Dental lavatory _ _ _ _ _ _ _ _ _	Dental lav

PLUMBING (continued)	
Plain kitchen sink _ _ _ _ _ _ _ _	S
Kitchen sink, R & L drain board _ _	
Kitchen sink, L H drain board _ _ _	
Combination sink & dishwasher _ _	
Combination sink & laundry tray _ _	S&T
Service sink _ _ _ _ _ _ _ _ _ _	SS
Wash sink (Wall type) _ _ _ _ _ _	
Wash sink _ _ _ _ _ _ _ _ _ _ _	
Laundry tray _ _ _ _ _ _ _ _ _ _	LT
Water closet (Low tank) _ _ _ _ _	
Water closet (No tank) _ _ _ _ _	
Urinal (Pedestal type) _ _ _ _ _ _	
Urinal (Wall type) _ _ _ _ _ _ _ _	
Urinal (Corner type) _ _ _ _ _ _ _	
Urinal (Stall type) _ _ _ _ _ _ _ _	
Urinal (Trough type) _ _ _ _ _ _	TU
Drinking fountain (Pedestal type) _	DF
Drinking fountain (Wall type) _ _ _	DF
Drinking fountain (Trough type) _ _	DF
Hot water tank _ _ _ _ _ _ _ _ _	HWT
Water heater _ _ _ _ _ _ _ _ _ _	WH
Meter _ _ _ _ _ _ _ _ _ _ _ _ _	M
Hose rack _ _ _ _ _ _ _ _ _ _ _	HR
Hose bibb _ _ _ _ _ _ _ _ _ _ _	HB
Gas outlet _ _ _ _ _ _ _ _ _ _ _	G
Vacuum outlet _ _ _ _ _ _ _ _ _	V
Drain _ _ _ _ _ _ _ _ _ _ _ _ _	D
Grease separator _ _ _ _ _ _ _ _	
Oil separator _ _ _ _ _ _ _ _ _ _	
Cleanout _ _ _ _ _ _ _ _ _ _ _ _	
Garage drain _ _ _ _ _ _ _ _ _ _	
Floor drain with backwater valve _ _	
Roof sump _ _ _ _ _ _ _ _ _ _ _	

Types of joints

Flanged   Screwed   Bell & spigot   Welded   Soldered

FIGURE 1-6   Standard symbols for plumbing, piping, and valves. (*Delmar/Cengage Learning*)

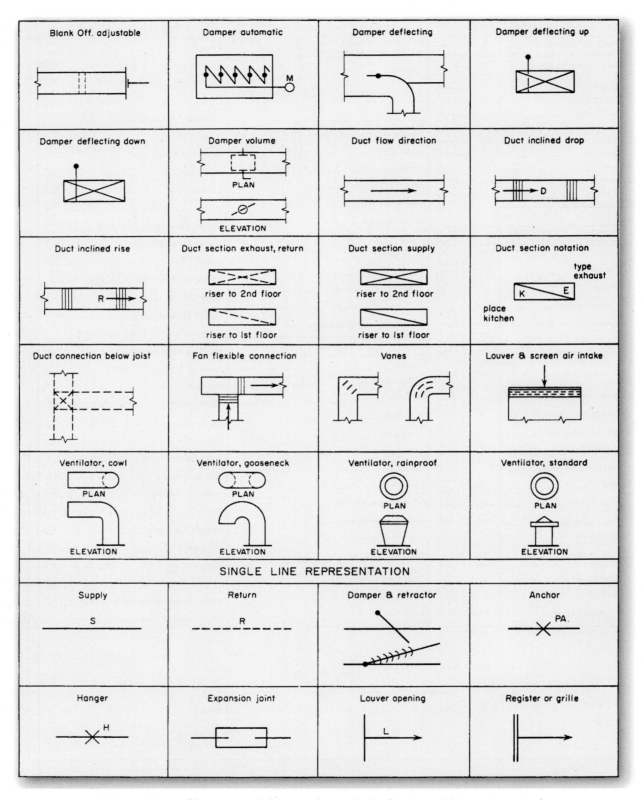

FIGURE 1-7 Sheet metal ductwork symbols. (*Delmar/Cengage Learning*)

details shall be provided prior to the installation of the feeders. Such a diagram shall show the area in square feet of the building or other structure supplied by each feeder, the total calculated load before applying demand factors, the demand factors used, the calculated load after applying demand factors, and the size and type of conductors to be used.

Construction Terms

As you will learn, *Electrical Wiring—Commercial* covers all aspects of typical commercial wiring. On construction sites, electricians work with others. Knowing construction terms and symbols is a key element to getting along with the other workers. A rather complete dictionary of construction terms can be found on http://www.constructionplace.com.

CODES AND ORGANIZATIONS

Many organizations such as cities and power companies develop electrical codes that they enforce within their areas of influence. These codes generally are concerned with the design and installation of electrical systems. It is important to verify which edition of the *NEC* has been adopted and is to be used as the basis for the local code. Some jurisdictions routinely adopt the latest edition of the *NEC* soon after it is published. Other jurisdictions may be operating on an edition of the *NEC* that is several years out of date. Consult these organizations before starting work on any project. The local codes may contain special requirements that apply to specific and particular installations. Additionally, the contractor may be required to obtain special permits and/or licenses before construction work can begin.

National Fire Protection Association

Organized in 1896, the National Fire Protection Association (NFPA) is an international, nonprofit organization dedicated to the twin goals of promoting the science of fire protection and improving fire protection methods. The NFPA publishes an eleven-volume series covering the national fire codes. These are available by subscription in 3-ring binder form, on CD, and by Internet download. The *NEC*[1] is a part of volume 3 of this series. The purpose and scope of this code are set forth in *NEC Article 90.*

Although the NFPA is an advisory organization, the codes, standards, and recommended practices contained in its published codes are widely used as a basis for local codes. Additional information concerning the publications of the NFPA and membership in the organization can be obtained by writing to

National Fire Protection Association
1 Batterymarch Park
PO Box 9101
Quincy, Massachusetts 02169-7471
617-770-3000
Fax: 617-770-0700
www.nfpa.org

National Electrical Code

The original *NEC* was developed in 1897. Sponsorship of the *Code* was assumed by the NFPA in 1911.

The *National Electrical Code* generally is the bible for the electrician. However, the *NEC* does not have a legal status until the appropriate authorities adopt it as a legal standard. In May 1971, the Department of Labor through OSHA adopted the *NEC* as a national consensus standard. Therefore, in the areas where OSHA is enforced, the *NEC* is the law.

Throughout this text, references are made to chapters, articles, sections, and tables of the *National Electrical Code*. The use of the term *section* has been removed from the *Code* for the most part. It is used extensively in this text to ensure proper identification of the *Code* references.

The student, and any other person interested in electrical construction, should obtain and use a copy of the latest edition of the *NEC*. Keep in mind the importance of determining which edition of the *NEC* is being enforced by the authority having jurisdiction (AHJ) where the work is being performed. To help the user of this text, relevant *Code* sections are paraphrased where appropriate. However, the

---

[1]*National Electrical Code*® and *NEC*® are Registered Trademarks of the National Fire Protection Association, Inc., Quincy, MA.

NEC must be consulted before any decision related to electrical installation is made.

The *NEC* is revised and updated every three years.

Who Writes the *Code*?

The process for revising the *NEC* is very comprehensive. The process begins, continues, and ends with involvement from the public, particularly from those who use or enforce the *NEC*.

For each *Code* cycle, the NFPA solicits proposals to make a change in the current *NEC* from anyone interested in electrical safety.

Anyone may submit a proposal to change the *NEC*, using the proposal form found in the back of the *NEC*. Proposal and Comment forms are also available for download on NFPA's Web site (www.nfpa.org) as well as on the Web sites of several organizations that are involved in the *NEC* process. Proposals received are then assigned to a specific Code-Making Panel (CMP) for action. The Code-Making Panel can take one of the following actions: accept, reject, accept in part, accept in principle, or accept in principle in part. These actions are published in the Report on Proposals (ROP). This document is available at no cost from NFPA in book form and can be downloaded from NFPA's Web site. More than 5000 proposals were submitted for the 2011 *NEC*.

The next phase in the process is the comment stage. After review of the Code-Making Panel's actions at the ROP meeting, individuals may send in their comments on the proposal actions using the Comment Form found in the ROP. Slightly more than 2900 comments were submitted for the 2011 *NEC*. The CMP meets again to review and take action on the comments received. These actions are published in the Report on Comments (ROC). Like the Report on Proposals, the ROC is available in printed form or electronically by downloading from NFPA's Web site.

The next step in the process is the final action (voting) on proposals and comments taken at the NFPA Annual Meeting.

After the Annual Meeting voting, should there be disagreement on the actions, there still is an opportunity for *Appeals* that are considered by the NFPA Standards Council, and/or *Petitions* that are considered by the NFPA Board of Directors.

After all of the final decisions are made, the *National Electrical Code* is published.

NEC ARRANGEMENT

NEC 90.3 contains important rules on the arrangement of the *NEC*. The *NEC* is divided into the introduction and nine chapters. The introduction is included in *Article 90*. This organization is shown in Figure 1-8. *Chapters 1*, *2*, *3*, and *4* apply generally. These chapters include the general requirements for all installations, wiring, and protection in *Chapter 2*; wiring methods and materials in *Chapter 3*; and equipment for general use in *Chapter 4*.

Chapters 5, *6*, and *7* apply to special occupancies, special equipment, or other special conditions. These three chapters supplement or modify the general rules in *Chapters 1–4*. *Chapters 1–4* apply to all of the requirements in the *NEC* unless they are amended by the rules in *Chapters 5*, *6*, or *7*.

Examples of this organization can be found in *Article 250* and *Article 680*. *Article 250* contains the general requirements for grounding and bonding of electrical systems and equipment. *Article 680* includes requirements for swimming pools, spas, and hot tubs. Many grounding and bonding rules are contained in *Article 680* and amend the rules found in *Article 250*.

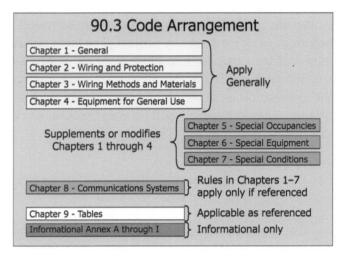

FIGURE 1-8 Organization of the *NEC*.
(*Delmar/Cengage Learning*)

The requirements in *Chapter 8* cover *Communications Systems* and are not subject to the requirements of *Chapters 1–7* unless a rule in *Chapter 8* specifically refers to a rule in *Chapters 1–7*.

Chapter 9 consists of tables that are very helpful and important in the proper application of the *NEC*.

Annexes are included in the back of the *NEC* and provide valuable information but are not enforceable.

⊸⋀⋁⊢ LANGUAGE CONVENTIONS

The *National Electrical Code* is intended for mandatory adoption by authorities having jurisdiction. As such, it is very important that the language used in the *Code* be suitable for mandatory enforcement. *NEC 90.5* provides an explanation of mandatory rules, permissive rules, and explanatory material. Other requirements for writing the *National Electrical Code* are contained in the *NEC Style Manual*. These rules help ensure uniformity throughout the *NEC*.

- Mandatory rules identify what is required or prohibited, and use the term *shall* or *shall not*.

- Permissive rules are actions that are allowed, but not required. Permissive rules use the term *shall be permitted* or *shall not be required*.

- Explanatory material is identified as an *Informational Note*. *Informational Notes* may make reference to other important rules or provide helpful information related to the *Code* itself. These *Informational Notes* are not intended to be enforceable. If more than one *Informational Note* is applicable to a *Code* rule, they are numbered sequentially.

Some articles or sections in the *NEC* include brackets at the end of the rule or figure. The information in these brackets, such as [33:6.5.1], identifies the source of the rule that is imported into the *NEC*. This is done under the NFPA extract policy. As identified in this example, the rule is extracted from *NFPA 33*, the *Standard for Spray Application Using Flammable and Combustible Materials*. This is an efficient manner in developing and maintaining electrical code requirements as the expertise for flammable and combustible materials resides in another NFPA committee and can be imported into the *NEC*.

Exceptions

The *NEC Manual of Style* gives instructions on how Exceptions are to be used in the *NEC*. Although there has been an effort in recent years to reduce the number of exceptions used in the *NEC*, in some cases they remain the best method of rule construction. When exceptions are used, general requirement is stated first, followed by one or more modifications of the general requirement. Often the exception contains a condition that must be met for the exception to apply.

- Exceptions are required to immediately follow the main rule to which they apply. If exceptions are made to items within a numbered list, the exception must clearly indicate the items within the list to which it applies. Exceptions containing the mandatory terms *shall* or *shall not* are to be listed first in the sequence. Permissive exceptions containing *shall be permitted* are to follow any mandatory exceptions and should be listed in their order of importance as determined by the Code-Making Panel.

- If used, exceptions are to convey alternatives or differences to a basic Code rule. The terms *shall* and *shall not* are used to specify a mandatory requirement that is either different from the rule or diametrically opposite to the rule. The term *shall be permitted* designates a variance from the main rule that is permitted but not required.

- See *250.110* for an example of three exceptions to the general rule. The exceptions present a different set of conditions for providing relief from the general rule.

Copies of the *NEC* are available from the NFPA, the International Association of Electrical Inspectors, and from many bookstores.

Citing *Code* References

Every time an electrician makes a decision concerning the electrical wiring, the decision should be checked by reference to the *Code*. Usually this is done from memory, without actually using the *Code*

TABLE 1-2

Citing the *NEC*.

Division	Designation	Example
Chapter	1–9	*Chapter 1*
Article	90 through 840	*Article 250*
Part	Roman numeral	*Article 250, Part II*
Section	Article number, a dot (period), plus one, two, or three digits	*250.20*
Paragraph	Section designation, plus uppercase letter in ( ), followed by digit in ( ), followed by a lowercase letter in ( ) as is required	*250.119(A)(1)*
List	Usually follows an opening paragraph or section	*285.23(B), (1), (2), (3), and (4)*
Exception to	Follows a rule that applies generally and applies under the conditions included in the Exception	*Exception No. 1 to 250.24(B)* or *250.61(B) Exception 3a*
Informative Annex	A, B, C, D, E, F, G, H, I (are not part of the *NEC* and are not enforceable)	*Informative Annex A*

book. If there is any doubt in the electrician's mind, then the *Code* should be referenced directly—just to make sure. When the *Code* is referenced, it is a good idea to record the location of the information in the *Code* book—this is referred to as "citing the *Code* reference." Electrical inspectors should always give a reference, preferably in writing, for any correction they ask be made. If they cannot cite the site of the rule, they should not cite the installation!

There is a very exact way that the location of a *Code* item is to be cited. The various levels of *Code* referencing are shown in Table 1-2. Starting at the top of the table, each step becomes a more specific reference. If a person references *Chapter 1*, this reference includes all the information and requirements that are set forth in several pages. When citing a specific *Section* or an *Exception*, only a few words may be included in the citation. The electrician and inspector should be as specific as possible when citing the *Code*. For the most part, the word *section* does not precede the section numbers in the *Code*.

DEFINED TERMS

Many terms used in the *NEC* have a meaning that is particular or unique and must be carefully followed and understood for proper application of the rules.

Standard dictionary terms do not apply to a term that is defined in the *NEC*. Although an exhaustive study of the rules often seems boring, the importance of understanding the meaning of the terms used in the *NEC* cannot be overstated. Terms that are used in more than one *NEC* article are included in *Article 100*. As you will find, *Article 100* is divided into two parts. *Part I* includes terms used throughout the *Code* and *Part II* includes terms used in parts of articles that apply to installations of equipment operating at over 600 volts.

Many articles in the *Code* have terms that are used in only that article and have a definition that is important to the proper application of requirements in the article. These terms are most often included near the beginning of the article in the XXX.2 location. For example, see *240.2, 250.2, 330.2, 517.2,* and *680.2*.

We will not review all the definitions at this point but suggest that you do that on your own. The following terms are defined in *NEC Article 100* and are used throughout the *Code* as well as in this text. We will review other definitions at the location where the term is used in this text. It is important to understand the meanings of these terms.

- **APPROVED:** *Acceptable to the authority having jurisdiction*(AHJ). Note that *90.4* of the *NEC*

---

*Reprinted with permission from NFPA 70-2011.

outlines several of the duties of the AHJ. Other duties, responsibilities, and authority of the AHJ are included in *Annex H* of the *NEC*. The local law or ordinance enacting an electrical installation and inspection program states the exact applicable requirements. These laws or ordinances often include requirements for licensing of contractors, electricians, and apprentices as well as for permits and inspections.

- **AUTHORITY HAVING JURISDICTION:** *The organization, office, or individual responsible for approving equipment, materials, an installation, or a procedure.* An extensive list of examples is included in an *Informational Note* that follows this definition in *Article 100.*

- **EQUIPMENT:** *A general term including material, fittings, devices, appliances, luminaires, apparatus, machinery, and the like used as a part of, or in connection with, an electrical installation.*

- **IDENTIFIED (as applied to equipment):** *Recognizable as suitable for the specific purpose, functions, use, environment, application, and so forth, where described in a particular Code requirement.*

- **LABELED:** *Equipment or materials to which has been attached a label, symbol, or other identifying mark of an organization that is acceptable to the authority having jurisdiction and concerned with product evaluation, that maintains periodic inspection of production of labeled equipment or materials, and by whose labeling the manufacturer indicates compliance with appropriate standards or performance in a specified manner.*

- **LISTED:** *Equipment, materials, or services included in a list published by an organization that is acceptable to the authority having jurisdiction and concerned with the evaluation of products or services, that maintains periodic inspection of the production of listed equipment or materials or periodic inspection of services, and whose listing states that either the equipment, material, or services meets appropriate designated standards or has been tested and found suitable for a specified purpose.*

Electrical Products Safety Standards

Recent harmonizing of standards has resulted in Underwriters Laboratories and several other testing laboratories now evaluating equipment for a manufacturer under both a UL standard as well as a Canadian Standards Association standard. This can save a manufacturer time and money because the manufacturer can have the testing and evaluation done in one location, acceptable to both U.S. and Canadian authorities enforcing the *Code*.

Nationally Recognized Testing Laboratories (NRTL)

When legal issues arise regarding electrical equipment, the courts must rely on the testing, evaluation, and product certification by qualified testing organizations. OSHA has recognized a number of organizations that meet the legal requirements found in OSHA 29 CFR 1910.7. Such an organization is referred to in the industry as a "NRTL" (pronounced "nurtle," as in *turtle*). The letters stand for *Nationally Recognized Testing Laboratory.* Visit the OSHA Web site at http://www.osha.gov/dts/otpca/nrtl/nrtllist.html for more details.

Many electrical inspection authorities accept the listing marks on equipment under the conditions stated in the OSHA recognition. Keep in mind that all electrical product testing laboratories are not created equal. Some can test any and all products. Others are limited by their testing equipment, facilities, or personnel. Some electrical inspection authorities have created and enforce their own electrical testing laboratory certification program.

Although there is no blanket requirement in the *NEC* that all electrical equipment used in installations must be listed and labeled to indicate conformity with a product safety standard, many electrical inspectors in fact require just that. Some inspection jurisdictions require blanket listing and labeling under authority of a local law, rule, or ordinance. Other inspectors require listing and labeling of electrical equipment as a condition of their acceptance of the installation. This authority is endorsed in *NEC 110.2*, which states, *Approval. The conductors* and equipment required or permitted by this Code shall be acceptable only if approved. The word *approved* is defined in *Article 100* as *Acceptable to the authority having jurisdiction*. So, the rule in *110.2* can be read with substitution of the definition of *approved* as, "The conductors and equipment required or permitted by this *Code* shall be acceptable only if acceptable to the authority having jurisdiction."

Other important terms or rules that are related are *90.4, 90.7, 110.3*: *Identified, Labeled,* and *Listed*.

Most electrical testing laboratories provide field evaluations and labeling for products that have been installed without a listing mark on the product. Some of these products have been produced in a foreign country without testing laboratory evaluation at the factory. Others are one-of-a-kind products, are unique in one or more features, or are complex and were produced at the factory without evaluation by a third-party testing laboratory.

Underwriters Laboratories, Inc.

Founded in 1894, Underwriters Laboratories (UL) is a nonprofit organization that produces a large number of product safety standards. UL also operates laboratories to investigate materials, devices, products, equipment, and construction methods and systems to determine whether the product being tested complies with the product safety standard. The organization provides a listing service to manufacturers. Any product authorized to carry an Underwriters Laboratories listing mark (label) has been evaluated with respect to all reasonable foreseeable hazards to life and property, and it has been determined that the product provides safeguards to these hazards to an acceptable degree. A listing by the Underwriters Laboratories does not mean that a product is approved by the *NEC*. However, many local agencies do make this distinction.

The NEC defines *approved* as *acceptable to the authority having jurisdiction*. *NEC 110.3(B)* stipulates that, *Listed or labeled equipment shall be installed and used in accordance with any instructions included in the listing or labeling.*

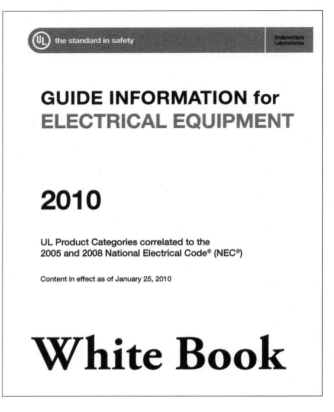

FIGURE 1-9 Underwriters Laboratories *White Book* (*Courtesy Underwriters Laboratories*)

Useful UL publications are

- *Electrical Construction Equipment Directory (Green Book)*
- *Electrical Appliances and Utilization Equipment Directory (Orange Book)*
- *Hazardous Locations Equipment Directory (Red Book)*
- *Guide Information for Electrical Equipment (White Book), Figure 1-9*
- *Recognized Component Directories (Yellow Books)* (a three-volume set usually distributed only to manufacturers)
- *Fire Resistance Directory (a four-volume set)*
- *Marking Guides:*
  - *Dead-Front Switchboards*
  - *Electrical Heating and Cooling Equipment*
  - *Luminaires*
  - *Molded Case Circuit Breakers*
  - *Panelboards*

- *Swimming Pool Equipment, Spas, Fountains and Hydromassage Bathtubs*

- *Wire and Cable*

Many inspection authorities continually refer to these books in addition to the *NEC*. For example, many guidelines on the proper installation of listed electrical equipment are found in the UL White Book. This information is so valuable, you could say without overstatement, "Don't leave home without it!" If the answer to a question cannot be found readily in the *NEC*, it generally can be found in the UL publications listed. An index of publications and information concerning the UL can be obtained by writing to

Public Information
Underwriters Laboratories, Inc.
333 Pfingsten Road
Northbrook, IL 60062-2096
847-272-8800
Fax: 847-272-9562
http://www.ul.com

UL Marking

The UL marking is required to be on the product! The UL marking will always consist of four elements: UL in a circle, the word *LISTED* in capital letters, the product identity, and a unique alphanumeric control or issue number, Figure 1-10. If the product is too small to have the UL marking on the product itself, or has a shape or is made of a material that will not accept the UL mark, the marking is permitted on the smallest unit carton or container that the product comes in. Marking on the carton or box is nice, but does not ensure that the product is UL listed!

The letters will *always* be staggered: U$_L$. They will *not* be side by side: UL.

The UL listing mark without any designation just outside the circle indicates the product has been evaluated to only U.S. product standards. A "US" at about the 4:00 o'clock location also indicates this. A "C" at about the 8:00 o'clock position indicates the product has been evaluated only to the Canadian standards. If both the "C" and "US" appear at their respective positions, the product has been evaluated to both the U.S. and Canadian standards.

Several other testing laboratories have the same or a similar marking scheme.

Intertek Testing Services (ITS)

Formerly known as Electrical Testing Laboratories (ETL), the Intertek Testing Service is a nationally recognized testing laboratory. Intertek Testing Services owns the copyright to several listing marks, including the ETL mark. The listing mark that indicates compliance with Canadian and United States product standards is shown as Figure 1-11. ITS provides testing, evaluation, labeling, listing, and follow-up service for the safety testing of electrical products. This is done in conformance to nationally recognized safety standards or to specifically designated requirements of jurisdictional authorities. Information can be obtained by writing to

Intertek Testing Services, NA, Inc.
ETL SEMKO
3933 U.S. Route 11
Cortland, NY 13045
607-753-6711
Fax: 607-756-9891
http://www.intertek-etlsemko.com

FIGURE 1-10 UL Listing mark indicates product compliance with Canadian and USA product standards. (*Courtesy Underwriters Laboratories*)

FIGURE 1-11 Listing mark for Intertek Testing Services for electrical products. (*Courtesy Intertek Testing Services N.A.*)

Canadian Standards Association (CSA)

The Canadian Electrical Code is significantly different from the *National Electrical Code.* It is considered Part 1 of the Canadian Electrical Code. Part 2 of the Canadian Electrical Code consists of electrical product safety standards similar to the standards produced in the United States by Underwriters Laboratories. Canadian product safety standards are produced by the Canadian Standards Association (CSA). In fact, many of the Canadian and U.S. standards have been harmonized. This allows a product to be evaluated and listed to the same requirements in both countries. Efforts are continuing to harmonize U.S. and Canadian standards with those from Mexico.

CSA also serves as a third-party independent electrical products testing laboratory. Manufacturers are permitted to use a listing mark to identify products that have been found by examination and testing to comply with the Canadian Electrical Code Part Two. CSA's listing mark is shown in Figure 1-12 and is arranged identically to UL's listing mark to designate suitability of the product for installation in the United States or Canada.

Those using this text in Canada must follow the Canadian Electrical Code. *Electrical Wiring— Commercial* (© Thomas Nelson Holdings) is available based on the Canadian Electrical Code.

The Canadian Electrical Code is a voluntary code suitable for adoption and enforcement by electrical inspection authorities. The Canadian Electrical Code is published by and available from

CSA International
178 Rexdale Boulevard
Toronto, Ontario, Canada
M9W 1R3
416-747-4000
Fax: 416-747-4149
http://www.csa-international.org

Counterfeit Products

Be on the lookout for counterfeit electrical products. Counterfeit electrical products can present a real hazard to life and property. These products have not been tested and listed by a recognized testing laboratory. Most counterfeit products as of this writing supposedly come from China. Counterfeit electrical products might be referred to as "black market products."

Look for unusual logos or wording. For example, the UL logo might be illustrated in an oval instead of a circle. The UL logo might not be encircled with anything. The wording might say "*approved*" instead of "LISTED." UL does not approve anything! They test the product. If the product meets its standards, the product is then LISTED. The UL marking might be on the carton and not on the product. Marking on the carton or box is nice but can be considered to be advertising! Absence of the manufacturer's name should raise the caution flag.

Currently, there is federal legislation underway to make it a criminal offense to traffic in counterfeit products and counterfeit trademarks. The legislation makes it mandatory that the counterfeit products and any tools that make the products or markings be seized and destroyed.

To learn more about counterfeits, check out http://www.ul.com, then search the word *counterfeit.* Also, check out http://www.nema.org, and search the word *counterfeit.*

FIGURE 1-12 Listing mark for the
Canadian Standards Association.
(Courtesy Canadian Standards Association)

National Electrical Manufacturers Association (NEMA)

NEMA is a nonprofit organization supported by the manufacturers of electrical equipment and supplies. NEMA develops standards that are designed to assist the purchaser in selecting and obtaining the

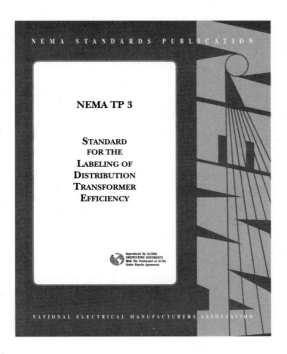

FIGURE 1-13 A typical NEMA standard. (*Courtesy National Electrical Manufacturers Association*)

correct product for specific applications. A typical standard is illustrated in Figure 1-13. Information concerning NEMA standards may be obtained by writing to

National Electrical Manufacturers Association
1300 North 17th Street, Suite 1752
Rosslyn, VA 22209
703-841-3200
Fax: 703-841-5900
http://www.nema.org

Free Standards: NEMA offers a number of its standards at no cost. These free standards are available for downloading. Check out the Web site.

American National Standards Institute (ANSI)

Various working groups in the organization study the numerous codes and standards. An American National Standard implies "a consensus of those concerned with its scope and provisions." The *National Electrical Code* is approved by ANSI and is numbered ANSI/NFPA 70-1999.

ANSI
25 West 43rd Street, 4th Floor
New York, NY 10036

212-642-4900
Fax: 212-398-0023
http://www.ansi.org

International Association of Electrical Inspectors (IAEI)

IAEI is a nonprofit organization whose membership consists of electrical inspectors, electricians, contractors, testing laboratories, electric utilities, and manufacturers throughout the United States and Canada. One goal of the IAEI is to improve the understanding of the *NEC*. Representatives of this organization serve as members of the various panels of the *NEC* Committee and share equally with other members in the task of reviewing and revising the *NEC*. The IAEI publishes a bimonthly magazine, the *IAEI News*. Additional information concerning the organization may be obtained by writing to

International Association of Electrical Inspectors
901 Waterfall Way, Suite 602
Richardson, TX 75080-7702
800-786-4234, 972-235-1455
Fax: 972-235-6858
http://www.iaei.org

Illuminating Engineering Society of North America (IESNA)

IESNA was formed in 1906. The objective of this group is to communicate information about all facets of good lighting practice to its members and to consumers. The IESNA produces numerous publications that are concerned with illumination.

The *IESNA Lighting Handbooks* are regarded as the standard for the illumination industry and contain essential information about light, lighting, and luminaires. Information about publications or membership may be obtained by writing to

Illuminating Engineering Society of North America
120 Wall St., 17th Floor
New York, NY 10005-4001
212-248-5000
Fax: 212-248-5017
http://www.iesna.org

American Society of Heating, Refrigerating and Air-Conditioning (ASHRAE)

ASHRAE was formed in 1894. ASHRAE states it is an international organization of some 51,000 persons and fulfills its mission of advancing heating, ventilation, air conditioning, and refrigeration to serve humanity and promote a sustainable world through research, standards writing, publishing, and continued education.

What has become increasingly important to ASHRAE and every other organization is energy conservation—all are concerned about "Green" technology. ASHRAE has a number of publications relating to energy conservation, for example, their *Standard 189.1* and *Standard 189.1P* with emphasis on the design of high performance "Green" buildings. ASHRAE standards even cover illuminance (lighting) calculations, which are identical to the illuminance calculations recommended by the Illuminating Engineers Society of North America (IESNA).

The *ASHRAE Handbook* is considered the Bible of the industry, as the *National Electrical Code* is to the electrical industry.

Those of us in the electrical arena are aware of the importance of promoting energy conservation through the use of energy efficient ballasts, low wattage fluorescent lamps (T8 instead of T12), less use of conventional incandescent lamps, increased use of compact fluorescent lamps (CFLs) and light emitting diodes (LEDs), along with lower current ratings for air-conditioning equipment.

Visiting their Web site www.ashrae.org will reveal a tremendous amount of technical information available to members of their organization. They have many publications, standards, and conduct on-line courses relating to energy conservation and sustainability.

Information may be obtained by contacting

ASHRAE
1791
Tullie Circle, N.E.
Atlanta, GA 30329
Phone: (404) 636-8400
Fax: (404) 321-5478

National Electrical Installation Standards (NEIS)

The *NEC* in *110.12* states that *Electrical equipment shall be installed in a neat and workmanlike manner.* Just what does that mean?

It means that unused openings shall be closed and that *Internal parts of electrical equipment, including busbars, wiring terminals, insulators, and other surfaces, shall not be damaged or contaminated by foreign materials such as paint, plaster, cleaners, abrasives, or corrosive residues. There shall be no damaged parts that may adversely affect safe operation or mechanical strength of the equipment such as parts that are broken; bent; cut; or deteriorated by corrosion, chemical action, or overheating.* *

To help the electrical industry understand what is required by *110.12*, the National Electrical Contractors Association (NECA) has been developing installation standards.

NEIS, an ongoing project of NECA, covers installation standards for the electrical trade. These installation standards are not about the *NEC* but rather cover those issues not covered in the *NEC*, such as housekeeping; how to properly handle, receive, and store electrical equipment; checking out the equipment before energizing; and so on. These are actual on-the-job issues that are not necessarily *Code*-related, but issues electricians need to know. In the past, this was hands-on, on-the-job training. Now, there are installation standards that can be followed by everyone in the electrical industry. These installation standards are also recognized by ANSI.

Installation Standards. At the moment, installation standards are available on the following subjects:

- *NECA 1, Good Workmanship in Electrical Contracting*
- *NECA 90, Commissioning Electrical Systems*
- *NECA 100, Electrical Symbols for Construction Drawings*
- *NECA 101, Steel Conduits*
- *NECA 102, Aluminum Rigid Metal Conduit*
- *NECA 104, Aluminum Wire and Cable*
- *NECA 105, Metal Cable Tray Systems*
- *NECA 111, Nonmetallic Raceways*

*Reprinted with permission from NFPA 70-2011.

- *NECA 120, Armored Cable (AC) and Metal Clad Cable (MC)*

- *NECA 121, Installing Type NM and Type UF Cables*

- *NECA 200, Temporary Power at Construction Sites*

- *NECA 202, Industrial Heat Tracing Systems*

- *NECA 230, Electric Motors and Controllers*

- *NECA 301, Installing and Testing Fiber Optic Cables*

- *NECA 303, Closed-Circuit Television (CCTV) Systems*

- *NECA 305, Fire Alarm System Job Practices*

- *NECA 331, Building and Service Entrance Grounding and Bonding*

- *NECA 400, Installing and Maintaining Switchboards*

- *NECA 402, Installing and Maintaining Motor Control Centers*

- *NECA 404, Installing Generator Sets*

- *NECA 406, Residential Generator Sets*

- *NECA 407, Installing and Maintaining Panelboards*

- *NECA 408, Installing and Maintaining Busways*

- *NECA 409, Installing and Maintaining Dry-Type Transformers*

- *NECA 410, Installing & Maintaining Liquid-Filled Transformers*

- *NECA 411, Installing and Maintaining Uninterruptible Power Supplies (UPS)*

- *NECA 420, Standard for Fuse Applications*

- *NECA 430, Installing Medium-Voltage Metal-Clad Switchgear*

- *NECA 500, Installing Indoor Lighting Systems*

- *NECA 501, Installing Exterior Lighting Systems*

- *NECA 502, Installing Industrial Lighting Systems*

- *NECA 503, Installing Fiber Optic Lighting Systems*

- *NECA 568, Installing Building Telecommunications Cabling*

- *NECA 600, Installing Medium-Voltage Cable*

- *NECA 605, Installing Underground Nonmetallic Utility Duct*

Check price and availability by contacting

National Electrical Contractors Association
3 Bethesda Metro Center, Suite 1100
Bethesda, MD 20814
301-657-3110
Fax: 301-215-4500
http://www.necanet.org

Registered Professional Engineer (PE)

Although the requirements may vary slightly from state to state, the general statement can be made that a registered professional engineer has demonstrated his or her competence by graduating from college and passing a difficult licensing examination. Following the successful completion of the examination, the engineer is authorized to practice engineering under the laws of the state. A requirement is usually made that a registered professional engineer must supervise the design of any building that is to be used by the public. The engineer must indicate approval of the design by affixing a seal to the plans.

Information concerning the procedure for becoming a registered professional engineer and a definition of the duties of the professional engineer can be obtained by writing the state government department that supervises licensing and registration.

ABANDONED CABLES

Abandoned cables might not be the cause of starting a fire, but they certainly are fuel for a fire.

Often overlooked when bidding and working on commercial and industrial installations is being confronted with a lot of existing unused conductors and cables. The specifications may not be clear on this issue. The electrical inspector will probably take a strong position on abandoned wiring. There are many places in the *NEC* that require the removal of abandoned conductors and cables. Here is the list:

- *372.13*: Cellular concrete floor raceways

- *374.7*: Cellular metal floor raceways

- *390.7*: Underfloor raceways
- *640.6(C)*: Audio signal processing, amplification, and reproduction equipment
- *645.5(F)*: Information technology cables
- *725.25*: Class 1, 2, and 3 remote-control, signaling, and power-limited circuits
- *760.3(A)*: Fire alarm cables
- *770.25*: Optical fiber cables
- *770.154(A)*: Optical fiber cables in ducts and plenums
- *800.25*: Telephone cables
- *800.154(A)*: Telephone cables in ducts and plenums
- *820.25*: CATV
- *820.154(A)*: CATV cables in ducts and plenums
- *830.25*: Network-powered broadband communications systems
- *840.25*: Premises-powered broadband communications systems

METRICS (SI) AND THE *NEC*

The United States is the last major country in the world not using the metric system as the primary system for weights and measurements. We have been very comfortable using English (United States Customary) values, but this is changing. Manufacturers are now showing both inch-pound and metric dimensions in their catalogs. Plans and specifications for governmental new construction and renovation projects started after January 1, 1994, have been using the metric system. You may not feel comfortable with metrics, but metrics are here to stay. You might just as well get familiar with the metric system.

Some common measurements of length in the English (Customary) system are shown with their metric (SI) equivalents in Table 1-3.

The *NEC* and other *NFPA Standards* are becoming international standards. All measurements in the 2011 *NEC* are shown with metrics first, followed by the inch-pound value in parentheses. For example, 600 mm (24 in.).

TABLE 1-3

Customary and metric comparisons.

Customary	*NEC* SI Units	SI Units
0.25 in.	6 mm	6.35 mm
0.5 in.	12.7 mm	12.7 mm
0.62 in.	15.87 mm	15.875 mm
1.0 in.	25 mm	25.4 mm
1.25 in.	32 mm	31.75 mm
2 in.	50 mm	50.8 mm
3 in.	75 mm	76.2 mm
4 in.	100 mm	101.6 mm
6 in.	150 mm	152.4 mm
8 in.	200 mm	203.2 mm
9 in.	225 mm	228.6 mm
1 ft	300 mm	304.8 mm
1.5 ft	450 mm	457.2 mm
2 ft	600 mm	609.6 mm
2.5 ft	750 mm	762 mm
3 ft	900 mm	914.4 mm
4 ft	1.2 m	1.2192 m
5 ft	1.5 m	1.524 m
6 ft	1.8 m	1.8288 m
6.5 ft	2.0 m	1.9182 m
8 ft	2.5 m	2.4384 m
9 ft	2.7 m	2.7432 m
10 ft	3.0 m	3.048 m
12 ft	3.7 m	3.6576 m
15 ft	4.5 m	4.572 m
18 ft	5.5 m	5.4864 m
20 ft	6.0 m	6.096 m
22 ft	6.7 m	6.7056 m
25 ft	7.5 m	7.62 m
30 ft	9.0 m	9.144 m
35 ft	11.0 m	10.668 m
40 ft	12.0 m	12.192 m
50 ft	15.0 m	15.24 m
75 ft	23.0 m	22.86 m
100 ft	30.0 m	30.48 m

In *Electrical Wiring—Commercial*, ease in understanding is of utmost importance. Therefore, inch-pound values are shown first, followed by metric values in parentheses. For example, 24 in. (600 mm).

Units of measurement are covered in *90.9* of the *NEC*. It permits both soft and hard conversion from the inch-pound system to or from the SI units of measurement. The accuracy of the conversion is just the opposite from what it might appear. Soft conversion is more precise than hard conversion; it is fairly easy to determine which dimension is a soft conversion and which has a hard conversion. Typically, soft conversion will result in two or more digits to the right of the decimal point. For example, *Table 110.26(A)(1)* gives the minimum depth of working spaces around electrical equipment operating at 600 volts, nominal, or less. The minimum distance for Condition 1 is 3 ft (914 mm), and for Condition 3, if the equipment operates from 151 to 600 volts to ground, the minimum clear working space is 4 ft, or 1.22 m.

A *soft metric conversion* is when the dimensions of a product already designed and manufactured to the inch-pound system have their dimensions converted to metric dimensions. The product does not change in size. Soft conversions are more precise than hard conversions.

A *hard metric measurement* is where a product has been designed to SI metric dimensions. No conversion from inch-pound measurement units is involved. A *hard conversion* is where an existing product is redesigned into a new size.

In the 2011 edition of the *NEC*, existing inch-pound dimensions did not change. Metric conversions were made, then rounded off. Please note that when comparing calculations made by both English and metric systems, slight differences will occur due to the conversion method used. These differences are not significant, and calculations for both systems are therefore valid. Where rounding off would create a safety hazard, the metric conversions are mathematically identical (a soft conversion).

For example, if a dimension is required to be 6 ft, it is shown in the *NEC* as 6 ft (1.8 m). Note that the 6 ft remains the same, and the metric value of 1.83 m has been rounded off to 1.8 m. This edition of *Electrical Wiring—Commercial* reflects these rounded-off changes. In this text, the inch-pound measurement is shown first, in other words, 6 ft (1.8 m).

TABLE 1-4	
Trade sizes of raceways versus actual inside diameters.	
Trade Size	**Inside Diameter (I.D.)**
½ Electrical Metallic Tubing	0.622 in.
½ Electrical Nonmetallic Tubing	0.560 in.
½ Flexible Metal Conduit	0.635 in.
½ Rigid Metal Conduit	0.632 in.
½ Intermediate Metal Conduit	0.660 in.

Trade Sizes

A unique situation exists. Strange as it may seem, what electricians have been referring to for years has not been correct!

Raceway sizes have always been an approximation. For example, there has never been a ½-in. raceway! Measurements taken from the *NEC* for a few types of raceways are shown in Table 1-4.

You can readily see that the cross-sectional areas, critical when determining conductor fill, are different. It makes sense to refer to conduit, raceway, and tubing sizes as *trade sizes*. The *NEC* in *90.9(C)(1)* states that *where the actual measured size of a product is not the same as the nominal size, trade size designators shall be used rather than dimensions. Trade practices shall be followed in all cases.* This edition of *Electrical Wiring—Commercial* uses the term *trade size* when referring to conduits, raceways, and tubing. For example, instead of referring to a ½-in. electrical metallic tubing (EMT), it is referred to as trade size ½ EMT. EMT is also referred to in the trade as "thinwall."

The *NEC* also uses the term *metric designator*. A ½-in. EMT is shown as *metric designator 16 (trade size 1/2)*. A 1-in. EMT is shown as *metric designator 27 (trade size 1)*. The numbers 16 and 27 are the *metric designator* values. The (½) and (1) are the *trade sizes*. The metric designator is based on the inside diameter of Rigid Metal Conduit—in rounded-off millimeters (mm). Table 1-5 shows some of the more common sizes of conduit, raceways, and tubing. A complete table is found in the *NEC, Table 300.1(C)*. Because of possible confusion, this text uses only the term *trade size* when referring to conduit and raceway sizes.

TABLE 1-5

Metric designators for raceways through trade size 3.

Metric Designator	Trade Size
12	⅜
16	½
21	¾
27	1
35	1¼
41	1½
53	2
63	2½
78	3

Conduit knockouts in boxes do not measure up to what we call them. Table 1-6 shows trade size knockouts and their actual measurements.

A word unique to the electrical industry is *device*. The *NEC* defines a device as *A unit of an electrical system that carries or controls electric energy as its principal function.* You will see this word often throughout this text.

Outlet boxes and device boxes use their nominal measurement as their *trade size*. For example, a 4 in. × 4 in. × 1½ in. does not have an internal cubic-inch area of 4 in. × 4 in. × 1½ in. = 24 cubic inches. *Table 314.16(A)* shows this size box as having a 21-cubic-inch area. This table shows *trade sizes* in two columns—millimeters and inches.

Table 1-7 provides the detailed dimensions of some typical sizes of outlet and device boxes in both metric and English units.

In this text, a square outlet box is referred to as 4 × 4 × 1½-inch square box, 4" × 4" × 1½" square box, or trade size 4 × 4 × 1½ square box.

TABLE 1-6

Trade size of a knockout compared to the actual measurement of the knockout.

Trade Size Knockout	Actual Measurement
½	⅞ in.
¾	1³⁄₃₂ in.
1	1⅜ in.

Similarly, a single-gang device box might be referred to as a 3 × 2 × 3-inch device box, a 3" × 2" × 3"-deep device box, or a trade size 3 × 2 × 3 device box. The box type should always follow the trade size numbers.

Trade sizes for construction material will not change. A 2 × 4 is really a *name*, not an actual dimension. A 2 × 4 stud will still be referred to as a 2 × 4 stud though its actual dimension is approximately 1½ × 3½ inches. This is its *trade size*.

In this text, measurements directly related to the *NEC* are given in both inch-pound and metric units. In many instances, only the inch-pound units are shown. This is particularly true for the examples of raceway calculations, box fill calculations, and load calculations for square foot areas, and on the plans (drawings). To show both English and metric measurements on a plan would certainly be confusing and would really clutter up the plans, making them difficult to read.

Because the *NEC* rounded off most metric conversion values, a calculation using metrics results in a different answer when compared to the same calculation done using inch-pounds. For example, load calculations for a residence are based on 3 volt-amperes per square foot or 33 volt-amperes per square meter.

For a 40 ft × 50 ft dwelling:

3 VA × 40 ft × 50 ft = 6000 volt-amperes

In metrics, using the rounded-off values in the *NEC*:

33 VA × 12 m × 15 m = 5940 volt-amperes

The difference is small, but nevertheless, there is a difference.

To show calculations in both units throughout this text would be very difficult to understand and would take up too much space. Calculations in either metrics or inch-pounds are in compliance with *NEC 90.9(D)*. In *90.9(C)(3)* we find that metric units are not required if the industry practice is to use inch-pound units.

It is interesting to note that the examples in *Annex D* of the *NEC* use inch-pound units, not metrics.

Guide to Metric Usage

The metric system is a base-10 or decimal system in that values can be easily multiplied or

TABLE 1-7

Box dimensions.

Box Dimensions		Box Type	Minimum Capacity	
mm	**in.**		**cm³**	**in.³**
100 × 32	4 × 1¼	round/octagonal	205	12.5
100 × 38	4 × 1½	round/octagonal	254	15.5
100 × 54	4 × 2⅛	round/octagonal	353	21.5
100 × 32	4 × 1¼	square	295	18.0
100 × 38	4 × 1½	square	344	21.0
100 × 54	4 × 2⅛	square	497	30.3
75 × 50 × 38	3 × 2 × 1½	device	123	7.5
75 × 50 × 50	3 × 2 × 2	device	164	10.0
75 × 50 × 57	3 × 2 × 2¼	device	172	10.5

See NEC Table 314.16(A) for complete listing.

divided by ten or powers of ten. The metric system as we know it today is known as the International System of Units (SI) derived from the French term *le Système International d'Unités.*

In the United States, it is the practice to use a period as the decimal marker and a comma to separate a string of numbers into groups of three for easier reading. In many countries, the comma has been used in lieu of the decimal marker and spaces are left to separate a string of numbers into groups of three. The SI system, taking something from both, uses the period as the decimal marker and the space to separate a string of numbers into groups of three, starting from the decimal point and counting in either direction. For example, 12 345.789 99. An exception to this is when there are four numbers on either side of the decimal point. In this case, the third and fourth numbers from the decimal point are not separated. For example, *2015.1415.*

In the metric system, the units increase or decrease in multiples of 10, 100, 1000, and so on. For instance, one megawatt (1,000,000 watts) is 1000 times greater than one kilowatt (1000 watts).

By assigning a name to a measurement, such as a watt, the name becomes the unit. Adding a prefix to the unit, such as *kilo-*, forms the new name *kilowatt*, meaning 1000 watts. Refer to Table 1-8 for prefixes used in the numeric systems.

TABLE 1-8

Numeric presentations.

Name	Exponential	Metric (SI)	Script	Customary
mega	(10^6)	1 000 000	one million	1,000,000
kilo	(10^3)	1 000	one thousand	1000
hector	(10^2)	100	one hundred	100
deka		10	ten	10
unit		1	one	1
deci	(10^{-1})	0.1	one-tenth	1/10 or 0.1
centi	(10^{-2})	0.01	one-hundredth	1/100 or 0.01
mili	(10^{-3})	0.001	one-thousandth	1/1000 or 0.001
micro	(10^{-6})	0.000 001	one-millionth	1/1,000,000 or 0.000,001
nano	(10^{-9})	0.000 000 001	one-billionth	1/1,000,000,000 or 0.000,000,001

TABLE 1-9

Useful conversions and their abbreviations.

inches (in.) × 0.0254 = meter (m)	square inches (in.²) × 6.452 = square centimeters (cm²)
inches (in.) × 0.254 = decimeters (dm)	square centimeters (cm²) × 0.155 = square inches (in.²)
inches (in.) × 2.54 = centimeters (cm)	square feet (ft²) × 0.093 = square meter (m²)
centimeters (cm) × 0.3937 = inches (in.)	square meters (m²) × 10.764 = square feet (ft²)
inches (in.) × 25.4 = millimeters (mm)	square yards (yd.²) × 0.8361 = square meters (m²)
millimeters (mm) × 0.039 37 = inches (in.)	square meters (m²) × 1.196 = square yards (yd.²)
feet (ft) × 0.3048 = meters (m)	kilometers (km) × 1 000 = meters (m)
meters (m) × 3.2802 = feet (ft)	kilometers (km) × 0.621 = miles (mi)
	miles (mi) × 1.609 = kilometers (km)

Certain prefixes shown in Table 1-8 have a preference in usage. These prefixes are *mega-*, *kilo-*, the unit itself, *centi-*, *milli-*, *micro-*, and *nano-*. Consider that the basic metric unit is a meter (one). Therefore, a kilometer is 1000 meters, a centimeter is 0.01 meter, and a millimeter is 0.001 meter.

The advantage of the SI metric system is that recognizing the meaning of the proper prefix lessens the possibility of confusion.

In this text, when writing numbers, the names are often spelled in full, but when used in calculations, they are abbreviated. For example: *m* for *meter*, *mm* for *millimeter*, *in.* for *inch*, and *ft* for *foot*. It is interesting to note that the abbreviation for inch is followed by a period (12 in.), but the abbreviation for foot is not followed by a period (6 ft). Why? Because *ft.* is the abbreviation for *fort.*

SUMMARY

As time passes, there is no doubt that metrics will be commonly used in this country. In the meantime, we need to take it slow and easy. The transition will take time. Table 1-9 shows useful conversion factors for converting English units to metric units. Appendix E of this text contains a comprehensive table showing the conversion values for inch-pound and metric units.

REVIEW

Refer to the *National Electrical Code* or the working drawings when necessary. Where applicable, responses should be written in complete sentences.

1. What section of the specification contains a list of contract documents?

   _____

   _____

2. The requirement for temporary light and power at the job site will be found in what portion of the specification? _____

3. The electrician uses the Schedule of Working Drawings for what purpose?

   _____

   _____

Complete the following items by indicating the letter(s) designating the correct source(s) of information:

4. _____ Ceiling height   A. Architectural floor plan
5. _____ Electrical receptacle style B. Details
6. _____ Electrical outlet location   C. Electrical layout drawings
7. _____ Exterior wall finishes   D. Electrical symbol schedule
8. _____ Grading elevations   E. Elevations
9. _____ Panelboard schedules   F. Sections
10. _____ Room width   G. Site plan
11. _____ Swing of door   H. Specification
12. _____ View of interior wall

Match the items on the left with those on the right by writing the letter designation of the appropriate organization from the list on the right.

13. _____ Electrical *Code*   A. IAEI
14. _____ Electrical inspectors   B. IESNA
15. _____ Fire codes   C. *NEC*
16. _____ Lighting information   D. NEMA
17. _____ Listing service   E. NFPA
18. _____ Manufacturers' standards   F. PE
19. _____ Seal   G. UL

Match the items on the left with those on the right by writing the letter designation of the proper level of *NEC* interpretation from the list on the right.

20. _____ Allowed by the *Code*   A. never
21. _____ May be done   B. shall
22. _____ Must be done   C. with special permission
23. _____ Required by the *Code*
24. _____ Up to the electrician

25. Find *NEC 250.52(A)(5)* and record the first four words.

_____

Show required conversion calculations for Question 26 and Question 27.

26. Luminaire style F is four feet long. The length in SI units, as specified by the *NEC*, is

_____ .

27. The gross area of the drugstore basement is 1395 square feet. The area in square meters is _____ .

Determine the following dimensions. Write the dimensions using unit names, not symbols (for example, 1 foot, not 1'), and indicate the source of the information.

28. What is the inside clear distance of the interior stairway to the drugstore basement?

_____

29. What is the gross square footage of each floor of the building?

_____

30. What is the distance in the drugstore from the exterior block wall to the block wall separating the drugstore from the bakery?

_____

31. What is the finished floor to finished ceiling height on the second floor? _____

_____

For Questions 32–34, cite the *NEC* source.

32. The standard ampere ratings for fuses and fixed trip circuit breakers. _____

33. The minimum bending radius for metal clad cable with a smooth sheath with an external diameter of 1 inch. _____

34. The permission to use splices in busbars as grounding electrode conductor. _____

Perform the following:

35. a) Where is the sump pump in the Commercial Building located? _____

    b) To what branch circuit is the sump pump connected? _____

36. Write a letter to one of the organizations listed in this unit requesting information about the organization and the services it provides.

_____

_____

_____

37. The temperatures from an arc flash can reach approximately _____ °F.

38. The Occupational Safety and Health Act (OSHA) Code of Federal Regulations (CFR) Number 29, Subpart S, in paragraph _____ discusses the training needed for those who face the risk of electrical injury.

39. The *NEC* defines a *Qualified person* as _____

40. According to *NFPA 70E Article 110,* a journeyman electrician is qualified to do any and all electrical wiring tasks. True _____ False _____

41. Live parts to which an employee may be exposed are required to be de-energized before the employee works on or near them, unless _____.

42. The OSHA regulations provide rules regarding _____ to make sure that the electrical equipment being worked on will not inadvertently be turned on while someone is working on the supposedly dead equipment.

43. Nearly _____ % of the electrocutions each year are from 120-volt systems.

44. *NEC 590.6(A)* and *(B)* require that ground-fault circuit-interrupter protection for personnel be provided for all 125-volt, single-phase, _____-ampere receptacle outlets used for temporary wiring.

45. Section _____ of the *NEC* contains important rules on the arrangement of the *NEC*.

Reading Electrical Working Drawings—Entry Level

OBJECTIVES

After studying this chapter, you should be able to

- read and interpret electrical symbols used in construction drawings.
- identify the electrical installation requirements for the Drugstore.
- identify the electrical installation requirements for the Bakery.

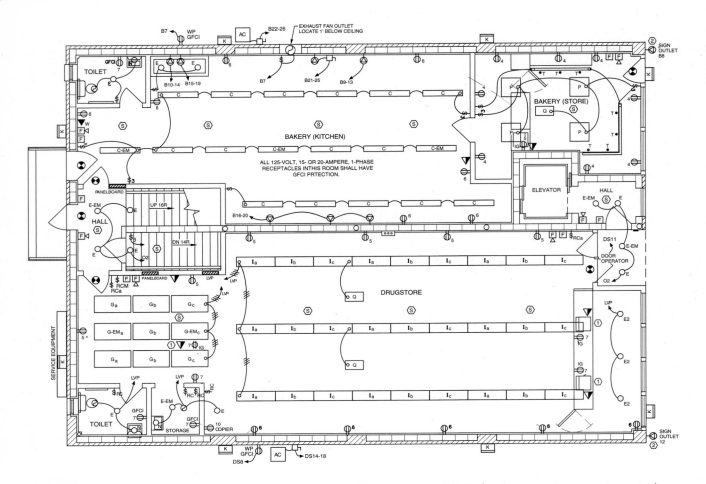

FIGURE 2-1 Electrical working drawings for the Drugstore and Bakery. (*Delmar/Cengage Learning*)

Electrical and architectural working drawings are the maps that the electrician must read and understand. Having this skill is essential to being able to install a complete electrical system and to coordinating related activities with those of workers in the other crafts. This chapter will provide the first step in developing the ability to read symbols that appear on the drawings and apply them to the electrical work.

The chapters in this text that address electrical working drawings will apply information presented in preceding chapters and will introduce special features of the building area being discussed. The questions at the end of each of these chapters will require that the student use the specifications, the drawings, and the 2011 *National Electrical Code.*

The user of this text is encouraged to study not only the electrical working drawings but also the architectural working drawings, the panelboard load calculation forms, the panelboard worksheets, and the panelboard directories. (The panelboard load calculation forms and panelboard worksheets are in Appendix A, and the panelboard schedules are on electrical drawing E-4 located in the back of the book.) Most of the information given on and in these items is yet to be discussed in detail, but much is self-evident. See Figure 2-1, which is a representation of the first floor of the building.

ELECTRICAL SYMBOLS

The electrical symbols you are most likely to find on electrical construction drawings are found in Figures 2-2 through 2-11 in this chapter. Knowing the special characteristics of these symbols will improve your ability to remember them and to interpret other symbols that are not used in these drawings. Other symbols you may find on working drawings are included in Appendix H-1 through H-13 of this textbook.

Symbols used in this book are widely understood in the electrical design and construction fields.

Other symbols may also be used, provided that a suitable explanation of their meaning is included on the drawings where the symbol is used or that a symbol legend sheet is provided. Most drawings today are produced with computer-aided design and drafting software (a CADD program). A symbols library is included with the CADD software. These symbols may vary slightly from one software package to another, but the experienced electrician can usually readily understand them.

Electrical plans are generally drawn to scale. However, graphic symbols indicate only the approximate locations of electrical equipment such as switches and receptacles and are not drawn to scale. Details are provided in the specifications or on the plans that will give mounting heights, dimensions above countertops, distances from doors, height above the floor, and so forth, for accurate locations of receptacles, luminaires, and other equipment.

If more than one symbol is located immediately adjacent to another, it usually means that a multigang box for multiple wiring devices and a single cover are to be installed. For example, if three switches are to be located in a common box and under a 3-gang cover plate, three "S" symbols will be drawn at the location. Obviously, the electrician will need to be certain a box with adequate size is selected to enclose the number of wires and devices required. The owner or architect will indicate in the specifications the maximum number of wiring devices permitted in a common box. Typically, this is three or four. Cover plates to accommodate more than four switches are often a special order item.

Several figures follow, with blueprint symbols grouped by type. Some symbols have an indicator such as the dimensions of pull boxes, size and number of conductors in a raceway or cable, whether the luminaire is supplied by a normal or emergency circuit, and any special circuitry or switching arrangement.

Raceway symbols are shown in Figure 2-2. Additional indicators are shown in Figure 2-3. For installations, an indication of conduit or cable routing on the drawing is diagrammatic—that is, conduit or cable is shown being routed from one location to another. However, actual routing of raceways or cables is left up to the electrician on the job. The electrician may want to install some conduit or cable under the floor and, in other cases, drill holes through joists or other framing members or even run the wiring method

RACEWAY SYMBOLS	
PREFERRED SYMBOL	**DESCRIPTION**
————————	Conduit concealed in finished areas, exposed in unfinished areas.
– – – – – – – – –	Conduit concealed in or under floor slab.
∿	Nonrigid raceway system.
—— N or E ——	N = Normal E = Emergency circuit.
– – – – – T – – – – –	Underfloor telecommunications raceway.
– – – – – PT – – – – –	Underfloor raceway for power and telecommunications.
– – – – – S – – – – –	Underfloor signal raceway.
– – – – – PTD – – – – –	Underfloor raceway for power, telephone, and data.
– – – – – UCP – – – – –	Undercarpet flat conductor cable (FCC) wiring system, power.
– – – – – UCT – – – – –	Undercarpet flat conductor cable (FCC) wiring system, telephone.
– – – – – UCD – – – – –	Undercarpet flat conductor cable (FCC) wiring system, data.

FIGURE 2-2 Raceway symbols.
(*Courtesy NECA 100-2006*)

above ceiling joists. Note that Type MC cable can most often be substituted for raceways such as electrical metallic tubing (EMT) for branch circuit and feeder wiring unless restricted by the local authority. The labor units for installing Type MC cable are often lower than for EMT, so there may be an economic advantage to using the cable for branch-circuit wiring in commercial buildings. Some smaller commercial buildings are permitted to be wired with nonmetallic sheathed cable. Specific occupancies may require specific wiring methods. For example, patient care areas of doctors' and dentists' offices are required to be wired in a metallic wiring method such as EMT or appropriate Type AC or Type MC cables, in compliance with *Article 517*.

Figure 2-3 shows raceway indicator symbols. These symbols supplement or modify the symbols

RACEWAY INDICATOR SYMBOLS	
PREFERRED SYMBOL	**DESCRIPTION**
	Conduit stub. Terminate with bushing or cap if underground.
 ——o	Conduit turning up.
 ——●	Conduit turning down.
SZ 2C, 4#1&1#6GND. OR SZ 53cm, 4#1&1#6GND.	Indicates trade size 2" or 53 mm conduit with (4) 1 AWG and (1) 6 AWG ground.
(2)SZ 2C , 4#1&1#6GND. OR (2)SZ 53cm, 4#1&1#6GND.	Indicates (2) trade size 2" or 53 mm conduits with (4) 1 AWG and (1) 6 AWG ground conductors in each conduit.
L211–1,3 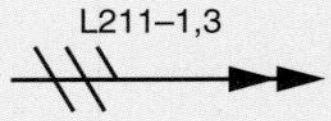	Homerun to panelboard. Number of arrows indicates number of circuits. (Example: Homerun to panel L211 CKTS. #1 and #3.)
	Flexible connection to equipment.
 ——●	Direct connection to equipment.
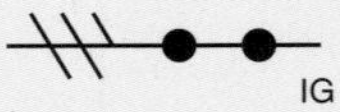 IG	Branch circuit, full hashes indicate ungrounded—"hot" (or switch-leg)—circuit conductors. Half hashes indicate grounded neutral circuit conductors. (No hashes indicates 1 hot and 1 neutral.) Dots indicate grounding conductors. Equipment bond size U.N.O. "IG" indicates an isolated grounding conductor.

FIGURE 2-3   Raceway indicator symbols. (*Courtesy NECA 100-2006*)

shown in Figure 2-2. When concealed raceways are drawn on plans, they usually are shown as curved lines. Hash marks drawn across raceway or cable lines indicate the *number* and *use* of the installed conductors. There are different acceptable ways of using hash marks. One way is to use full slashes to indicate "hot" (or switch leg) conductor(s), and half slashes to indicate grounded neutral conductor(s). No slashes indicate one "hot" and one grounded conductor. A dot(s) indicates an equipment grounding conductor(s). The letters "IG" are added near the dot to indicate an isolated-insulated grounding conductor(s). Another way is to use long

hash marks to indicate a neutral (white) conductor and short hash marks to indicate "hot" ungrounded conductors. Check the plans and/or specifications for a Symbol Schedule to be sure you fully understand the meaning of the hash marks indicated on the plans you are working with. An arrowhead at the end of a branch-circuit symbol indicates that the raceway goes from this point to the panelboard but will no longer be drawn on the plans. This symbol is used to avoid the graphic congestion created if all the lines were to come into a single point on the plans. The small numbers indicate which branch circuits are to be installed in the raceway. As the overcurrent devices in a panelboard are usually numbered with odd numbers on the left and even on the right, it is common to see groups of odd or even numbers.

When raceways are installed vertically in the building, from one floor to another, the vertical direction may be represented by the arrow symbol inside a circle. A dot represents the head of the arrow and indicates that the raceway is headed upward. A cross represents the tail of the arrow and that the raceway is headed downward. When both are shown, it means the raceway passes through. This symbol should be shown in the exact same location on the next, or previous, floor plan but should indicate the opposite direction.

If the raceway is for telephone system use, the line will be broken and an uppercase *T* will be inserted.

Architects and engineers do not often indicate the size or dimensions of pull or junction boxes that are required but leave it up to the electrician to follow the *Code* rules for proper sizing of these boxes. The location of specific raceways, such as underfloor raceway systems, are often shown on the scaled drawings, so the electrician must carefully follow the layout. The location of power, telephone, and data outlets will also be shown on the scaled floor drawings. It is obviously critical to get these locations correctly positioned before the concrete floor is poured. Symbols for pull and junction boxes and busways are located in Figure H-1 of Appendix H.

The location of luminaires (often referred to in trade jargon as "fixtures") is often specifically identified on the construction drawings. See Figure 2-4 for typical symbols used on construction drawings. The specific type of luminaire to be installed as well as the manufacturer and model is often identified in the specifications or in a luminaire schedule. Luminaires must be designed and installed to be

LUMINAIRE (LIGHTING FIXTURE) SYMBOLS	
PREFERRED SYMBOL	**DESCRIPTION**
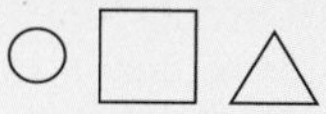 ○ □ △	Luminaire: (drawn to approximate shape and to scale or large enough for clarity).
	Luminaire strip type (length drawn to scale).
	Fluorescent strip luminaire.
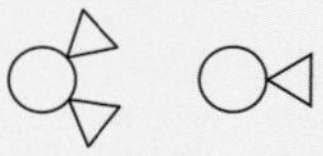	Fixture—double- or single-head spotlight or floodlight.
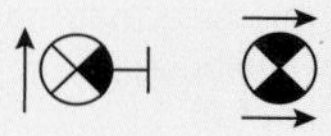	Exit luminaire fixture. Arrows and exit face as indicated on drawings (mounting heights to be determined by job specifications).
 ▽ ▽ ▽	Light track. Length as indicated on the drawings, with number of fixtures as indicated on drawings, and indicated in the fixture schedule.
	Emergency battery remote luminaire heads.
	Emergency battery unit with luminaire heads.
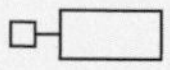	Single-luminaire pole-mounted site luminaire fixture.
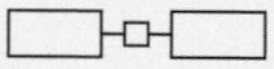	Twin-luminaire pole-mounted site luminaire fixture.
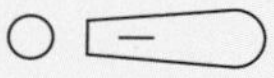	Roadway luminaire—cobra head.
	Bollard-type site luminaire.
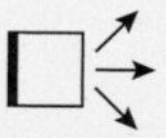	Outdoor wallpack.

FIGURE 2-4   Luminaire (lighting fixture) symbols. (*Courtesy NECA 100-2006*)

appropriate with all environmental conditions where the luminaire is installed. For example, luminaires installed at a wet location must be identified for those conditions. The luminaire manufacture often includes instructions for installing the luminaire to prevent the entrance of water, such as gasketing or caulking. These instructions must be followed. Carefully follow any mounting height specified for the luminaire.

An indication of luminaire mounting or orientation is provided, as shown in Figure 2-5.

LUMINAIRE (LIGHTING FIXTURE) BASIC AND EXTENDED MODIFIERS	
PREFERRED SYMBOL	**DESCRIPTION**
 ○ ▭	Surface-mounted fixture.
 ∅ ▭	Recessed fixture.
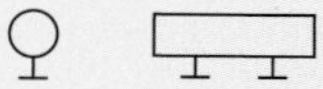	Wall-mounted fixture.
 ⊙	Suspended, pendant, chain, stem, or cable-hung fixture.
	Pole mounted with arm.
⊙	Pole mounted on top.
▣	In-ground or floor mounted. (Box around symbol.)
	Accent/directional arrow, with or without tail. (Drawn from photometric center in direction of optics or photometric orientation.)
 ○—	Directional aiming line. (Drawn from photometric center and may be extended to actual aiming point if required.)
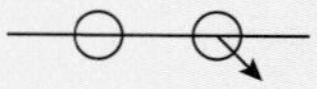	Track mounted; length, luminaire types and quantities as shown. (Track length drawn to scale.)
	Luminaire providing emergency illumination. (Filled in.)
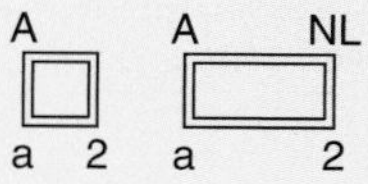 A A NL a 2 a 2	Standard designations for all luminaire fixtures. "A" = Fixture type, refer to fixture schedule "NL" = Unswitched night light "2" = Circuit number "a" = Switch control
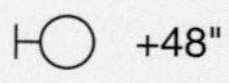 ⊢○ +48"	Mounting height.
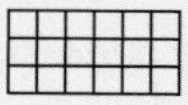	Louvers.
	Recessed, emergency fixture.

NOTE: Modifiers are shown with specific base symbols for clarity. Each modifier can be used with any of the base symbols.

FIGURE 2-5   Luminaire (lighting fixture) basic and extended modifiers. (*Courtesy NECA 100-2006*)

	OUTLET AND RECEPTACLE SYMBOLS AND NOTATIONS		
PREFERRED SYMBOL	**DESCRIPTION**	**PREFERRED SYMBOL**	**DESCRIPTION**
	Floor duplex receptacle. F = flush MTD. S = surface MTD.		Multioutlet assembly with outlets on centers as indicated on the drawings and in the specifications, mounted 6" above counter or at height as directed. A—indicates type.
	Duplex convenience receptacle. 20A 125V.	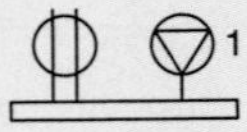	Multioutlet assembly, devices as indicated.
EP–2 CKT.1 	Duplex convenience receptacle on emergency/standby circuit. Specify panelboard and circuit.	 	Special receptacle—typical notation: 1—indicates example "1" = _A,_/_V, _Pole,_Wire, _NEMA__ "2" = _A,_/_V, _Pole,_Wire, _NEMA__ "3" = _A,_/_V, _Pole,_Wire, _NEMA__
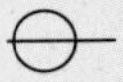	Single convenience receptacle.	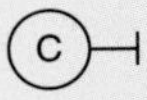	Clock hanger outlet recessed mounted 8'-0" AFF or 8" below ceiling as appropriate and as directed.
EP–2 CKT.3 	Single convenience receptacle on emergency/standby circuit. Specify panelboard and circuit.		Flush-mounted floor box, adjustable, with both power and voice/data receptacles.
	Double duplex convenience receptacle.		Duplex receptacle ceiling mounted 15 or 20A 125V.
EP–2 CKT.5 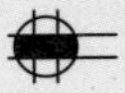	Double duplex convenience receptacle on emergency/standby circuit. Specify panelboard and circuit.		Double duplex receptacle—ceiling mounted.

Receptacles And Outlets

Typical Outlet Notations:

"a"	=	Switched outlet, "a"—indicates switch control.
"C"	=	Mounted 6" above counter or 42" AFF. Coordinate exact mounting height with architectural drawings.
"CLG"	=	Ceiling mounted.
"D"	=	Dedicated device on individual branch circuit.
"E"	=	Emergency.
"EXIST."	=	Existing device/equipment.
"F"	=	Flush floor box with fire/smoke rated penetration.
"GFCI"	=	Ground fault circuit interupter, personnel protection.
"GFPE"	=	Ground fault protection of equipment.
"H"	=	Horizontally mounted.
"IG"	=	Isolated ground receptacle with separate green ground conductor to isolated ground bus in panel.
"M"	=	Modular furniture service—provide flexible connection; coordinate exact location with furniture plans.
"PT"	=	Poke thru with 2-hour fire/smoke-rated penetration.
"S"	=	Surface-mounted floor box.
"SP"	=	Surge protection receptacle.
"T"	=	Tamper-resistant safety receptacle.
"TL"	=	Twist-lock.
"W"	=	Wall-mounted device at 48" AFF unless otherwise indicated.
"WP"	=	Weatherproof receptacle with listed coverplate for wet location with plug installed. MTD. 48" AFF unless otherwise indicated.
+XX	=	Dimensioned height.

FIGURE 2-6 Outlet and receptacle symbols and notations. (*Courtesy NECA 100-2006*)

As can be seen, these modifiers are important as they indicate the type of luminaire as well as orientation.

Figure 2-6 shows typical symbols for outlets and receptacles. Note that the term *outlet* can refer to a variety of receptacles or connection points for utilization equipment. Outlets include receptacle outlets and outlets for motors, ceiling paddle fans, and luminaires. The location of outlets and receptacles is shown on the drawings. Sometimes, these locations are approximate, and the outlet or receptacle can be moved slightly to accommodate framing members such as studs or joists. At other

SWITCHES AND SENSOR SYMBOLS	
PREFERRED SYMBOL	**DESCRIPTION**
$ or S	Single-pole switch.
$_2$ or S$_2$	Double-pole switch.
$_3$ or S$_3$	Three-way switch
$_4$ or S$_4$	Four-way switch.
$_a$ or S$_a$	Switch control of luminaire or outlet (lowercase letter).
$_{CB}$ or S$_{CB}$	Circuit-breaker switch.
$_{DT}$ or S$_{DT}$	Single-pole/double-throw switch.
$_G$ or S$_G$	Glow switch toggle, glows in off position.
$_H$ or S$_H$	Horizontally mounted—with on position to the left.
$_K$ or S$_K$	Key-operated switch.
$_{KP}$ or S$_{KP}$	Key-operated switch with pilot light on when switch is on.
$_{LV}$ or S$_{LV}$	Low-voltage switch.
$_{LM}$ or S$_{LM}$	Low-voltage master switch.
$_{MC}$ or S$_{MC}$	Momentary contact switch.
$_P$ or S$_P$	Switch with pilot light on when switch is on.
$_T$ or S$_T$	Timer switch.
$_{WP}$ or S$_{WP}$	Weatherproof single-pole switch.
$_D$ or S$_D$	Dimmer switch rated 1000W, unless otherwise indicated. "LV" = low voltage "FL" = fluorescent
 (M symbol)	Occupancy sensor, wall mounted with off-auto override switch.
 (M P symbol)	Occupancy sensor—ceiling mounted. "P"—indicates multiple switches wired in parallel.
$_{PROJ}$ or S$_{PROJ}$	Motorized projection screen raise/lower switch.

FIGURE 2-7 Switches and sensor symbols.
(*Courtesy NECA 100-2006*)

times, the location identified on the drawings is expected to be precise. The mounting height of these outlets is often included in the specifications or is shown as notations on the drawings. Carefully follow these instructions, including those for the specific height above counters or for special equipment. Having a problem positioning the box for the receptacle or other outlet? Bring this to the attention of the responsible supervisor or agent of the owner for resolution. Note that typical outlet notations provide specific information on type of outlet, mounting instructions, or switching indication.

Typical symbols for switches and sensors are shown in Figure 2-7. Carefully review the specifications or notations on the construction drawings for the height above finished floor or for location in proximity to doors. Switches are not usually permitted on the hinged side of the door but are required to be located on the strike side. In some cases, the switch is required to be located on the outside of the door to the room or area.

Special switches such as keyed type and occupancy sensors are indicated by symbols on the drawings. See Figure 2-7.

The lighting in some buildings will be controlled by computers or lighting relays. Occupancy sensors and low-voltage switching can be used to facilitate these controls. Other switching designs call for 3-way or 4-way controls. Three-way switches allow one or more luminaires to be controlled from two locations. Four-way switching involves inserting one or more 4-way switches between a 3-way switch that is located at each end of the switch legs. Additional information on wiring 3-way and 4-way switches is included in Chapter 5 of this text.

Figure 2-8 shows symbols used to identify the location of disconnecting means and other components of circuits for motorized and HVAC equipment. These symbols often include the ampere or other rating of the equipment used. For example, the symbol for a fusible-type disconnect switch indicates the frame (switch) ampere rating as well as the fuse rating.

Special control schemes are often provided on the drawings. Carefully follow the requirements in *Article 430* for motors and in *Article 440* for air-conditioning equipment.

MOTORIZED AND HVAC EQUIPMENT—CONTROLS, SYMBOLS	
PREFERRED SYMBOL	**DESCRIPTION**
xxA ▢	Disconnect switch, unfused type, size as indicated on drawings. "xxA" indicates amperage.
$\frac{xxAF}{yyAT}$ ⊏F⊐	Disconnect switch, fused type, size as indicated on drawings. "xxAF" indicates frame size. "yyAT" indicates fuse rating.
$\frac{xxAF}{yyAT}$ ⊏CB⊐	Enclosed circuit breaker, size as indicated. "xxAF" indicates frame size. "yyAT" indicates trip size.
⊏C⊐	Magnetic contactor; size as indicated on drawings.
RV ⊠ NEMA x	Magnetic motor starter. "RV" indicates reduced voltage. Starter size as indicated.
F ⊠ NEMA x $\overline{xxA-xP}$	Combination magnetic starter and disconnect switch. Starter size and fuse rating as indicated.
ATC	Automatic temperature control panel.
CP	Equipment control panel.
Ⓡ	Relay.
Ⓐ	Aquastat.
Ⓕ	Firestat.
Ⓗ	Humidistat.
Ⓣ L	Line voltage thermostat.
Ⓣ LV	Low-voltage thermostat.
Ⓣ	Thermostat.
Ⓢⓥ	Solenoid valve.
TS	Time switch.
AF	Airflow switch.
EP	Electric/pneumatic switch.
FS	Flow switch.
IC	Irrigation control.
LS	Limit switch.
PE	Pneumatic/electric switch.
PC	Photo cell or photo control.
PS	Pressure switch.

FIGURE 2-8 Motorized and HVAC equipment controls and symbols. (*Courtesy NECA 100-2006*)

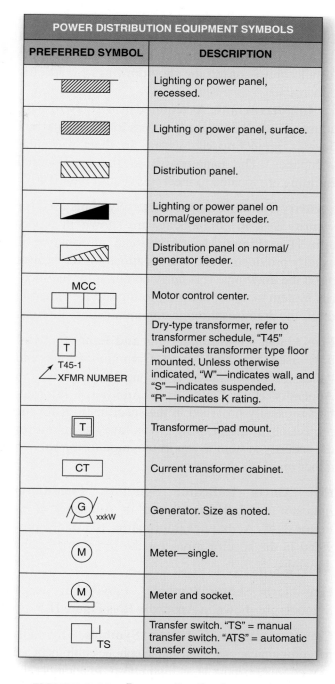

MOTORIZED AND HVAC EQUIPMENT SYMBOLS	
PREFERRED SYMBOL	DESCRIPTION
(3)	Motor "3" —indicates horsepower.
(D)	Motorized damper.
BBH	Baseboard heater.
BBH	Baseboard heater with receptacle (on different circuit).
	Unit-type heater.
	Ceiling exhaust fan.
	Paddle fan.
	Wall fan.
WH	Water heater.

FIGURE 2-9 Motorized and HVAC equipment symbols. (*Courtesy NECA 100-2006*)

POWER DISTRIBUTION EQUIPMENT SYMBOLS	
PREFERRED SYMBOL	DESCRIPTION
	Lighting or power panel, recessed.
	Lighting or power panel, surface.
	Distribution panel.
	Lighting or power panel on normal/generator feeder.
	Distribution panel on normal/generator feeder.
MCC	Motor control center.
T / T45-1 / XFMR NUMBER	Dry-type transformer, refer to transformer schedule, "T45" —indicates transformer type floor mounted. Unless otherwise indicated, "W"—indicates wall, and "S"—indicates suspended. "R"—indicates K rating.
T	Transformer—pad mount.
CT	Current transformer cabinet.
G xxkW	Generator. Size as noted.
M	Meter—single.
M	Meter and socket.
TS	Transfer switch. "TS" = manual transfer switch. "ATS" = automatic transfer switch.

FIGURE 2-10 Power distribution equipment symbols. (*Courtesy NECA 100-2006*)

Included in Figure 2-9 are the symbols for equipment such as motors, baseboard heaters, ceiling (paddle) fans, and water heaters.

Figure 2-10 includes symbols for distribution equipment such as panelboards, motor control centers, transformers, and transfer switches. It is particularly important to review the spacing required for distribution equipment so that the working space requirements of *NEC 110.26* are complied with. The power distribution symbols shown in this figure are supplemented by a one-line riser diagram that indicates the source and destination of feeders as well as the size of conduits, conductors to be installed, and the required overcurrent protection. Panelboard schedules are typically provided to detail the circuits and overcurrent protection required for distribution equipment.

Panelboards are distribution points for electrical circuits. They contain circuit protective devices. See *NEC Article 100* for the official definition. The previous two basic classes of panelboards, power and lighting and appliance, have been deleted. Panelboards are available to accommodate from 2 to 84 branch circuits and from 30 to several hundred amperes rating. The requirements for construction and application of panelboards are set forth in *NEC Article 408*. In the building used as an example in this text, a panelboard is located in each occupancy, so the tenant will have ready access to the overcurrent devices. The symbols are not drawn to scale, and the installer must consult the shop drawings for specific dimensions.

Important note: As required by *NEC 240.24(B)*, all tenants shall have ready access to the overcurrent devices protecting the wiring in their particular occupancy. The panelboards in the Commercial Building are located to meet this requirement.

Security System Components. Symbols used for security system components are shown in Appendix H as Figure H-2. Additional information is usually provided in the specifications for the project or on the drawings. It is fairly common for the security system to report to a monitoring service either on or off the premises through communications circuits.

Fire-Alarm Communications and Panels. Symbols used for Fire-Alarm Communications and Panels security system components are shown in Appendix H as Figure H-3. Additional information is usually provided in the specifications for the project or on the drawings. It is fairly common for the security system to report to a monitoring service either on or off the premises.

Fire Alarm Indicator Symbols. Fire-Alarm Indicator Symbols are shown in Appendix H as Figure H-4. Additional information is usually provided in the specifications for the project or on the drawings.

Fire Alarm Sensor Symbols. Fire-Alarm Sensor Symbols are shown in Appendix H as Figure H-5.

Communications—Teledata Symbols. Symbols for communications, data, and telephone equipment are located in Appendix H as Figure H-6. In addition, the plans or specifications may specify installing an empty raceway from the vicinity of the outlet to the location above a suspended ceiling. This facilitates the installation of communications cables later in the project.

Communications—Audio/Visual Symbols. Symbols for communications-audio/visual are located in Appendix H as Figure H-7.

Communications—Equipment. Symbols for communications equipment are located in Appendix H as Figure H-8.

SITE WORK	
PREFERRED SYMBOL	**DESCRIPTION**
------- UF -------	Underground feeder.
------- UT -------	Underground telephone.
------- UFA -------	Underground fire alarm.
------- UTV -------	Underground television (CATV).
—| MH |—	Manhole.
—| HH |—	Handhole.
—| E | T |—	Combination prefabricated manholes for power and tel/data systems. "E" = denotes power, "T" = denotes tel/data.

FIGURE 2-11 Site work. (*Courtesy NECA 100-2006*)

Site Work Symbols. Symbols for site work are shown in Figure 2-11. You might find an additional variation in the form of underground service lateral. Some plans will require conduit from the building to the utility transformer. Others will permit conduit to be stubbed into the ground and direct burial cable to be run to the transformer. In yet other variations, conduit is required to be installed on power poles to protect the service lateral as it rises to the pole-mounted transformers.

Schematic and One-Line Diagram Symbols. Note that these symbols that are commonly used in schematic and on one-line drawings are located in Appendix H in Figure H-9(A), (B), and (C). Schematic symbols for switches are included in Appendix H as Figure H-10.

Miscellaneous Symbols. Miscellaneous symbols are included in Appendix H as Figure H-11. These include symbols used on drawings to indicate mechanical equipment, feeders that appear on feeder schedules and sections or details.

Abbreviations. Abbreviations commonly used on electrical construction drawings are included in Appendix H as Figure H-12.

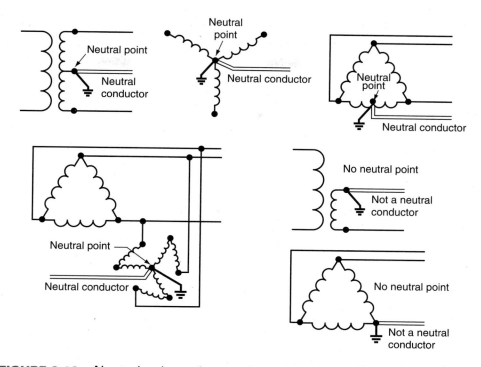

FIGURE 2-12 *Neutral point and neutral conductor. (Delmar/Cengage Learning)*

Nurse Call Systems. Nurse call systems are located in Appendix H as Figure H-13.

Conductor Designations

The terms *neutral* and *hot* are often used in trade jargon. The terms *neutral* and *grounded conductor* are defined terms but *hot*, or *hot wire*, is not. The term *ungrounded* is defined in the *NEC*. These terms are found many times in the *NEC* and in this text. These terms are often misunderstood. The *NEC* contains the following definitions:

- **Neutral Conductor.** *The conductor connected to the neutral point of a system that is intended to carry current under normal conditions.** See Figure 2-12. It is important to note that in this definition, the neutral conductor is expected to carry current under normal conditions, whereas the equipment-grounding conductor is expected to carry current only under abnormal conditions. Both connect to the neutral terminal bar at the service or source of a separately derived system.

- **Neutral Point.** *The common point on a wye-connection in a polyphase system or midpoint on a single-phase, 3-wire system; or midpoint of a single-phase portion of a 3-phase delta* *system; or midpoint of a 3-wire, direct-current system.** See Figure 2-12.

- **Ungrounded.** *Not connected to ground or a conductive body that extends the ground connection.**

Electricians might use the term *neutral* or they might use the term *neutral conductor* when referring to the grounded circuit conductor. The lower two drawings on the right in Figure 2-12 show the location of a grounded conductor that is not a neutral.

THE DRUGSTORE

A special feature of the Drugstore wiring is the low-voltage remote-control system. See Chapter 20, "Low-Voltage Remote Control," for a complete discussion. This system offers control flexibility that is not available in the traditional control system. The switches used in this system operate on 24 volts, and the power wiring, at 120 volts, goes directly to the electrical load. This reduces branch-circuit length and voltage drop. A switching schedule gives details on the system operation, and a wiring diagram provides valuable information to the installer.

*Reprinted with permission from NFPA 70-2011.

One of the reasons for the low popularity of this system is the scarcity of electricians who are prepared to install a low-voltage control system. The system specified for the Drugstore has been around a long time and is still being used. It is the most basic type of low-voltage control available today and is discussed in Chapter 20.

The student is encouraged to request manufacturer's literature from any electrical distributor. Check out the manufacturer's Web sites by browsing the Web. Several companies manufacture low-voltage control systems.

Different types of illumination systems have been selected for many of the spaces in the building. The student should observe the differences in the wiring requirements.

In the merchandise area, several luminaires are installed in a continuous row. It is necessary to install electrical power to only one point of a continuous row of luminaires. From this point, the conductors are installed to supply other luminaires in the wiring channel of each luminaire. In the pharmacy area, a luminous ceiling is shown. This illumination system consists of rows of strip fluorescents and a ceiling that will transmit light. The installation of the ceiling, in many jurisdictions, is the work of the electrician. For this system to be efficient, the surfaces above the luminous ceiling must be highly reflective (white).

THE BAKERY

For the production area of the Bakery, a special luminaire is selected to prevent contamination of the food products. These units are totally enclosed and require a separate electrical connection to each luminaire. The luminaires may be supplied by installing a conduit in the upper-level slab or on the ceiling surface. In the sales area, more attractive luminaires have been selected.

A conventional control system is to be installed for the lighting in the Bakery. The goal of the system is to provide control at every entry point so that a person is never required to walk through an unlighted space. Often this requires long switching circuits, such as the three-point control of the main lighting in the work area.

The electrician may be responsible for making changes in conductor size to compensate for excessive voltage drop. This requires the electrician to be alert for high loads on long circuits, such as the control circuit on the bakery work area lighting.

The panelboard worksheets and feeder load calculations for the Drugstore, Bakery, Insurance Office, Real Estate Office, Beauty Salon, and the owner's loads are found in Appendix A of this text. The panelboard schedules are found on electrical drawing E-4. The drawings are located in the envelope inside the back cover of the book.

Table B-1 in Appendix B is a table of useful electrical formulas. The Watts Wheel is included in Appendix B as Table B-2.

You will want to refer to these tables and formulas often as you study *Electrical Wiring— Commercial*.

REVIEW

Refer to the *National Electrical Code* or the working drawings when necessary. Where applicable, responses should be written in complete sentences.

Answer Questions 1–7 by identifying the symbol and the type of installation (wall, floor, or ceiling) for the boxes.

1. ⊖ _____

2. ⬤ _____

3. S_{KP} or S_{KP} _____

4. ○ _____

5. ▼F _____

6. ⊡⊣ _____

7. –○– _____

8. How many duplex receptacle outlets are to be installed in the Drugstore basement storage area? _____

9. The duplex receptacle outlets in the Drugstore basement storage area are supplied from which panelboard? _____

10. The duplex receptacle outlets in the Drugstore basement storage area are to be connected to which branch circuit(s)? _____

11. How many lighting outlets are to be installed in the Drugstore basement storage area? _____

12. What style(s) of luminaires is/are to be installed in the Drugstore basement storage area? _____

13. How may the luminaires for the Drugstore basement storage area be installed? _____

_____

14. List the luminaires required for the Drugstore including the basement storage area. Give the style, count, and mounting method for each style.

Style _____	Count _____	Mounting method _____
Style _____	Count _____	Mounting method _____
Style _____	Count _____	Mounting method _____
Style _____	Count _____	Mounting method _____
Style _____	Count _____	Mounting method _____
Style _____	Count _____	Mounting method _____
Style _____	Count _____	Mounting method _____
Style _____	Count _____	Mounting method _____
Style _____	Count _____	Mounting method _____
Style _____	Count _____	Mounting method _____
Style _____	Count _____	Mounting method _____

15. From the information you have studied about the various electrical symbols, identify the following symbols, some of which are not shown in the symbol schedule.

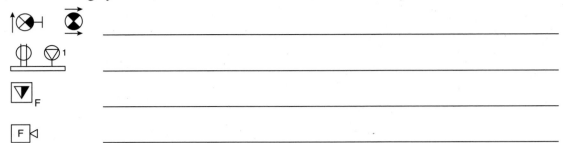

_____

_____

_____

_____

16. What is the source of supply and circuit numbers for the air-conditioner compressor unit for the Drugstore? _____

17. In the spaces, draw in by hand the symbol for each of the following:

   a. Wall-mounted exit luminaire _____

   b. Bollard-type site luminaire _____

   c. A double duplex receptacle _____

   d. A key-operated switch _____

   e. Enclosed circuit breaker with a 400 A frame size and 350 A trip _____

   f. A fluorescent dimmer switch _____

   g. A 3-HP motor _____

   h. A surface-mounted panelboard _____

   i. A flush panelboard on normal and generator feeder _____

   j. A capacitor _____

   k. One-line diagram symbol for a corner-grounded delta transformer _____

   l. One-line diagram symbol for a system- or equipment grounding connection _____

   m. One-line diagram symbol for a start push button _____ and a stop push button _____

Calculating the Electrical Load

OBJECTIVES

After studying this chapter, you should be able to

- determine the minimum lighting loading for a given area.
- determine the minimum receptacle loading for a given area.
- determine the minimum equipment loading.
- determine a reasonable calculated and connected load.
- tabulate the unbalanced or neutral load.
- apply the factors for continuous loads where appropriate.
- apply the factors for noncoincident loads where appropriate.

⎯⎯⎯⎯⎯| INTRODUCTION

Note: The first two chapters in this text covered the basics. You will now be introduced to specific *National Electrical Code* requirements as they become important to the specific topic being discussed.

The first sentence in the *NEC* sets the stage for electrical installations. The wording is extremely important.

90.1 Purpose.

(A) Practical Safeguarding. The purpose of this Code is the practical safeguarding of persons and property from hazards arising from the use of electricity.

(B) Adequacy. This Code contains provisions that are considered necessary for safety. Compliance therewith and proper maintenance results in an installation that is essentially free from hazard but not necessarily efficient, convenient, or adequate for good service or future expansion of electrical use.

Informational Note: Hazards often occur because of overloading of wiring systems by methods or usage not in conformity with this Code. This occurs because initial wiring did not provide for increases in the use of electricity. An initial adequate installation and reasonable provisions for system changes provide for future increases in the use of electricity.*

What about Future Growth?

As you study *Electrical Wiring—Commercial,* you will learn how to make load calculations in conformance to the *NEC.* You will also note that the authors have added additional load values in certain instances. This provides for future growth. The *NEC* talks about future growth in *90.8.* As you read through this section, you'll find that the text gives a recommendation, not a requirement. Providing spare capacity in the design of the electrical system, such as leaving room in conduits and other raceways, along with unused spaces in panelboards, accommodates future growth in the electrical system and should be practiced. This is above and beyond the minimum *NEC* requirements but is considered good practice by electricians, contractors, and consulting engineers.

Important: Electrical calculations always involve certain values that come about from rounding up, rounding down, or following certain *Code* requirements.

For example, we might solve for amperes using the following formula:

$$I = \frac{kVA \times 1000}{E \times 1.732}$$

The value of 1.732 is the square root of 3…$\sqrt{3}$. Using a calculator to find $\sqrt{3}$, we find 1.732050808. To be practical, the electrical industry uses 1.732, although some may shorten this to 1.73.

In the above formula, when E equals 208 volts, the result of E $\times$ 1.732 = 360.256. Again, to be practical, we use 360 in many calculations.

The value of 208 is the product of 120 $\times$ 1.732. This comes from 208Y/120 volt 3-phase systems. These systems are created by connecting three, 120-volt transformers in a Y configuration with one end of each transformer connected in the center. The voltage between outer points of the single-phase transformers is about 208 volts. This is another example of rounding off. Actually, 120 $\times$ 1.732 = 207.84. Multiply 120 by 1.732050808 and we get 207.846097.

Because electrical systems may operate at slightly varying voltages, *220.5(A)* contains the following rule to simplify calculations by telling us,

Voltage. Unless other voltages are specified, for purposing of calculating branch-circuit and feeder loads, nominal system voltages of 120, 120/240, 240, 208Y/120, 240, 347, 480Y/277, 480, 600Y/347 and 600 volts shall be used.*

The word *nominal* means "in name only, not in reality."

A branch-circuit nominal voltage might be 120 volts, but the actual operating voltage be 122, 117, 115, 110, or just about anything. As an example, the *Code* decided that for consistency, calculations will use the value of 120 volts.

---

*Reprinted with permission from NFPA 70-2011.

Fractions of an Ampere. Except where the calculations result in a major fraction of an ampere (0.5 or larger), such fractions are permitted to be dropped.*

So as you work your way through this text, don't get "hung up" and waste time trying to pin down precise calculation results.

THE ELECTRICAL LOAD

To plan any electrical wiring project, the first step is to determine the load the electrical system is to serve. Only with this information can components for the branch circuits, the feeders, and the service be properly selected. The *NEC* provides considerable guidance in determining the minimum loading that is appropriate, for a given occupancy, for feeders and the service. Often the electrician or electrical contractor is asked to generate this information. It is quite common for electrical inspection departments to require load calculation information to be submitted before the permit will be issued for the project. This unit of this text provides a foundation for performing load calculations so proper selection of electrical circuit components can be accomplished.

NEC Article 220 establishes the procedure that is to be used to calculate electrical loads for services, feeders, and branch circuits. *Section 220.40* states the overall concept for load calculations: *The calculated load of a feeder or service shall not be less than the sum of the loads on the branch circuits supplied, as determined by Part II of this article, after any applicable demand factors permitted by Part III or IV or required by Part V have been applied.*

Load calculations for many types of equipment such as for air-conditioning equipment, cranes and hoists, fire pumps, and electric welders are provided in other articles in the *NEC*. See *Table 220.3* for this information.

As we will see, lighting loads in commercial occupancies are generally considered to be continuous, that is, in operation for 3 hours or more. To prevent overheating conductors and terminals for equipment such as circuit breakers and connections for fusible switches, additional capacity not less

than 125% of the continuous load is required to be added to any noncontinuous load. See the following table for application of this rule.

Component	*NEC* Section	Continuous Load	Noncontinuous Load
Branch-circuit conductor	*210.19 (A)(1)*	125%	100%
Branch-circuit overcurrent device	*210.20(A)*	125%	100%
Feeder conductor	*215.2(A) (1)*	125%	100%
Feeder overcurrent device	*215.3*	125%	100%
Service conductors	*230.42(A)*	125%	100%

These factors must be applied where appropriate so that the conductors and overcurrent protection as well as the equipment that contains the overcurrent protection will be sized properly. An exception is permitted to be applied to these rules for equipment that is rated at 100%, or continuous duty, as well as for grounded (often referred to as "neutral") conductors that do not connect to the overcurrent device. Most overcurrent protective devices in the up to 600-volt class of equipment are not rated for continuous, or 100%, duty and must be increased in ampacity to compensate for continuous loads.

Lighting loads can either be a calculated load based on the area in square feet (square meters), using a unit load as determined from *NEC Table 220.12*, or may be the actual connected load, obtained by referring to the nameplate on the equipment. Always use the larger load determined by the methods.

Using the Drugstore in this text as an example case, the application of these procedures will be illustrated.

- The phrase *calculated load* will be used to designate when the value is in compliance with the requirements of *NEC Article 220*. This calculated load is often determined by selecting an appropriate unit load and multiplying this by the number of units.

---

*Reprinted with permission from NFPA 70-2011.

TABLE 3-1

Table 220.12 General Lighting Loads by Occupancy

| Type of Occupancy | Unit Load | |
	Volt-Amperes/ Square Meter	Volt-Amperes/ Square Foot
Armories and auditoriums	11	1
Banks	39[b]	3½[b]
Barber shops and beauty parlors	33	3
Churches	11	1
Clubs	22	2
Court rooms	22	2
Dwelling units[a]	33	3
Garages — commercial (storage)	6	½
Hospitals	22	2
Hotels and motels, including apartment houses without provision for cooking by tenants[a]	22	2
Industrial commercial (loft) buildings	22	2
Lodge rooms	17	1½
Office buildings	39[b]	3½[b]
Restaurants	22	2
Schools	33	3
Stores	33	3
Warehouses (storage)	3	¼
In any of the preceding occupancies except one-family dwellings and individual dwelling units of two-family and multifamily dwellings:		
Assembly halls and auditoriums	11	1
Halls, corridors, closets, stairways	6	½
Storage spaces	3	¼

[a]See 220.14(J).
[b]See 220.14(K).

Reprinted with permission from NFPA 70-2011.

- The phrase *connected load* will be used to designate the value of the load as it actually exists and is to be connected to the electrical system.

If the results of the load study differ, use the larger of these two methods for sizing the electrical system components. The unit loads provided for calculating general lighting loads are shown in Table 3-1.

ENERGY CODE CONSIDERATIONS

Several local jurisdictions have adopted a building or energy code that limits the maximum lighting load an occupancy is permitted to have, in an effort to force the use of more energy-efficient lighting equipment. Although the *NEC* does not recognize this limitation, the authority having jurisdiction for enforcing the code should be consulted before the design is completed, as reduced size feeders or the service may be permitted.

LIGHTING LOADING CALCULATIONS

The following guidelines should be followed when calculating the lighting load:

- Use *NEC Table 220.12*, shown in this text as Table 3-1, to find the value of volt-amperes per square foot (or per square meter) for general lighting loads, or use the actual volt-amperes if that value is higher.
- For show window lighting, allow 200 volt-amperes per linear foot (660 volt-amperes per linear meter), or use the connected load if that value is higher; see *NEC 220.14(G)*.

The *NEC* defines a show window in *Article 100* as: *Any window used or designed to be used for the display of goods or advertising material, whether it is fully or partly enclosed or entirely open at the rear and whether or not it has a platform raised higher than the street floor level.**

At least one receptacle outlet is required to be installed above each 12 ft (3.7 m) linear section of show window. To ensure the required receptacle(s) is not a great distance above the show window, which would necessitate the use of extension cords, the receptacle(s) is required to be installed within 18 inches (450 mm) of the top of the show window; see *NEC 210.62*.

If lighting track is installed, 150 volt-amperes is to be allowed for every 2 feet (600 mm) of track or any fraction thereof; see *NEC 220.43(B)*. An exception provides that if the track lighting is supplied through a device that limits the current to the track, the load is permitted to be calculated based on the rating of the device used to limit the current.

The application of these requirements is illustrated in Table 3-2.

You will often see the term "nonlinear" load. The high harmonic currents associated with nonlinear loads cause havoc by overheating the neutral conductor. Neutral conductors need to be increased in size to safely handle these high harmonic currents. This subject is covered in detail in Chapter 8 of this text.

*Reprinted with permission from NFPA 70-2011.

TABLE 3-2

Luminaires connected in the Drugstore.

Luminaire Type and Load	Continuous Load	Noncontinuous Load
Style I @ 61 VA	27 = 1647 VA	0 VA
Style G @ 80 VA	7 = 560 VA	0 VA
Style G-EM @ 86 VA	2 = 172 VA	0 VA
Style E2 @ 26 VA	3 = 78 VA	0 VA
Style E @ 14 VA	2 = 28 VA	0 VA
Style E-EM @ 20 VA	0	1 = 20 VA
Style EXIT @ 4.6 VA	2 = 9 VA	0 VA
Style N @ 34 VA	0	2 = 68 VA
Style Q @ 24 VA	0	2 = 48 VA
Total connected load	2494 VA	136 VA

General Lighting

The minimum lighting load to be included in the calculations for a given type of occupancy is determined from *NEC Table 220.12,* which is shown in Table 3-1. For a store occupancy, the table indicates that the unit load per square foot is 3 volt-amperes and that the unit load per square meter is 33 volt-amperes. Therefore, for the first floor of the Drugstore, the lighting load allowance is

60 ft × 23.25 ft × 3 VA per sq ft = 4185 VA
(18 m × 7 m × 33 VA per sq m = 4158 VA)

Because this value is a minimum, it is also necessary to determine the connected load. The greater of the two values becomes the lighting load, in accordance with the requirements of *NEC Article 220.* These loads for lighting, whether calculated or connected loads are used, are considered to be continuous and are placed in the Continuous Loads column of Table 3-2.

As presented in detail in a later chapter, the illumination in the sales area of the Drugstore is provided by fluorescent luminaires equipped with

two lamps and a ballast. Previously, ballasts had a significant effect on the volt-ampere requirement for a luminaire, as they were typically low-power-factor loads. Modern ballast constructions with energy-saving features include ballasts that are of high power factor and are near unity, or 1 for 1. As a result, we are including the lamp wattage rating as the ballast load. A schedule of luminaires is included in Appendix A and in Chapter 15 of this text.

The *NEC* addresses luminaires with ballasts as *Inductive Lighting Loads* in *220.18(B).* It requires that the ballast load be used in load calculations, rather than the rating of the lamps. Be sure to verify the ballast load current when doing load calculations, as well as for actual branch-circuit layout. Watts can be used to determine the cost of operation; volt-amperes are used to determine the size of the conductors and overcurrent devices.

The connected lighting load for the first floor of the Drugstore is shown in Table 3-2.

Note that due to the fact that we are using high-power-factor ballasts or LED power supplies, we are treating the lamp watts as equal to volt-amperes. Incandescent lamps are not being used in the luminaires for the building, to achieve the highest efficiency and to comply with Federal Department of Energy use reductions. It is important that the connected load be tabulated as accurately as possible. The values will not only be used to select the proper electrical components, but they may also be used to predict the cost of energy for operating the building.

Storage Area Lighting

According to *NEC Table 220.12,* the minimum load for basement storage space is 0.25 volt-amperes per square foot (3 volt-amperes per square meter). These loads for lighting are not considered to be continuous, as they will not likely be operated for 3 hours or more, and are placed in the Noncontinuous Loads column of Table 3-3.

The load allowance for this space is

858 sq. ft × 0.25 VA per sq. ft = 216 VA
(79.7 sq. m × 3 VA per sq. m = 239 VA)

The connected load for the storage space for the Drugstore is shown in Table 3-3.

TABLE 3-3	
Luminaires connected in Drugstore storage area.	
5 Style B luminaires @ 64 VA	320 VA
4 Style B-EM luminaires @ 70 VA	280 VA
Total connected load	600 VA

Note: Luminaire loads in the storage spaces are considered noncontinuous as it is unlikely the luminaires will operate for 3 hours or more.

OTHER LOADS

The remaining loads consist of

- miscellaneous receptacle outlets both in the basement storage area and the main floor;
- a receptacle for servicing the outdoor AC unit;
- a receptacle for the copy machine;
- receptacle outlets for the cash registers;
- an outlet for the electric sign;
- an outlet for the electric door operators;
- exhaust fans for the toilet room and workroom;
- a water heater located in the basement storage area;
- an unit heater for the basement storage area; and
- indoor and outdoor air-conditioning equipment;

Receptacle Outlets

- A single piece of equipment consisting of a multiple receptacle comprised of four or more receptacles shall be calculated at not less than 90 volt-amperes per receptacle. As shown in Figure 3-1, the first outlet has one receptacle on a strap or yoke for a calculated load of 180 VA. The second outlet has 2 receptacles on a strap or yoke for a calculated load of 180 VA. The third outlet shows two straps or yokes for a total of 4 receptacles resulting in a calculated load of 360 VA. The fourth outlet shows a single piece of equipment or wiring device with four receptacles. This represents a calculated load of 360 VA. See *NEC 220.14(I)*.

- The actual rating is used for specific loads such as receptacles for a copy machine and cash registers, *NEC 220.14(A)*.

The allowance for the receptacle outlets for nonspecific loads is

15 on the first floor and 6 in the basement storage area for a total of 21 outlets @ 180 VA = 3780 VA

Receptacle for Servicing AC Equipment

The receptacle outlet is required to be located at the same level as the HVAC equipment and within 25 ft (7.56 m) of the equipment, by *NEC 210.63*. *NEC 210.8(B)(3)* requires that the receptacle be GFCI protected. No load allowance is stipulated other than the 180 VA from *NEC 220.14(I)*. The arbitrary allowance is

1 receptacle outlet for servicing the HVAC @ 1500 VA per outlet = 1500 VA

Enclosures for receptacles in damp locations shall be weatherproof when an attachment plug cap is not inserted, *NEC 406.8(A)*. Examples of this would be under an overhanging roof or a porch that has a roof over it, and similar locations that will be protected from a pounding rain.

Enclosures for receptacles in wet locations shall be weatherproof whether or not the attachment plug cap is inserted, *NEC 406.8(B)*.

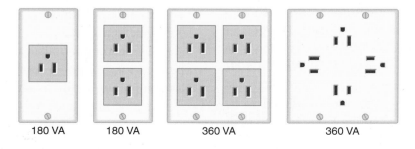

| 180 VA | 180 VA | 360 VA | 360 VA |

FIGURE 3-1 Minimum receptacle outlet allowance.

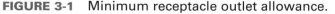

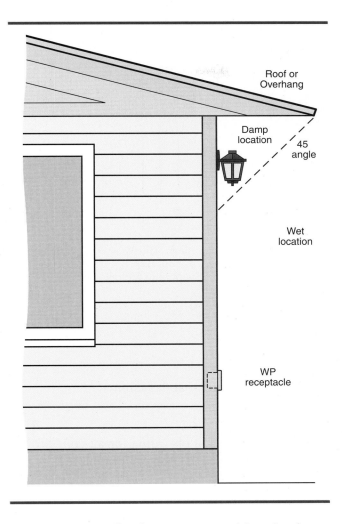

FIGURE 3-2   Outdoor areas considered to be damp or wet locations.

Some inspectors consider the area created by a 45° angle from the edge of an overhanging roof to the wall to be a damp location. Below the point where the 45° line meets the wall is considered to be a wet location. This is a judgment call. See Figure 3-2.

All straight blade receptacles installed in wet locations rated 15- and 20-ampere, 125- and 250-volts, installed in the previously described weatherproof enclosures shall be listed as weather-resistant type, *NEC 406.8(B)(1)*.

Sign Outlet

Article 600 in the *NEC* covers the installation of signs and outline lighting. Here, we shall discuss the required load for a sign outlet and the very important safety requirement—the disconnecting means.

A sign outlet

- is a requirement of *NEC 600.5(A)*.
- is assigned a minimum allowance of 1200 volt-amperes in *NEC 220.14(F)* for each branch circuit required in *NEC 600.5(A)*.
- branch circuit(s) shall be rated at least 20 amperes and shall not supply other loads.
- has a 125% factor is applied in the calculation, as the sign is considered to be a continuous load.

1 sign outlet @ 1200 VA per outlet = 1200 VA.

In accordance with *NEC 600.6*, each sign and outline lighting system is required to have a disconnecting means (switch or circuit breaker) that

- is externally operable.
- opens all ungrounded (hot) conductors.
- is within sight of the sign or is capable of being locked in the off position.
- does not rely on portable means of locking the disconnect off.

Exception: Indoor signs that are cord and plug connected do not require another disconnect. The cord and plug connection serves as the required disconnecting means.

MOTORS AND APPLIANCES

- *NEC Article 422* sets forth the installation requirement for appliances. In accordance with *NEC 220.14(A)*, outlets for specific appliances are required to be calculated on the ampere rating of the appliance or load served. The panelboard load calculation includes loads for the cash registers, copy machine, exhaust fans, electric door operator, and water heater.
- *NEC Article 424* applies to fixed electric space-heating equipment. The load for the unit heater in the basement storage space is shown in the panelboard load calculation form.
- *NEC Article 430* applies to motors, including those that are a part of motor-operated appliances unless "specifically amended" by *NEC Article 422*.

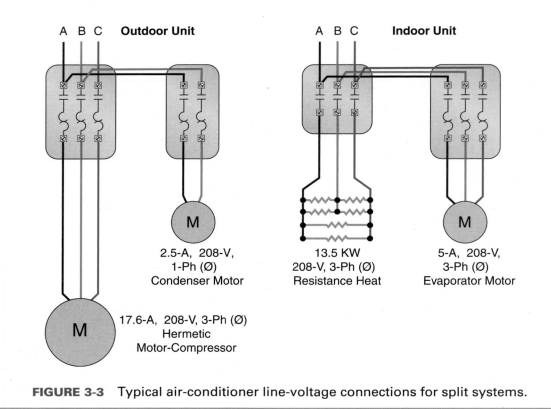

FIGURE 3-3 Typical air-conditioner line-voltage connections for split systems.

- *NEC Article 440* addresses air-conditioning and refrigeration equipment that incorporates hermetic refrigeration motor compressor(s), including those that are a part of appliances unless "specifically amended" by *NEC Article 422*.

Hermetic Motor-Compressor

Included as a part of the Specification in Appendix A of this text are the electrical characteristics of the air-conditioning equipment to be installed in the Commercial Building. Note there are two types of equipment specified. For tenants on the second floor, roof-mounted equipment is provided that includes the heating, cooling, and air-handling functions in one packaged unit. A single branch circuit is run to this equipment. For the two tenants on the first floor, a split system is specified. The hermetic compressors are mounted on concrete pads outside the building, and air handlers are located in the basement. A branch circuit is installed for the outside equipment and another branch circuit for the indoor equipment. The load requirements for the Drugstore are copied here for convenience. The connection scheme is shown in Figure 3-3, and the nameplates for the HVAC equipment, in Figure 3-4.

Additional information on wiring for air-conditioning equipment is included in Chapter 21 of this text.

Supply voltage:

208-volt, 3-phase, 3-wire, 60 hertz

Hermetic refrigeration compressor-motor:

Rated load current 17.6 amperes @ 208-volt, 3-phase

Condenser motor:

Full-load current 2.5 amperes @ 208-volt, single-phase

An evaporator motor is supplied by a separate branch circuit to the air handler that is located in the basement. Electric heating coils are installed in the air handler.

It is necessary to convert these loads to volt-ampere values before they can be added to the other loads. For a 208-volt, 3-phase supply, the rated load current is to be multiplied by 360. (This is the result of multiplying 208 volts by 1.732 [the square root of 3, because there are three working conductors] = 360.256. This is considered the effective voltage and is sometimes shown as 360_{EF}.) For the 208-volt loads for the small motors that are connected single-phase, the rated load current is to be multiplied by 208.

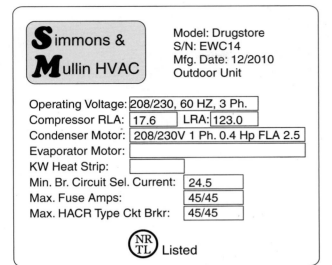

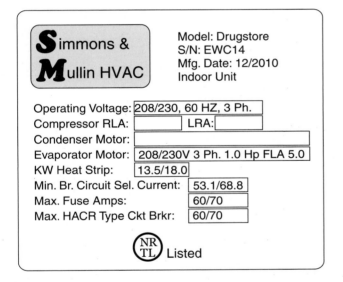

FIGURE 3-4 Nameplate information for the air-conditioning equipment for the Drugstore.

The air-conditioning load is shown in the following table:

Compressor: 3 ph., 208 V, 17.6 A		6336 VA
Condenser: 1 ph., 208 V, 2.5 A		520 VA
Total air-conditioning load		6856 VA

This is a load calculation and does not indicate the current that will occur in the conductors supplying the equipment. A typical connection scheme is shown in Figure 3-3. It can be seen that the current in the three phases would not be equal. This will be discussed in detail in later chapters.

So far as electricians are concerned, the outdoor equipment represents a single load with a single branch circuit and the indoor equipment is the same. The electrician will use the minimum branch-circuit selection current for installing the minimum-size branch-circuit conductors. The maximum fuse or HACR circuit-breaker information is used for the overcurrent protection. Both of these values are provided on the unit nameplate.

Note that the compressor motor will often be the largest motor load on the panelboard and service. This load is subject to the 125% rule of *NEC 430.24*.

SUMMARY OF DRUGSTORE LOADS

General Lighting

The connected load (2546 volt-amperes) is less than the *NEC* load allowance (4185 volt-amperes), so the *NEC* load allowance of 4185 volt-amperes is used. Lighting is considered a continuous load, so the load is shown in the Continuous Load column.

Storage

The connected load (600 volt-amperes) is greater than the *NEC* load allowance (215 volt-amperes), so the actual load is used. Lighting is considered a continuous load, so the load is shown in the Continuous Load Column.

Receptacle Loads

Receptacle loads are not generally considered continuous unless the receptacle is for dedicated loads that are continuous in nature. A calculated load of 180 VA is used for convenience receptacles for cord-and-plug-connected loads.

Other Loads

A specific allowance is provided for

- cash registers,
- electric door operators,
- copy machine,
- exhaust fans,
- water heater, and
- unit heater for storage space.

TABLE 3-4

Drugstore feeder load calculation.

	Count[1]	VA/Unit[2]	*NEC* Load[3]	Connected Load[4]	Continuous Load[5]	Noncontinuous[6]	Neutral Load[7]
General Lighting:							
NEC 220.12 Connected loads from Table 3-2	1395	3	4185	2630			
Totals:			4185	2630	4185		4185
Storage Area:							
NEC 220.12 Connected load from Table 3-3	858	0.25	215	600			
Totals:			215	600	600		600
Other Loads:							
Receptacle outlets	21	180	3780			3780	3780
Copy Machine	1	1440		1440		1440	1440
Receptacle for HVAC	1	1500		1500		1500	1500
Sign Outlet	1	1200	1200	1200	1200		1200
Cash registers	2	240		480		480	480
Electric doors	2	360		720		720	720
Water heater	1	4500		4500	4500		
Unit htr (storage)	1	7500		7500	7500		
Motors & appliances	Amperes	VA/Unit					
AC outdoor unit		360		6856			
AC Indoor unit[8]	42.5	360		15,300	15,300		
Exhaust fans	2	120		240		240	
Total Loads				33,285	8160	13,905	
Minimum Feeder load 125% of continuous loads plus 100% of noncontinuous:						49,766	13,905
Minimum Ampacity:						138.2	38.6

[1]Count: Square footage of given area, the number of a given type load, or the ampere rating of the given type load.

[2]VA/Unit: Specific unit load values required by *NEC Article 220*, or actual load values for the given type load.

[3]*NEC* Load: Count × *NEC* VA/Unit. Also the "calculated load."

[4]Connected Load: Actual load connected to the electrical system.

[5]Continuous Loads: The maximum current is expected to continue for 3 hours or more. Column also used to apply 125% to the largest motor load.

[6]Noncontinuous: Loads that are not continuous.

[7]Neutral Load: The maximum unbalanced load calculated according to *NEC 220.61*.

[8]Because the heating and cooling loads are noncoincident, only the larger load is included.

HVAC Equipment

The connected motor loads are used. The load for the compressor would normally be shown in the Continuous Load column to satisfy the requirement in *NEC 430.24* that 125% of the largest motor load be used in the load calculation. However, in this feeder the heating load is shown rather than the air-conditioning equipment, and only one of the loads will operate at any one time. The *NEC* considers this as *noncoincident loads*; see *NEC 220.60*. Note also that the heating load is considered *continuous*, so it will be calculated at 125%.

Refer to the *National Electrical Code* or the working drawings when necessary. Where applicable, responses should be written in complete sentences.

For Questions 1–4, indicate the general lighting unit load that would be included in the branch-circuit calculations as set forth in *NEC Table 220.12*.

1. A restaurant _____

2. A schoolroom _____

3. A corridor in a school _____

4. A hallway in a dwelling _____

5. The load to be included for general-purpose convenience receptacles is _____ VA.

6. The load to be included for an outlet for a specific appliance or load is _____.

For questions 7–9, indicate the effect, if any, a change in loads can have on the feeder serving the panelboard for a given area.

7. An increase in the calculated load

_____

8. An increase in the balanced load

_____

9. An increase in the nonlinear load

_____

Answer Questions 10, 11, and 12 using the following information:

An addition is being planned to a school building. You have been asked to determine the load that will be added to the panelboard that will serve this addition.

The addition will be a building 80 ft × 50 ft. It will consist of four classrooms, each 40 ft × 20 ft, and a corridor that is 10 ft wide.

The following loads will be installed:

Each classroom:

12 fluorescent luminaires, 2 ft × 4 ft @ 85 VA each
20 duplex receptacles
AC unit, 208-volt, 1-phase @ 5000 VA

Corridor:

5 fluorescent luminaires, 1 ft × 8 ft @ 85 VA each
8 duplex receptacles

Exterior:

4 wall-mounted luminaires @ 125 VA each
4 duplex receptacles

Load	Count	VA/Unit	NEC Load	Connected Load	Calculated Load	Minimum Branch-Circuit	Neutral Load
Classrooms *NEC 220.12* Luminaires Subtotal							
Corridor *NEC 220.12* Luminaires Subtotal							
Exterior Luminaires Subtotal							
Receptacles *NEC 220.14(1)* Subtotal							
Motors *NEC 220.14(C)* Subtotal							
Totals							

10. The calculated load is _____ VA.

11. The connected load is _____ VA.

12. The neutral load is _____ VA.

13. Why does the *NEC* require the overcurrent device to be rated not less than 125% of continuous loads? _____

_____

14. Explain the concept of noncoincident loads for load calculations. _____

_____.

Branch Circuits

OBJECTIVES

After studying this chapter, you should be able to

- determine the required number of branch circuits for a set of loads.

- apply adjustment and correction factors as appropriate.

- apply factors for continuous, motor, and heating loads.

- determine the correct rating for branch-circuit protective devices.

- determine the preferred type of wire for a branch circuit.

- determine the required minimum size conductor for a branch circuit.

CONDUCTOR SELECTION

When called on to connect an electrical load such as lighting, motors, heating, or air-conditioning equipment, the electrician must have a working knowledge of how to select the proper type and size of conductors to be installed. Installing a conductor of the proper type and size will ensure that the voltage at the terminals of the equipment is within the minimums as set forth by the *NEC,* and that the circuit will have a long uninterrupted life.

Manufacturers of electrical equipment may have a lower and upper voltage range at which their equipment is designed to operate. Our job is to design and install the electrical system so the proper voltage is supplied at the line terminals of the equipment. Failing to do so can result in voiding the manufacturer's warranty and equipment failure. Installing and operating equipment to comply with the manufacturer's warranty is also required by *NEC 110.3(B).*

The first step in understanding conductor selection and installation requirements is to refer to *NEC Article 310.* This article contains such topics as insulation types; conductors in parallel, wet, and dry locations; marking; maximum operating temperatures; permitted use, trade names; direct burial; allowable ampacity tables; adjustment factors; correction factors; and derating factors.

Selecting the proper conductor for the circuit can be one of the most complicated tasks you will be called on to perform. Some of the issues that must be known and considered are listed here:

- The temperature rating of the terminals from the beginning to the end of the circuit
- The temperature rating of the conductor insulation
- The conductor material (copper or aluminum)
- The size of the conductor in American Wire Gage (AWG) or kcmil
- The length of the conductor
- The number of current-carrying conductors in the raceway or cable
- If the neutral conductor counts as a current-carrying conductor

- The ambient temperature where the conductor actually operates
- If the conductor is in a circular raceway (conduit or EMT) that is exposed to direct sunlight exposure on rooftops
- Determining if the load is considered continuous and, if so, making appropriate accommodation
- Applying correct factors for electric heating loads and for water heaters
- Selecting correct conductors and overcurrent protection for motor loads as well as for air-conditioning equipment

What's the Big Deal with Ampacity?

Understanding the concepts included in the term *ampacity* is most important. The term is defined in *NEC Article 100* as ▶*The maximum current, in amperes, that a conductor can carry continuously under the conditions of use without exceeding its temperature rating.*◀

We can observe the components of the definition as follows:

- It is the maximum current.
- The current is expressed in amperes.
- It is a continuous (long-time) rating.
- Conditions of use must be observed.
- The conductor insulation has a temperature rating that must not be exceeded.

We note that the current rating is *continuous,* unlike most other components involved in the circuit, such as terminals and equipment ratings. The conductor insulation often has a better or higher rating than terminals, for example. Note that the temperature of concern here is that of the insulation system and not the conductor. The conductor material can withstand much higher temperature without damage than the insulation.

The conditions of use must be known and observed. For most all the work we do, we use the allowable ampacities from *Table 310.15(B)(16)* (formerly *Table 310.16*). The conditions of use are included in the table heading. It reads:

Allowable Ampacities of Insulated Copper Conductors Rated Up to and Including 2000 Volts, 60°C Through 90°C, Not More Than Three Current-Carrying Conductors in Raceway, Cable, or Earth (Directly Buried), Based on Ambient Temperature of 86°F (30°C).

If our installation varies from the conditions of use stated in the table heading, either an adjustment (for more than three current-carrying conductors) or correction (for higher ambient temperatures) must be made to the current-carrying capability of the conductor. You will hear us say more than once that heat is the great enemy of electrical systems. Our installation must meet the *NEC* rules to be considered safe.

Be sure you fully understand the meaning, because the word *ampacity* is used throughout this text whenever conductors are discussed. We will explore its meaning throughout this chapter.

Conductor Type Selection

For installation of conductors in raceways or cables, an important step in choosing a conductor is the selection of an insulation type appropriate for the installation. An examination of *NEC Table 310.15(B)(16)* will reveal that although several different types of insulation are listed, there are only three different temperature ratings: 60°C, 75°C, and 90°C. We refer to conductor insulation temperature ratings throughout this text by only their °C values as this is standard electrical industry practice. It should also be noted that the higher the temperature rating, the higher the ampacity for a given conductor size. As the current through the conductor increases, the temperature of the conductor also increases. The insulation on the conductor must be suitable for this elevated temperature.

Should the operational temperature of a conductor become excessive, the insulation may soften or melt, causing ground faults or short circuits and possible equipment damage and personal injury. Heat comes from the following sources:

- Surrounding (ambient) temperature where the conductor is located

- Current flowing through the conductor due to the resistance of the conductor material

- Adjacent current-carrying conductors

- The sun where exposed to direct sunlight

Heat generated in a conductor as a result of current flow over a long period of time is expressed in watts, and is calculated

$$W = I^2R$$

Where W = watts, I represents current, and R represents the resistance of the conductor material.

Heat generated in a conductor over a short period of time (subcycle) as in the case of a short circuit or ground fault is expressed in ampere-square seconds, and is calculated:

$$H = I^2t$$

Ampere-square seconds are discussed in detail in Chapter 19.

The resistance of a conductor is dependent on the size, material, and length. Resistance values are listed in *NEC Chapter 9, Table 8*. Conductor resistance may be reduced by selecting copper instead of aluminum or by increasing the size. (The circuit length is usually established by other factors.) The resistance of a conductor does not change appreciably in response to current or ambient conditions.

The current has a dramatic effect on the heat. A 25% load change, from 16 to 20 amperes, will result in a 56% change in heat. This is the result of the current being squared in the calculation. The *NEC* incorporates many features to address this issue. The most notable of these is in the selection of the conductor type. Certain types of conductor insulation can tolerate higher heat levels, thus permitting a higher current with no change in wire size. The *NEC* also restricts the connected load, thus limiting the current. In addition, it recognizes the heating effect of bundling conductors (installing conductors without maintaining spacing) and requires the application of an adjustment factor.

Many factors should be considered when selecting a conductor. *NEC Table 310.104(A)* provides very valuable information on the properties of conductor insulations (see Table 4-1 in this text for an example of this information):

- Column 1 provides the technical names for the conductors, which are seldom used in the trade.

TABLE 4-1

Conductor specification and application.

Trade Name	Type Letter	Maximum Operating Temperature	Application Provisions	Insulation	Insulation Thickness			Outer Covering
					AWG-kcmil	mm	mils	
Heat-resistant thermoplastic	THHN	90°C (194°F)	Dry and damp locations	Flame-retardant, heat-resistant thermoplastic	14–12	0.38	15	Nylon jacket or equivalent
					10	0.51	20	
					8–6	0.76	30	
					4–2	1.02	40	
					1–4/0	1.27	50	
					250–500	1.52	60	
					501–1000	1.78	70	
Moisture- and heat-resistant thermoplastic	THWN	75°C (167°F)	Dry and wet locations	Flame-retardant, moisture- and heat-resistant thermoplastic	14–12	0.38	15	Nylon jacket or equivalent
					10	0.51	20	
	THWN-2	90°C (194°F)			8–6	0.76	30	
					4–2	1.02	40	
					1–4/0	1.27	50	
					250–500	1.52	60	
					501–1000	1.78	70	
Moisture-resistant thermoset	XHHW	90°C (194°F)	Dry and damp locations	Flame-retardant, moisture-resistant thermoset	14–10	0.76	30	None
					8–2	1.14	45	
					1–4/0	1.40	55	
		75°C (167°F)	Wet locations		213–500	1.65	65	
					501–1000	2.03	80	
					1001–2000	2.41	95	
Moisture-resistant thermoset	XHHW-2	90°C (194°F)	Dry and wet locations	Flame-retardant, moisture-resistant thermoset	14–10	0.76	30	None
					8–2	1.14	45	
					1–4/0	1.40	55	
					213–500	1.65	65	
					501–1000	2.03	80	
					1001–2000	2.41	95	

See *NEC Table 310.104(A)* for complete listing.

- Column 2 gives the letter-type designation. These are used extensively in the trade, and there is some logic to their origin. In general, R formerly referred to a rubber-based covering and now indicates thermoset insulation; H indicates a high temperature rating. Each H indicates a 15°C temperature rating above 60°C; W is used when the conductor is usable in wet locations; T indicates a thermoplastic covering; N is used when there is a nylon outer covering; U indicates suitability for underground use; and X indicates crosslinked synthetic polymer insulation.

- Column 3 lists the maximum operating temperature ratings. The 60°C, 75°C, and 90°C ratings are used extensively in building wiring.

- Column 4 is highly used as it lists the application provisions where the conductor is approved for

use. Turning to *NEC Article 100, Definitions,* and finding *Location* will provide complete definitions for dry, damp, and wet. For example, if a circuit is run underground in a conduit, conductors rated for wet locations must be used.

- Column 5 gives a brief, technical description of the insulation.

- Columns 6, 7, and 8 give the insulation thickness for various sizes and types of conductors. These values become very important when calculating conduit fill. For example, for electrical metallic tubing (EMT) trade size ¾, *NEC Annex C, Table C1* shows that 16 size 12 AWG conductors Type THHN/THWN, 15 Type TW, or 11 Type THW, may be installed.

- Column 9 provides a description of the outer covering. Popular types THHN and THWN are listed as having a nylon jacket. This jacket tends to make installation easier but is not critical to the insulation value.

Conductor Insulation

The *NEC* requires that unless permitted elsewhere in the *NEC*, all conductors be insulated, *310.106(D)*. There are a few exceptions, such as the permission to use a bare neutral conductor for services, and bare equipment grounding conductors. *NEC Table 310.104(A)* shows many types of conductors and their applications and insulations. The most commonly used insulations are of the thermoplastic and thermoset type.

What Is Thermoplastic Insulation?

Thermoplastic insulation is the most common. Like chocolate, it will soften and melt if heated above its rated temperature. It can be heated, melted, and reshaped. Thermoplastic insulation will stiffen at temperatures colder than 14°F. Typical examples of thermoplastic insulation are Types THHN, THHW, THW, THWN, and TW.

What Is Thermoset Insulation?

Thermoset insulation can withstand higher and lower temperatures. Like a baked cake, once the ingredients have been mixed, heated, and formed, it can never be reheated and reshaped. If heated above its rated temperature, it will char and crack. Typical examples of thermoset insulation are Types RHH, RHW, XHH, and XHHW.

Wire Selection

NEC Table 310.15(B)(16) places the wires available for building construction into two categories: copper and aluminum or copper-clad aluminum. Copper-clad aluminum wires have not been produced for many years.

Copper wire is usually the wire of choice partly because the allowable ampacity of a given size of conductor is greater for copper than for the aluminum wire.

For example, the allowable ampacity of a 12 AWG conductor with Type THHN insulation is 30 amperes for copper and 25 amperes for the aluminum. These ampacities are suitable for applying adjustment or correction (derating) factors, but not for terminations.

In addition, serious connection problems exist with the aluminum conductors:

- If moisture is present, there may be corrosive action when dissimilar metals come in contact, such as an aluminum conductor in a copper connector.

- The surface of aluminum oxidizes when exposed to air. If the oxidation is not broken, a poor connection will result. Thus, when installing larger aluminum conductors, an inhibitor is brushed onto the conductor and then it is scraped with a stiff brush. The scraping breaks through the oxidation, thus allowing the inhibitor to cover the wire surface. This prevents air from coming in contact with the aluminum surface. Pressure-type aluminum connectors usually have a factory-installed inhibitor in the connector.

- Aluminum wire expands and contracts to a greater degree than copper wire for a given current. Referred to as "cold flow," this action can loosen connections that were not correctly made. Thus, it is extra important that an aluminum connection be tightened properly in compliance with the manufacturer's instructions.

Conductor Insulations with Multiple Ratings

Some conductors have an insulation with a dual or even a triple rating. For example, a conductor with Type XHHW insulation is rated 90°C in a dry and damp location and 75°C in a wet location. Conductors with a −2 insulation system are rated 90°C in a dry or wet location. Conductors available with −2 insulations include THW-2, THWN-2 and XHHW-2. See Table 4-2 for an example of conductor ampacity with −2 insulation.

Conductors are also available with two or more insulation types such as THWN/THHN and USE/RHW/RHH. For example, THWN/THHN insulated conductors are suitable for use at 75°C in a wet location and 90°C in a dry location. Triple-rated

TABLE 4-2

Conductor ampacity with −2 insulation.

250 kcmil CU	Allowable Ampacity		
	Dry	Damp	Wet
XHHW	255	255	255
XHHW-2	290	290	290

conductor insulations like USE/RHH/RHW are permitted for direct burial (USE) in or on buildings or other structures in dry locations at 90°C ampacities (RHH), and in or on buildings or other structures in wet locations at 75°C ampacities (RHW).

What Are Wet Locations?

Underground wiring, raceways in direct contact with the earth, installations subject to saturation of water or other liquids, and unprotected locations exposed to the weather are wet locations per the definition in the *NEC*. Insulated conductors and cables in these locations shall be listed for wet locations, *300.5(B)*.

A similar wet location requirement found in *300.9* requires that *Where raceways are installed in wet locations above grade, the interior of these raceways shall be considered to be a wet location. Insulated conductors and cables installed in raceways in wet locations above grade shall comply with 310.10(C).* Note the importance of the words "interior of these raceways."

Suffice it to say that insulated conductors and cables in wet locations must have a "W" designation in their type lettering. See *NEC 310.10(C)*.

As always, check the markings on the conductor and refer to the appropriate *NEC* tables.

Size Selection

NEC Table 310.104(A) (Table 4-1) and *NEC Table 310.15(B)(16)* (Table 4-3) are referred to regularly by electricians, engineers, and electrical inspectors when information about wire sizing is needed. *Table 310.15(B)(16)* shows the allowable ampacities of insulated conductors. The temperature ratings, the types of insulation, the material the wire

is made of, the conductor size in American Wire Gauge (AWG) or kcmil are also shown.

Of the ampacity tables in the *NEC, Table 310.15(B)(16)* is used most often because it is always referred to when conductors are to be installed in raceways or when cables are installed. It is important to read and understand all the notes and footnotes to these tables.

Two footnotes now provide helpful information for the user of the *NEC*. The first refers to *Table 310.15(B)(2)*, where correction factors are found for application if the ambient temperature where the conductors are located is above or below 86°F. The second footnote refers to *NEC 240.4(D)*, which covers overcurrent protection of small conductors.

This covers 14, 12, and 10 AWG conductors. The overcurrent protection rules in *240.4(D)* for the small conductors matches the allowable ampacity of these conductors in the 60°C column of *Table 310.15(B)(16)*. *NEC 240.4(E)*, *(F)*, and *(G)* should be consulted for exceptions to this rule.

Heat Is a Problem for Conductor Insulation

NEC 310.15(A)(3) tells us that *no conductor shall be used in such a manner that its operating temperature exceeds that designated for the type of insulated conductor involved.*

Excessive heat degrades and destroys conductor insulation. The insulation might melt or become brittle and break off. In either case, the conductors could touch one another or a grounded surface. Metal raceways not properly grounded could become "live" and present a shock hazard. In any event, the stage is set for short circuits, ground faults, and/or possible electrocution. You need to pay close attention to the requirements in *Article 310* of the *NEC* that pertain to temperature limitations of conductor insulation.

Correction Factors for Ambient Temperature

It is virtually impossible to calculate how the various elements contribute to and affect the heat buildup in a conductor. We simply turn to the *NEC*.

TABLE 4-3

Allowable ampacities for copper conductors.

▶*Allowable Ampacities of Insulated Copper Conductors Rated Up to and Including 2000 Volts, 60°C Through 90°C (140°F Through 194°F), Not More Than Three Current-Carrying Conductors in Raceway, Cable, or Earth (Directly Buried). Based on Ambient Temperature of 30°C (86°F)*◀

SIZE	▶TEMPERATURE RATINGS OF CONDUCTORS [SEE *TABLE 310.104(A)*]◀		
	60°C (140°F)	75°C (167°F)	90°C (194°F)
AWG or kcmil	Types: TW, UF	Types: RHW, THHW, THW, THWN, XHHW, USE, ZW	Types: TBS, SA, SIS, FEP, FEPB, MI, RHH, RHW-2, THHN, THHW, THW-2, THWN-2, USE-2, XHH, XHHW, XHHW-2, ZW-2
18			14
16			18
14**	15	20	25
12**	20	25	30
10**	30	35	40
8	40	50	55
6	55	65	75
4	70	85	95
3	85	100	115
2	95	115	130
1	110	130	140
1/0	125	150	170
2/0	145	175	195
3/0	165	200	225
4/0	195	230	260
250	215	255	290
300	240	285	320
350	260	310	350
400	280	335	380
500	320	380	430
600	350	420	475
700	385	460	520
750	400	475	535
800	410	490	555
900	435	520	585
1000	455	545	615
1250	495	590	665
1500	525	625	705
1750	545	650	735
2000	555	665	750

*Refer to *310.15(B)(2)* for the ampacity correction factors where the ambient temperature is other than 86°F (30°C).
**Refer to *240.4(D)* for conductor overcurrent protection limitations.
For aluminum or copper-clad aluminum values, consult the *NEC Table 310.15(B)(16)*.
Reprinted with Permission from NFPA 70-2011

TABLE 4-4

Correction factors for ambient temperature.

For ambient temperatures other than 86°F, (30°C) multiply the allowable ampacities specified in *Table 310.15(B)(16)* by the appropriate correction factor shown below

Ambient Temperature	TEMPERATURE RATING OF CONDUCTOR		
	(60°C)	(75°C)	(90°C)
50°F (10°C) or less	1.29	1.20	1.15
51–59°F (11–15°C)	1.22	1.15	1.12
60–68°F (16–20°C)	1.15	1.11	1.08
69–77°F (21–25°C)	1.08	1.05	1.04
78–86°F (26–30°C)	1.00	1.00	1.00
87–95°F (31–35°C)	0.91	0.94	0.96
96–104°F (36–40°C)	0.82	0.88	0.91
105–113°F (41–45°C)	0.71	0.82	0.87
114–122°F (46–50°C)	0.58	0.75	0.82
123–131°F (51–55°C)	0.41	0.67	0.76
132–140°F (56–60°C)	—	0.58	0.71
141–149°F (61–65°C)	—	0.47	0.65
150–158°F (66–70°C)	—	0.33	0.58
159–167°F (71–75°C)	—	—	0.50
168–176°F (76–80°C)	—	—	0.41
177–185°F (81–85°C)	—	—	0.29

The *NEC* contains a number of tables that show correction factors and adjustment factors. These might be referred to as "penalty" factors. In this chapter, Table 4-3 [*NEC Table 310.15(B)(16)*] shows the allowable ampacity of conductors commonly used in commercial and industrial installations. Table 4-4 [formerly located below *NEC Table 310.16*, now *NEC Table 310.15(B)(2)(a)*] and Table 4-5 [*NEC Table 310.15(B)(3)(a)*] in this chapter show the correction and adjustment factors required by the *NEC*. Let's discuss how to apply these adjustment or correction (penalty) factors.

The insulation on a conductor will collect heat from the ambient (surrounding) temperature, from current flowing through the conductor, and from adjacent current-carrying conductors. Some of the collected heat will be dissipated to whatever the conductor is touching (e.g., inside of the raceway, adjacent insulation) and from convection of the heat into the surrounding air inside the raceway.

TABLE 4-5

Table 310.15(B)(3)(a) Adjustment Factors for More Than Three Current-Carrying Conductors in a Raceway or Cable

Number of Conductors[1]	Percent of Values in Table 310.15(B)(16) through Table 310.15(B)(19) as Adjusted for Ambient Temperature if Necessary
4–6	80
7–9	70
10–20	50
21–30	45
31–40	40
41 and above	35

[1]Number of conductors is the total number of conductors in the raceway or cable adjusted in accordance with 310.15(B)(5) and (6).

After the accumulated heat minus the dissipated heat settles down, we are faced with how to determine the safe allowable current-carrying capability of the conductor.

FIGURE 4-1   An outdoor installation of service, feeders, and branch circuits is often subject to high ambient temperatures. (*Delmar/Cengage Learning*)

The correction factors are given in Table 4-4. This set of factors applies to both copper and aluminum or copper-clad aluminum and is given in new *NEC Table 310.15(B)(2)(a)*. This table previously appeared at the bottom of both *Tables 310.16* and *310.17*.

If the ambient temperature anywhere along the length of a raceway is less than 78°F or higher than 86°F, the ampacity of the conductors in the raceway, or cable, must be modified by a correction factor. A discussion of the conditions that require this action is set forth in *NEC 310.15(A)(3)*, including *Informational Note 1*. The selection is dependent on the conductors' insulation temperature rating. The conductors shown in Figure 4-1 are subject to the outdoor ambient temperatures and may be subject to correction.

As Type THHN/THWN is the conductor of choice for the Commercial Building, the temperature correction factor will be applied to the 90°C column of *NEC Table 310.15(B)(16)* for dry locations and the 75°C column for wet locations.

 EXAMPLE

From Table 4-3 [*NEC Table 310.15(B)(16)*], the allowable ampacity of a 4 AWG THHN conductor is 95 amperes in a dry location. If installed in an ambient temperature of 96–104°F, such as might be the case in a boiler room or outdoors in some climates, the corrected ampacity of the conductor is found by applying the correction factor of 0.91 from Table 4-4.

$$95 \times 0.91 = 86.5 \text{ amperes}$$

Conduits on Roofs Exposed to Direct Sunlight

Conduits, raceways, and cables run on or above rooftops where exposure to direct sunlight can be very hot. The temperature inside a raceway is considerably greater than the temperature on the outside of the raceway. The raceways absorb the rays of the sun. Figure 4-2 illustrates a location where the ambient temperature and exposure to the rays of the sun must be compensated for to avoid overheating conductors. Think about how hot the metal on an automobile gets after standing out in the blazing sun for a long time. In addition, the interior of the vehicle gets much hotter than the exterior temperature. Some questions that seem to require an answer are presented here:

- Is the raceway (metal or nonmetallic conduit or EMT) its original manufactured color or has it been painted a dark or light color?
- Is the rooftop dark or light in color?
- What is the expected temperature above the roof?

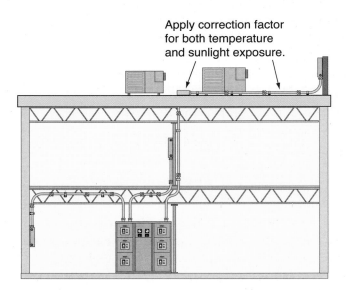

Apply correction factor for both temperature and sunlight exposure.

FIGURE 4-2   Application of correction factors for both temperature and sunlight exposure. (*Delmar/Cengage Learning*)

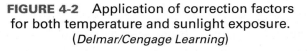

TABLE 4-6

Application of *Correction Factors* for sunlight exposure.

Location	Outdoor Ambient	Adder for Rooftop	Temperature for Application of Correction Factor	Allowable Ampacity of 4 AWG Conductor	Ambient Correction Factor	Corrected Allowable Ampacity
Phoenix, AZ	107°F	30°F	137°F	XHHW-2	0.71	95 × 0.71 = 67.5 A
				XHHW	0.58	85 × 0.58 = 49.3 A
Chicago, IL	91°F	30°F	121°F	XHHW-2	0.82	95 × 0.82 = 77.9 A
				XHHW	0.75	85 × 0.58 = 63.8 A

The *NEC* does not answer these questions. Instead, *NEC Table 310.15(B)(3)(c)* provides temperature values, based on the distance above the roof to the bottom of the circular raceway, that must be added to the expected ambient temperature.

EXAMPLE ————————————

From Table 4-3 [*NEC Table 310.15(B)(16)*], the allowable ampacity of a 4 AWG conductor with XHHW-2 insulation is 95 amperes. These conductors will be in a raceway run across the top of a roof on supports 8 inches above the roof. From Table 4-5 [*NEC Table 310.15(B)(3)(c)*], we find that we must add 30°F to the expected ambient design temperature. Checking Table A-30 in the Appendix of this text, we can find the expected design temperature for many cities across the country. Let's assume we are in Phoenix, Arizona, see Table 4-6. The calculation is

$$30°F + 107°F = 137°F$$

We then look at Table 4-5 [*NEC Table 310.15(B)(2)(a)*] and find that the correction factor for 137°F is 0.71. The corrected ampacity of the 4 AWG conductor with XHHW-2 insulation is

$$95 × 0.71 = 67.45 \text{ amperes}$$

————————————————————

EXAMPLE ————————————

Using the same 4 AWG conductors with XHHW-2 insulation on or above a rooftop in Chicago, we find the expected design temperature in Table A-30 in the Appendix of this text is 91°F, see Table 4-6. The calculation is

$$30°F + 91°F = 121°F$$

We then look at *NEC Table 310.15(B)(2)(a)* and find that the correction factor for 121°F is 0.82. The corrected ampacity of the 4 AWG XHHW-2 conductor is

$$95 × 0.82 = 78 \text{ amperes}$$

————————————————————

After looking at the above examples but substituting a conductor with an insulation that is rated for 75°C in the wet location, the value of selecting a conductor that is suitable for 90°C becomes obvious. Not only do we begin with a greater allowable ampacity, the correction factor is lower.

The full report of the studies made regarding conduits on rooftops can be found on the International Association of Electrical Inspectors Web site: http://www.iaei.org. Search for *Effect of Rooftop Exposure on Ambient Temperatures Inside Conduits*. This article contains temperature tables for many cities in the United States. This information is also found in Table 30 in the Appendix of this text.

Adjustment Factors for More Than Three Current-Carrying Conductors

The use of adjustment factors is another response to excessive heat affecting the current-carrying capacity of a conductor. This is the collective heat generated by a group of conductors, such as shown

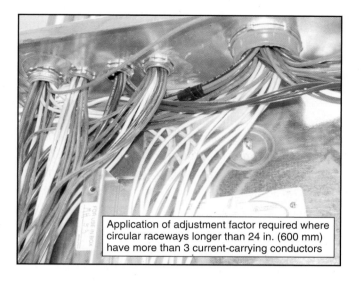

Application of adjustment factor required where circular raceways longer than 24 in. (600 mm) have more than 3 current-carrying conductors

FIGURE 4-3 Application of *Adjustment Factors* required where there are more than three current-carrying conductors longer than 24 in. (600 mm) in raceways or cables. (*Delmar/Cengage Learning*)

in Figure 4-3. Those conductors are placed in a conduit or cable or otherwise stacked, grouped, or bundled without maintaining spacing in a way that prevents airflow from dissipating the heat. Whenever a grouping of more than three current-carrying conductors exists, an adjustment factor from *NEC Table 310.15(B)(3)(a)* must be applied. A huge challenge is created when electricians install more than three conductors in raceways, as shown in Figure 4-3. First, the electrician must determine whether the neutral conductors are considered by the NEC as current carrying. Following that, the electrician counts the ungrounded (hot) conductors and goes to the table to apply the correct adjustment factor. In some cases, larger conductors may be required to provide the needed ampacity after application of the *Adjustment Factors*. This may require larger raceways. A practical approach is to install fewer conductors in a raceway than will require increasing the size of conductors after application of the *Adjustment Factors*.

💡 **EXAMPLE** ────────────────

From Table 4-3 [*NEC Table 310.15(B)(16)*], the allowable ampacity of a 4 AWG THHN/THWN conductor is 95 amperes. If more than three current-carrying conductors are installed in a raceway, the allowable ampacity must be adjusted (reduced).

TABLE 4-7

Table 310.15(B)(3)(c) Ambient Temperature Adjustment for Circular Raceways Exposed to Sunlight on or Above Rooftops

Distance Above Roof to Bottom of Conduit	Temperature Adder	
	°C	°F
0–13 mm (½ in.)	33	60
Above 13 mm (½ in.)–90 mm (3½ in.)	22	40
Above 90 mm (3½ in.)–300 mm (12 in.)	17	30
Above 300 mm (12 in.)–900 mm (36 in.)	14	25

Reprinted with Permission from NFPA 70-2011

Let's say that we have four 4 AWG THHN/THWN conductors. From Table 4-7 [*NEC Table 310.15(B)(3)(a)*], the adjustment factor is 0.80. Another way of looking at the application of this factor is, because of each current-carrying conductor contributing heat, the current on each conductor must be reduced to prevent overheating.

The adjusted ampacity of the 4 AWG THHN/THWN conductor is

$$95 \times 0.80 = 76 \text{ amperes}$$

Heat and More Than Three Current-Carrying Conductors

If high ambient temperatures *and* more than three current-carrying conductors are installed in the same raceway, both adjustment and correction factors must be applied.

💡 **EXAMPLE** ────────────────

From Table 4-3 [*NEC Table 310.15(B)(16)*], the allowable ampacity of a 4 AWG THHN/THWN conductor is 95 amperes. Let's say that four 4 AWG THHN/THWN current-carrying conductors are installed in a raceway and the ambient temperature is expected to be in the range of 96–104°F (36–40°C). The correction factor from Table 4-4 [*NEC Table 310.15(B)(2)(a)*] is 0.91 and the adjustment factor from Table 4-7 [*NEC Table 310.15(B)(3)(a)*] is 0.80.

Applying both adjustment and correction factors, the allowable ampacity of the 4 AWG THHN/THWN conductor is

$$95 \times 0.91 \times 0.80 = 69 \text{ amperes}$$

In the Commercial Building, five current-carrying conductors are installed in a raceway from the panelboard in the Insurance Office, Beauty Salon, and Real Estate Office to the air-conditioning equipment and the receptacle located on the roof. The adjustment factor for five current-carrying conductors, found in Table 4-7 [*NEC Table 310.15(B)(3)(a)*], is 80%, or 0.8. Depending on how the raceway is brought out over the roof and how it is run to the air-conditioning unit, additional temperature adjustments will have to be made in accordance with Table 4-6 [*NEC Table 310.15(B)(3)(c)*]. In other words, you might have to apply three adjustment or correction factors.

Derating, Where the 10% Rule Might Apply

There is an exception to the ampacity derating rule where two different temperatures are encountered for the same conductors. An example is where a portion of the conduit is inside a building in normal condition spaces, and a portion of the conduit crosses the rooftop in full sunlight exposure. This exception permits the application of the higher ampacity (lower temperature) if the length of the conductors that enter a higher temperature is not more than 10 ft (3.0 m) or not more than 10% of the total circuit length. See *310.15(A)(2), Exception*. This exception might minimize the drastic derating required by *NEC Table 310.15(B)(3)(c)*. Where and how a raceway is brought out onto a rooftop should be given a lot of consideration. Check this out with the local electrical inspector.

Let's Review the Requirements for Derating Conductors

Conductors must be derated according to factors from *NEC 310.15(B)* that apply:

- If conductors are installed in an ambient temperature higher than 86°F (30°C), the ampacity must be corrected in accordance with *NEC Table 310.15(B)(2)(a)*.

- If more than three current-carrying conductors are installed in a raceway or cable for a continuous length longer than 24 in. (600 mm), apply the adjustment factor from *NEC 310.15(B)(3)(a)*. See Figure 4-3.

- If more than three current-carrying conductors in cable assemblies are installed without maintaining spacing for lengths greater than 24 in. (600 mm), apply the adjustment factor from *NEC 310.15(B)(3)(a)*. See Figure 4-4 and *NEC 310.15(B)(3)(a)*.

- If conductors are installed in conduits that are exposed to direct sunlight above rooftops, add the factors from *NEC Table 310.15(B)(3)(c)* to *NEC Table 310.15(B)(2)(a)* and apply the temperature correction factors.

The *NEC* has moved away from using the term *derated* or *derating* in favor of applying *adjustment* or *correction* factors. The term *derated* has been used in the trade for many years but will probably be relegated to "trade jargon."

It is important to note that the *NEC* refers to current-carrying conductors for the purpose of derating when more than three current-carrying conductors are installed in a raceway or cable. Following are the basic rules:

- DO count all current-carrying conductors.

- DO count the neutral conductor of a 3-wire, single-phase circuit when the circuit is taken from a 4-wire, 3-phase, wye-connected system. See Figure 4-5 and *NEC 310.15(B)(5)(b)*.

Table 4-8 gives terminal markings that shall be followed to ensure proper connection.

How Equipment Affects Conductor Selection

Connection and termination requirements for equipment often limit the current a conductor is otherwise permitted to carry. This is done to prevent overheating and damaging the equipment. *NEC 110.14(C)(1)* indicates that for equipment rated 100 amperes or less (or marked for 14 AWG through 1 AWG), the allowable ampacity values found in the 60°C column of *NEC Table 310.15(B)(16)* are generally to be used. See

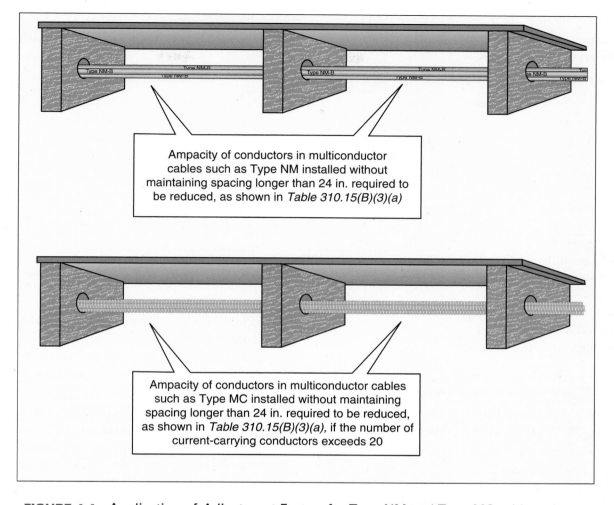

FIGURE 4-4 Application of *Adjustment Factors* for Type NM and Type MC cables where separation is not maintained. (*Delmar/Cengage Learning*)

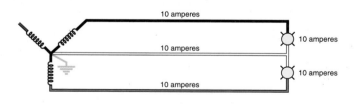

FIGURE 4-5 Four-wire wye system, 3-wire circuit. (*Delmar/Cengage Learning*)

Figure 4-6. For example, when installing Type THHN conductors that have 90°C insulation, the allowable ampacity values found in the 60°C column are used unless all termination markings indicate suitability for 75°C ampacities. The reasoning is rather simple: a wire has two ends! A receptacle is typically marked with a wire size and not with a terminal temperature rating. This connection is typically listed for 60°C. The other end of the conductor might be connected to a circuit breaker listed for 75°C. However, the current the branch circuit conductors are permitted to carry is selected from the 60°C column of *Table 310.15(B)(16)* and not from the 75°C or 90°C columns. Table 4-8 provides details on the type of device, markings for terminations, and the conductors that are permitted.

NEC 110.14(C)(1) says that for equipment rated over 100 amperes (or marked for conductors larger than 1 AWG), the allowable ampacity values found in the 75°C column of *NEC Table 310.15(B)(16)* are generally to be used, Figure 4-7. For example, when installing Type THHN conductors that have 90°C insulation, the allowable ampacity values

TABLE 4-8

Devices and conductors.

Type of Device	Markings	Conductors Permitted
15- or 20-ampere receptacles and switches	CO/ALR	Copper, aluminum, copper-clad aluminum
15- and 20-ampere receptacles and switches	NONE	Copper, copper-clad aluminum
30-ampere and greater receptacles and switches	AL/CU	Copper, aluminum, copper-clad aluminum
30-ampere and greater receptacles and switches	None	Copper only
Screwless pressure terminal connectors of the push-in type	None	Solid 14 AWG copper only
Wire connectors	AL	Aluminum
Wire connectors	AL/CU	Copper, aluminum, copper-clad aluminum
Wire connectors	CC/CU	Copper or copper-clad aluminum
Wire connectors	CU or None	Copper only

found in the 75°C column are used. The reasoning is the same as that given previously.

Why even have insulation rated 90°C? Boiler rooms, attics, and outside installations in the hot sun are examples of where the high-temperature rating might be needed. The big advantage of high-temperature insulations is they permit the use of the allowable ampacity found in the 90°C column for applying adjustment and correction (derating) factors. In addition, some equipment may require conductor insulation rated higher than required for ampacity issues to safely withstand the temperature internal to the equipment. An example

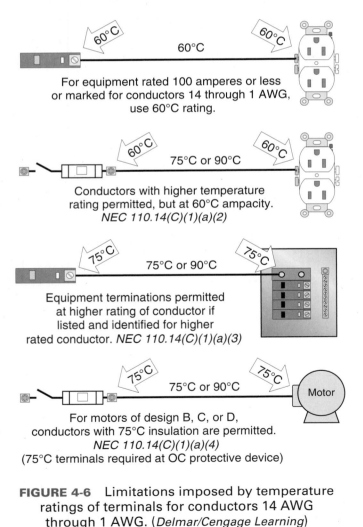

FIGURE 4-6 Limitations imposed by temperature ratings of terminals for conductors 14 AWG through 1 AWG. (*Delmar/Cengage Learning*)

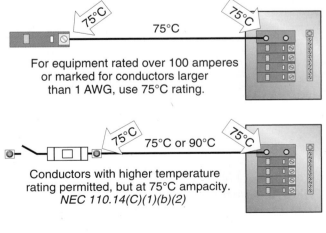

FIGURE 4-7 Limitations imposed by temperature ratings of terminals for conductors larger than 1 AWG. (*Delmar/Cengage Learning*)

is luminaires that may require 90°C or even 105°C conductor insulation for termination.

- DO count each current-carrying conductor when conductors are run in parallel. Parallel conductors are two or more conductors that are connected at each end to more economically create a set of conductors with an increased ampacity. Rules for installing conductors in parallel are found in *NEC 310.10(H)* and *310.15(B)(3)(a)*. Branch circuits are not usually installed in parallel as 1/0 AWG is the minimum size conductor generally permitted to be installed in parallel.

- DO count the neutral conductor of a 4-wire, 3-phase, wye-connected circuit when the major portion of the load is electric discharge lighting (fluorescent, mercury vapor, high-pressure sodium, etc.), data processing, and other loads where the neutral conductor carries third harmonic current. See Figure 4-8 and *NEC 310.15(B)(5)(c)*. In branch circuits and feeders that supply major nonlinear loads, the current in the neutral conductor can be as great as two times that of the current in the ungrounded (hot) conductors. In these situations, design engineers often specify that the neutral conductor be double sized. Special K-rated transformers are recommended for these applications.

- You are NOT required to count the neutral conductor of a 3-wire, single-phase circuit if the neutral conductor carries only the unbalanced current. See Figure 4-9 and *NEC 310.15(B)(5)(a)*.

- You are NOT required to count equipment-grounding conductors that are run in the same raceway or cable with circuit conductors, Figure 4-10. Equipment grounding conductors must be included when calculating raceway fill. See *NEC 310.15(B)(6)*.

- You are NOT required to apply an adjustment factor to conductors in sections of raceway 24 in. (600 mm) or less in length. See Figure 4-11 and *NEC 310.15(B)(3)(a)(2)*.

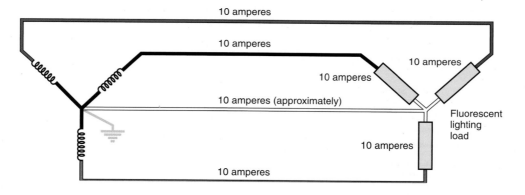

FIGURE 4-8 Four-wire wye system, nonlinear loads. (*Delmar/Cengage Learning*)

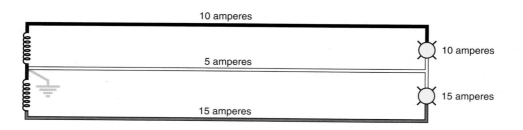

FIGURE 4-9 Neutral current in single-phase, 3-wire circuit. (*Delmar/Cengage Learning*)

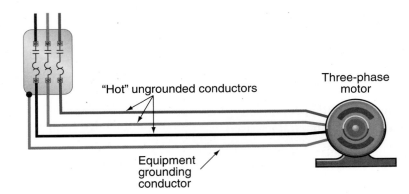

FIGURE 4-10 Count equipment-grounding conductors for raceway fill, but not as current-carrying conductor. (*Delmar/Cengage Learning*)

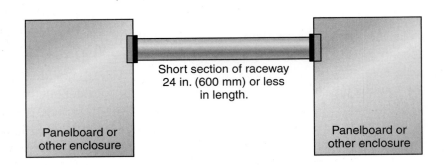

FIGURE 4-11 Ampacity adjustment is not required if conductors are installed without maintaining spacing for 24 in. (600 mm) or less, *310.15(B)(3)(a)*. (*Delmar/Cengage Learning*)

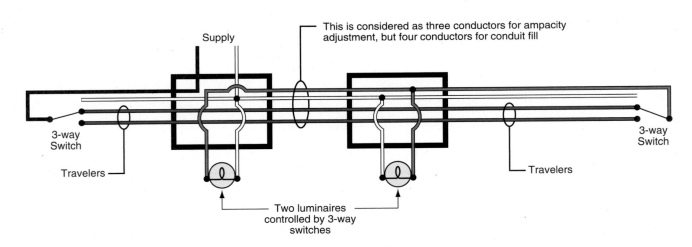

FIGURE 4-12 Counting 3-way switch travelers as current-carrying conductors. (*Delmar/Cengage Learning*)

- You are NOT required to count the noncurrent-carrying "traveler" conductor on 3-way or 4-way switching arrangements. Only one of the travelers is carrying current at any one time, so only one conductor need be counted when determining ampacity adjustment factor, Figure 4-12. Both travelers are counted when computing raceway fill.

Keep All Conductors of the Same Circuit in the Same Raceway

Several factors or conditions can cause conductors to overheat, such as the number of conductors in a raceway or cable and the ambient temperature. Often overlooked is the possibility of induction heating. The *NEC* in *300.3(B)* requires that *All conductors of the same circuit and, where used, the grounded conductor and all equipment grounding conductors and bonding conductors shall be contained within the same raceway, auxiliary gutter, cable tray, cablebus assembly, trench, cable, or cord*. Why is this requirement important?

When all conductors of the same alternating-current circuit are in the same raceway, the magnetic flux around each conductor cancels out the magnetic flux of the other circuit conductors, Figure 4-13. If the flux does not cancel, induction heating of a metal raceway or metal-jacketed cable will occur. Also, if the conductors of a circuit were in different raceways, in the event of a fault, the impedance of the circuit would be higher than if all of the circuit conductors of the circuit were in the same raceway. Ohm's Law proves that in a given circuit, the higher the impedance, the lower the current flow. Inversely, the lower the impedance, the higher the current flow. We want the overcurrent device ahead of the circuit to open as fast as possible. Overcurrent devices open fast when they "see" high values of current flow. They open slowly when they "see" low values of current flow.

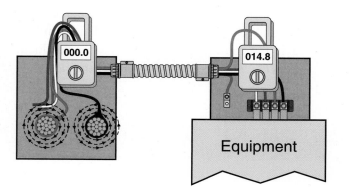

FIGURE 4-13 The magnetic flux cancels if all circuit conductors are in the same raceway or cable. (*Delmar/Cengage Learning*)

DETERMINING CIRCUIT COMPONENTS

As you study from this point on, you will learn a very important requirement for sizing conductors and sizing overcurrent protection. Conductors and overcurrent protection are key components of an electrical system. Every portion of the total length of the circuit must be evaluated individually to compensate for any condition of use that applies to that section. One portion of the circuit length may have excessive numbers of conductors in the raceway, another part of the run might pass through a high-temperature environment, and finally a portion of the circuit might be exposed to sunlight. Every condition of use that is outside of the allowable ampacity table conditions must be compensated for.

Continuous Loads

Several *NEC* sections contain the very important requirement to size conductors and overcurrent devices at 100% of the noncontinuous loads plus 125% of the continuous loads. This takes into account heat buildup resulting from the current flowing through the conductors and overcurrent devices for an extended period of time. These requirements are found in *NEC 210.19(A)(1)* for branch circuits, *210.20(A)* for branch circuit overcurrent devices, *215.2(A)(1)* for feeder conductors, *215.3* for a feeder overcurrent protection, *230.42(A)(1)* for service conductors, *409.20* and *409.21* for industrial control panels, and *625.21* for electric vehicle charging system overcurrent protection.

It is not necessary to apply the 125% factor for grounded (often neutral) conductors that are not connected to an overcurrent device, *NEC 210.19(A) (1) Exception No. 2*, *215.2(A)(1) Exception No. 2*, and *230.42(A)(2)*. The logic of this *Code* exception is that the heat developed in the neutral conductor will not contribute to the possible nuisance tripping of a circuit breaker or the opening of a fuse because the neutral conductor is terminated on a neutral bus, not to the terminal of an overcurrent device. A second exception provides that if both the overcurrent device and its assembly are listed for operation at 100% of their rating, it is not necessary to increase the ampacity of the conductors and overcurrent

TABLE 4-9

Continuous and noncontinuous loads and minimum rating of overcurrent device.

Noncontinuous Loads		Continuous Loads		Minimum Rating of Overcurrent Device and Conductor
Load	Factor	Load	Factor	
40 A	100%	40 A	125%	$40 + (40 \times 1.25 = 50) = 90$

device by 25%. Be very cautious with this exception. Most electrical equipment in the 600 volt class is not rated for continuous operation at 100% of its rating.

Store lighting is an example of continuous loads, whereas receptacle outlets typically are not considered continuous loads.

There are two ways to compensate for continuous loads. One way is to apply a 125% factor to the continuous load plus 100% of the noncontinuous load, and this becomes the minimum rating of the conductor and the overcurrent device. This rule is illustrated in Table 4-9.

The second method is to limit the continuous load on the circuit to not more than 80% of the rating of the overcurrent device and the conductor. This concept is shown in Table 4-10.

Conductor Sizing

- The conductor and insulation type must be selected to meet the criteria of *NEC 110.14(C)(1)*. If the circuit (i.e., the overcurrent protective device [OCPD]) rating is 100 amperes or less, the conductor may be selected from any of the three columns in *NEC Table 310.15(B)16*. However, the allowable ampacity is limited by the rating of the terminals. If the circuit rating is greater than 100 amperes, the conductor

insulation must either be 75°C or 90°C. Note that equipment of the 600-volt class does not have terminals rated 90°C. As a result, the 90°C allowable ampacities of *Table 310.15(B)(16)* are permitted for ampacity adjustment and correction (derating) purposes only. As you will see, using these higher ampacities for derating can be very valuable.

- The conductor type must comply with the requirements of the location, for example, dry, damp, or wet.

- Branch-circuit conductors shall have an ampacity not less than the load to be served, *NEC 210.19(A)*.

- In compliance with *NEC 210.19(A)*, the allowable ampacity [the value given in *NEC Table 310.15(B)(16)*] of the conductor must be equal to or greater than the noncontinuous load plus 125% of the continuous load.

- For circuits rated 100 amperes or less, or if any of the terminations are marked for size 1 AWG conductors or less, use the 60°C column of *NEC Table 310.15(B)(16)* to identify the allowable ampacity that is equal to or next greater than the OCPD ampere rating. This conductor is the minimum size allowed for the load. See *NEC 110.14(C)(1)* for details on electrical

TABLE 4-10

Limiting the continuous load.

Ampere Rating of OCPD and Conductor	Continuous Load		
	Load	Factor	Result
90 Amperes	72 Amperes	125%	90 Amperes

TABLE 4-11					
Equipment termination limitation on conductor allowable ampacity [Table 310.15(B)(16)].					
110.14(C)(1)	**60°C**	**75°C**		**90°C**	
(a) OCPD 100 A or less or wire size 14 through 1 AWG	Generally required	Permitted at 60°C ampacity	Permitted at 75°C ampacity if all terminals are rated for 75°C	Permitted at 60°C ampacity	Permitted to use this ampacity for derating
(b) OCPD greater than 100 A or wire size larger than 1 AWG		Generally required		Permitted at 75°C ampacity	Permitted to use this ampacity for derating

connections. A conductor with 75°C or 90°C insulation is acceptable so long as the conductor size is selected from the 60°C column. See Table 4-11.

- For circuits with a rating greater than 100 amperes, or if all the terminations are rated for 75°C or higher, use the 75°C column of *NEC Table 310.15(B)(16)* to identify the allowable ampacity that is equal to or next greater than the OCPD ampere rating. The size of this conductor is the minimum size allowed for the load. A conductor with 90°C insulation is acceptable so long as the conductor size is selected from the 75°C column. See Table 4-11.

- The allowable ampacity of the conductor must be sufficient to allow the use of the required OCPD and not less than the calculated load (noncontinuous load plus 125% of the continuous load).

- Storage-type water heaters with a capacity of 120 gallon (450 L) or less are to be considered continuous loads for purposes of sizing branch circuits. As a result, the branch circuit conductors are to have a rating of 125% of the nameplate load. See *NEC 422.13*.

- Fixed electric space-heating equipment is required to be considered a continuous load. As a result, the branch circuit conductors are to have a rating of 125% of the nameplate load. See *NEC 424.3(B)*.

- For motors, *NEC 430.6(A)(1)* requires that conductor rating, switch rating, and branch-circuit short-circuit and ground-fault protection sizing shall be based on *NEC Tables 430.247, 430.248, 430.249,* and *430.250*. In addition, *430.22* requires that conductors for a motor shall have an ampacity not less than 125% of the motor's current rating as determined using the tables given above.

However, the running overload protection for a motor shall be based on the motor's nameplate current rating, *430.6(A)(2)*.

- When serving an air-conditioning unit, the conductors must have an ampacity not less than the *Minimum Circuit Ampacity* as marked on the nameplate of the unit. This marking on the nameplate makes it easy to size the conductors. UL Standard 1995 requires that the manufacturer provide this data. *Article 440* also provides the basis for this requirement as well as for sizing the disconnecting means for the unit.

Overcurrent Protection

The purpose of an overcurrent protective device (OCPD) is to protect the circuit wiring and devices and to some extent the equipment served by the circuit. The following *NEC* references must be consulted when selecting an OCPD. The student should read these references with great care because only a project-specific synopsis is presented.

- A continuous load is defined as one where the load is expected to operate for a period of 3 hours or more. Office and store lighting are common examples of continuous loads. See *NEC Article 100*.

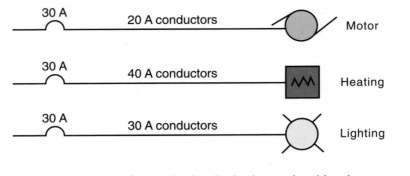

FIGURE 4-14 The rating of branch circuits is determined by the overcurrent device. (*Delmar/Cengage Learning*)

- *NEC 210.20(A)* states that the OCPD shall have a rating not less than the noncontinuous load plus 125% of the continuous load.

- The rating of the OCPD becomes the rating of the branch circuit regardless of the conductor size or load type. See Figure 4-14 and *NEC 210.3*.

- In general, branch-circuit conductors shall be protected in accordance with their ampacities. See *NEC 240.4*.

- A footnote to *NEC Table 310.15(B)(16)* must be observed when selecting 14, 12, or 10 AWG conductors. This footnote refers to *NEC 240.4(D) Small Conductors*. In this section, the OCPD ratings for copper conductors of 14, 12, and 10 AWG are limited to 15, 20, and 30 amperes, respectively. For aluminum and copper-clad aluminum conductors of 12 and 10 AWG, the ratings are limited to 15 and 25 amperes. *NEC 240.4(E)*, *(F)*, and *(G)* should be consulted for exceptions. For example, conductors installed for motors under *Article 430* and for air-conditioning equipment under *Article 440* are exempt from these small conductor rules.

- If the ampacity of the conductor is between standard ampere ratings for overcurrent devices as shown in *NEC 240.6(A)*, the next higher ampere rating fuse or circuit breaker is permitted under specific conditions. These conditions include the rating if the circuit is 800 amperes or less and the circuit is not a multioutlet branch circuit supplying cord-and plug-connected portable loads. See *NEC 240.4(B)*.

- Multioutlet branch-circuit ratings are limited to 15, 20, 30, 40, and 50 amperes except on indus-trial premises, where they may have a higher rating. See *NEC 210.3*.

- OCPDs are available in the standard ampere ratings listed in *NEC 240.6(A)*.

- A storage-type water heater with a capacity of 120 gallons (450 L) or less is to be considered a continuous load for purposes of sizing branch circuits. As a result, the branch circuit conductors and the overcurrent device are to have a rating of 125% of the nameplate load. See *NEC 422.13*.

- Fixed electric space-heating equipment is required to be considered a continuous load. As a result, the branch circuit conductors and the overcurrent device are to have a rating of 125% of the nameplate load. See *NEC 424.3(B)*.

- For motor loads of less than 100 amperes, the rating of an inverse-time circuit breaker (as used in the Commercial Building) shall not exceed 400% of the load. A rating of 250% is considered appropriate. See *NEC 430.52(C)* and *Table 430.52*.

- When a branch circuit supplies an air-conditioning system that consists of a hermetic motor-compressor and other loads (as occurs in the Commercial Building), the circuit rating shall not exceed 225% of the hermetic motor load plus the sum of the additional loads. See *NEC 440.22(B)(1)*. The nameplate on an AC unit will be marked with the *maximum overcurrent protection size and type*.

Selection Criteria

Abbreviated circuit component selection criteria for motors, motor-compressor units, and other circuits are shown in Table 4-12. The designations C, H, N,

TABLE 4-12

Circuit component selection.

Size/Rating Type Circuit	Conductor Minimum Allowable Ampacity	Minimum OCPD Ampere Rating	Maximum OCPD Ampere Rating
C – Continuous	Not less than 125% of load	Not less than 125% load	Next higher over conductor ampacity
H – Hermetic Motor-Compressor	Not smaller than *Minimum Circuit Ampacity* on nameplate	Not lower than the *Minimum OCPD* ampere rating marked on nameplate	Not greater than *Maximum OCPD* ampere rating marked on the nameplate
N – Noncontinuous	Not less than load	Not less than load	Next higher over conductor ampacity
M – Motor	125% of motor loads plus other loads	Varies based on type of OCPD. See *430.52*	Varies based on type of OCPD. See *430.52*
R – Cord and Plug Branch-Circuit	Not less than OCPD ampere rating	Not less than load	Not higher than conductor ampacity

M, and R, as shown in Table 4-12, are also used in the panelboard worksheet. For motors, the OCPDs are inverse-time circuit breakers. Keep in mind that the conductor must always be adequate for the load to be supplied. Under the conditions explained earlier, the overcurrent device is sometimes permitted to exceed the ampacity of the conductor.

DEFINING THE BRANCH CIRCUITS

In Chapter 3, the electrical loads to be installed in the Drugstore were identified and quantified. The next step usually would be to draw, or sketch, a layout of the space. For this exercise, review working drawing E2, observing the location of the various electrical outlets.

The information in this chapter concentrates on the branch circuits, including the selection of the overcurrent devices and the conductor type and sizes. It will be useful to refer to the panelboard worksheet shown in this chapter. Much of this information is taken from the loading schedule.

Laying out an electrical system usually starts with a sketch or drawing of the space(s). Symbols, as discussed in Chapter 2, are placed on the plans representing the location of the equipment. Then lines are drawn showing the circuiting of the equipment. Following this, branch circuits can be assigned. The actual number of branch circuits to be installed in the Drugstore is determined by referring to the plans or specifications. The following discussion illustrates the process employed to arrive at the final number.

The specifications (see Appendix) for this Commercial Building direct the electrician to install copper conductors of a size not smaller than 12 AWG. The maximum noncontinuous load on a circuit is limited by the rating of the overcurrent protective device. For other than motor branch circuits, the maximum overcurrent protection for a 12 AWG copper conductor is 20 amperes, per *NEC 240.4(D)*. On a 120-volt branch circuit, this limits the load to

$$120 \text{ V} \times 20 \text{ A} = 2400 \text{ VA}$$

The maximum continuous load on a circuit is limited to 80% of the rating of the overcurrent protective device. For 120-volt branch circuits, this limits the load to

$$120 \text{ V} \times 20 \text{ A} \times 0.8 = 1920 \text{ VA}$$

Minimum Number Branch Circuits

Referring to the loading schedule developed in Chapter 3, we will use Table 4-13 to determine the minimum number of branch circuits for the general lighting load and general purpose receptacles. This table will allow us to apply 125% for the portion of the lighting load that is considered continuous.

Actual Number of Branch Circuits

To the four branch circuits required for general lighting and receptacles we will add additional branch circuits shown in the Drugstore panelboard worksheet, Table 4-14. Other factors, such as

TABLE 4-13

Devices and conductors.

Load	Value in VA	Factor	Result in VA
Calculated load for first floor	4185	125%	5231
Calculated load for basement storage area	600	100%	600
Receptacles on first floor and basement storage area	3780	100%	3780
		Total Load	9611
9611 ÷ (120 V × 20 A branch circuits = 2400 VA) = 9611 ÷ 2400			4 branch circuits

switching arrangements and the convenience of installation, are additional important considerations in determining the number of circuits to be used.

At this point, we will consider an example that is unrelated to the Commercial Building in this text to illustrate the concepts we have discussed.

Example 1

Ten luminaires are to be installed in a 1000 sq. ft (92.9 sq. m) industrial loft where the summer temperature will be as high as 97°F (36.11°C). Each luminaire is designed for one 300-watt, 120-volt incandescent lamp, although a 100-watt lamp will be installed. The activities in the room will be continuous from 7 AM to 6 PM. The power system is 120/240-volt, single-phase. In compliance with *NEC 210.23(A)*, branch circuits with a 20-ampere rating shall be used. The conductors shall be no smaller than 12 AWG and shall be Type THHN/THWN.

For this problem, the circuit component selection procedure is divided into three steps. The first step is to determine the number of circuits and the load per circuit.

1a. First, calculate the *minimum load* by multiplying the area by the volt-ampere value given in *NEC Table 220.12* for the occupancy.

1b. Next, determine the *connected load* by multiplying the number of luminaires by the luminaire rating in volt-amperes.

1c. The greater of the *calculated load* and *connected load* in volt-amperes is used to determine the number of branch circuits required.

1d. The load in volt-amperes from 1a. or 1b. is divided by the circuit voltage to determine the calculated load in amperes.

1e. The load in amperes is then divided by the branch-circuit rating. That quotient, rounded up if required to a whole number, indicates the number of circuits.

In the second step, the rating of the OCPD is determined. As this is a branch circuit, *NEC Article 210* is consulted.

2a. First, establish the load that it is expected to operate continuously for three hours or more. The remainder of the load is classified as non-continuous.

2b. The *minimum circuit ampacity* is determined by adding together 125% of the continuous load to 100% of the noncontinuous load.

2c. The rating of the OCPD is selected from *NEC 240.6(A) Standard Ampere Ratings*. It is the standard rating equal to, or next higher than, the OCPD selection ampacity.

2d. The conductor size and type is selected. It shall have an ampacity not less than the *minimum circuit ampacity* and of a type consistent with the problem statement.

The third step is to verify the conductor selection.

3a. Read the allowable ampacity for the selected size and type of conductor from *NEC Table 310.15(B)(16)*.

3b. Record the *maximum ambient temperature*.

3c. Read the correction factor for the selected conductor type and the *maximum ambient temperature* from *NEC Table 310.15(B)(2)(a)*.

3d. Count the number of current-carrying conductors that are to be installed in the raceway.

3e. Read the adjustment factor, for the number of *current-carrying conductors*, from *NEC Table 310.15(B)(3)(a)*.

3f. Calculate the ampacity by multiplying the allowable ampacity by the adjustment and correction factors. If it is not less than the minimum circuit ampacity (2c), the conductor size is verified. If it fails, increase the conductor size and re-verify.

⊸⊸ APPLICATION OF PROCEDURE

Step 1: Determine the number of circuits and load per circuit.

1a. Minimum load = 1000 square feet × 2 volt-amperes per square foot = 2000 volt-amperes (91 square meters × 22 volt-amperes per square meter = 2000 volt-amperes)

1b. Connected load = 300 volt-amperes per luminaire × 10 luminaires = 3000 volt-amperes (Note that 300 VA per luminaire is used as that is the capacity of the luminaire rather than the 100 VA lamp per luminaire the owner intends to install. See *NEC 220.14(D).*)

1c. Use the connected load as it is greater than the calculated load = 3000 volt-amperes

1d. Connected load amperes = 3000 volt-amperes ÷ 120 volts = 25 amperes

1e. Number of circuits = 25 amperes ÷ 20 amperes per circuit = 2 circuits

1f. Connected load per circuit = 25 amperes ÷ 2 circuits = 12.5 amperes per circuit

Step 2: Determine the OCPD (branch-circuit) rating.

2a. Continuous load = 12.5 amperes
     Noncontinuous load = 0 ampere

2b. OCPD selection ampacity = (12.5 amperes × 1.25) + 0 = 15.6 amperes

2c. OCPD rating = 20 amperes

2d. Conductor size and type = 12 AWG, Type THHN/THWN

Step 3: Verify the conductor size.

3a. Conductor allowable ampacity = 30 amperes

3b. 97°F (36.1°C)

3c. Correction factor = 0.88 (for 97°F (36.1°C))

3d. Current-carrying conductors = 3

3e. Adjustment factor = 1.0

3f. Conductor ampacity = 30 × 0.88 × 1.0 = 26.4 amperes. The conductor ampacity is not less than the minimum circuit ampacity (2c), thus the conductor size is verified.

⊸⊸ USING THE PANELBOARD WORKSHEET

The information studied in this unit is now applied to the Drugstore and is illustrated in the Drugstore panelboard worksheet.

Following is a discussion of the purpose and function performed by each of the columns. If there is a question, students are encouraged to review the information presented at the beginning of this chapter.

Load/Area Served (Column A)

This column is a listing of the loads and the areas of the occupancy. Some of the information in this column comes from the Drugstore feeder load calculation form prepared in the preceding chapter.

Quantity (Column B)

This column is used to count the quantity of the various loads to be supplied by the panelboard. Note that in some cases the result will be a calculated load where a unit load is used such as for general-purpose receptacles or where a factor is applied due to continuous loads. The result will be a connected load if actual load values are used such as for the copy machine.

Load Each in VA (Column C)

Here, the load for each unit is entered. Note that due to the very high power factor ballasts for the fluorescent luminaires and for ease of calculation, the lamp wattage is used for the volt-amperes in the worksheet. The load of the air-conditioning equipment is converted to volt-amperes for calculation in the worksheet.

Calculated or Connected Volt-Amperes (Column D)

This column is simply the product or result of multiplying the load in column C by the number of units in column B.

Load Type (Column E)

The following load types will appear in this column:

- **C (Continuous Loads)**

  When the load is continuous, the OCPD ampere rating and conductor must not be less than 125%

TABLE 4-14

Drugstore panelboard worksheet.

A Load/Area Served	B Quantity	C Load Each VA[1]	D Calculated or Connected VA	E Load Type	F Load Modifier	G Ambient Temp. (°F/°C)	H Correction Factor[2]	I Current-Carrying Conductors	J Adjustment Factor[3]	K Minimum Allowable Ampacity	L Conductor Size (AWG)	M Allowable Ampacity[4]	N Overcurrent Protection	O Circuit Number
Luminaires														
Style I	27	61	1647	C	1.25	86/30	1	2	1	17.2	12	20	20	1
Style G	7	80	560	C	1.25	86/30	1	2	1	5.8	12	20	20	2
Style G-EM	2	86	172	C	1.25	86/30	1	2	1	1.8	12	20	20	2
Style E2 Cash Reg	3	26	78	C	1.25	86/30	1	2	1	0.8	12	20	20	2
Style E	2	14	28	C	1.25	86/30	1	2	1	0.3	12	20	20	3
Style E-EM	1	20	20	N	1	86/30	1	2	1	0.2	12	20	20	3
Style N	2	34	68	N	1	86/30	1	2	1	0.6	12	20	20	3
Style Q	2	24	48	N	1	86/30	1	2	1	0.4	12	20	20	3
Exit signs	2	5	9	C	1.25	86/30	1	2	1	0.1	12	20	20	3
Style B Storage	5	64	320	N	1	86/30	1	2	1	2.7	12	20	20	4
Style B-EM Storage	4	70	280	N	1	86/30	1	2	1	2.3	12	20	20	4
Receptacles First Floor, North	6	180	1080	R	1	86/30	1	2	1	9.0	12	20	20	5
Receptacles First Floor, South	5	180	900	R	1	86/30	1	2	1	7.5	12	20	20	6
Receptacles, Cash Registers	2	230	460	R	1	86/30	1	2	1	3.8	12	20	20	7
Receptacles, Floor,	4	180	720	R	1	86/30	1	2	1	6.0	12	20	20	7
Toilet & Storage Areas														
Receptacle, for AC	1	1500	1500	R	1	86/30	1	2	1	12.5	12	20	20	8
Receptacles Storage	8	180	1440	R	1	86/30	1	2	1	12.0	12	20	20	9
Receptacle, Copier	1	1250	1250	R	1	86/30	1	2	1	10.4	12	20	20	10
Exhaust fan, Toilet & Storage	2	120	240	C	1	86/30	1	2	1	2.0	12	20	20	3
Electric Doors	2	360	720	N	1	86/30	1	2	1	6.0	12	20	20	11
Sign	1	1200	1200	C	1.25	86/30	1	2	1	12.5	12	20	20	12
Water heater	1	4500	4500	C	1.25	86/30	1	3	1	15.6	12	20	20	13-17
Air Conditioner, Outdoor Unit[5]	1	10725	10725	H	1	100/38	0.88	3	1	33.9	8	45	60	14-18
Air Conditioner, Indoor Unit	1	15300	15300	H	1.25	86/30	1	3	1	53.1	6	65	60	19-23
Space heater storage	1	7500	7500	H	1.25	86/30	1	3	1	26.0	10	30	30	20-24

[1]Loads of fluorescent luminaires figured at lamp wattage due to very high power factor ballasts.
[2]Correction factor is based upon temperature where the conductors are actually installed.
[3]The adjustment factor is determined by the number of current-carrying conductors in the same raceway or cable.
[4]Assume 75°C terminals
[5]The "Load Each VA" includes 125% of largest motor

of the load or 20 amperes as required by the contract specifications. If combination loads are supplied, the OCPD and conductor are required to be not less than 125% of the continuous loads and 100% of the noncontinuous loads.

- **N (Noncontinuous Loads)**

When the load is noncontinuous, the OCPD ampere rating shall be not less than the load, and the ampacity of the conductor shall not be less than the load and not less than the rating of the OCPD unless the round-up provisions of *NEC 240.4(B)* are permitted and used.

- **H (Hermetic Refrigeration and Heating Units)**

When the circuit serves a hermetic refrigeration unit, the calculated load and minimum branch circuit conductor size can be determined from the branch circuit selection current or minimum branch circuit ampacity given on the unit nameplate. Other loads such as the air-handler and unit or wall mounted heaters are classified as heating loads.

- **M (Motors)**

Motor loads, other than air-conditioning equipment are classed as motor loads. The minimum ampacity of the circuit conductors shall not be less than 125% of the motor load. See *NEC 430.22*.

The motor leads and controller terminals have a 75°C temperature rating, thus that column in *NEC Table 310.15(B)(16)* shall be used when determining the minimum size conductor regardless of the load or the conductor size. See *NEC 110.14(1)*.

See *NEC 430.52(C)(1)* and *Table 430.52* for the selection of the maximum rating of the overcurrent device to be used for branch-circuit short circuit and ground-fault protection. If an inverse-time circuit breaker is used and the motor load is 100 amperes or less, the preferred maximum rating is 250%, but it may be increased to 400% if required for starting the motor. If a dual-element time–delay fuse is used, it can be rated not more than 175% of the motor full load current from the appropriate table in *NEC Article 430*. See Chapter 7 of this text for more information on motor loads.

- **R (Receptacles Supplying Cord- and Plug-Connected Portable Loads)**

If a branch circuit supplies cord- and plug-connected equipment that is not fastened in place, the equipment shall not exceed 80% of the branch-circuit ampere rating, *210.23(A)(1)*. If the cord-and-plug-connected equipment is fastened in place, and the circuit also supplies cord-and-plug-connected equipment not fastened in place, or lighting, or both, the load of the fastened in place equipment shall not exceed 50% of the branch-circuit ampere rating, *210.23(A)(2)*. Receptacles may be placed in C, N, or R class depending on the load they serve.

Load Modifier (Column F). For certain types of loads, the NEC requires that the basic load value be increased. This might be to account for such things as the starting current of a motor or to limit the heat buildup in overcurrent devices where continuous loads are involved. Calculations for the panelboard worksheets in this text refer to this increase as a *load modifier*.

If the load type is C, the load is continuous and is to be increased by 25%, in accordance with *NEC 210.19(A)(1)*.

If the load type is M, the load for the motor is to be increased by 25% in accordance with *NEC 430.22(A)*.

If the load is an H, it shall comply with *NEC 440.22*. Fixed electric space heating loads are also required to be considered continuous for the branch circuit (see *NEC 424.3(B)*) but are not considered continuous for the feeder. *See NEC 220.51*. This section indicates that the authority having jurisdiction is permitted to allow a demand factor below 100% if reduced loading occurs.

If the load is a water heater, it shall be included as a continuous load. See *NEC 422.13*. Other loads remain unchanged.

Ambient Temperature (°F) (Column G). The ambient temperature in degrees Fahrenheit (°F) where the conductor actually operates as discussed previously and in *NEC 310.15(A)(2)* is recorded in this column. A value of 86°F (30°C) is used for most areas of the Commercial Building other than outdoors where higher temperatures can be expected.

Correction Factor (Column H). This factor is taken from ►*NEC Table 310.15(B)(2)(a)*.◄ The information in this table was formerly located below *NEC Table 310.16*. The factor from the table is based on the ambient temperature where the circuit is installed. *NEC 310.15(B)* sets forth the requirements for the use of correction factors if the ambient temperature is outside the norms of *NEC Table 310.15(B)(16)* (formerly *Table 310.16*).

Number of Current-Carrying Conductors (Column I). The number of current-carrying conductors in a specific raceway or cable is recorded in this column. If in doubt, review the earlier discussion on how to determine if a conductor is considered to be current carrying. This is particularly important when considering if it is required to count neutral conductors.

Adjustment Factor (Column J). This factor is taken from *NEC Table 310.15(B)(3)(a)* and is used to compensate for the increase in temperature caused by grouping current-carrying conductors in a raceway or cable.

Minimum Allowable Ampacity (Column K). The calculated or connected load in column D is multiplied by the load modifier in column F, is divided by the temperature correction factor in column H and finally is divided by the adjustment factor in column J. Finally, the volt-amperes determined as explained in the previous sentence is divided by the circuit voltage or, in the case of a 3-phase circuit, by 360. This gives the minimum conductor ampacity.

Conductor Size (AWG) (Column L). For conductors sized 14 AWG through 1 AWG or having overcurrent protection rated 100 amperes or less, the default or standard rating of terminations is 60°C. Thus, the conductor size is selected from the 60°C column of *Table 310.15(B)(16)* that exceeds the minimum allowable ampacity from column K. Conductors with 75°C or 90°C insulation are permitted if used at 60°C ampacities.

For conductors larger than 1 AWG or for overcurrent devices rated greater than 100 amperes, the default rating of terminals is 75°C. The conductor size is selected from the 75°C column of *Table 310.15(B)(16)* that exceeds the minimum allowable ampacity from column K. Conductors with 90°C insulation are permitted if used at 75°C ampacities.

Allowable Ampacity (Column M). This is the allowable ampacity from *NEC Table 310.15(B)(16)* for the size of conductor selected in Column L. ▶ Note that the allowable ampacity for 14 and 12 AWG conductors has been reduced in *NEC Table 310.15(B)(16)* to match the overcurrent protection rules of *NEC 240.4(D)(3) and (5)*.◀ Note as well that the 10 AWG conductor has an allowable ampacity of 26.4 amperes due to the application of a correction factor of 0.88 for the high outside ambient temperature. It can continue to be used for the branch circuit as it exceeds the minimum allowable ampacity of 24.5 amperes.

Overcurrent Protection (Column N). The overcurrent protection for all the circuits with 12 AWG copper conductors is selected based on the allowable ampacity of the conductor and in compliance with *NEC 240.4(D)(5)*. The overcurrent protection for the air-conditioning circuits is selected based on the information on the nameplate for the equipment.

Circuit Number (Column O). The circuit number assigned here will have to be coordinated with the panelboard schedule. If the viewer is facing the panelboard, the circuits connected on the left side are given odd numbers with the even numbers being assigned to the circuits on the right side. Thus, circuits 1, 3, and 5 would be the top three circuits on the left side, and 2, 4, and 6 on the right side. If it is a 3-phase panelboard, two or three circuits can be grouped together with a single neutral conductor and the conductors installed in a raceway, or cable, to serve two or three single-phase 120-volt loads.

The circuit breakers for the 3-phase circuits are numbered with odd numbers on the left and even numbers on the right.

REVIEW

Refer to the *National Electrical Code* or the working drawings when necessary. Where applicable, responses should be written in complete sentences.

Complete the table using the following circuits:

1. Eleven 120-volt duplex receptacles for general use. Receptacle loads are considered to be noncontinuous and are not subject to the 80% load limitation as applied for continuous loads.

2. Ten 120-volt fluorescent luminaires, rated at 150 volt-amperes each.

3. Eight 120-volt incandescent luminaires, rated at 200 watts each.

4. One 240-volt exhaust fan, rated at 5000 watts.

5. One 120/240-volt single-phase, 3-wire feeder, 40-kVA continuous load, originating in a room with an ambient temperature of 100°F (38°C).

Row	Circuit Number	1	2	3	4	5
1	Calculated Load in Volt-Amperes					
2	Calculated·Load in Amperes					
3	Load Type					
4	Load Modifier					
5	OCPD Selection Amperes					
6	OCPD Ampere Rating					
7	Minimum Allowable Conductor Ampacity					
8	Minimum Conductor Size Type THHN/THWN					
9	Ambient Temperature					
10	Temperature Correction Factor					
11	Number of Current-Carrying Conductors					
12	Raceway Fill Adjustment Factor					
13	Allowable Conductor Ampacity Before Correcting/ Adjusting					
14	Allowable Conductor Ampacity After Correcting/ Adjusting					

Row 1:   Calculated load in volt-amperes from problem statements.

Row 2:   Calculated load in amperes divided by voltage factor.

Row 3:   Load type: C = continuous; N = noncontinuous; M = motor, R = receptacle,

Row 4:   Load modifier, 1.25 for C and M load types, 1.0 for R and N load types.

Row 5:   Calculated load amperes (Row 2) multiplied by load modifier (Row 4).

Row 6:   OCPD rating equal to or next higher standard ampere rating than OCPD selection amperes (Row 5).

Row 7:   Ampacity of conductor equal to or greater than calculated load amperes (Row 2).

Row 8:   Conductor size with ampacity equal to or greater than calculated load amperes (Row 2). Select from *Table 310.15(B)(16)* using the appropriate 60°C for circuits rated 100 amperes or less, and the 75°C column for loads over 100 amperes.

Row 9:   From problem statements.

Row 10: From *NEC® Tables 310.15(B)(2)(a)* and *310.15(B)(2)(b)*

Row 11: From problem statements.

Row 12: From *NEC® Table 310.15(B)(3)(a)*.

Row 13: From *NEC® 310.15(B)(16)* for conductor type specified in problem statement: (THHN/THWN).

Row 14: Value from Row 13 multiplied by correction and/or adjustment factors. This value must be equal to or greater than the minimum allowable conductor ampacity value in Row 7.

- Circuits 1, 2, 3, and 4 are installed in the same raceway for a distance of 10 ft in an environment where the maximum ambient temperature will be 86°F (30°C).

- Type THHN/THWN conductors shall be used for all circuits.

- Minimum conductor size 12 AWG.

Switches and Receptacles

OBJECTIVES

After studying this chapter, you should be able to

- select switches and receptacles with the proper rating for a particular application.

- install various types of receptacles correctly.

- connect single-pole, 3-way, 4-way, and 2-pole switches to control circuits.

Electricians select and install numerous receptacles and switches. Therefore, it is essential that electricians know the important characteristics of these devices and how they are to be connected into the electrical system.

RECEPTACLES

The National Electrical Manufacturers Association (NEMA) has developed standards for the configuration of locking and nonlocking plugs and receptacles. The differences in the plugs and receptacles are based on the ampacity and voltage rating of the device. The configuration also ensures receptacles of a lower voltage or ampere rating are not interchangeable with receptacles having a higher rating. For example, the two most commonly used receptacles are the NEMA 5-15R, Figure 5-1, and the NEMA 5-20R, Figure 5-2. The NEMA 5-15R receptacle has a 15-ampere, 125-volt rating and has two parallel slots and a ground pinhole. This receptacle will accept the NEMA 5-15P plug only, Figure 5-3. The NEMA 5-20R receptacle

(A)   (B)

FIGURE 5-2   (A & B) NEMA Figure 5-20R.
(*Photo provided by Leviton*)

has two parallel slots and a T slot. This receptacle is rated at 20 amperes, 125 volts. The NEMA 5-20R will accept either a NEMA 5-15P or 5-20P plug, Figure 5-4. A NEMA 6-20R receptacle is shown in

FIGURE 5-3   NEMA Figure 5-15P.
(*Delmar/Cengage Learning*)

FIGURE 5-4   NEMA Figure 5-20P.
(*Delmar/Cengage Learning*)

(A)   (B)

FIGURE 5-1   (A & B) NEMA Figure 5-15R.
(*Photo provided by Leviton*)

(A)   (B)

FIGURE 5-5 (A) NEMA 6-20R (B) NEMA 5-20R.
(*Photo provided by Leviton*)

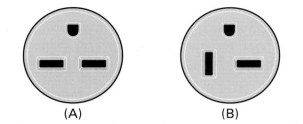

FIGURE 5-6 (A) NEMA 6-15R (B) NEMA 6-20P.
(*Delmar/Cengage Learning*)

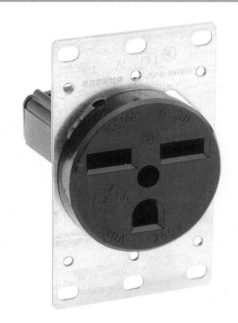

FIGURE 5-7 NEMA 6-30R.
(*Photo provided by Leviton*)

Figure 5-5. This receptacle has a rating of 20 amperes at 250 volts and will accept either the NEMA 6-15P or 6-20P plug, Figure 5-6. The acceptance of 15- and 20-ampere plugs complies with *NEC Table 210.21(B)(3)*, which permits the installation of either the NEMA 5-15R or 5-20R receptacle on a 20-ampere branch circuit. Another receptacle, which is specified for the Commercial Building, is the NEMA 6-30R, Figure 5-7. This receptacle is rated at 30 amperes, 250 volts.

The NEMA standards for general-purpose non-locking and locking plugs and receptacles are shown in Tables 5-1, 5-2, and 5-3. A special note should be made of the differences between the 125/250 (NEMA 14)

TABLE 5-1	
NEMA terminal identification for receptacles and plugs.	
Green-Colored Terminal (marked G, GR, GN, or GRND)	This terminal has a hexagon shape. Connect equipment grounding conductor ONLY to this terminal. (Green, bare, or green with yellow stripe.)
White (Silver)-Colored Terminal (marked W)	Connect white or gray grounded circuit conductor ONLY to this terminal.
Brass-Colored Terminal (marked X, Y, Z)	Connect HOT conductor to this terminal (black, red, blue, etc.). There is no NEMA standard for a specific color conductor to be connected to a specific letter. It is good practice to establish a color code and stick to it throughout the installation.

TABLE 5-2

NEMA configurations for general-purpose nonlocking plugs and receptacles.

		15 AMPERE		20 AMPERE		30 AMPERE		50 AMPERE		60 AMPERE	
		RECEPTACLE	PLUG	RECEPTACLE	PLUG	RECEPTACLE	PLUG	RECEPTACLE	PLUG	RECEPTACLE	PLUG
2-POLE 2-WIRE	125 V (1)	1-15R	1-15P		1-20P		1-30P				
	250 V (2)		2-15P	2-20R	2-20P	2-30R	2-30P				
	277 V AC (3)			(RESERVED FOR FUTURE CONFIGURATIONS)							
	600 V (4)			(RESERVED FOR FUTURE CONFIGURATIONS)							
2-POLE 3-WIRE GROUNDING	125 V (5)	5-15R	5-15P	5-20R	5-20P	5-30R	5-30P	5-50R	5-50P		
	250 V (6)	6-15R	6-15P	6-20R	6-20P	6-30R	6-30P	6-50R	6-50P		
	277 V AC (7)	7-15R	7-15P	7-20R	7-20P	7-30R	7-30P	7-50R	7-50P		
	347 V AC (24)	24-15R	24-15P	24-20R	24-20P	24-30R	24-30P	24-50R	24-50P		
	480 V AC (8)			(RESERVED FOR FUTURE CONFIGURATIONS)							
	600 V AC (9)			(RESERVED FOR FUTURE CONFIGURATIONS)							
3-POLE 3-WIRE	125/250 V (10)			10-20R	10-20P	10-30R	10-30P	10-50R	10-50P		
	3 Ø 250 V (11)	11-15R	11-15P	11-20R	11-20P	11-30R	11-30P	11-50R	11-50P		
	3 Ø 480 V (12)			(RESERVED FOR FUTURE CONFIGURATIONS)							
	3 Ø 600 V (13)			(RESERVED FOR FUTURE CONFIGURATIONS)							
3-POLE 4-WIRE GROUNDING	125/250 V (14)	14-15R	14-15P	14-20R	14-20P	14-30R	14-30P	14-50R	14-50P	14-60R	14-60P
	3 Ø 250 V (15)	15-15R	15-15P	15-20R	15-20P	15-30R	15-30P	15-50R	15-50P	15-60R	15-60P
	3 Ø 480 V (16)			(RESERVED FOR FUTURE CONFIGURATIONS)							
	3 Ø 600 V (17)			(RESERVED FOR FUTURE CONFIGURATIONS)							
4-POLE 4-WIRE	3 Ø 208Y/120 V (18)	18-15R	18-15P	18-20R	18-20P	18-30R	18-30P	18-50R	18-50P	18-60R	18-60P
	3 Ø 480Y/277 V (19)			(RESERVED FOR FUTURE CONFIGURATIONS)							
	3 Ø 600Y/347 V (20)			(RESERVED FOR FUTURE CONFIGURATIONS)							
4-POLE 5-WIRE GROUNDING	3 Ø 208Y/120 V (21)			(RESERVED FOR FUTURE CONFIGURATIONS)							
	3 Ø 480Y/277 V (22)			(RESERVED FOR FUTURE CONFIGURATIONS)							
	3 Ø 600Y/347 V (23)			(RESERVED FOR FUTURE CONFIGURATIONS)							

TABLE 5-3

NEMA configurations for general-purpose locking plugs and receptacles.

			15 AMPERE		20 AMPERE		30 AMPERE		50 AMPERE		60 AMPERE	
			RECEPTACLE	PLUG	RECEPTACLE	PLUG	RECEPTACLE	PLUG	RECEPTACLE	PLUG	RECEPTACLE	PLUG
2-POLE 2-WIRE	125 V	L1	L1-15R	L1-15P								
	250 V	L2			L2-20R	L2-20P						
	277 V AC	3		(RESERVED FOR FUTURE CONFIGURATIONS)								
	600 V	4		(RESERVED FOR FUTURE CONFIGURATIONS)								
2-POLE 3-WIRE GROUNDING	125 V	5	L5-15R	L5-15P	L5-20R	L5-20P	L5-30R	L5-30P	L5-50R	L5-50P	L5-60R	L5-60P
	250 V	6	L6-15R	L6-15P	L6-20R	L6-20P	L6-30R	L6-30P	L6-50R	L6-50P	L6-60R	L6-60P
	277 V AC	7	L7-15R	L7-15P	L7-20R	L7-20P	L7-30R	L7-30P	L7-50R	L7-50P	L7-60R	L7-60P
	347 V AC	24			L24-20R	L24-20P						
	480 V AC	8			L8-20R	L8-20P	L8-30R	L8-30P	L8-50R	L8-50P	L8-60R	L8-60P
	600 V AC	9			L9-20R	L9-20P	L9-30R	L9-30P	L9-50R	L9-50P	L9-60R	L9-60P
3-POLE 3-WIRE	125/250 V	10			L10-20R	L10-20P	L10-30R	L10-30P				
	3 Ø 250 V	11	L11-15R	L11-15P	L11-20R	L11-20P	L11-30R	L11-30P				
	3 Ø 480 V	12			L12-20R	L12-20P	L12-30R	L12-30P				
	3 Ø 600 V	13					L13-30R	L13-30P				
3-POLE 4-WIRE GROUNDING	125/250 V	14			L14-20R	L14-20P	L14-30R	L14-30P	L14-50R	L14-50P	L14-60R	L14-60P
	3 Ø 250 V	15			L15-20R	L15-20P	L15-30R	L15-30P	L15-50R	L15-50P	L15-60R	L15-60P
	3 Ø 480 V	16			L16-20R	L16-20P	L16-30R	L16-30P	L16-50R	L16-50P	L16-60R	L16-60P
	3 Ø 600 V	17					L17-30R	L17-30P	L17-50R	L17-50P	L17-60R	L17-60P
4-POLE 4-WIRE	3 Ø 208Y/ 120 V	18			L18-20R	L18-20P	L18-30R	L18-30P				
	3 Ø 480Y/ 277 V	19			L19-20R	L19-20P	L19-30R	L19-30P				
	3 Ø 600Y/ 347 V	20			L20-20R	L20-20P	L20-30R	L20-30P				
4-POLE 5-WIRE GROUNDING	3 Ø 208Y/ 120 V	21			L21-20R	L21-20P	L21-30R	L21-30P	L21-50R	L21-50P	L21-60R	L21-60P
	3 Ø 480Y/ 277 V	22			L22-20R	L22-20P	L22-30R	L22-30P	L22-50R	L22-50P	L22-60R	L22-60P
	3 Ø 600Y/ 347 V	23			L23-20R	L23-20P	L23-30R	L23-30P	L23-50R	L23-50P	L23-60R	L23-60P

devices and the 3-phase, 250-volt (NEMA 15) devices. The connection of the 125/250-volt receptacle requires a neutral conductor, an equipment grounding conductor, and 2-phase connections. For the 3-phase, 250-volt receptacle, an equipment grounding conductor and 3-phase connections are required.

Hospital-Grade Receptacles

In locations where severe abuse or heavy use is expected, hospital-grade receptacles are recommended. These are high-quality products and meet UL requirements. These receptacles are marked with a small green dot, Figure 5-8.

Electronic Equipment Receptacles

To reduce "electrical noise" (electromagnetic interference), circuits with receptacle outlets that serve loads such as cash registers, computers, or other sensitive electronic equipment are often supplied by *isolated ground* receptacles,

Figures 5-9 and 5-10(B). In addition to having an isolated ground feature, these receptacles may also have transient voltage surge protection or be hospital grade, or both.

Isolated Ground Receptacles

Isolated ground receptacles are covered in *250.146(D)*.

Where isolated grounding is desired, a green, or green with one or more yellow stripes, insulated equipment grounding conductor is run from the neutral terminal at the main service entrance or the source of a separately derived system to the isolated ground terminal on the receptacle.

Isolated ground receptacles are required by *NEC 406.2(D)* to have an orange triangle on their face, Figures 5-9 and 5-10(B). The orange triangle might be on an ivory face, or the entire face might be orange, with the triangle outlined using black lines. This allows the orange to show through the outlined triangle.

"Electrical noise," as referred to when discussing electronic equipment, is any extraneous signal

FIGURE 5-8   Hospital grade receptacle.
(*Photo provided by Leviton*)

FIGURE 5-9   Isolated ground receptacle.
(*Photo provided by Leviton*)

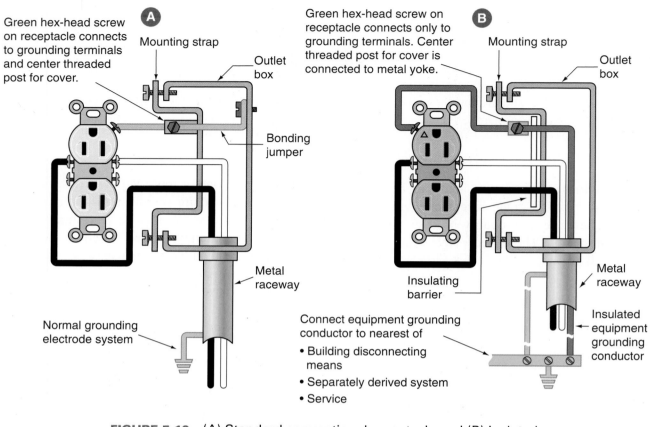

Green hex-head screw on receptacle connects to grounding terminals and center threaded post for cover.

A

Mounting strap

Outlet box

Bonding jumper

Metal raceway

Normal grounding electrode system

Green hex-head screw on receptacle connects only to grounding terminals. Center threaded post for cover is connected to metal yoke.

B

Mounting strap

Outlet box

Insulating barrier

Metal raceway

Insulated equipment grounding conductor

Connect equipment grounding conductor to nearest of
• Building disconnecting means
• Separately derived system
• Service

FIGURE 5-10 (A) Standard conventional receptacle and (B) isolated ground receptacle. (*Delmar/Cengage Learning*)

that tends to interfere with the signal normally present in or passing through a system. Induced noise might be EMI (electromagnetic interference) or RFI (radio frequency interference). It might be heard as a hum, buzzing, whining, or static. The cause might be traced to such things as an electric motor, ignition on a furnace, fluorescent and other types of electronic discharge luminaires, nearness to a power line, arc welders, power-supply switching circuits, cellular phones, other audio/video equipment, office electronic equipment, and nearness of computer cables to electric light and power wiring. These things are capable of creating unwanted waveforms that can scramble or distort digital signals.

In a standard conventional receptacle outlet, Figure 5-10(A), the green hexagonal grounding screw, the grounding contacts, the yoke (strap), and the metal wall box are all "tied" together to the building's equipment grounding system. Thus, the many receptacles in a building create a multiple-ground path situation.

In an isolated ground receptacle, Figures 5-9 and 5-10(B), the green hexagonal grounding screw and the grounding contacts of the receptacle are isolated from the metal yoke (strap) of the receptacle and also from the wiring method's equipment grounding conductor system. Wiring methods typically used that provide an equipment grounding conductor include EMT, Type MC, having the armor that qualifies as an equipment grounding conductor and Type AC cable. Type MC cable is available with two equipment grounding conductor paths for wiring isolated ground circuits. It is often referred to as Type MC-IG.

Permission is given in *NEC 250.146(D)* to install a separate green insulated equipment grounding conductor from the green hexagonal screw on the receptacle all the way back to the main service disconnect or source of a separately derived system (see Figure 5-11). This separate green equipment grounding conductor is not required to be connected to any panelboards, load centers, junction or pull boxes, or other ground reference points in between.

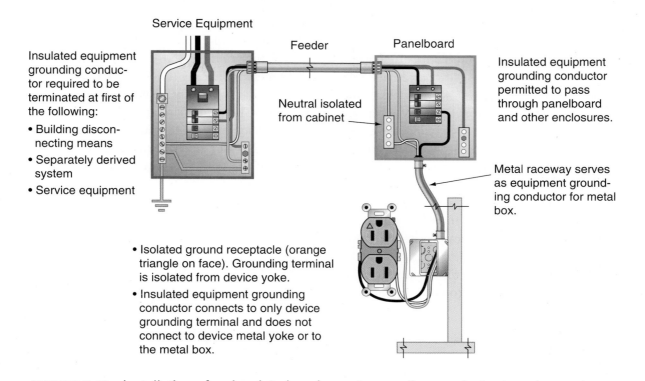

Insulated equipment grounding conductor required to be terminated at first of the following:

• Building disconnecting means
• Separately derived system
• Service equipment

Insulated equipment grounding conductor permitted to pass through panelboard and other enclosures.

Neutral isolated from cabinet

Metal raceway serves as equipment grounding conductor for metal box.

• Isolated ground receptacle (orange triangle on face). Grounding terminal is isolated from device yoke.
• Insulated equipment grounding conductor connects to only device grounding terminal and does not connect to device metal yoke or to the metal box.

FIGURE 5-11   Installation of an insulated equipment grounding conductor to reduce noise on circuits supplying computers. The receptacle is of the isolated ground type. (*Delmar/Cengage Learning*)

The isolated system grounding is now "clean," and the result is less transient noise (disturbances) transmitted to the connected load.

Equipment grounding conductors and isolated equipment grounding conductors are two different items. The *Informational Note* to *NEC 250.146(D)* calls attention to the fact that metal raceways and outlet boxes may still have to be grounded with an equipment grounding conductor. An additional isolated equipment grounding conductor may then be installed to minimize noise as discussed earlier.

When nonmetallic boxes are used in conjunction with an isolated ground receptacle, nonmetallic faceplates should be used because the metal yoke on the receptacle is not connected to the receptacle's equipment grounding conductor terminal. A metallic faceplate would not be properly grounded.

Electronic Equipment Grounding

Most commercial buildings today contain electronic equipment such as computers, copy machines, fax machines, data-processing equipment, telephone systems, security systems, medical diagnostic instruments, HVAC and similar electronic controls, and electronic cash registers.

Electronic equipment is extremely sensitive and susceptible to line disturbances caused by such things as

• voltage sags, spikes, and surges;
• static charges;
• lightning strikes and surges on the system;
• electromagnetic interference (EMI);
• radio frequency interference (RFI); and
• improper grounding.

Metal raceways in a building act like a large antenna and can pick up electromagnetic interference (electrical noise) that changes data in computer/data-processing equipment.

Electric utilities are responsible for providing electricity within certain voltage limits. Switching surges or lightning strikes on their lines can cause problems with electronic equipment. Static, lightning-energy, radio interference, and electromagnetic interference can cause voltages that exceed the tolerance limits of the electronic equipment. These can cause equipment computing problems. Electronic data-processing equipment will put out garbage that is unacceptable in the business world. Electronic equipment must be protected from these disturbances.

Quality voltage provides a clean sine wave that is free from distortion, Figure 5-12. Poor voltage (dirty power) might show up in the form of sags, surges, spikes or impulses, notches or dropouts, or total loss of power, Figure 5-13. Disturbances that can change the sine wave may be caused by other electronic equipment operating in the same building.

Proper grounding of the ac distribution system and the means by which electronic equipment is grounded are critical, as these can affect the operation of electronic equipment. The ungrounded (HOT) conductor, the grounded (NEUTRAL) conductor, and the equipment grounding (GROUND) conductor must be properly sized, tightly connected, and correctly terminated. The equipment grounding conductor serves a vital function for a computer in that the computer's dc logic has one side directly connected to the metal frame of the computer, thus using ground as a reference point for the processing of information. An acceptable impedance for the grounding path associated with normal equipment grounding for branch-circuit wiring is 1 to 2 ohms. Applying Ohm's law:

$$\text{Amperes} = \frac{\text{Volts}}{\text{Ohms}} = \frac{120}{1} = 120 \text{ amperes}$$

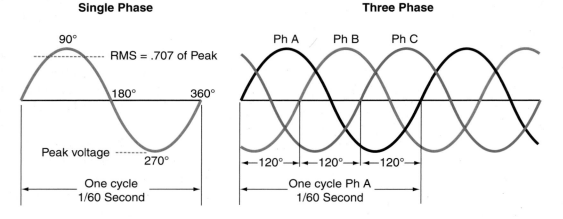

FIGURE 5-12 Quality voltage and sine wave. (*Delmar/Cengage Learning*)

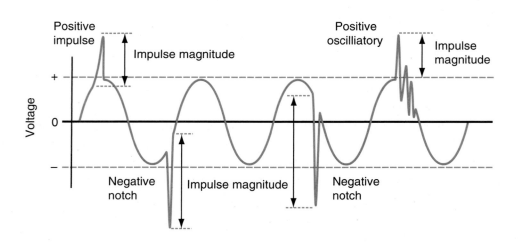

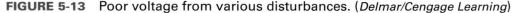

FIGURE 5-13 Poor voltage from various disturbances. (*Delmar/Cengage Learning*)

A ground-fault value of 120 amperes will easily cause a 15- or 20-ampere branch-circuit breaker to trip off. *NEC 250.118* lists the different items that are considered acceptable equipment grounding conductors.

The acceptable impedance of the ground path for branch-circuit wiring is *not* acceptable for grounding electronic equipment such as computer/data-processing equipment. The acceptable impedance of a ground path for grounding electronic equipment is much more critical than the safety equipment ground impedance for branch-circuit wiring. The maximum impedance for grounding electronic equipment is 0.25 ohms. That is why a separate insulated isolated equipment grounding conductor is often used for the grounding of computer/data-processing equipment to ensure a low-impedance path to ground. Furthermore, stray currents on the equipment grounding conductor can damage the electronic equipment or destroy or alter the accuracy of the information being processed. Proper equipment grounding minimizes these potential problems. A separate insulated isolated equipment grounding conductor is sized according to *NEC Table 250.122*. For longer runs, a larger size equipment grounding conductor is recommended to keep the impedance to 0.25 ohms or less. The electrical engineer designing the circuits and feeders for computer rooms and data-processing equipment considers all of this. *NEC Article 645* covers information technology equipment and systems. The National Fire Protection Association publication *NFPA-75* titled *Protection of Electronic Computer/ Data Processing Equipment* contains additional information.

For circuits supplying computer/data-processing equipment, *never* use the grounded metal raceway or metal armor of Type MC cable as the only equipment ground for the electronic data-processing equipment. Instead of allowing the grounded metal raceway system or Type MC cable to serve as the equipment ground for sensitive electronic equipment, use a separate *insulated* equipment grounding conductor that has green, or green with one or more yellow stripes, insulation that

- connects to the isolated equipment grounding terminal of an isolated type receptacle.
- does not connect to the metal yoke of the receptacle.

- does not connect to the metal outlet or switch box.
- must be isolated from the metal raceway system.
- does not touch anywhere except at the connection at its source and at the grounding terminal of the isolated receptacle.
- must be run in the same raceway as the HOT and NEUTRAL conductors.
- provides a ground path having an impedance of 0.25 ohms or less.

This separate insulated equipment grounding conductor is permitted by the *NEC* to pass through one or more panelboards, boxes, wireways, or other enclosures without a connection to the panelboard grounding terminal bar. It can be carried all the way back to the main service or to the transformer, such as a 480-volt primary, 208/120-volt wye connected secondary step-down transformer. See Figure 5-11, *NEC 250.146(D)*, and *NEC 408.40, Exception.*

To summarize grounding when electronic computer/data-processing equipment is involved, generally there are two equipment grounds: (1) the metal raceway system or Type MC cable that serves as the safety equipment grounding means for the metal outlet boxes, switch boxes, junction boxes, and similar metal equipment that is part of the premise wiring; and (2) a second insulated isolated equipment ground that serves the electronic equipment. This is sometimes referred to as a "dedicated equipment grounding conductor."

NEC 250.96(B), Informational Note, and *250.146(D), Informational Note,* state that the use of an isolated equipment grounding conductor does not relieve the requirement for grounding the raceway system and outlet box.

Figure 5-13 illustrates one way that receptacles serving computer-type equipment can be connected. The panelboard enclosure might have been provided with an equipment ground bus insulated from the panelboard. This would provide a means of connecting one or more separate insulated isolated equipment grounding conductors running from the ground pin of the isolating-type receptacles. One larger size insulated isolated equipment grounding conductor would then be run from the source (in this figure, the ground bus in the main switch) to the equipment grounding terminal bar in the panelboard.

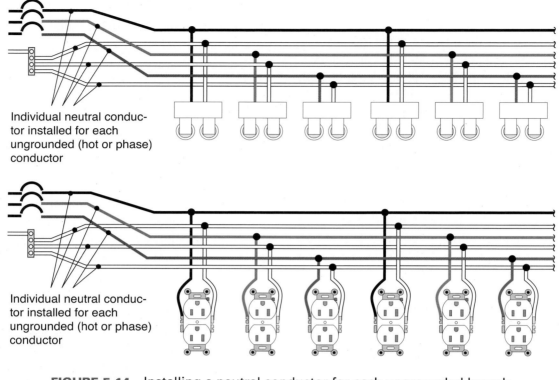

Individual neutral conductor installed for each ungrounded (hot or phase) conductor

Individual neutral conductor installed for each ungrounded (hot or phase) conductor

FIGURE 5-14 Installing a neutral conductor for each ungrounded branch circuit conductor. (*Delmar/Cengage Learning*)

Another problem that occurs where there is a heavy concentration of electronic equipment is that of overheated neutral conductors. To minimize this problem on branch circuits that serve electronic loads such as computers, it is highly recommended that separate neutral conductors be run for each phase conductor of the branch-circuit wiring, rather than running a common neutral conductor for multiwire branch circuits. This is done to eliminate the problem of harmonic currents overheating the neutral conductor. See Figure 5-14 for an illustration of installing a neutral conductor for each ungrounded branch circuit conductor.

If surge protection is installed, it will absorb high-voltage surges on the line and further protect the equipment, Figure 5-15. These are highly recommended in areas of the country where lightning strikes are common. When a comparison is made between the cost of the equipment, or the cost of re-creating lost data, and the cost of the receptacle, probably under $50, this is inexpensive protection. Surge protection can be thought of as a system with protection installed at the service, at a panelboard, and finally, at the receptacle or equipment.

FIGURE 5-15 Surge suppressor receptacle. (*Photo provided by Leviton*)

Ground-Fault Circuit-Interrupter Protection

Electrocutions and personal injury have resulted because of electrical shock from defective tools and appliances and other defective electrical equipment. This shock hazard exists whenever a person inadvertently touches a hot ungrounded conductor or defective equipment and a conducting surface, such as a water pipe, metal sink, sheet metal ducts, or any conducting material that is grounded. The injury might be from the electrical shock itself or from reacting to an electrical shock by falling off a ladder or into moving parts of equipment. To protect against the possibility of shock, 15- and 20-ampere branch circuits can be protected with GFCI receptacles or GFCI circuit breakers.

NEC 210.8(B) sets forth the requirements for the installation of ground-fault circuit-interrupter protection (GFCI) for buildings other than dwellings. Note that a new requirement in the opening paragraph of *NEC 210.8* states ▶ *The ground-fault circuit-interrupter shall be installed in a readily accessible location.* *◀ The term *readily accessible* is defined in *NEC Article 100 as Capable of being reached quickly for operation, renewal, or inspections without requiring those to whom ready access is requisite to climb over or remove obstacles or to resort to portable ladders, and so forth.* Be very mindful of this rule when installing receptacle type GFCIs.

In commercial buildings, GFCI protection is required for all 125-volt, single-phase, 15- and 20-ampere receptacles installed

1. in bathrooms,
2. in kitchens,
3. on rooftops,
4. outdoors,
5. within 6 ft (1.8 m) *from the outside edge of any sink,*
6. ▶ in indoor wet locations, ◀
7. ▶ in locker rooms with adjacent showering facilities, ◀ and
8. ▶ in garages, service bays, and similar areas. ◀

The *NEC* defines a bathroom in *Article 100* as ▶ *An area including a basin with one or more of the* following: a toilet, a urinal, a tub, a shower, a bidet, or similar plumbing fixtures.* ◀

The *NEC* defines a kitchen, also in *Article 100,* as ▶ *An area with a sink and permanent provisions for food preparation and cooking.* *◀ Some employee break areas meet the definition of a kitchen and thus require GFCI protection of 15- and 20-ampere, 125-volt receptacles.

GFCI protection for the outdoor and rooftop receptacles in the Commercial Building can be provided by installing GFCI circuit breakers on the circuit(s) supplying the receptacles or by installing GFCI receptacles. This is a matter of choice. If GFCI circuit breakers are used, a tripping requires going to the panelboard to reset the GFCI circuit breaker. In the case of tripping a GFCI receptacle, the tripping is right there at the GFCI receptacle—easy to reset. A GFCI receptacle opens both grounded and ungrounded circuit conductors to deal with issues like reverse-polarity. A circuit breaker GFCI opens only the ungrounded conductor.

Exception No. 1 exempts the GFCI requirement for rooftop and outdoor receptacles that are

- supplied by a branch circuit dedicated to electric snow-melting, deicing, or pipeline and vessel heating equipment, and
- not readily accessible.

Elevator Receptacles. Receptacles of the 125-volt, single-phase, 15- and 20-ampere type that are installed in elevator pits as well as in elevator equipment rooms are required to be of the GFCI type. See *NEC 620.85* for these requirements. Note that GFCI protection of the branch circuit does not satisfy this requirement.

Drinking Fountains. Electric drinking fountains are required to have GFCI protection. See *NEC 422.52.* This requirement does not extend to water heater/coolers that typically have an exchangeable water bottle.

Cord-and-Plug-Connected Vending Machines. Cord-and-plug-connected vending machines (new and remanufactured) are required to have integral GFCI protection either in the attachment plug cap or within 12 in. (300 mm) of the attachment plug cap. If the

---

*Reprinted with permission from NFPA 70-2011.

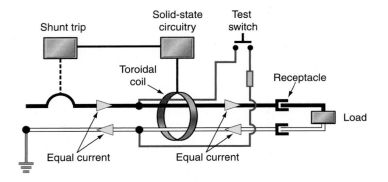

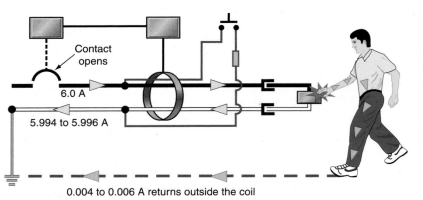

No current is induced in the toroidal coil because both circuit wires are carrying equal current. The contacts remain closed.

An imbalance of from 4 to 6 milliamperes in the coil will cause the contacts to open. The GFCI must open in approximately 25 milliseconds. Receptacle-type GFCIs have a switching contact in each circuit conductor.

0.004 to 0.006 A returns outside the coil

FIGURE 5-16   Basic principle of ground-fault circuit-interrupter operation. (*Delmar/Cengage Learning*)

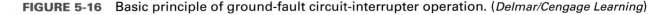

vending machine does not have integral GFCI protection, *NEC 422.51* requires that the outlet into which the vending machine will be plugged be GFCI protected.

Replacing Receptacles in Existing Buildings

Be very careful when replacing receptacles in existing installations. No doubt you will be replacing a grounding-type or a nongrounding-type receptacle. *NEC 406.4(D)(3)* contains a retroactive requirement that states, *Ground-fault circuit-interrupter protected receptacles shall be provided where replacements are made at receptacle outlets that are required to be so protected elsewhere in this Code.**

The Underwriters Laboratories require that Class A GFCIs do not trip on ground-fault currents of 4 milliamperes or less and must trip on ground-fault currents of 6 milliamperes or greater (0.004 to 0.006 ampere). Figure 5-16 illustrates the principle of how a GFCI

operates. The GFCI could have been appropriately called an unbalanced or leakage current detector, which would have accurately described its function.

Receptacles in Electric Baseboard Heaters

Electric baseboard heaters are available with or without receptacle outlets. Figure 5-17 shows the relationship of an electric baseboard heater to a receptacle outlet. *NEC Article 424* of the *Code* states the requirements for fixed electric space-heating equipment. See *NEC 210.52*. This receptacle shall not be connected to the heater's branch circuit.

SNAP SWITCHES

The term *snap switch* is rarely used today except in the *NEC* and the Underwriters Laboratories standards. Most electricians refer to snap switches as "toggle switches" or "wall switches." Typical switches are shown in Figure 5-18.

*Reprinted with permission from NFPA 70-2011.

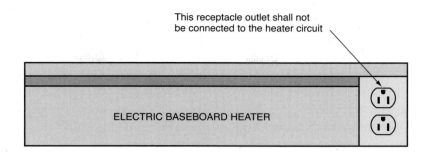

This receptacle outlet shall not
be connected to the heater circuit

ELECTRIC BASEBOARD HEATER

FIGURE 5-17   Receptacles in a permanently installed electric baseboard heater. Listed baseboard heaters include instructions that may not permit their installation below receptacle outlets. See *210.52 Informational Note*. (*Delmar/Cengage Learning*)

FIGURE 5-18   General use snap switches. (*Photo provided by Leviton*)

Switches are divided into two categories. *NEC 404.14(A)* covers ac general-use snap switches that are used to control:

- alternating-current (ac) circuits only.

- resistive and inductive loads not to exceed the ampere rating of the switch at the voltage involved. This includes electric-discharge lamps that involve ballasts, such as fluorescent, HID, mercury vapor, and sodium vapor.

- tungsten-filament lamp loads not to exceed the ampere rating of the switch at 120 volts.

- motor loads not to exceed 80% of the ampere rating of the switch at rated voltage.

Ac general-use snap switches may be marked ac only, or they may be marked with the current and voltage rating markings. A typical switch marking is 15A, 120-277V ac. The 277-rating is required on 480Y/277-volt systems.

See *NEC 210.6* for additional information pertaining to maximum voltage limitations. *NEC 404.14(B)* covers ac–dc general-use snap switches that are used to control:

- ac or direct-current (dc) circuits.

- resistive loads not to exceed the ampere rating of the switch at applied voltage.

- inductive loads not to exceed one-half the ampere rating of the switch at applied voltage.

- motor loads not to exceed the ampere rating of the switch at applied voltage only if the switch is marked in horsepower.

- tungsten-filament lamp loads not to exceed the ampere rating of the switch at applied voltage when marked with the letter T. A tungsten-filament lamp draws a very high current at the instant the circuit is closed. T-rated switch contacts are capable of handling arcing associated with the high inrush current surge.

An ac–dc general-use snap switch normally is not marked ac–dc. However, it is always marked with the current and voltage rating, such as 10A-125V or 5A-250V-T.

Terminals of switches rated at 20 amperes or less, when marked CO/ALR, are suitable for use with aluminum, copper, and copper-clad aluminum conductors, *NEC 404.14(C)*. Switches not marked CO/ALR are suitable for use with copper and copper-clad aluminum conductors only.

Screwless pressure terminals of the conductor push-in type may be used with copper and copper-clad aluminum conductors only. Push-in type terminals are not suitable for use with ordinary aluminum conductors.

Full information on snap-switch ratings and their use is found in *NEC 404.14* and in the Underwriters Laboratories *White Book: Guide Information for Electrical Equipment* in the category *Snap Switches (WJQR)*.

Grounding of Snap Switches and Metal Faceplates

In *404.9(B)* we find the requirements for grounding snap switches, dimmers, and similar types of switches. The metal yoke on snap switches shall be grounded. All snap switches have a green hexagon-shaped equipment grounding screw secured to the metal yoke of the switch for attaching an equipment grounding conductor.

Let's look at how proper grounding is accomplished.

A metal box is considered adequately grounded provided the metal raceway, Type AC or Type MC cable, is properly grounded at its origin, *NEC 250.118*. This being the case, the metal yoke of the snap switch is considered grounded when secured to the metal box by metal screws. In turn, a metal faceplate is considered grounded when secured to the metal yoke of a snap switch. The effective ground-fault current path is a multistep situation: first the metal raceway or armor must be grounded; then the metal outlet box must be grounded; then the yoke on the snap switch is grounded; and finally the metal faceplate is grounded by the No. 6-32 metal screws that secure the faceplate to the switch.

When there is direct metal-to-metal contact of the metal yoke between the switch and the surface-mounted metal wall box or metal cover, or if the switch has a self-grounding clip on the yoke, there is no need to connect a separate equipment grounding conductor to this grounding screw.

Without these requirements, a metal faceplate could unknowingly become energized, resulting in a lethal shock to anyone touching the metal faceplate and a grounded surface of any kind.

Although not likely found in other than dwellings, there is an exception to *NEC 404.9(B)* that addresses older buildings that might be wired with nonmetallic-sheathed cable that does not have an equipment grounding conductor, or in which the wiring might be knob and tube. In this situation, if you are replacing a switch that is located within reach of ground, earth, conducting floors, or another conducting surface, you have two choices: (1) install a proper equipment grounding conductor as defined in *NEC 250.118*, or (2) install a GFCI on that particular branch circuit.

CONDUCTOR COLOR CODING

Some color coding is mandated by the *National Electrical Code*. These are the grounded conductors (often referred to as "neutral") and equipment grounding conductors. Other color coding choices are left up to the electrician. Reference to color coding is found in *NEC 110.15*, *200.6*, *200.7*, *210.5*, *215.12*, *230.56*, *250.119*, and *310.110*

Ungrounded Conductor

An ungrounded "hot" conductor must have an outer finish that is other than green, white, or gray, or with three continuous white stripes. This conductor is commonly called the hot conductor. You will get a shock if this hot conductor and a grounded conductor or grounded surface, such as a metal water pipe, are touched at the same time.

Popular colors are black, white, red, blue, green, yellow, brown, orange, and gray. Conductors are now available in these insulation colors through 500 kcmil or larger.

Grounded Conductor

There are many important terms used in the electrical industry. The following will help clarify two of these terms.

Electricians often use the term *neutral* whenever they refer to a white grounded circuit conductor. The correct definition is found in the *NEC*:

- **Neutral Conductor.** *The conductor connected to the neutral point of a system that is intended to carry current under normal conditions.* * See Figure 5-19.

- **Neutral Point.** *The common point on a wye connection in a polyphase system, midpoint on a single-phase, 3-wire system; midpoint of a single-phase portion of a 3-phase delta system; or midpoint of a 3-wire, direct-current system.* * See Figure 5-19.

For alternating-current circuits, the *NEC* requires that the grounded (identified) conductor have an outer finish of either a continuous white or gray color, or have an outer finish (not green) that has three continuous white stripes along the conductor's entire length. In multiwire circuits, the grounded conductor is also called a *neutral* conductor.

A grounded conductor is often referred to as a neutral conductor. A neutral conductor is always grounded under the requirements of *NEC 250.20*, but a grounded conductor is not always a neutral

conductor, such as in the case of the grounded phase conductor of a 3-phase, grounded B-phase system.

NEC 200.6(A) requires that a grounded conductor, sizes 6 AWG or smaller, be identified by a continuous white or gray color. *NEC 200.6(B)* permits grounded conductors larger than 6 AWG to be identified at its terminations. For example, this would allow a black conductor to be identified in a disconnect switch or panelboard with white paint or marking tape.

When more than one system is installed in the same raceway, such as combinations of 208Y/120-volt wye and 480Y/277-volt wye circuits, you must be able to distinguish which grounded conductor belongs to which set of ungrounded conductors. Generally, white is used for the grounded conductor of 208Y/120-volt circuits, and gray is used for the grounded conductor of 480Y/277-volt circuits. The color chosen for identifying the grounded conductors of the different systems shall be posted at each branch-circuit panelboard, *NEC 200.6(D)*.

A white or gray conductor shall not be used as an ungrounded (hot) phase conductor except when reidentified according to *NEC 200.7*, which requires it to be a part of a cable or flexible cord.

Three Continuous White Stripes.   In commercial and industrial wiring, some conductors have three continuous white stripes. This method of identifying conductors is done by the cable manufacturer and is found on larger size conductors and some service-entrance cables, such as those used for a mobile-home feeder (two black, one black that has three white stripes, and one green).

Equipment Grounding Conductor

An equipment grounding conductor may be bare, green, or green with one or more yellow stripes, *NEC 250.119* and *310.110(B)*. Never use an insulated conductor that is green or green with yellow stripes for an ungrounded (hot) or grounded (often a neutral) phase conductor. An exception is allowed for low-voltage circuit such as control circuits for air-conditioning systems.

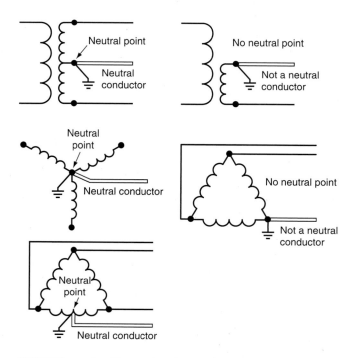

FIGURE 5-19 Neutral point and neutral conductors.
(*Delmar/Cengage Learning*)

*Reprinted with permission from NFPA 70-2011.

Typical Color Coding (for Conduit Wiring)

The following are some examples of color coding quite often used for the conductors of branch circuits, feeders, and services. These color coding recommendations are not code mandated, other than for the white (or gray) grounded conductor. Certainly, other color combinations are permitted. Just be sure that you do not use white, gray, green, and bare conductors other than for their permitted uses discussed previously.

- 120 volts, single-phase, 2-wire: black, white
- 120/240 volts, single-phase, 3-wire: black, white, red
- 208Y/120 volts, 3-phase, 4-wire wye: black, white, red, blue
- 240 volts, 3-phase, 3-wire delta, corner grounded B-phase: black, white (B-phase), red
- 120/240 volts, 3-phase, 4-wire delta with "high leg": black, white, red, orange (the high leg). You could also use black, white, blue, or orange. *NEC 110.15* and *230.56* requires the high leg conductor to be orange. See *NEC 408.3(E)* for phase arrangement of the high leg.
- 480Y/277 volts, 3-phase wire, 4-wire wye: brown, gray, orange, yellow
- 480 volts, 3-phase, 3-wire delta: brown, orange, yellow

CAUTION: For new work, the grounded conductor is permitted to be white or gray. See *NEC 200.6* and *200.7*. In old installations, a conductor having gray insulation was sometimes used as an ungrounded "hot" conductor. This practice is no longer permitted. Be very careful when working on any circuit that has a gray insulated conductor. It might be a hot conductor—and could be lethal.

Two Electrical Systems in One Building

For installations having more than one electrical system, such as 480Y/277 volts and 208Y/120 volts, use white as the grounded conductor for one system and use gray as the grounded conductor for the other system.

Typical Color Coding (for Cable Wiring)

When wiring with cable (Type NM, AC, or MC), the choice of colors is limited to those that are furnished with the cable unless the cable is special ordered with different colors:

- **Two-wire cable:** black, white, bare, or green equipment grounding conductor
- **Three-wire cable:** black, white, red, bare, or green equipment grounding conductor
- **Four-wire cable:** black, white, red, blue, bare, or green equipment grounding conductor
- **Four-wire cable:** black, white, red, white with red stripe, or bare equipment grounding conductor
- **Five-wire cable:** black, white, red, blue, yellow, bare, or green equipment grounding conductor

Several of the wiring diagrams in this chapter illustrate how the white wire is correctly used in switching circuits.

Home-run cables are available that have multiple sets of multiwire branch circuits, with a neutral for each ungrounded conductor and with oversized neutral conductors where harmonic currents may be a problem.

Conductor Insulation Colors for Connecting Switches

Before we discuss switch connections, let's discuss conductor insulation colors. Color selection is easy when the wiring method is conduit or tubing because so many colors are available. When the wiring method is cable, color coding becomes a bit more complicated. You are stuck with black, white, and red.

Small conductors are available with insulation colors of black, blue, brown, gray, green, orange, pink, purple, red, tan, white, and yellow. You might find other colors. Some conductors are also available that have a stripe on the insulation for further identification for the more complex installation.

Insulation color coding has been pretty well established over the years for phase conductors,

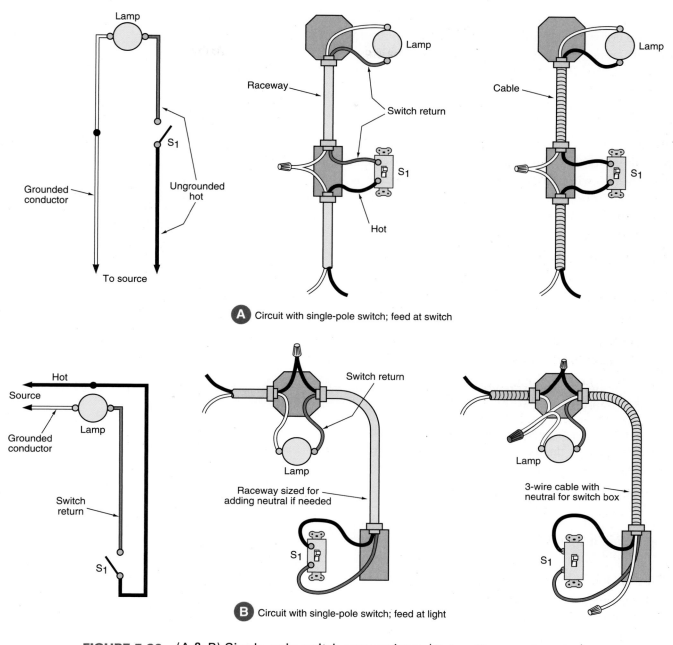

A Circuit with single-pole switch; feed at switch

B Circuit with single-pole switch; feed at light

FIGURE 5-20   (A & B) Single-pole switch connections. (*Delmar/Cengage Learning*)

switch returns, and travelers of electrical systems. For this Commercial Building, the electrical system is 208Y/120, 3-phase, 4-wire wye using: Phase A—black, white (neutral), Phase B—red, and Phase C—blue. That leaves brown, orange, pink, purple, tan, and yellow for switch returns (switch legs) and travelers.

Switch Returns.   To keep confusion to a minimum, select a color that is not the same as the phase conductors. Over the years, different localities seem to have established color coding by tradition. Don't use white, gray, or green for switch returns. See Figures 5-20, 5-22, and 5-24.

Travelers. Again to keep confusion to a minimum, select a color that is different from the switch return and different from the phase conductors. Don't use white, gray, or green for travelers (Figs. 5-21, 5-22, 5-23, and 5-24).

When using cable for the wiring of 3-way switches, *NEC 200.7(C)(1)* permits the use of a conductor with insulation that is white or gray or has three continuous white stripes (not too common) as a switch supply *only* if the conductor is permanently reidentified by painting or other effective means such as marking tape at each location where the conductor is visible and accessible. A white or gray insulated conductor is not permitted to be used as the switch return (Figs. 5-20, 5-22, and 5-24).

Snap Switch Types and Connections

Snap switches are readily available in four types: single-pole, 3-way, 4-way, and double-pole.

Single-Pole Switches. A single-pole switch is used where it is desired to control a light or group of lights, or other load, from one switching point. This type of switch is connected in series with the ungrounded (hot) wire feeding the load. Figure 5-20 shows typical applications of a single-pole switch controlling a light from one switching point, either at the switch or at the light. Note that orange has been used for the switch return (switch leg/switch loop) in the raceway (conduit) diagrams. Any color other than white, gray, or green could have been used for the switch return.

To repeat, *NEC 200.7(C)(1)* makes it very clear that the white conductor is *not to be used as a return conductor from the switch to the switched outlet.* Note the use of the words *permanently reidentified* in *200.7(C)(1)*. Some inspectors do not consider plastic tape as permanent. Instead, they prefer the use of permanent marking pens or heat shrinkable tubing, both available in a multitude of colors and sizes. For the more complex commercial and industrial applications, there are thermal printers for use with heat shrinkable tubing to mark wire and cable. Find out what means of "permanently reidentified" is acceptable in your area.

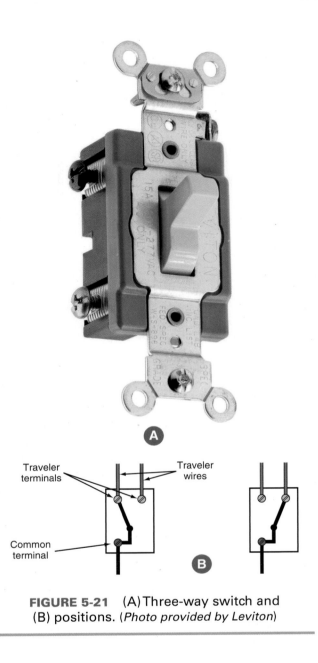

FIGURE 5-21 (A) Three-way switch and (B) positions. (*Photo provided by Leviton*)

Permanent marking pens are a great way to reidentify conductor insulation. These are readily available with a wide felt tip (⅝ inch). Cut a notch or drill a small hole in the felt tip. Slide the notched marker up and down the conductor insulation or slide the marker with the small hole over the conductor and slide it up and down. This makes the job of reidentifying a conductor very easy and fast.

Three wiring diagrams are shown in Figure 5-20 for each of the switching connections discussed. The first diagram is a schematic drawing and is valuable when visualizing the current path. The second

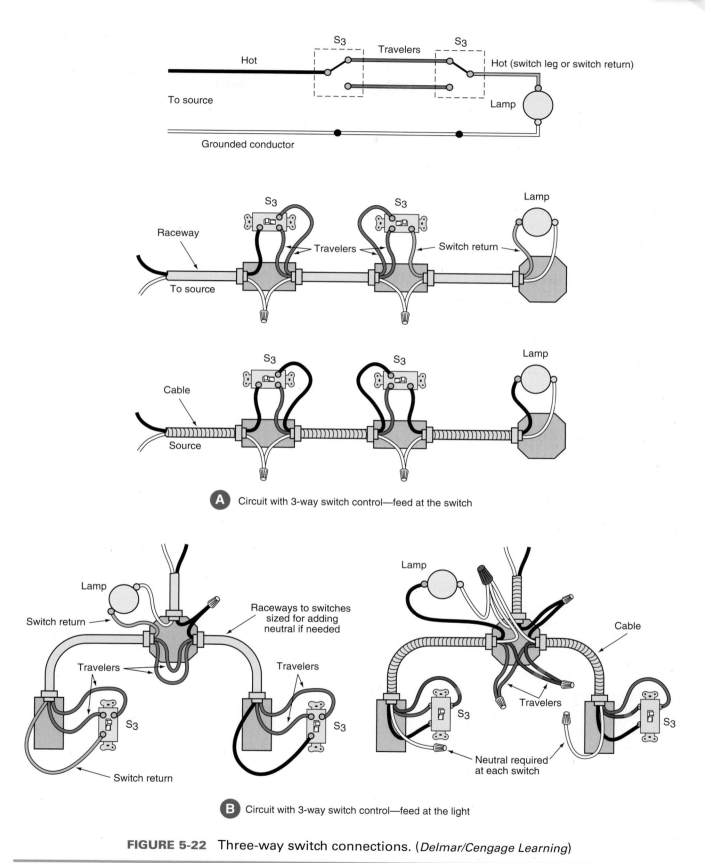

A Circuit with 3-way switch control—feed at the switch

B Circuit with 3-way switch control—feed at the light

FIGURE 5-22 Three-way switch connections. (*Delmar/Cengage Learning*)

diagram represents the situation where a raceway will be available for installing the conductors. The third diagram illustrates the connection necessary when using a cable such as armored cable (AC), metal-clad cable (MC), or nonmetallic-sheathed cable (NM). Equipment grounding conductors and connections are not shown in order to keep the diagrams as simple as possible.

Three-Way Switches. A 3-way switch should really be called a 3-terminal switch. Three-way switches provide switch control from two locations. A 3-way switch, Figure 5-21, has a *common terminal* to which the switch blade is always connected. The other two terminals are called the *traveler terminals*. In one position, the switch blade is connected between the common terminal and one of the traveler terminals. In the other position, the switch blade is connected between the common terminal and the second traveler terminal. A 3-way switch can easily be identified because it has no On or Off position. Note that On and Off positions are not marked on the switch handle in Figure 5-21A. A 3-way switch is also identified by its three terminals. The common terminal is darker in color than the two traveler terminals, which have a brass color. Figure 5-22 shows 3-way switch connections, one with the source feeding at the light, the other with the source feeding at one of the 3-way switches.

In Figures 5-22(A) and (B), orange has been used for the switch return (switch leg/switch loop) in the raceway (conduit) diagrams. Any color other than white, gray, or green could have been used for the switch return.

For the *travelers*, purple was selected. Other colors could have been used. Always use a color that is different from the switch return.

As with single-pole switch connections, when using cable for the wiring of 3-way switches, *NEC 200.7(C)(1)* permits the use of a conductor with insulation that is white or gray or has three continuous white stripes (not too common) as a switch supply *only* if the conductor is permanently reidentified by painting or other effective means at each location where the conductor is visible and accessible.

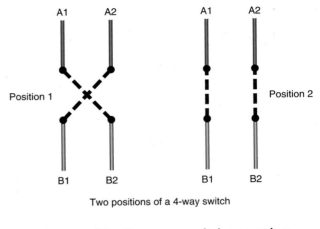

Two positions of a 4-way switch

FIGURE 5-23 Four-way switch operation. (*Delmar/Cengage Learning*)

Four-Way Switches. A 4-way switch should really be called a 4-terminal switch. A 4-way switch is used when a light, a group of lights, or other load must be controlled from more than two switching points.

A 4-way switch is similar to a 3-way switch in that it does not have On and Off positions. A 4-way switch is always connected between two 3-way switches. Three-way switches are connected to the source and to the load. At all other control points, 4-way switches are used.

A 4-way switch has four terminals. Two of these terminals are connected to the traveler conductors from one 3-way switch, and the other two terminals are connected to traveler conductors from the other 3-way switch, Figure 5-22.

In Figure 5-23, terminals A1 and A2 are connected to one 3-way switch and terminals B1 and B2 are connected to the other 3-way switch. In position 1, the switch connects A1 to B2 and A2 to B1. In position 2, the switch connects A1 to B1 and A2 to B2.

Figure 5-24 illustrates a typical circuit in which a lamp is controlled from any one of three switching points.

Note that orange has been used for the switch return (switch leg/switch loop) in the raceway (conduit) diagrams. Any color other than white, gray, or green could have been used for the switch return. For the *travelers*, select a color that is different from the switch return. In these diagrams, purple and pink

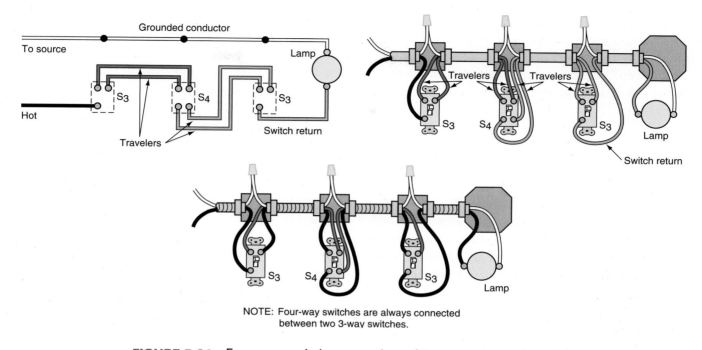

NOTE: Four-way switches are always connected between two 3-way switches.

FIGURE 5-24 Four-way switch connections. (*Delmar/Cengage Learning*)

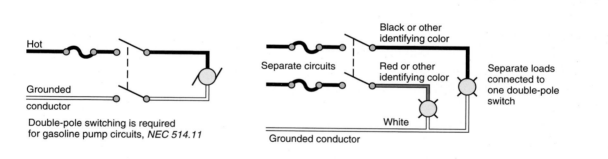

Double-pole switching is required for gasoline pump circuits, *NEC 514.11*

FIGURE 5-25 Double-pole switch connections. (*Delmar/Cengage Learning*)

have been used for the travelers. Other color combinations could have been used.

For cable wiring, the white conductor is used as the supply to a switch and shall be permanently reidentified as being an ungrounded conductor. The white conductor shall not be used as the switch return, *200.7(C)(2)*.

In Figure 5-24, the arrangement is such that white conductors were spliced together. Care must be used to ensure that the traveler wires are connected to the proper terminals of the 4-way switch. That is, the two traveler wires from one 3-way switch must be connected to the two terminals on

one end of the 4-way switch. Similarly, the two traveler wires from the other 3-way switch must be connected to the two terminals on the other end of the 4-way switch.

Double-Pole Switch. A double-pole switch is rarely used on lighting circuits. As shown in Figure 5-25, a double-pole switch can be used for those installations where two separate circuits are to be controlled with one switch. However the switch must be marked "2-circuit" to indicate the switch has been investigated for this purpose. See the UL *Guide Information for Electrical Equipment (White Book)* under the category *Snap Switches* (*WJQR*). All conductors of circuits supplying gasoline-dispensing pumps, or running through such pumps, must have a disconnecting means. Thus, the lighting mounted on gasoline-dispensing islands may require 2-pole switches, *NEC 514.11.*

OCCUPANCY SENSORS

Occupancy sensors come in several types and configurations. At times, they are ceiling mounted so they can "see" the area better. These may be wired directly to the luminaires they control or be the control circuit to a lighting controller. All are installed to reduce the use of energy and are considered to be a "green" control.

Figure 5-26 shows a simple form of occupancy sensor. It is contained in a switch-type wiring device that will fit into a standard device box. These are

FIGURE 5-26 Light-switch type of occupancy sensor. (*Photo provided by Leviton*)

often supplied with an on/off override switch to give nonautomatic control.

A new requirement was added to the 2011 *NEC* in *404.2(C)* to require a neutral (grounded) conductor to be provided at the switch location. The rule reads as follows:

▶*(C) Switches Controlling Lighting Loads. Where switches control lighting loads supplied by a grounded general purpose branch circuit, the grounded circuit conductor for the controlled lighting circuit shall be provided at the switch location.*

Exception: The grounded circuit conductor shall be permitted to be omitted from the switch enclosure where either of the conditions in (1) or (2) apply:

(1) Conductors for switches controlling lighting loads enter the box through a raceway. The raceway shall have sufficient cross-sectional area to accommodate the extension of the grounded circuit conductor of the lighting circuit to the switch location whether or not the conductors in the raceway are required to be increased in size to comply with 310.15(B)(2)(a).

(2) Cable assemblies for switches controlling lighting <u>*loads*</u> *enter the box through a framing cavity that is open at the top or bottom on the same floor level, or through a wall, floor, or ceiling that is unfinished on one side.* * ◀

The purpose of the rule is stated in the *Informational Note* that follows. It reads,

▶*Informational Note: The provision for a (future) grounded conductor is to complete a circuit path for electronic lighting control devices.* * ◀

SWITCH AND RECEPTACLE COVERS

The cover that is placed on a recessed box containing a receptacle or a switch is called a faceplate, and a cover placed on a surface-mounted box, such as a 4-in. square, is called a raised cover.

Faceplates come in a variety of colors, shapes, and materials, but for our purposes, they are placed in two categories: metal and insulating. Metal

*Reprinted with permission from NFPA 70-2011.

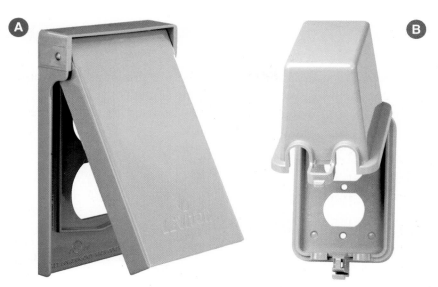

FIGURE 5-27 (A) Cover suitable for damp locations and (B) cover required for wet locations. (*Photo provided by Leviton*)

faceplates can become a hazard because they can conduct electricity. Metal faceplates are considered to be effectively grounded through the No. 6-32 screws that fasten the faceplate to the grounded yoke of a receptacle or switch. If the box is nonmetallic, the yoke must be grounded, and both switches and receptacles are available with a grounding screw for grounding the metal yoke.

In the past, it was common practice for a receptacle to be fastened to a raised cover (installed on a surface-mounted box such as a 4 in. [100 mm] square) by a single 6–32 screw. *NEC 406.4(C)* now prohibits this practice and requires that two screws or another approved method be used to fasten a receptacle to a raised (surface-mounted) cover.

Receptacles installed outdoors where protected from the weather, or in any damp location, shall remain weatherproof when the receptacle cover is closed, *NEC 406.9(A)*. See Figure 5-27 for an example of weatherproof receptacle covers.

A box and cover installed in any wet location shall remain weatherproof:

- when an attachment plug cap is inserted into the receptacle, or
- when no attachment plug cap is inserted into the receptacle.

Receptacles rated 15- and 20-ampere, 125- through 250-volts that are installed in these

weatherproof enclosures shall be listed as weather-resistant type.

Wet location requirements for receptacles are found in *NEC 406.9(B)*. See *NEC 404.4* for requirements for switches installed in damp or wet locations.

Taken in part from *NEC Article 100*, here are definitions important to the understanding of locations:

> **Damp.** *Locations protected from weather and not subject to saturation with water or other liquids but subject to moderate degrees of moisture.**
>
> **Dry.** *A location not normally subject to dampness or wetness.**
>
> **Wet.** *Installations underground or in concrete slabs or masonry in direct contact with the earth; in locations subject to saturation with water or other liquids, such as vehicle washing areas; and in unprotected locations exposed to weather.**

See *NEC Article 100* for the complete definitions of the above terms.

---

*Reprinted with permission from NFPA 70-2011.

REVIEW

Refer to the *National Electrical Code* or the working drawings when necessary. Where applicable, responses should be written in complete sentences.

Indicate which of the following switches may be used to control the loads listed in Questions 1–7.

a. Ac/dc   10 A–125 V/5 A–250 V

b. Ac only   10 A–120 V

c. Ac only   15 A–120/277 V

d. Ac/dc   20 A–125 V–T/10 A–250 V–T

1. _____ A 120-volt incandescent lamp load (tungsten filament) consisting of ten 150-watt lamps

2. _____ A 120-volt fluorescent lamp load (inductive) of 1500 volt-amperes

3. _____ A 277-volt fluorescent lamp load of 625 volt-amperes

4. _____ A 120-volt motor drawing 10 amperes

5. _____ A 120-volt resistive load of 1250 watts

6. _____ A 120-volt incandescent lamp load of 2000 watts

7. _____ A 230-volt motor drawing 2.5 amperes

Using the information in Tables 5-2 and 5-3, select by number the correct receptacle for the conditions described in Questions 8–11.

8. A 20-ampere, 120/208-volt, single-phase load _____

9. A 50-ampere, 230-volt, 3-phase load _____

10. A 30-ampere, 208-volt, single-phase load _____

11. A 20-ampere, hospital-grade, 120-volt, single-phase _____

12. The metal yoke of an isolated grounding-type receptacle (is) (is not) connected to the green equipment grounding terminal of the receptacle. (Circle the correct answer.)

For Questions 13–15, check the correct statement.

13. ☐ The recommended method to ground computer/data-processing equipment is to use the grounded metal raceway as the equipment grounding conductor.

     ☐ The recommended method to ground computer/data-processing equipment is to install a separate insulated equipment grounding conductor in the same raceway in which the branch-circuit conductors are installed.

14. When a separate insulated equipment grounding conductor is installed to minimize line disturbances on the circuit supplying an isolated grounding-type receptacle, it

     ☐ shall terminate in the panelboard where the branch circuit originates.

     ☐ may run through the panelboard where the branch circuit originates, back to the main disconnect (the source) for the building.

15. To eliminate the overheated neutral problem associated with branch-circuit wiring that supplies computer/data-processing equipment,

    ☐ install a separate neutral conductor for each phase conductor.

    ☐ install a common neutral conductor on multiwire branch circuits.

For the following questions, wiring has been installed for 3-way control of a light. The power source is available at one of the 3-way switches; the load is located between the 3-way switches. Be sure to study the requirements of *404.2(C)*.

16. Draw a wiring diagram illustrating the connection of the conductors.

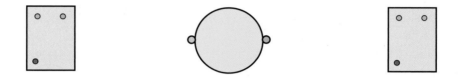

17. Draw the conductors in the raceway diagram that follows. Draw the connections to the switches and the light. Label the conductors for color and function.

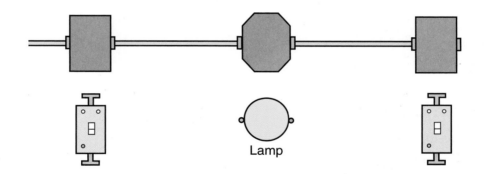

Lamp

18. Following is the same arrangement but using nonmetallic sheathed cable, Type AC or Type MC cable. Draw connections to the switches and the light. Label the conductors for color and function. A 2-wire cable would contain black and white conductors; a 3-wire cable would contain red, black, and white conductors. A 4-wire cable has black, white, red, and blue colored insulations.

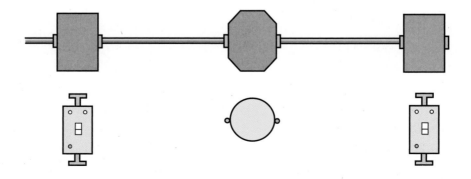

19. This diagram consists of two lamps and two 3-way switches. Draw a wiring diagram of the connections. Bring the circuit in from the left.

20. Following is a diagram showing the installation of a metallic raceway system connecting two lamps and two 3-way switches. Draw the necessary conductors and show their connections.

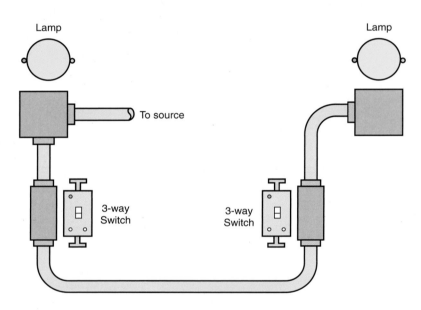

21. Following is a diagram showing the installation of a nonmetallic sheathed cable system connecting two lamps and two 3-way switches. Draw the necessary conductors and show their connections.

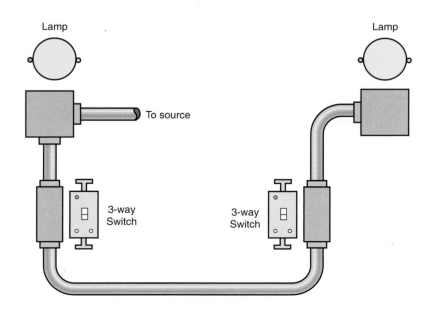

22. This diagram consists of one lamp, two 3-way switches, and one 4-way switch. Draw a wiring diagram of the connections. Bring the circuit in from the top (lamp location first).

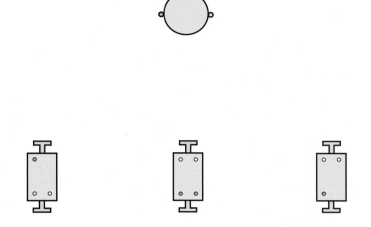

23. Following is a diagram showing the installation of a metallic raceway system connecting a lamp outlet to a 4-way switch system. Draw the necessary conductors and show their connections.

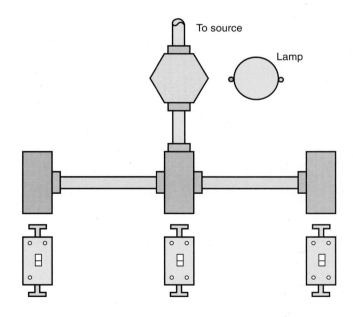

24. Following is a diagram showing the installation of a nonmetallic sheathed cable system connecting a lamp and a 4-way switch system. Draw the necessary conductors and show their connections.

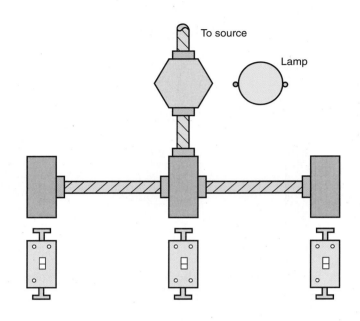

Wiring Methods

OBJECTIVES

After studying this chapter, you should be able to

- select the proper raceway or cable for the conditions.
- identify the installation requirements for a raceway or cable.
- select the proper raceway size, depending on the conductors to be installed.
- understand the term *Metric Designator*.
- select the proper size of outlet, pull, and junction boxes based on the applicable requirements.
- select the proper size of box, depending on entering raceways and orientation.

In a commercial building, the major part of the electrical work is the installation of the branch-circuit wiring. The electrician must have the ability to select and install the correct materials to ensure a successful job. For some jobs, the specifications or bid documents will dictate a specific wiring method that is required. For other installations, the owner will specify the wiring method to be used.

The term *raceway*, which is used in this chapter as well as others, is defined by the *NEC* as a channel that is designed and used expressly for the purpose of holding wires, cables, or busbars. The term *circular raceway* is finding its way into the *NEC* to distinguish between raceways that are circular in configuration from raceways that are rectangular such as wireways.

The following paragraphs describe several types of materials classified as a raceway. The *NEC* has introduced acronyms for most raceways as shown. The *NEC* recognizes intermediate metal conduit (IMC), rigid metal conduit (RMC), flexible metal conduit (FMC), liquidtight flexible metal conduit (LFMC), rigid polyvinyl chloride conduit (PVC), high density polyethylene conduit (HDPE), nonmetallic underground conduit with conductors (NUCC), reinforced thermosetting resin conduit (RTRC), liquidtight flexible nonmetallic conduit (LFNC), electrical metallic tubing (EMT), flexible metallic tubing (FMT), and electrical nonmetallic tubing (ENT).

RACEWAY SIZING IN THE *NEC*

In keeping with the emphasis of the metric system of measurements in the *NEC*, a Metric Designator has been introduced to provide an equivalency to the inch system of measurement for circular raceways used for many years as shown in Table 6-1.

You may have noticed the *NEC* uses the term "trade size" rather than simply "size" or "size in inches" to indicate the size of circular raceways. There is a simple explanation for this designation. Trade size 2 conduit is not two inches! This is shown in *Table 4* of *Chapter 9*. The nominal internal diameters in inches and millimeters as well as the square-inch (square millimeter) areas of trade size 2 (metric designator 53) circular raceways are as shown in Table 6-2.

You should know that the circular raceway with the largest internal size in trade sizes ½ through 2 (metric designator 16 through 53) size is IMC and from trade size 2½ through 4 (metric designator 63

TABLE 6-1

Raceway trade sizes and metric designators.

Trade Size	Metric Designator	Raceways Permitted in Sizes						
		IMC	RMC	FMC	LFMC	PVC	EMT	ENT
⅜	12	N	Y[1]	Y	Y	N	Y[1]	N
½	16	Y	Y	Y	Y	Y	Y	Y
¾	21	Y	Y	Y	Y	Y	Y	Y
1	27	Y	Y	Y	Y	Y	Y	Y
1¼	35	Y	Y	Y	Y	Y	Y	Y
1½	41	Y	Y	Y	Y	Y	Y	Y
2	53	Y	Y	Y	Y	Y	Y	Y
2½	63	Y	Y	Y	Y	Y	Y	N
3	78	Y	Y	Y	Y	Y	Y	N
3½	91	Y	Y	Y	Y	Y	Y	N
4	103	Y	Y	Y	Y	Y	Y	N
5	129	N	Y	N	N	Y	N	N
6	155	N	Y	N	N	Y	N	N

[1]Limited to enclosing the leads of motors as stated in *430.245(B)*.

TABLE 6-2

Dimensions of trade size 2 raceways.

Trade Size 2	Diameter		100% Area	
	Inches	Millimeters	Inches²	Millimeters²
EMT	2.067	52.5	3.356	2165
ENT	2.020	51.3	3.205	2067
FMC	2.040	51.8	3.269	2107
IMC	2.150	54.6	3.630	2341
RMC	2.083	52.9	3.408	2198
PVC (Schedule 40)	2.047	52.0	3.291	2124

through 103) is EMT. Thus, these raceways allow a larger number of conductors in some cases.

It will be useful to review some of the terms used in the *NEC* concerning raceways:

- A ferrous conduit is made of iron or steel; a nonferrous conduit is made of a metal other than iron or steel.

- Common nonferrous raceways include aluminum, brass and stainless steel.

- Metal or metallic would include both ferrous and nonferrous.

- Nonmetallic raceways would include PVC, ENT, and fiberglass.

- Couplings are used to connect sections of a raceway.

- Locknuts, metal bushings, or connectors are used to connect raceways to boxes or fittings.

- Integral couplings are formed into some raceways and cannot be removed.

- Associated fittings such as couplings and connectors are separate items.

- A running thread is a longer-than-standard-length thread cut on one conduit and is sometimes referred to as "continuous-thread" or "all-thread." Connection to another conduit is achieved by screwing a coupling on the running thread, butting the conduits together, and then backing the coupling onto the second conduit. As shown in Figure 6-1, running threads are not permitted for connecting conduits together with a coupling as a tight connection is not assured,

Running thread not permitted for connection at couplings

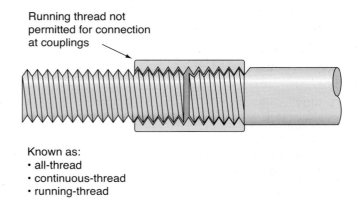

Known as:
- all-thread
- continuous-thread
- running-thread

FIGURE 6-1 Running threads.
(*Delmar/Cengage Learning*)

NEC 344.42(B). Running thread can be used to connect two enclosures by cutting the conduit to length, installing the outer locknuts on the running thread, and sliding the conduit into one enclosure, then into the other. Caution: If running thread is created by cutting threads in the field, no galvanizing will remain on the conduit, and protection against rusting should be provided. An option to installing running thread between enclosures is to connect two conduit nipples together with a split coupling.

RIGID METAL CONDUIT (RMC)

Rigid metal conduit, Figure 6-2, is of heavy-wall construction to provide a maximum degree of physical protection to the conductors that run through it.

FIGURE 6-2 Rigid metal conduit.
(*Delmar/Cengage Learning*)

FIGURE 6-3 Rigid metal conduit fittings.
(*Courtesy Appleton*)

Rigid metal conduit is available in steel, aluminum, brass, and stainless steel. The conduit can be threaded on the job with a standard pipe die, or nonthreaded fittings may be used where permitted. See *NEC Article 344* and Figure 6-3.

NEC 300.6 covers the need for protection of metal raceways against corrosion. When rigid metal conduit is threaded in the field (on the job), the thread must be coated with an approved electrically conductive, corrosion-resistant compound if corrosion protection is necessary.

RMC bends of 45° and 90° can be purchased, or they can be made using special bending tools. Bends in trade sizes 1⁄2, 3⁄4, and 1 RMC can be made using hand benders or hickeys, Figure 6-4. See the Ideal Bending Guide in Appendix I of this text for additional information on using a hand bender. Hydraulic benders must be used to make bends in larger sizes of conduit.

Uses permitted for RMC are found in *NEC 344.10. Uses permitted* include these (see the *NEC* for specifics):

A. Atmospheric Conditions and Occupancies

   1. Galvanized steel and stainless steel

FIGURE 6-4 (A) "Hickey" for bending conduit (B) Bender for EMT and smaller conduit. Both, without handles. (*Photos Courtesy Ideal*)

   2. Red brass

   3. Aluminum

   4. Ferrous raceways and fittings

B. Corrosive Environments

   1. Galvanized steel, stainless steel, and red brass RMC, elbows, couplings, and fittings

   2. Supplementary protection of aluminum RMC

C. Cinder Fill

D. Wet Locations

Similar *uses permitted* are found in each of the circular raceway articles in the 3XX.10 section. If *uses not permitted* are provided, they are located in the 3XX.12 section.

INTERMEDIATE METAL CONDUIT (IMC)

Intermediate metal conduit has a wall thickness that is between that of rigid metal conduit (RMC) and electrical metallic tubing (EMT). As a result, this circular raceway is often permitted to contain more conductors than RMC.

This type of conduit can be installed using either threaded or nonthreaded fittings. *NEC Article 342* defines IMC and describes the uses permitted, the installation requirements, and the construction specifications. It is permitted to be installed in the same locations as RMC.

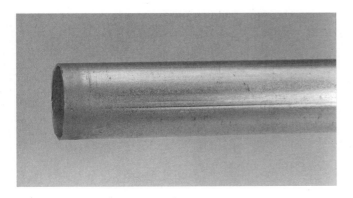

FIGURE 6-5 Electrical metallic tubing (EMT). (*Delmar/Cengage Learning*)

ELECTRICAL METALLIC TUBING (EMT)

EMT is a thin-wall metal raceway that is not permitted to be threaded, Figure 6-5. EMT is also referred to in the trade as *thinwall*. The specifications for the Commercial Building in this text permit the use of EMT for all branch-circuit wiring. *NEC Article 358* should be consulted for exact installation requirements.

Fittings

Electrical metallic tubing is a nonthreaded, thin-wall raceway. Because EMT is not permitted to be threaded, sections are joined together and connected to boxes, other fittings, or cabinets by couplings and connectors. EMT fittings available include the set-screw, compression and indenter styles.

Set-screw. When used with this type of fitting, the EMT is pushed into the coupling or connector to the stop and is secured in place by tightening the set-screws, Figure 6-6. This type of fitting is classified as concrete-tight but is not raintight. Some fittings may be concrete-tight only when

FIGURE 6-6 EMT connector and coupling, set-screw type. (*Delmar/Cengage Learning*)

FIGURE 6-7 EMT connector and coupling, compression type. (*Photo courtesy Appelton Electric LLC, Emerson Industrial Automation*)

wrapped with tape. If so, the carton containing the fitting is marked "Concrete-tight when taped."

Compression. EMT is secured in these fittings by tightening the compression nuts with a wrench or pliers, Figure 6-7. These fittings are classified as raintight and concrete-tight types.

When installing EMT outdoors where exposed to rain, or in any other wet location, compression EMT fittings must be listed for this use. Check the carton for wording that indicates the suitability of these fittings for use in wet locations.

Indenter. The indenter (crimp) type of fitting is a thing of the past, although an Underwriters Laboratories standard continues to have such a listing available. A special crimping tool was used to secure the fitting to the EMT. The tool places an indentation in both the fitting and the EMT. It was a standard wiring practice to make two sets of indentations at each connection, approximately 90° apart. This type of fitting is classified as concrete-tight. Figure 6-8 shows a straight indenter-type connector. Couplings were also available.

Installation of EMT

Due to numerous failures of EMT in underground installations, many electrical inspection authorities will not permit EMT to be installed underground

FIGURE 6-8 EMT connector, indenter type. ((*Photo courtesy Appelton Electric LLC, Emerson Industrial Automation*)

or in concrete slabs that are on or below grade. It is also common for electrical consulting engineers to require the use of PVC conduits in these and other corrosive locations. Be sure to consult the authority having jurisdiction as well as the project specifications before beginning the electrical installation.

The efficient installation of EMT requires the use of a bender, Figure 6-9. This tool is commonly available in hand-operated models for EMT in trade sizes ½, ¾, and 1 and in power-operated models for EMT larger than trade size 1. See the Ideal Bending Guide in Appendix I of this text for additional information on using a hand bender.

Several kinds of bends can be made with the use of the bending tool. The stub bend, the back-to-back bend, and the angle bend are shown in Figure 6-10. Other types of bends include offset, saddles and kicks. The manufacturer's instructions that accompany each bender indicate the method for making each type of bend.

Installation of Metallic Raceway

IMC, RMC, and EMT are to be installed according to the requirements of *NEC Articles 342, 344, and 358.* The following points summarize the contents of these articles. All conduit runs should be level, straight, plumb, and neat, showing good workmanship. Do not do sloppy work.

It is quite common for the authority having jurisdiction or the electrical engineer who designs

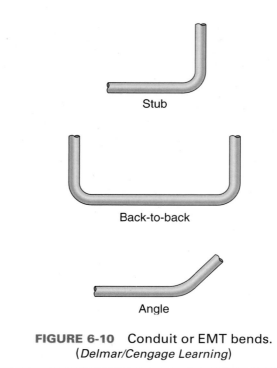

FIGURE 6-10 Conduit or EMT bends.
(*Delmar/Cengage Learning*)

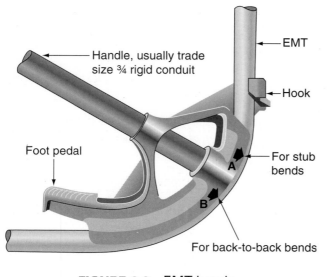

FIGURE 6-9 EMT bender.
(*Photo Courtesy Ideal*)

the project to require an equipment grounding conductor of the wire type to be installed through every metal raceway of the circular type. The reason for this rule is the numerous failures of the metal raceways to provide a continuous ground-fault return path. The failure can be the result of the conduit or EMT being loose at fittings, having a broken fitting, or rusting through. Ask yourself, "How many loose or broken fittings does it take to lose the ground-fault return path?" The answer is obvious!

The rigid types of metal conduit and IMC

- may be installed in concealed and exposed work.

- may be installed in or under concrete when of the type approved for this purpose.

- must not be installed in or under cinder concrete or cinder fill that is subject to permanent moisture, unless the conduit is encased in at least 2 in. (50 mm) of noncinder concrete, is at least 2 in. (450 mm) under the fill, or is of corrosion-resistant material suitable for the purpose.

- must not be installed where subject to severe mechanical damage. (An exception to this is rigid metal conduit, which may be installed in a location where it is subject to damaging conditions.)

- may contain up to four quarter bends (for a total of 360°) in any run.

- shall be fastened within 3 ft (900 mm) of each outlet box, junction box, cabinet, or conduit body but may be extended to 5 ft (1.5 m) to accommodate structural members.

- shall be securely supported at least every 10 ft (3 m) except if threaded couplings are used as given in *NEC Table 344.30(B)(2)*.

- may be installed in wet or dry locations if the conduit is of the type approved for this use.

- shall have the ends reamed to remove rough edges.

- are considered adequately supported when run through drilled, bored, or punched holes in framing members, such as studs or joists.

- have the following requirement for grounding: all listed fittings for metallic raceways are tested for a specified amount of current for a specified length of time, making the fittings acceptable for grounding.

Wet Location and Draining of Raceways

Raceways provide a path for water to run into electrical equipment. This is a hazard. The *NEC* addresses this issue as follows:

- *NEC 225.22* states, *Raceways on exteriors of buildings or other structures shall be arranged to drain and shall be suitable for use in wet locations.**

- *NEC 230.53* states, *Where exposed to the weather, raceways enclosing service-entrance conductors shall be suitable for use in wet locations and arranged to drain. Where embedded in masonry, raceways shall be arranged to drain.**

RACEWAY SEALS

If a raceway enters a building or structure from an underground distribution system it is required to be sealed in accordance with *NEC 300.5(G)*. Spare or unused raceways are also required to be sealed. The sealants used are required to be identified for use with the cable insulation, shield or other components. See Figure 6-11.

These requirements are found in *NEC 225.27, and 230.8*. Note that this requirement does not contemplate sealing the raceways in the manner required for hazardous (classified) locations. The sealant is required to be *identified* for the use and must not be deleterious to the insulation or shield. Note that the word *identified* is defined in *NEC Article 100*. The definition reads, *Identified (as applied to equipment). Recognizable as suitable for the specific purpose, function, use, environment, application, and so forth, where described in a particular Code requirement.** Though permitted, this rule does not require that the sealing compound be listed by a qualified electrical testing laboratory.

The sealing compound contemplated by this rule is often referred to as "duct seal." It consists of a putty-like substance that can be molded, shaped and placed by hand. It can often be removed and reused as desired.

The purpose behind sealing raceways that enter a building or structure from underground is found in *NEC 300.5(G)*. This section states *Conduits or raceways through which moisture may contact live parts shall be sealed or plugged at either or both ends.** An informational note states *Presence of hazardous gases or vapors may also necessitate sealing of underground conduits or raceways entering buildings.** No doubt, and electrical inspector would require seal-offs two be suitable for a hazardous location if the presence of a hazardous gas or vapor is known.

FLEXIBLE CONNECTIONS

The installation of certain equipment requires flexible connections, both to simplify the installation and to stop the transfer of vibrations.

There are three basic wiring methods used for flexible connections:

- Flexible metal conduit (FMC), Figures 6-12, 6-14, and 6-17. See *NEC Article 348*.

- Liquidtight flexible metal conduit (LFMC), Figures 6-13 and 6-15. See *NEC Article 350*.

- Liquidtight flexible nonmetallic conduit (LFNC), Figure 6-16. See *NEC Article 356*.

Grounding: All listed fittings for flexible metallic raceways are tested for a specified amount of current for a specified length of time, making the fittings acceptable for grounding.

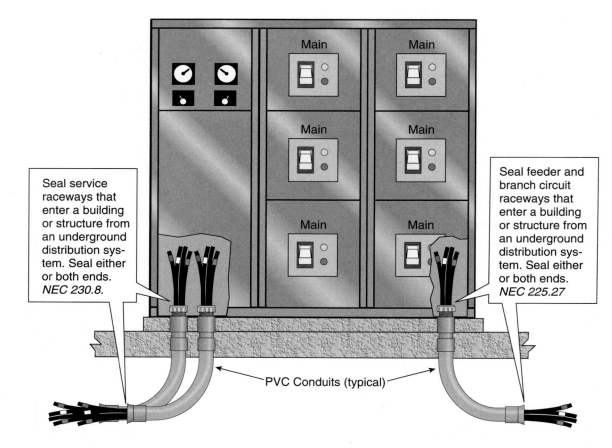

Seal service raceways that enter a building or structure from an underground distribution system. Seal either or both ends. *NEC 230.8.*

Seal feeder and branch circuit raceways that enter a building or structure from an underground distribution system. Seal either or both ends. *NEC 225.27*

PVC Conduits (typical)

FIGURE 6-11 Raceways that enter a building or structure from an underground distribution system are required to be sealed at one or both ends. See *225.27* for outside feeders and branch circuits and *230.8* for services. (*Delmar/Cengage Learning*)

Flexible Metal Conduit (FMC)

NEC Article 348 regulates the use and installation of FMC. FMC is similar to armored cable, except that the conductors are installed by the electrician. For Type AC and Type MC cable, the cable armor is wrapped around the conductors at the factory to form a complete cable assembly.

NEC 348.10 covers *Uses Permitted* and *348.12* gives *Uses Not Permitted* for FMC. Note that *NEC 348.12(1)* now prohibits the use of FMC in wet locations. This prohibition reduces the likelihood that water will enter enclosures through FMC. LFMC and LFNC are commonly used to connect equipment in wet locations.

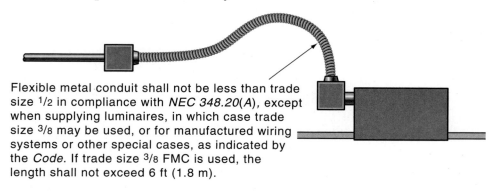

Flexible metal conduit shall not be less than trade size 1/2 in compliance with *NEC 348.20(A)*, except when supplying luminaires, in which case trade size 3/8 may be used, or for manufactured wiring systems or other special cases, as indicated by the *Code*. If trade size 3/8 FMC is used, the length shall not exceed 6 ft (1.8 m).

FIGURE 6-12 Flexible metal conduit. (*Delmar/Cengage Learning*)

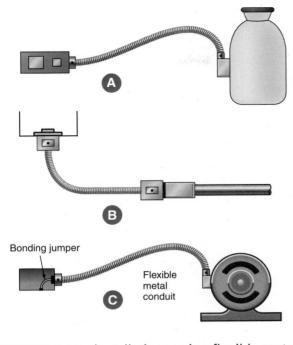

FIGURE 6-13 Installations using flexible metal conduit. (*Delmar/Cengage Learning*)

Some of the more common installations using FMC are shown in Figure 6-13. Note that the flexibility required to make the installation is provided by the FMC.

Flexible metal and nonmetallic conduit are commonly used to connect recessed or lay-in luminaires. When used in this manner, electricians refer to it as a "fixture whip."

FMC is permitted to be used as an equipment grounding conductor if

- it is listed;
- it is not more than 6 ft (1.8 m) in length;
- it is connected with listed fittings; and
- the overcurrent device protecting the contained conductors is rated at 20 amperes or less, Figure 6-13(A) and (B).

348.60 and 250.118(5): ▶If used to connect equipment where flexibility is necessary to minimize the transmission of vibration from equipment or to provide flexibility for equipment that requires movement after installation, . . . install an equipment grounding conductor sized in compliance with *NEC 250.122* through the FMC.◀

For other rules on use as an equipment grounding conductor, install FMC in compliance with

NEC 250.134(B). For bonding requirements, install in compliance with *NEC 250.102.*

FMC of any size may be installed with an equipment bonding jumper

- outside the conduit if it is not more than 6 ft (1.8 m) in length.
- inside the conduit of unlimited length, Figure 6-14.

Liquidtight Flexible Metal Conduit (LFMC)

The use and installation of LFMC is described in *NEC Article 350.* LFMC has a tighter and flatter fit of its spiral metallic tape or armor as compared to FMC. LFMC has a thermoplastic outer jacket that is liquidtight and is commonly used as a flexible connection to central air-conditioning units located outdoors, Figure 6-14.

LFNC is permitted to be installed

- in exposed or concealed locations.
- if flexibility or protection from liquids is required.
- in hazardous (classified) locations where specifically approved or permitted such as by *NEC 501.10(B), 502.10, 503.10,* and *504.20.*
- for direct burial, if so listed and marked.
- if not subject to physical damage.
- if the combination of ambient and conductor temperature does not exceed that for which the LFMC is approved.
- with fittings identified for use with the LFMC.
- in trade sizes ½ to 4. For enclosing motor leads, trade size ⅜ is suitable.
- for "fixture whips" not less than 18 in. (450 mm) or over 6 ft (1.8 m) in length permitted or required by *NEC 410.117(C),* or for flexible connection to equipment such as a motor.
- under raised floors in information technology rooms, as provided in *NEC 645.5(E)(2).*

See *NEC 350.10* for uses permitted for LFNC and 350.12 for uses not permitted.

LFNC is not permitted in ducts or plenums used for environmental air [*NEC 300.22(B)*] or in other

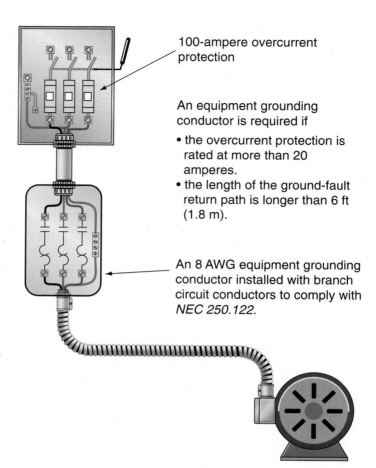

100-ampere overcurrent
protection

An equipment grounding
conductor is required if

• the overcurrent protection is
  rated at more than 20
  amperes.
• the length of the ground-fault
  return path is longer than 6 ft
  (1.8 m).

An 8 AWG equipment grounding
conductor installed with branch
circuit conductors to comply with
NEC 250.122.

FIGURE 6-14   Equipment grounding conductor in flexible metal conduit. (*Delmar/Cengage Learning*)

spaces used for environmental air *[NEC 300.22(C)]*. These environmental air spaces are sometimes referred to as "air-handling ceilings."

LFMC of trade size ½ or smaller may be used as an equipment grounding conductor if

• it is listed;

• it is not more than 6 ft (1.8 m) in length;

• it is connected with listed fittings; and

• the circuit is rated at 20 amperes or less, Figure 6-14.

LFMC of trade size ¾, 1, or 1¼ may be used as an equipment grounding conductor if

• it is listed;

• it is not more than 6 ft (1.8 m) in length;

• it is connected with listed fittings; and

• the overcurrent device is rated at 60 amperes or less, Figure 6-14.

NEC 350.60 and *250.118(6)*: ▶If LFMC is used to connect equipment where flexibility is necessary to minimize the transmission of vibration from equipment or to provide flexibility for equipment that requires movement after installation, . . . install an equipment grounding conductor sized in compliance with *NEC 250.122.*◀

• For use as an equipment grounding conductor, install in compliance with *NEC 250.134(B)*.

• For bonding purposes, make the installation in compliance with *NEC 250.102*.

LFMC of any size or length may be installed with a required bonding jumper subject to the inside/outside rules of *NEC 250.102(E)*, Figure 6-15.

Regarding fittings for LFMC, the UL *Guide Information for Electrical Equipment (White Book)* states in *Conduit Fittings (DWTT)* "Liquidtight flexible metal conduit fittings in the 1¼ inch and

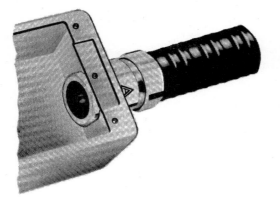

FIGURE 6-15 Liquidtight flexible metal conduit and fitting. (*Delmar/Cengage Learning*)

smaller trade sizes are considered suitable for grounding for use in circuits over and under 250 volts and where installed in accordance with the *NEC*." Because all of these listed fittings are suitable for grounding, it is not required that they be marked as such.

Liquidtight Flexible Nonmetallic Conduit (LFNC)

Liquidtight flexible nonmetallic conduit (LFNC) is covered in *Article 356* of the *NEC*.

The three types of LFNC are marked as follows:

- LFNC-A for a smooth seamless inner core and cover bonded together and having one or more reinforcement layers between the core and covers for layered conduit
- LFNC-B for a smooth inner surface with integral reinforcement within the conduit wall
- LFNC-C for a corrugated internal and external surface without integral reinforcement within the conduit wall

Both metal and nonmetallic fittings listed for use with the various types of LFNC will be marked with the "A," "B," or "C" designations.

LFNC types A and C are generally restricted to a length of 6 ft (1.8 m). Type B LFNC is permitted in lengths exceeding 6 ft (1.8 m) if secured as required.

See *NEC 356.10* for *Uses Permitted* and *356.12* for *Uses Not Permitted*.

ARMORED (TYPE AC) AND METAL-CLAD (TYPE MC) CABLES

Two increasingly popular wiring systems are presented in *NEC Articles 320* and *330*. As they are similar and different in many ways, the information on these systems is presented in the following parallel comparative format, Figure 6-18 and Table 6-3.

FIGURE 6-16 An application of liquidtight flexible metal conduit. (*Delmar/Cengage Learning*)

FIGURE 6-17 An application for liquidtight flexible nonmetallic conduit.
(*Delmar/Cengage Learning*)

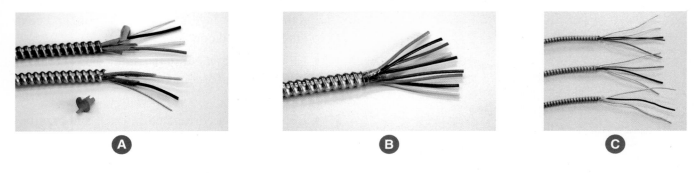

FIGURE 6-18 (A) Type AC cable construction, (B) traditional Type MC construction, (C) Type MC cable with 10 AWG aluminum grounding/bonding conductor. (*Delmar/Cengage Learning*)

TABLE 6-3	
Type AC and Type MC cables.	

Type AC (*NEC Article 320*)	Type MC (*NEC Article 330*)
Definition: ▶ Type AC cable is a fabricated assembly of insulated conductors in a flexible interlocked metallic armor. ◀	Definition: Type MC cable is a factory assembly of one or more insulated circuit conductors with or without optical fiber members enclosed in an armor of interlocking metal tape, or a smooth or corrugated metallic sheath.
Number of conductors: Two to four current-carrying conductors plus a bonding wire. It may also have a separate equipment grounding conductor.	Number of conductors: Any number of current-carrying conductors. At least one manufacturer has a "Home Run Cable" with conductors for more than one branch circuit plus the equipment grounding conductors.
Conductor size and type: Copper conductors 14 AWG through 1 AWG; aluminum conductors 12 AWG through 1 AWG.	Conductor size and type: Copper conductors 18 AWG through 2000 kcmil; aluminum or copper-clad aluminum 12 AWG through 2000 kcmil.
Color coding: Two conductors—one black and one white. Three conductors—one black, one white, and one red. Four conductors—one black, one white, one red, and one blue. These are in addition to any equipment grounding and bonding conductors.	Color coding: Two conductors—one black and one white. Three conductors— one black, one white, and one red. Four conductors—one black, one white, one red, and one blue. Or one brown, one orange, one yellow, and one gray. These are in addition to any equipment grounding and bonding conductors. Other color insulations are available from some manufacturers or by special order.
Bonding and grounding: Type AC cable shall provide an adequate path for fault current. It has a 16 AWG aluminum bare bonding wire in continuous contact with the metal armor. These act together to serve as an acceptable equipment ground. See *NEC 250.118(8)* and *320.108*. The bonding wire does not need to be terminated; it may be folded back over the armor or cut off. It must not be used as a neutral conductor or an equipment grounding conductor.	Bonding and grounding (interlocked armor type): Recent product developments have complicated the issue of Type MC cables and equipment grounding. The variations being produced today include 1. the "traditional" Type MC cable with a spiral interlocked metal armor and the conductors wrapped in an overall plastic (Mylar) tape, and 2. Type MC cable having a spiral interlocked metal armor with a 10 AWG bare aluminum conductor in constant contact with the armor and the circuit conductors having an overall plastic (Mylar) tape wrap. One variety of this Type MC cable includes an additional layer of plastic material extruded over the conductor insulation and does not have an overall mylar wrap around the circuit conductors. A green insulated equipment grounding conductor is always included in the "traditional" product. It may contain another equipment grounding conductor that has green with one or more yellow stripes. This conductor is often used for isolated ground branch circuits. This variety of Type MC

TABLE 6-3

(*Continued*)

Type AC (*NEC Article 320*)	Type MC (*NEC Article 330*)
	cables is not permitted for wiring patient care areas of health care facilities. Type MC with the 10 AWG aluminum grounding/bonding conductor may have an additional 12 AWG or larger insulated copper equipment grounding conductor. This variety is suitable for wiring patient care areas of health care facilities as well as for isolated ground circuits. Some manufacturers' installation instructions state that the bare bonding wire need not be terminated or folded (wrapped) back over the jacket of the cable. Merely cutting off the bonding wire where it comes out of the jacket is acceptable. The sheath of smooth or corrugated tube-Type MC is typically "listed" as an acceptable equipment grounding conductor.

Insulation:
    Typical Type AC cables have copper conductors with THHN insulation. Type ACTH cables have 75°C thermoplastic insulation; Type ACTHH have 90°C thermoplastic insulation; or Type ACHH 90°C thermosetting insulations.
	Insulation:
	    Typical Type MC cables have copper or aluminum conductors with THHN, THHN/THWN, or XHHW insulation for circuits up to 600 volts. Type MC is available with TFN, TFFN, or THHN insulations for fire alarm applications.

Wrapping:
    Conductors are individually wrapped with flame retardant, light brown kraft paper.
	Wrapping:
	    No individual paper wrapping. A polyester (Mylar) tape is placed over all the conductors or may be extruded over the individual conductors.

Covering over armor:
    None available.
	Covering over armor:
	    A PVC outer covering is available for use in wet, corrosive, and underground locations, and in concrete.

Locations permitted:
    Dry locations only.
	Locations permitted:
	    Dry. Also wet locations, direct burial in direct contact with earth, or embedded in concrete if identified for direct burial. May be installed in a raceway.

Insulating bushings:
    Required to protect conductor insulation from damage. These keep sharp edges of the armor from cutting the conductor insulation. Usually provided with the cable.
	Insulating bushings:
	    Recommended but not mandatory. Provides additional protection for conductor insulation. Usually provided with the cable.

Minimum radius of bends:
    Five times the diameter of the cable.
	Minimum radius of bends:
	    Smooth sheath:
	        Ten times for sizes not more than ¾ in. (19 mm) diameter.
	        Twelve times for sizes more than ¾ in. (19 mm) and not more than 1½ in. (38 mm) in diameter.
	        Fifteen times for sizes more than 1½ in. (38 mm) in diameter.
	    Corrugated or interlocked:
	        Seven times diameter of cable.

Oversized neutral conductor:
    Not available.
	Oversized neutral conductor:
	    Available. May be needed where harmonics are high, such as branch circuits for computers. Example: a 10 AWG neutral conductor with three 12 AWG phase conductors.

Fiber-optic cables:
    Not available.
	Fiber-optic cables:
	    Available with power conductors.

Multiple neutral conductors:
    Not available.
	Multiple neutral conductors:
	    Available. Suitable for eliminating harmonic currents in branch circuits and to eliminate requirement for simultaneous disconnecting means for multiwire branch circuits.

(continues)

TABLE 6-3

(*Continued*)

Type AC (*NEC Article 320*)	Type MC (*NEC Article 330*)

Supports:
    Not more than 4 ft 6 in. (1.4 m) apart.
    Not more than 1 ft (300 mm) from box or fitting.
    Not required when fished through wall or ceiling
        or run through holes in studs, joists, or rafters.
    Not required for lengths of no more than
        6 ft (1.8 m) when used as a fixture whip in an
        accessible ceiling.
	Supports:
	    Not more than 6 ft (1.8 m) apart.
	    Not more than 1 ft (300 mm) from box or fitting for cable
	        having not more than four 10 AWG or smaller conductors.
	    Not required when fished through wall or ceiling or
	        run through holes in studs, joists, or rafters.
	    Not required for lengths of no more than 6 ft (1.8 m)
	        when used as a fixture whip in an accessible ceiling.

Voltage rating:
    Not higher than 600 volts.
	Voltage rating:
	    Type MC cables typically used for services, feeders and branch
	    circuits have conductors insulated for 600 volts. Type MC
	    cable rated 2400 to 35,000 volts is covered as Medium-voltage
	    Power Cable and may be marked "Type MV or Type MC."

Connectors:
    Set-screw-type connectors not permitted with
    aluminum armor. "Listed" connectors are suitable
    for grounding purposes.
	Connectors:
	    Set-screw-type connectors not permitted with aluminum
	    armor. "Listed" connectors are suitable for grounding
	    purposes.

Ampacity:
    Follow *NEC 310.15*. For installations in thermal
    insulation, use the 60°C column of *Table 310.15(B)16*.
    Note: See *NEC 310.15(B)(3)(a)(4)* for special
    adjustment factors.
	Ampacity:
	    Follow *NEC 310.15*. Use the 60°C column of *Table 310.15(B)(16)*.
	    For 1/0 and larger conductors, OK to use the 75°C column.
	    Note: See *NEC 310.15(B)(3)(a)(4)* for special adjustment
	    factors.

Armor:
    Galvanized steel or aluminum. Armored cable
    with aluminum armor is marked *Aluminum
    Armor*.
	Armor:
	    Galvanized steel, aluminum, or copper.
	Three types:
	    Interlocked (most common). May have a separate equipment
	        grounding conductor. The armor qualifies as an acceptable
	        equipment grounding conductor with a 10 AWG aluminum
	        grounding/bonding conductor. Steel or aluminum armor.
	    Smooth tube. Does not require a separate equipment
	        grounding conductor. Aluminum only.
	    Corrugated tube. Does not typically require a separate
	        equipment grounding conductor. Copper or aluminum armor.

⌁〰️ RIGID POLYVINYL CHLORIDE CONDUIT (PVC) (*NEC ARTICLE 352*)

The most commonly used PVC (formerly referred to as rigid nonmetallic conduit) is made of polyvinyl chloride, which is a thermoplastic polymer. Solvent-type cement is used for PVC connections and terminations.

Overview. The two most popular types of PVC are as follows:

- Schedule 40: Permitted underground (direct burial or encased in concrete) and aboveground (indoors and outdoors exposed to sunlight) where not subject to physical damage. Some Schedule 40 is marked for underground use only. Schedule 40 has a thinner wall than Schedule 80.

- Schedule 80: Permitted underground (direct burial or encased in concrete) and aboveground (indoors and outdoors in sunlight) where subject to physical damage. Schedule 80 has a thicker wall than Schedule 40.

- The main differences between aboveground and underground listed PVC are its fire resistance rating and its resistance to sunlight (UV).

- Some rigid nonmetallic conduit is made of fiberglass and is included in *NEC Article 355* as Reinforced Thermosetting Resin Conduit Type RTRC.

PVC conduit is permitted (see *NEC 352.10*)

- concealed in walls, floors, and ceilings.

- in corrosive areas if the PVC is suitable for the chemicals to which it is exposed.

- in cinder fill.

- in wet and damp locations.

- for exposed work where not subject to physical damage unless identified for such use. (Schedule 80 is so listed.)

- for underground installations.

- in places of assembly if the requirements of *NEC 518.4(B), 518.4(C),* and *520.5(C)* are met.

PVC conduit is not permitted (see *NEC 352.12*)

- in hazardous (classified) locations except for limited use as set forth in *NEC Chapter 5*.

- for support of luminaires.

- where exposed to physical damage unless so identified. (Schedule 80 is so listed.)

- where subject to ambient temperatures exceeding 122°F (50°C).

- for conductors whose insulation temperature limitations would exceed those for which the conduit is listed. (Check the marking or listing label to obtain this temperature rating.)

- for electrical circuits serving patient care areas in hospitals and health care facilities, as the raceway does not provide an equipment grounding conductor as required by *NEC 517.13(A)*.

- in places of assembly and theaters except as provided in *NEC 518.4(B)* and *520.5(C)*.

PVC conduit expands and contracts due to temperature changes about three times greater than aluminum and six times greater than steel. On an outdoor exposed run of PVC conduit, it is important that expansion fittings be installed to comply with the requirements set forth in *NEC 352.44*, where expansion characteristics of PVC conduit are given. See Figure 6-19 for an example of the installation of an expansion fitting. Expansion joints or fittings are required if the expected length change is to be 0.25 in. (6 mm) or greater in a straight run between securely mounted items such as boxes, cabinets, elbows, or other conduit terminations. Refer to *NEC Table 352.44* for the expansion characteristics of PVC conduit.

There you will find that PVC conduit increases in length 4.06 in. (103.124 mm) per 100°F of temperature change. This means that a 10-ft (3.04-m) length would increase/decrease about 0.4 in. (10.16-mm) for a 10-ft (3.04-m) length.

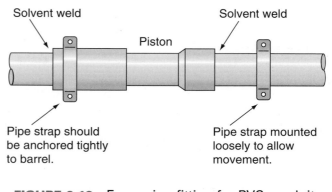

FIGURE 6-19 Expansion fitting for PVC conduit. (*Delmar/Cengage Learning*)

This obviously exceeds the 0.25-in. (6-mm) length change requiring an expansion fitting.

An informational bulletin on design and installation criterion is available for free download from the National Electrical Manufacturers Association (NEMA at http://www.nema.org/stds/expansionjoints-pvc.cfm).

Also consider expansion fittings where earth movement will be a problem, *300.5(J)*. An equipment grounding conductor must be installed in PVC conduit, and if metal boxes are used, the grounding conductor must be connected to the metal box. See *NEC Article 352* for additional information for the installation of PVC conduit.

PVC Conduit Fittings and Boxes

A complete line of fittings, boxes, and accessories are available for PVC conduit. Where PVC conduit is connected to a box or enclosure, a box adapter is recommended over a male connector because the box adapter is stronger and provides a gentle radius for wire pulling.

ELECTRICAL NONMETALLIC TUBING (ENT) (*NEC ARTICLE 362*)

ENT is a pliable, corrugated raceway made of polyvinyl chloride, a plastic material that is also used to make PVC conduit. ENT is hand-bendable and is available in coils and reels to facilitate a single-length installation between pull points. ENT is designed for use within a building or encased in concrete and is not suitable for outdoor use. The

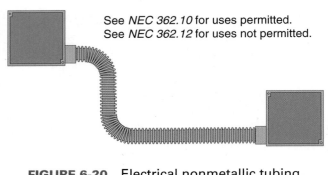

FIGURE 6-20 Electrical nonmetallic tubing. (*Delmar/Cengage Learning*)

See *NEC 362.10* for uses permitted.
See *NEC 362.12* for uses not permitted.

specification for the Commercial Building allows the use of ENT for feeders, branch circuits, and telephone wiring where a raceway trade size 2 or smaller is required. ENT is available in three colors, which allows the color coding of the various systems: blue for electrical power wiring . . . yellow for communication wiring . . . red for fire alarm wiring. See Figure 6-20 for an example of ENT.

ENT, according to *NEC 362.10*, is permitted to be used

- exposed, where not subject to physical damage, in buildings not more than three floors above grade.
- concealed, without regard to building height; in walls, floors, and ceilings of a material with a 15-min finish rating (½ in. [12.7 mm] UL classified gypsum board has a 15-min finish rating). Check with the manufacturer for further information.
- subject to corrosive influences (check with manufacturer's recommendations).
- in damp and wet locations, with fittings identified for the use.
- above a suspended ceiling if the ceiling tile has a 15-min finish rating. The ceiling tile does not need a finish rating if the building is not more than three floors above grade.
- embedded in poured concrete, including slab-on-grade or below grade, with fittings identified for the purpose.

ENT shall not be used (see *NEC 362.12*):

- in hazardous (classified) locations except as permitted by other Articles of the *NEC*.

- for luminaire or equipment support.
- subject to ambient temperature in excess of 122°F (50°C).
- for conductors whose insulation temperature would exceed those for which ENT is listed. (Check the UL label for the actual temperature rating.)
- for direct earth burial.
- where the voltage is more than 600 volts.
- in places of assembly and theaters except as provided in *NEC Articles 518* and *520*.
- if exposed to the direct rays of the sun unless identified as sunlight resistant.
- if subject to physical damage.

See *NEC Article 362* for other ENT installation requirements.

ENT Fittings

Two styles of listed mechanical fittings are available for ENT: one-piece snap-on and a clamshell variety. Solvent cement PVC fittings for PVC conduit are also listed for use with ENT. The package label will indicate which of these fittings are concrete-tight without tape.

Note that the listing of ENT, as well as manufacturer instructions, requires a special "quick-set" solvent cement for fittings that are secured with a cement. This cement is faster acting than the solvent cement for Schedule 40 and 80 PVC so as to not damage the thinner walls of the ENT. This "quick-set" solvent cement is permitted to be used on Schedule 40 and 80 PVC.

ENT with mechanical fittings identified for the purpose or with cemented-on fittings is suitable for use in poured concrete.

ENT Boxes and Accessories

Wall and ceiling boxes, with knockouts for trade sizes ½, ¾, and 1 (metric designator 16, 21, and 27), are available. ENT can also be connected to any PVC box. ENT mud boxes listed for luminaire support are available, with knockouts for trade sizes ½, ¾, and 1 (metric designator 16, 21, and 27).

RACEWAY SIZING

The conduit and tubing size required for an installation depends on three factors:

1. The number of conductors to be installed
2. The cross-sectional area of the conductors
3. The permissible raceway fill

The relationship of these factors is defined in *NEC Chapter 9, Table 1*, including *Notes 1* through *9* and *Tables 4, 5, 5A, 8*, and *Annex C* of the *NEC*. After examining the working drawings and determining the number of conductors to be installed to a certain point, either of the following procedures can be used to find the conduit size. Refer to Tables 6-4, 6-5, and 6-6.

If all of the conductors are the same size and have the same insulation, the raceway size can be determined directly by referring to *NEC Annex C*.

For example, assume that three 8 AWG, Type THWN conductors are to be installed in electrical metallic tubing to an air-conditioning unit. Referring to *NEC Annex C, Table C1*, it will be determined that a trade size ½ (metric designator 16) raceway is acceptable.

Next, assume that a short section of the EMT is replaced by liquidtight flexible nonmetallic conduit. To comply with *NEC 356.60*, this change requires the installation of an equipment grounding conductor. *NEC Table 250.122* specifies the use of a 10 AWG equipment grounding conductor if the overcurrent device is from 30 amperes through 90 amperes. The equipment grounding conductor must be included in the conductor fill calculations in compliance with *NEC Chapter 9, Tables, Note 3*.

- One 10 AWG Type THWN @ 0.0211 in.² (13.61 mm²)
- Three 8 AWG, Type THWN @ 3 × 0.0366 in.² = 0.1098 in.² (3 × 23.61 mm² = 70.83 mm²)
- Total cross-sectional area = 0.0211 in.² + 0.1098 in.² = 0.1309 in.² (13.61 mm² + 70.83 mm² = 84.44 mm²)

Referring to *NEC Chapter 9, Table 4*:

- for electrical metallic tubing, the allowable 40% fill for more than two conductors in trade size ½ (metric designator 16) is 0.122 in.² (78 mm²).

- for liquidtight flexible nonmetallic conduit, the allowable 40% fill for more than two conductors in trade size ½ (metric designator 16) is 0.125 in.² (81 mm²).

As both fill values must be 0.1309 in.² (86 mm²) or greater, the use of a trade size ½ (metric designator 16) is not permitted. Trade size ¾ (metric designator 21) is the minimum size.

NEC Chapter 9, Table 1, is based on common conditions of proper cabling and alignment of conductors where the length of the pull and the number of bends are within reasonable limits. It should be recognized that, for certain conditions, a larger size conduit or a lesser conduit fill should be considered.

When pulling three conductors or cables into a raceway, if the ratio of the raceway diameter to the conductor diameter is between 2.8 and 3.2, jamming can occur. With four or more conductors or cables, jamming is highly unlikely.

The limitations of this table apply only to complete conduit or tubing systems and are not intended to apply to sections of conduit or tubing used to protect exposed wiring from physical damage.

Where conduit or tubing not exceeding 24 in. (600 mm) in length is installed between boxes, cabinets, and similar enclosures, it is permitted to be filled to 60% of its total cross-sectional area. See *NEC Chapter 9, Table 1, Note (4)*. This note indicates that the adjustment factors for an excessive number of conductors in *NEC 310.15(B)(3)(a)* need not be applied.

Before any installation of conductors or cables, the complete set of notes to *NEC Chapter 9, Table 1*, should be consulted.

Wireways and Auxiliary Gutters

It is common to use wireways and auxiliary gutters when wiring commercial occupancies. Although permitted as a general wiring method, it is more common to see wireways installed at service equipment to facilitate the connection of the service disconnecting means to the service-lateral conductors coming from the utility transformer.

Metal wireways are covered in the *NEC* in *Article 376* and auxiliary gutters in *Article 366*. They are identical pieces of equipment. How they are to be installed determines which article applies.

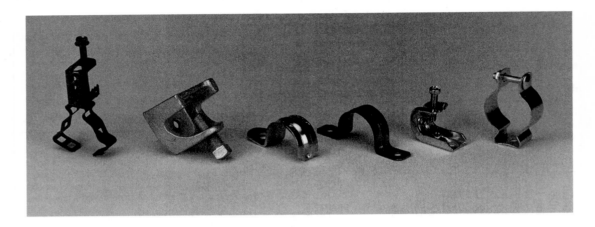

FIGURE 6-21 Raceway support devices. (*Delmar/Cengage Learning*)

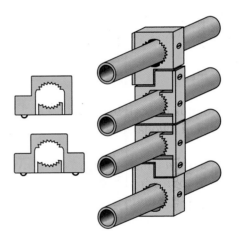

FIGURE 6-22 Nonmetallic conduit and supports. (*Delmar/Cengage Learning*)

This magnetic heating effect is considerably less in aluminum raceways and zero in PVC raceways. Losses in ferrous raceways and enclosures in an alternating current circuit are caused by

- eddy currents—the current induced in ferrous metals, and

- hysteresis losses—power loss that creates heat in a ferrous metal as a result of the molecules continually reversing.

The bottom line is that eddy currents and hysteresis losses create heat. We know that heat damages conductor insulation. Circuits where all of the conductors are close together have lower inductive reactance than circuits where the conductors are spaced farther apart.

Proof of the flux cancellation principle can be demonstrated using a clamp-on ammeter. You get a zero reading if you clamp the ammeter around all of the circuit conductors. You get an appropriate current reading if you put the clamp-on ammeter around the conductor(s) of one phase only, Figure 6-24.

In the real world, should a fault occur in a circuit, the greater the impedance of the circuit, the lower the current flow.

Inversely, the lower the impedance of the circuit, the greater the current flow.

Overcurrent devices open faster when they see high values of current flow. Overcurrent devices open more slowly when they see low values of current flow. This is referred to as "inverse time." When an overload or fault current occurs, we want the overcurrent device protecting the circuit to open as fast as possible. The conclusion is that we want the impedance of a circuit to be as low as possible so as to have the value of fault current as high as possible.

For conductors installed in a raceway, there is nothing an electrician can do to decrease the spacing between conductors. However, when installing conductors in a trench, it is important to keep conductors as close together as possible to minimize circuit impedance and voltage loss.

SPECIAL CONSIDERATIONS

Following the selection of the circuit conductors and the branch-circuit protection, there are a number of special considerations that generally are factors for each installation. The electrician is usually given

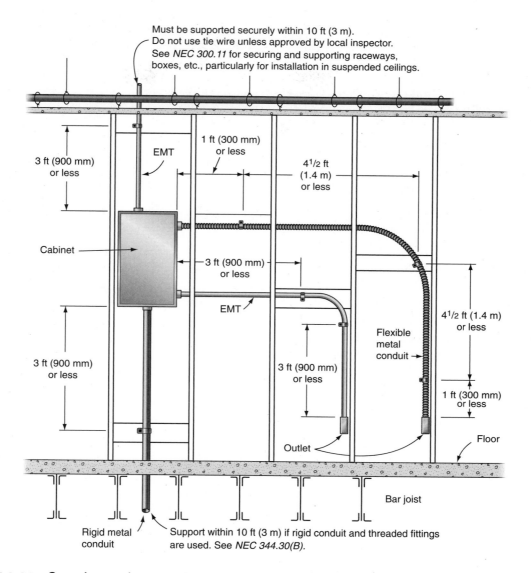

Must be supported securely within 10 ft (3 m).
Do not use tie wire unless approved by local inspector.
See *NEC 300.11* for securing and supporting raceways,
boxes, etc., particularly for installation in suspended ceilings.

3 ft (900 mm) or less

EMT

1 ft (300 mm) or less

4¹/₂ ft (1.4 m) or less

Cabinet

3 ft (900 mm) or less

EMT

3 ft (900 mm) or less

Flexible metal conduit

4¹/₂ ft (1.4 m) or less

3 ft (900 mm) or less

3 ft (900 mm) or less

1 ft (300 mm) or less

Outlet

Floor

Bar joist

Rigid metal conduit

Support within 10 ft (3 m) if rigid conduit and threaded fittings are used. See *NEC 344.30(B)*.

FIGURE 6-23 Securing and supporting requirements for raceways, tubing, and cable are found in .30 section of the specific wiring method article. (*Delmar/Cengage Learning*)

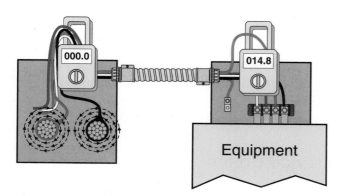

000.0

014.8

Equipment

FIGURE 6-24 Magnetic field around conductors cancelled by proper installation. (*Delmar/Cengage Learning*)

the responsibility of planning the routing of the conduit or cable to ensure the outlets are connected properly. For raceways such as EMT, the electrician must determine the length and the number of conductors in each raceway. In addition, the electrician must apply adjustment and correction factors to the conductors' ampacity as required, make allowances for voltage drops, recognize the various receptacle types, and be able to install these receptacles correctly on the system.

Surface Raceways

The *NEC* recognizes surface metal raceways in *Article 386* and nonmetallic surface raceways in *Article 388*. These are discussed in detail in Chapter 9.

Insulating, Grounding, and Bonding Bushings

Grounding and bonding bushings, bonding wedges, grounding conductors, and bonding jumpers are installed to ensure a low-impedance fault-return path if a fault occurs. Bonding jumpers are required where the ground-fault return path is impaired by concentric or eccentric knockouts, or reducing washers. *NEC 300.4(G)* states that if 4 AWG or larger ungrounded circuit conductors enter a cabinet, box enclosure, or raceway, an insulating bushing or equivalent must be used. Similar insulating bushing requirements are found in *NEC312.6(C), 314.17(D), 314.42, 354.46, and 362.46*. Insulating bushings protect the conductors from abrasion where they pass through the bushing. Combination metal/insulating bushings can be used. Some EMT connectors have an insulated throat. Insulating sleeves are also available that slide into a bushing between the conductors and the throat of the bushing.

Reducing Washers

Reducing washers are installed where the size of the conduit, EMT, or the cable connector is smaller than the knockout in the box, cabinet, or other enclosure. This is often the result of a smaller conduit, EMT, or cable replacing a larger item or where the electrician has removed more eccentric or concentric knockouts than required for the size of conduit.

The *NEC* contains little information on the installation of reducing washers. We find installation instructions in the *UL White Book* in "Outlet Bushings and Fittings (QCRV)". It reads, "Metal reducing washers are considered suitable for grounding for use in circuits over and under 250 V and where installed in accordance with ANSI/NFPA 70, *National Electrical Code.*" Reducing washers are intended for use with metal enclosures having a minimum thickness of 0.053 in. (1.35 mm) for nonservice conductors only. Reducing washers may be installed in enclosures provided with concentric or eccentric knockouts, only after all of the concentric and eccentric rings have been removed. However, those enclosures containing concentric and eccentric

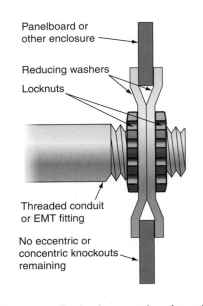

FIGURE 6-25 Reducing washer installation requirements. (*Delmar/Cengage Learning*)

knockouts that have been listed for bonding purposes may be used with reducing washers without all knockouts being removed. See Figure 6-25 for an example of the installation of reducing washers.

It should be noted that no enclosures such as cabinets, wireways, auxiliary gutters, or pull and junction boxes have knockouts "that have been listed for bonding purposes." This phrase refers to smaller metal boxes such as 4 ×1½ in. (120 × 38 mm) square boxes that have combination trade size ½ and ¾ in. knockouts (metric designator 16 and 21). The larger knockout has been securely retained during the manufacturing process by spot welding or a similar process.

BOX STYLES AND SIZING

The style of box required on a building project is usually established in the specifications. However, the sizing of the boxes is usually one of the decisions made by the electrician, Tables 6-8 and 6-9.

Device Boxes

Device boxes, Figure 6-26, are available with depths ranging from 1½ to 3½ in. (38 to 90 mm). These boxes can be purchased for trade size ½ or ¾ conduit or with cable clamps. Each side of the

TABLE 6-8

Table 314.16(A) Metal Boxes

Box Trade Size			Minimum Volume		Maximum Number of Conductors* (arranged by AWG size)						
mm	in.		cm³	in.³	18	16	14	12	10	8	6
100 × 32	(4 × 1¼)	round/octagonal	205	12.5	8	7	6	5	5	5	2
100 × 38	(4 × 1½)	round/octagonal	254	15.5	10	8	7	6	6	5	3
100 × 54	(4 × 2⅛)	round/octagonal	353	21.5	14	12	10	9	8	7	4
100 × 32	(4× 1¼)	square	295	18.0	12	10	9	8	7	6	3
100 × 38	(4 × 1½)	square	344	21.0	14	12	10	9	8	7	4
100 × 54	(4 × 2⅛)	square	497	30.3	20	17	15	13	12	10	6
120 × 32	(4¹¹⁄₁₆ × 1¼)	square	418	25.5	17	14	12	11	10	8	5
120 × 38	(4¹¹⁄₁₆ × 1½)	square	484	29.5	19	16	14	13	11	9	5
120 × 54	(4¹¹⁄₁₆ × 2⅛)	square	689	42.0	28	24	21	18	16	14	8
75 × 50 × 38	(3 × 2 × 1½)	device	123	7.5	5	4	3	3	3	2	1
75 × 50 × 50	(3 × 2 × 2)	device	164	10.0	6	5	5	4	4	3	2
75× 50 × 57	(3× 2 × 2¼)	device	172	10.5	7	6	5	4	4	3	2
75 × 50 × 65	(3 × 2 × 2½)	device	205	12.5	8	7	6	5	5	4	2
75 × 50 × 70	(3 × 2 × 2¾)	device	230	14.0	9	8	7	6	5	4	2
75 × 50 × 90	(3 × 2 × 3½)	device	295	18.0	12	10	9	8	7	6	3
100 × 54 × 38	(4 × 2⅛ × 1½)	device	169	10.3	6	5	5	4	4	3	2
100 × 54 × 48	(4 × 2⅛ × 1⅞)	device	213	13.0	8	7	6	5	5	4	2
100 × 54 × 54	(4 × 2⅛ × 2⅛)	device	238	14.5	9	8	7	6	5	4	2
95 × 50 × 65	(3¾ × 2 × 2½)	masonry box/gang	230	14.0	9	8	7	6	5	4	2
95 × 50 × 90	(3¾ × 2 × 3½)	masonry box/gang	344	21.0	14	12	10	9	8	7	4
min. 44.5 depth	FS — single cover/gang (1¾)		221	13.5	9	7	6	6	5	4	2
min. 60.3 depth	FD — single cover/gang (2⅜)		295	18.0	12	10	9	8	7	6	3
min. 44.5 depth	FS — multiple cover/gang (1¾)		295	18.0	12	10	9	8	7	6	3
min. 60.3 depth	FD — multiple cover/gang (2⅜)		395	24.0	16	13	12	10	9	8	4

*Where no volume allowances are required by 314.16(B)(2) through (B)(5).

Reprinted with permission from NFPA 70-2011.

box has holes through which a 20-penny nail can be inserted for nailing to wood studs. The boxes can be ganged by removing the common sides of two or more boxes and connecting the boxes together. Plaster ears may be provided for use on plasterboard or for work on old installations.

Securing Boxes with Screws

There are many ways to secure a box to framing members, such as boxes with brackets, nails, screws, bolts, and so forth. Make sure if you use screws that will pass through the box, the exposed threads on the screws are protected in an approved manner so as not to damage conductor insulation, *NEC 314.23(B)(1)*.

TABLE 6-9

Table 314.16(B) Volume Allowance Required per Conductor

Size of Conductor (AWG)	Free Space Within Box for Each Conductor	
	cm³	in.³
18	24.6	1.50
16	28.7	1.75
14	32.8	2.00
12	36.9	2.25
10	41.0	2.50
8	49.2	3.00
6	81.9	5.00

Reprinted with permission from NFPA 70-2011.

FIGURE 6-26 Device (switch) box.
(*Photo courtesy Appelton Electric LLC, Emerson Industrial Automation*)

Approved means *Acceptable to the authority having jurisdiction.* See *NEC Article 100.*

Masonry Boxes

Masonry boxes are designed for use in masonry block or brick. These boxes do not require an extension cover but will accommodate devices directly, Figure 6-27. Masonry boxes are available with depths of 2½ in. (65 mm) and 3½ in. (90 mm). Boxes of this type are available with knockout sizes up to trade size 1.

Handy Boxes

Handy boxes, Figure 6-28, are generally used in exposed or surface installations and are available with trade size ½, ¾, and 1 knockouts. These boxes range in depth from 1¼ to 2½ in. (32 to 58 mm) and can accommodate most devices without the use of an extension cover.

Trade Size 4 × 4 Square Boxes

A trade size 4 × 4 square box, Figure 6-29, is used for surface or concealed installations. Mud or plaster rings are available for different thicknesses of finish materials. If marked with their volume, the plaster rings add to the capacity of the box to contain conductors. Covers of various configura-

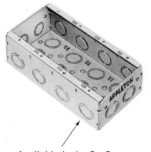

Available in 1-, 2-, 3-, 4-, and 5-gang sizes.

FIGURE 6-27 Four-gang masonry box.
(*Photo courtesy Appelton Electric LLC, Emerson Industrial Automation*)

FIGURE 6-28 Handy box and receptacle cover.
(*Photo courtesy Appelton Electric LLC, Emerson Industrial Automation*)

tions such as for switches or receptacles are available to accommodate devices where the box is surface mounted. This type of box is available with knockouts up to trade size 1.

Octagonal Boxes

Boxes of this type are used primarily to install ceiling outlets. Octagonal boxes are available either for mounting in concrete or for surface or concealed mounting, Figure 6-30. Extension covers are available but are not always required. Octagonal boxes are commonly used in depths of 1¼, 1½, and 2⅛ in. (32, 38, and 54 mm). These boxes are available with knockouts up to trade size 1.

Trade Size 4¹¹⁄₁₆ Square Boxes

These spacious boxes are used where the larger size is required, usually the result of larger size

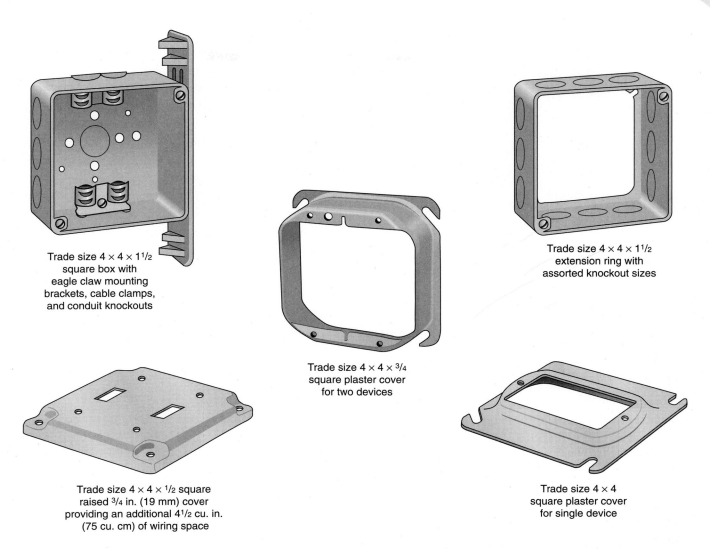

Trade size 4 × 4 × 1¹/₂
square box with
eagle claw mounting
brackets, cable clamps,
and conduit knockouts

Trade size 4 × 4 × ³/₄
square plaster cover
for two devices

Trade size 4 × 4 × 1¹/₂
extension ring with
assorted knockout sizes

Trade size 4 × 4 × ¹/₂ square
raised ³/₄ in. (19 mm) cover
providing an additional 4¹/₂ cu. in.
(75 cu. cm) of wiring space

Trade size 4 × 4
square plaster cover
for single device

FIGURE 6-29 Trade size 4 square boxes and covers. (*Delmar/Cengage Learning*)

FIGURE 6-30 Octagonal box on telescopic hanger and box extension. (*Delmar/Cengage Learning*)

conductors, many conductors, or wiring devices in the box. They are available in 1½ in. and 2⅛ in. deep sizes, with trade size ½ and ¾ knockouts. These boxes require an extension ring or a raised plaster cover to permit the attachment of wiring devices, Figure 6-31.

Box Sizing

NEC 314.16 states that outlet boxes, switch boxes, and device boxes must be large enough to provide ample room for the wires in that box without having to jam or crowd the wires in the box.

When conductors are the same size and metal boxes are used, the proper box size can be determined by referring to *NEC Table 314.16(A)*. *Table 314.16(A)* applies to metal boxes as they are produced in standard

FIGURE 6-31 Trade size 4¹¹⁄₁₆ square box. (*Photo courtesy Appelton Electric LLC, Emerson Industrial Automation*)

sizes. Nonmetallic boxes are produced in nonstandard sizes, with their internal volume marked inside the box. When conductors of different sizes are installed, refer to *NEC Table 314.16(B)* to determine the minimum size of device and outlet boxes.

NEC Table 314.16(A) does not consider the space taken up by fittings or devices such as studs for luminaires, cable clamps, hickeys, switches, pilot lights, or receptacles that may be in the box. An adjustment must be made for this equipment or wiring device. The box may have to be increased in size to accommodate this equipment without overcrowding. Table 6-10 provides a

TABLE 6-10

Box fill according to *NEC 314.16*.

When a box contains no fittings, devices, studs for luminaires, cable clamps, hickeys, switches, receptacles, or equipment grounding conductors.	Refer directly to *NEC Table 314.16(A)* or *Table 314.16(B)*.
When a box contains one or more internal cable clamps.	Add a single-volume based on the largest conductor in the box.
When a box contains one or more fixture studs for luminaires or hickeys.	Add a single-volume for each type based on the largest conductor in the box.
When a box contains one or more wiring devices on a yoke.	Add a double-volume for each yoke based on the largest conductor connected to a device on that yoke. Some large wiring devices, such as a 30-ampere, 3-pole, 4-wire receptacle, will not fit into a single gang box. It requires a 2-gang box. See text for details.
When a box contains one or more equipment grounding conductors.	Add a single-volume based on the largest conductor in the box.
When a box contains one or more additional sets of isolated (insulated) equipment grounding conductors as permitted by *NEC 250.146(D)* for noise reduction.	Add another volume based on the largest equipment grounding conductor in the additional set.
For conductors less than 12 in. (300 mm) long between the raceway entries for the conductor that is looped or coiled through the box without being spliced.	Add a single-volume for each conductor that is looped or coiled through the box.
For conductors 12 in. (300 mm) or longer between the raceway entries for the conductor that is looped or coiled through the box without being spliced.	Add a double-volume for each conductor that is looped or coiled through the box.
When conductors originate outside of the box and terminate inside the box.	Add a single-volume for each conductor based on the size of the conductor.
When no part of the conductor leaves the box—for example, a jumper wire used to connect three wiring devices on one yoke, or pigtails as illustrated in Figure 6-34.	Do not count this (these). No additional volume is required.
An equipment grounding conductor or not more than four fixture wires smaller than 14 AWG that originate from a luminaire canopy or similar canopy (like a fan) and terminate in the box.	Not required to count this (these). No additional volume is required.
When small fittings such as locknuts and bushings are present.	Not required to count this (these). No additional volume is required.

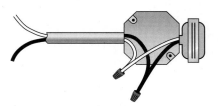

FIGURE 6-32   Box containing four conductors. Transformer leads 18 AWG or larger are to be included, *NEC 314.16(B)(1)*. (*Delmar/Cengage Learning*)

list of conditions that may arise when evaluating box fill and the appropriate response to each of those conditions. Figure 6-32 covers the special case presented when transformer leads enter a box. Figure 6-33 addresses the common situation of luminaire wires in a box. If there are not more than four luminaire wires and they are smaller than 14 AWG, they may be omitted from the box fill calculations. See *NEC 314.16(B)(1), Exception.*

Boxes to Support Ceiling-Suspended Paddle Fans

The twisting and vibrating nature of a ceiling-suspended paddle fan can lead to personal injury should the fan fall. In locations where a ceiling-suspended paddle fan is to be installed,

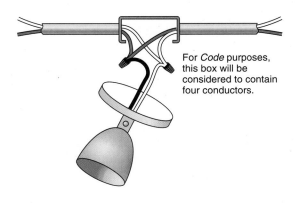

For *Code* purposes, this box will be considered to contain four conductors.

FIGURE 6-33   Box containing four conductors. Not more than four luminaire wires that are smaller than 14 AWG may be omitted from the box fill calculations, *NEC 314.16(B)(1), Exception.* (*Delmar/Cengage Learning*)

NEC 314.27(C) states that the box or outlet box system shall be listed for the support of ceiling-suspended paddle fans. Watch this closely when roughing in ceiling outlet boxes.

Unused Openings

To keep dirt, bugs, and varmints out of electrical equipment, *NEC 110.12(A)* and *NEC 408.7* require that unused openings be closed. Openings for mounting the equipment, openings necessary for the operation of the equipment, and openings permitted by the design for listed equipment do not have to be closed.

SELECTING THE CORRECT SIZE BOX

Box Fill When All Conductors Are the Same Size

A box contains one stud for luminaire and two internal cable clamps. Four 12 AWG conductors and an equipment grounding conductor enter the box.

Four 12 AWG conductors	4
One stud for luminaire	1
Two cable clamps (count only one)	1
One equipment grounding conductor	1
Total	7

Referring to *NEC 314.16(A)*, we find that a $4 \times 2^{1}/_{8}$ in. (100×54 mm) octagonal box is suitable for this example.

Box Fill When the Conductors Are Different Sizes

When a box contains different size wires, refer to *NEC 314.16* and do the following:

- Determine the minimum volume requirement based on the volume required for the conductors, clamps, support fittings, device or equipment, and equipment grounding conductors, according to *NEC 314.16(B)(1)* through *(B)(5)*.

- When conductors of different sizes are connected to a device(s) on one yoke or strap, the volume allowance of the largest conductor connected to the device(s) shall be used in computing the box fill.
- When more than one size of equipment grounding conductor is in a box, the volume allowance of the largest equipment grounding conductor shall be used in computing the box fill.

Box Fill When a Box Contains Devices

A box is to contain two devices—a duplex receptacle and a toggle switch. Two 12 AWG conductors are connected to the receptacle, and two 14 AWG conductors are connected to the toggle switch. Two equipment grounding conductors are in the box, a 12 AWG and a 14 AWG. Two cable clamps are in the box. See *NEC Table 314.16(B)* or Table 6-6 in this text for volume allowance per conductor.

The minimum box size is calculated as follows:

Item	Customary Volume	SI Volume
Two 12 AWG (2.25 in.³ each)	4.50 in.³	73.8 cm³
Two 14 AWG (2.0 in.³ each)	4.00 in.³	65.6 cm³
One cable clamp (2.25 in.³ each based on largest conductor)	2.25 in.³	36.9 cm³
One equipment grounding conductor (2.25 in.³ each based on largest conductor)	2.25 in.³	36.9 cm³
One duplex receptacle (2.25 in.³ × 2, based on largest conductor connected to device)	4.50 in.³	73.8 cm³
One toggle switch (2.0 in.³ × 2, based on largest conductor connected to device)	4.00 in.³	65.6 cm³
Minimum volume	21.50 in.³	352.6 cm³

Therefore, select a box having a *minimum* volume of 21.5 in.³ (352.6 cm³). The volume may be marked on the box; otherwise, refer to the second column of *NEC Table 314.16(A)*, titled *Minimum Volume*.

The *Code* requires that all boxes *other* than those listed in *NEC Table 314.16(A)* be durable and legibly marked by the manufacturer with their cubic-inch capacity. When sectional boxes are ganged together, the volume to be filled is the total cubic-inch volume of the assembled boxes. Fittings may be used with the sectional boxes, such as plaster rings, raised covers, and extension rings.

When these fittings are marked with their volume in cubic inches or have dimensions comparable to those boxes shown in *NEC Table 314.16(A)*, their volume may be considered in determining the total cubic-inch volume to be filled. See *NEC 314.16(A)(1)*. Figure 6-34 illustrates how the volume of a plaster ring, or of a raised cover, is added to the volume of the box to increase the total volume.

Depth of Box—Watch Out!

Determining the cubic volume of a box is not the end of the story. Outlet and device boxes shall be deep enough so that conductors will not be damaged when installing the wiring device or equipment into the box. Merely calculating and providing the proper volume of a box is not always enough. The volume calculation might prove adequate, yet the size (depth) of the wiring device or equipment might be such that conductors behind it may possibly be damaged, *NEC 314.24*.

Width of Box—Watch Out!

Large wiring devices, such as a 30-ampere, 3-pole, 4-wire receptacle, will not fit into a single gang box that is 2 in. (50.8 mm) wide. These receptacles measure 2.10 in. (53.3 mm) in width. Likewise, a 50-ampere, 3-pole, 4-wire receptacle measures 2.75 in. (69.9 mm) in width. Consider using a 4-in. (101.6-mm) square box with a 2-gang plaster device ring for flush mounting or a 2-gang raised cover for surface mounting, or use a 2-gang device box. The center-to-center mounting holes of both the 30-ampere and the 50-ampere receptacle are 1.81 in. (46.0 mm) apart, exactly matching the center-to-center holes of a 2-gang plaster ring, raised cover, or 2-gang device box, *NEC 314.16(B)(4)*.

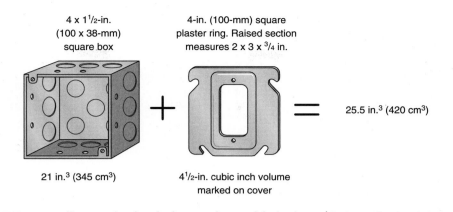

4 x 1½-in.
(100 x 38-mm)
square box

4-in. (100-mm) square
plaster ring. Raised section
measures 2 x 3 x ¾ in.

25.5 in.³ (420 cm³)

21 in.³ (345 cm³)

4½-in. cubic inch volume
marked on cover

FIGURE 6-34   **Box and raised plaster ring cubic inches.** (*Delmar/Cengage Learning*)

Box Fill When There Is a Plaster Ring

How many 12 AWG conductors are permitted in this box and raised plaster ring, as shown in Figure 6-33?

See *NEC 314.16* and *NEC Tables 314.16(A)* and *314.16(B)* for the volume required per conductor. This box and cover will accommodate

$$\frac{25.5 \text{ in.}^3 \text{ (420 cm}^3)}{2.25 \text{ in.}^3 \text{ (36.9 cm}^3)} = \text{Eleven 12 AWG conductors maximum, less the required deductions per } \textit{NEC 314.16(B)(2)}$$

Many devices such as GFCI receptacles, dimmers, and timers are much larger than the conventional receptacle or switch. The *Code* has recognized this problem by requiring that one or more devices on one strap require a conductor reduction of two for standard boxes or a volume allowance of two conductors for nonstandard boxes when determining maximum box capacity or minimum size. It is always good practice to install a box that has ample room for the conductors, devices, and fittings rather than forcibly crowding the conductors into the box.

The sizes of equipment grounding conductors are shown in *NEC Table 250.122*. The equipment grounding conductors are the same size as the circuit conductors in cables having 14, 12, or 10 AWG circuit conductors. Thus, box sizes can be calculated using *NEC Table 314.16(A)* for standard boxes.

Another example of the procedure for selecting the box size is the installation of a 4-way switch in the Bakery. Four 12 AWG conductors and one device are to be installed in the box. Therefore, the box selected must be able to accommodate six conductors; a 3 × 2 × 2¼ in. (75 × 50 × 57 mm) switch box is adequate.

When the box contains fittings such as luminaire studs, clamps, or hickeys, the number of conductors permitted in the box is one less for *each* type of device. For example, if a box contains two clamps, deduct one conductor from the allowable number shown in *NEC Table 314.16(A)* (Figure 6-35).

An example of this situation occurs in the Bakery at the box where the conduit or Type MC cable leading to the 4-way switch is connected.

- There are two sets of travelers (one from each of the 3-way switches) passing through the box on the way to the 4-way switch (four conductors).

- A neutral conductor from the panelboard is spliced to a neutral conductor that serves the luminaire attached to the box and to neutral conductors entering the two conduits leaving the box that go to other luminaires (four conductors).

- A switch return is spliced in the box to a conductor that serves the luminaire that is attached to the box and to a switch return that enters the conduits leaving the box to serve other luminaires (four conductors).

- A luminaire stud-stem assembly is used to connect the luminaire directly to the box (one conductor).

This is a total of 13 conductors.

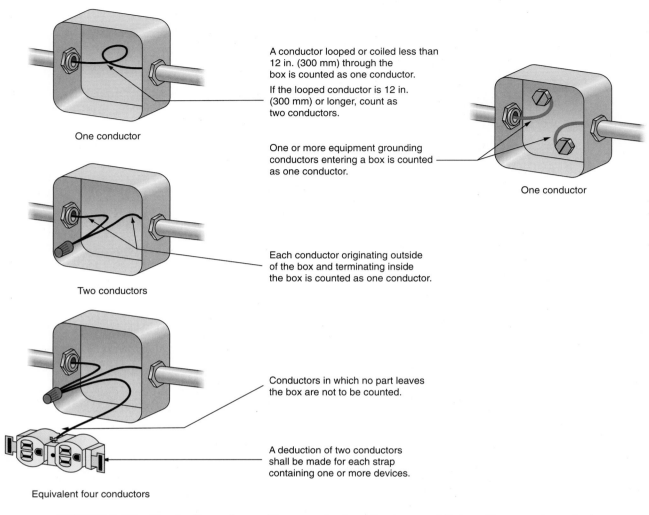

One conductor

A conductor looped or coiled less than 12 in. (300 mm) through the box is counted as one conductor.

If the looped conductor is 12 in. (300 mm) or longer, count as two conductors.

One conductor

One or more equipment grounding conductors entering a box is counted as one conductor.

Two conductors

Each conductor originating outside of the box and terminating inside the box is counted as one conductor.

Conductors in which no part leaves the box are not to be counted.

Equivalent four conductors

A deduction of two conductors shall be made for each strap containing one or more devices.

FIGURE 6-35 Illustration of counting conductors for boxes. (*Delmar/Cengage Learning*)

From *NEC Table 314.16(A)*, a 4 × 2⅛ in. (100 × 54 mm) square or a 4¹¹⁄₁₆ × 1½ in. (100 × 38 mm) square box would qualify. Because these boxes require plaster rings, a smaller size box could be used if the plaster ring volume added to the box volume were sufficient for the 13 conductors.

Pull and Junction Boxes and Conduit Bodies

When necessary, conduit bodies (Figure 6-36) or pull boxes (Figure 6-37) are used. They must be sized according to *NEC 314.28* when they contain conductors 4 AWG or larger that are required to be insulated.

FIGURE 6-36 Conduit bodies. (*Delmar/Cengage Learning*)

Conduit bodies are often referred to in the trade as "Condulets" and "Unilets." The terms *Condulets* and *Unilets* are copyrighted trade names of Cooper Crouse-Hinds and Appleton Electric, respectively, for their line of conduit bodies.

Conduit bodies are excellent for turning corners and changing directions of a conduit run. They are available with threaded hubs for threaded conduit and threaded fittings, set screw hubs for direct insertion of EMT without needing an EMT connector, gasketed covers, built-in rollers to make it easy to pull wires, and in jumbo sizes to provide more room for pulling, splicing, and making taps. The following chart shows how typical conduit bodies are identified.

When Looking at the Front Opening with the Inlet Up, If the Direction Is:	The Conduit Body Is Called a(n):
To the back	LB
Left	LL
Right	LR
Straight through	C
Straight through in both directions	X
Straight through and to the left or right	T
Straight through and to the back	TB
One conduit entry only	E

- For straight pulls, the pull box, junction box, or conduit body must be at least eight times the diameter of the largest raceway, Figures 6-37 and 6-38.
- For angle pulls or U pulls or where splices are made, the distance between the raceways and the opposite wall of the box or conduit body must be at least six times the diameter of the largest raceway. To this value, it is necessary to add the diameters of all other raceways entering in any one row of the same wall of the box, Figures 6-39, 6-40, and 6-41. When more than

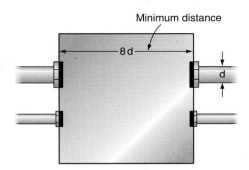

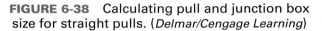

FIGURE 6-38 Calculating pull and junction box size for straight pulls. (*Delmar/Cengage Learning*)

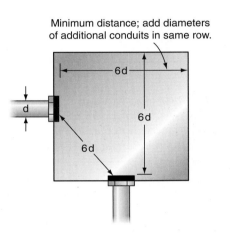

FIGURE 6-39 Calculating pull and junction box size for angle pull. (*Delmar/Cengage Learning*)

one row exists, the row with the maximum diameters must be used in sizing the box.

- Boxes may be smaller than the preceding requirements when approved and marked with the size and number of conductors permitted.
- Conductors must be racked or cabled if any dimension of the box exceeds 6 ft (1.83 m).
- Boxes must have an approved cover.
- Boxes must be accessible.

Depth of Pull Boxes. The depth of a pull box when a simple angle pull is made is not clearly addressed in the *NEC*.

In reality, the depth would have to be at least the diameter of the bushing for the raceway entering the side of the pull box *plus* a little more to be able to install the bushing and tighten it.

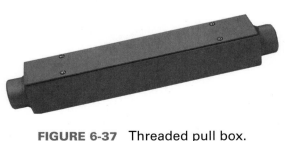

FIGURE 6-37 Threaded pull box. (*Delmar/Cengage Learning*)

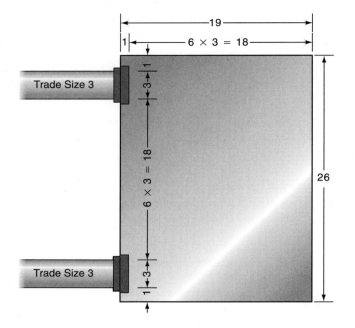

FIGURE 6-40 Example of sizing pull and junction boxes for a U pull for two trade size 3 raceways in accordance with *NEC 314.28(A)(2).* (*Delmar/Cengage Learning*)

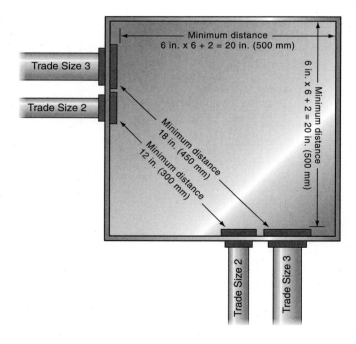

FIGURE 6-41 Calculating pull and junction boxes for angle pulls with side-by-side conduits. (*Delmar/Cengage Learning*)

For example, a locknut and bushing for a 3-in. rigid metal conduit has a diameter of approximately 3.8 to 4 in. (97 to 102 mm). If this were a grounding/bonding bushing, the diameter (including the huge grounding/bonding lug) would be much more than that of a simple locknut and bushing. To install the grounding/bonding lug, 5 to 6 in. (127 to 152 mm) may be needed to turn the bushing onto the conduit. With a little space to spare, the pull box would need to be at least 8 in. (203 mm) deep.

For multiple raceways, you have to take into consideration the diameters of the bushings, as discussed previously.

Power Distribution Blocks

Power distribution blocks are permitted in metal pull and junction boxes over 100 in.^{3} (1650 cm^3) in size, Figure 6-42. The installation must comply with the following five requirements:

1. Power distribution blocks are required to be listed by a qualified electrical products testing organization.

2. In addition to complying with the size requirements for pull and junction boxes in the first sentence of *NEC 314. 28(A)(2),* the metal box must have the minimum dimensions specified

FIGURE 6-42 Power distribution blocks. (*Photo Courtesy Cooper Bussmann*)

in the installation instructions for the power distribution block.

3. Wire bending space at the terminals of the power distribution block is required to comply with *NEC 312.6*.

4. Power distribution blocks are not permitted to have uninsulated live parts exposed within the box whether or not the box cover is installed.

5. If the pull or junction boxes are used for conductors that do not terminate on the power distribution block, through conductors are required to be arranged so the power distribution block terminals are not obstructed following the installation.

Similar provisions are included in *NEC 376.56(B)* for power distribution blocks installed in metal wireways. Also, the UL White Book contains the requirement that power distribution blocks be located on the load side of overcurrent protection. As a result, power distribution blocks are not permitted on the supply side of the service disconnecting means.

REVIEW

Refer to the *National Electrical Code* or the working drawings when necessary. Where applicable, responses should be written in complete sentences.

For Questions 1–4, select from the following list of raceway types (check the letter(s) representing the correct response(s)):

a. electrical metallic tubing

b. electrical nonmetallic tubing

c. flexible metallic tubing

d. rigid metal conduit

e. intermediate metal conduit

f. rigid nonmetallic conduit

1. Which raceway(s) may be used where cinder fill is present?
   ☐ a.   ☐ b.   ☐ c.   ☐ d.   ☐ e.   ☐ f.

2. Which raceway(s) may be used in hazardous locations?
   ☐ a.   ☐ b.   ☐ c.   ☐ d.   ☐ e.   ☐ f.

3. Which raceway(s) may be used in wet locations?
   ☐ a.   ☐ b.   ☐ c.   ☐ d.   ☐ e.   ☐ f.

4. Which raceway(s) may be used for service entrances?
   ☐ a.   ☐ b.   ☐ c.   ☐ d.   ☐ e.   ☐ f.

Give the raceway size required for each of the following conductor combinations:

5. Five, 10 AWG, Type THW in Schedule 80 PVC _____

6. Three, 1/0 AWG, Type THHN in EMT _____

7. Three, 1/0 AWG and one, 1 AWG, Type THW in EMT _____

For Questions 8 and 9, determine the correct box size for each of the following conditions:

8. Two nonmetallic sheathed cables with two 12 AWG conductors, a grounding conductor, and a switch in a metal box.

9. A conduit run serving a series of luminaires, connected to a total of three circuits. The luminaires are supplied by 120 volts from a 3-phase, 4-wire system. Each box will contain two circuits running through the box and a third circuit connected to a luminaire, which is hung from a luminaire stud. Use 12 AWG Type THHN conductors.

10. What is the minimum cubic in. volume that is permissible for the following outlet box? Show your calculations.

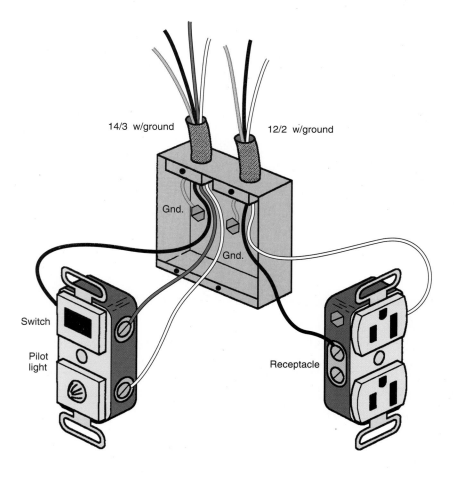

Two trade size 3 raceways enter a box directly across from each other. No other raceways enter the box. What are the minimum dimensions of the box?

11. Length _____

12. Width _____

13. Depth _____

Two trade size 3 raceways enter a box at right angles to each other. No other raceways enter the box. What are the minimum dimensions of the box?

14. Length _____

15. Width _____

16. Depth _____

17. If you had the freedom of choice, what type of raceway would you select for installation under the sidewalks and driveways of the Commercial Building? Support your selection. _____

_____

_____

_____

Determine the required EMT raceway size for the following combination of conductors:

18. Four 8 AWG Type THW and four 12 AWG Type THW

19. Three 350 kcmil and one 250 kcmil Type XHHW conductors and a 4 AWG bare conductor

Determine the required box size for the following situations:

20. In a nonmetallic sheathed cable installation, a 10/3 with ground is installed in a metal octagonal box to supply two 12/2 with ground branch-circuit cables. What is the minimum size box? The box contains cable clamps.

Two 3-in. raceways enter a box, one through a side and the other in the back. Four 500 kcmil, type THHN conductors will be installed in the raceway. No other raceways enter the box. What are the minimum dimensions of the box?

21. Length   ——————————

22. Width   ——————————

23. Depth   ——————————

Calculate the answers using information shown in the drawing.

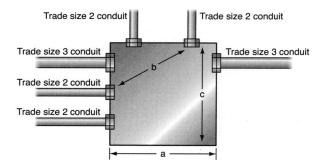

24. Dimension *a* must be at least ——————— inches.

25. Dimension *b* must be at least ——————— inches.

26. Dimension *c* must be at least ——————— inches.

27. Do you foresee any difficulties in installing the conductors in these raceways? Explain.

CHAPTER

7

Motor and Appliance Circuits

OBJECTIVES

After studying this chapter, you should be able to

- use and interpret the word *appliance*.
- use and interpret the term *utilization equipment*.
- choose an appropriate method for installing electrical circuits to appliances.
- properly locate the disconnecting means for appliances.
- identify appropriate grounding for appliances.
- determine branch-circuit ratings, conductor sizes, and overcurrent protection for appliances and motors.
- understand the meaning of the terms *Type 1* and *Type 2* protection.

∿ APPLIANCES

Most of the *Code* rules that relate to the installation and connection of electrical appliances are found in *NEC Article 422*. For motor-operated appliances, reference will be made to *NEC Article 430 (Motors, Motor Circuits, and Controllers)*. If the appliance is equipped with hermetic refrigerant motor compressor(s), such as refrigeration and air-conditioning equipment, *NEC Article 440 (Air-Conditioning and Refrigeration Equipment)* applies.

The *National Electrical Code* defines an appliance as utilization equipment, generally other than industrial, of standardized sizes or types, that is installed or connected as a unit to perform such functions as air-conditioning, washing, food mixing, and cooking, among others.

The *Code* defines utilization equipment as equipment that uses electric energy for mechanical, chemical, heating, lighting, or similar purposes.

In commercial buildings, one of the tasks performed by the electrician is to make the electrical connections to appliances. The branch-circuit sizing may be specified by the consulting engineer in the specifications or on the working drawings. In other cases, it is up to the electrician to do all of the calculations, and/or to make decisions based on the nameplate data of the appliance.

The Bakery in the Commercial Building provides examples of several different types of electrical connection methods that can be used to connect electrical appliances.

Appliance Branch-Circuit Overcurrent Protection

- For those appliances that are marked with a maximum size and type of overcurrent protective device, this rating shall not be exceeded, *NEC 422.11(A)*.

- If the appliance is not marked with a maximum size overcurrent protective device, *NEC 422.11(A)* stipulates that *NEC 240.4* applies. This section states that the fundamental rule is to protect conductors at their ampacities, as specified in *NEC Table 310.15(B)(16)*.

There are several variations allowed to this fundamental rule, three of which relate closely to motor-operated appliances, air-conditioning, and refrigeration equipment.

NEC 240.4(B) allows the use of the next standard higher ampere rating overcurrent device when the ampacity of the conductor does not match a standard ampere rating overcurrent device. There are two exceptions to this: if the conductors are part of a multioutlet branch-circuit supplying cord- and plug-connected portable loads or if the next standard rating is greater than 800-ampere, then the next lower standard rating overcurrent device must be chosen. Standard ratings of overcurrent devices are listed in *NEC 240.6(A)*.

NEC 422.11 lists overcurrent protection requirements for several types of appliances. Commercial appliances include

- infrared lamp heating,
- surface heating elements,
- single nonmotor-operated appliances,
- resistance-type heating elements,
- sheathed-type heating elements,
- water heaters and steam boilers, and
- motor-operated appliances.

For a single nonmotor-operated appliance, *NEC 422.11(E)* states that the overcurrent protection shall not exceed

a. the overcurrent device rating if so marked on the appliance.

b. 150% of the appliance's rated current, if the appliance is rated at more than 13.3 amperes. If the 150% results in a nonstandard ampere overcurrent device value, then it is permitted to use the next higher standard rating overcurrent device.

For example, an appliance is rated at 4500 volt-ampere, 240 volts, single-phase. What is the maximum size fuse permitted?

$$\frac{4500 \text{ VA}}{240 \text{ V}} = 18.75 \text{ A}$$

$$18.75 \text{ A} \times 1.5 = 28.125 \text{ A}$$

Therefore, the *Code* permits the use of a 30-ampere fuse or circuit breaker for the overcurrent protection.

Storage-Type Water Heaters

Fixed storage water heaters that have a capacity of 120 gal (450 L) or less are considered a continuous load for the purposes of sizing branch circuits, *NEC 422.13*.

This means the nameplate load of the water heater must have a 125% factor applied in keeping with *NEC 422.10(A)* and *210.19(A)(1)*. You also need to explore the design the of the water heater controls. Some controllers allow all the heating elements to operate at the same time, so all the load must have the 125% factor applied. Other controls

interlock the elements so only one or two of the elements can be operated at the same time. Obviously, we need to be informed about the controls used on the water heater to apply the calculation to determine branch circuit, feeder, and disconnecting means ratings. These issues are explored in Table 7-1.

Appliance Grounding

- When appliances are to be grounded, the requirements are given in *NEC Article 250*. Grounding is discussed in several places in this text. With few exceptions, an equipment grounding conductor that is sized in accordance with *NEC 250.122* is required to be installed with the branch circuit and be connected to the

TABLE 7-1

Electrical values for the electric water heaters installed in the Commercial Building.

Tenant	Size	kW	Volts	Phase	Amperes	125% Factor	Branch Circuit Rating	Conductor Size/AWG THHN/THWN	How Connected
Women's Toilet, 2nd floor	2.5 gal	1.5 kW	120	1	12.5	15.6	20	12	Cord & Plug
Men's Toilet, 2nd floor	2.5 gal	1.5 kW	120	1	12.5	15.6	20	12	Cord & Plug
Insurance Office	6 gal	1.65 kW	120	1	13.75	17.2	20	12	Hard wired. Nonfused disconnect required adjacent to water heater
Real Estate Office	6 gal	1.65 kW	120	1	13.75	17.2	20	12	Hard wired. Nonfused disconnect required adjacent to water heater
Bakery	80 gal	30 kW	208	3	83.3	104.1	110	2	Hard wired. Nonfused 200 A disconnect required adjacent to water heater
Beauty Salon	120 gal	24 kW	208	3	66.6	83.3	90	4	Hard wired. Nonfused 100 A disconnect required adjacent to water heater
Drugstore	80 gal	18 kW	208	3	50	62.5	70	4	Hard wired. Nonfused 100 A disconnect required adjacent to water heater
Owner (in Basement)	50 gal	6 kW	208	3	16.7	20.9	25	10	Hard wired. Circuit breaker in panelboard serves as disconnect.

appliance. It is suggested that the reader consult the *Code Index* in the *Appendix* where all *NEC* references are listed.

Appliance-Disconnecting Means

- Permanently connected appliances, *NEC 422.31:*

  a. If the appliance rating does not exceed ⅛ horsepower or does not exceed 300 volt-amperes, the branch-circuit overcurrent device may serve as the appliance's disconnecting means, *NEC 422.31(A).*

  b. If the appliance rating exceeds 300 volt-amperes or exceeds ⅛ horsepower, the branch-circuit disconnect switch or breaker can serve as the appliance's disconnecting means, but only if it is within sight of the appliance or is capable of being locked in the off position. The lock-off provision must be permanently installed on or at the switch or circuit breaker, *NEC 422.31(B).* Figure 7-1 shows an accessory that is available from circuit breaker manufacturers and can be installed between the panelboard circuit breakers and the deadfront or cover. To implement the lockout procedure, turn the circuit breaker off and install a control mechanism such as a padlock in the opening in the accessory.

- For motor-driven, permanently connected appliances rated at more than ⅛ horsepower, the disconnecting means must be within sight of the motor controller, *NEC 422.32.*

- Some appliances have an integral unit switch. If this unit switch has a marked off position and disconnects all of the ungrounded conductors, it can be considered the appliance's disconnecting means. This applies only in nondwelling occupancies if the branch-circuit switch or circuit breaker supplying the appliance is readily accessible, *NEC 422.34(D).* The term *readily accessible* is covered in the *Definitions* in the *NEC.* Simply stated, this means that there must be easy access to the disconnect switch without the need to *climb over or remove obstacles or to resort to portable ladders, and so forth.*

- The term *in sight* is covered in the *Definitions* in the *NEC.* Further discussion is found in *NEC 430.102.*

- *NEC 422.33* permits the use of a cord- and plug-connected arrangement to serve as the appliance's disconnecting means. Such cord-and-plug assemblies may be furnished and attached to the appliance by the manufacturer to meet the requirements of specific Underwriters Laboratories standards.

THE BASICS OF MOTOR CIRCUITS

Motor failures happen for many reasons. Abuse, lack of oil, bad bearings, insulation failure (end-of-normal life, oil, moisture), dirt or other foreign material (reduced ventilation), single-phasing, unbalanced voltage (more about this a little later), and jammed V-belts are a few of the conditions that can cause a motor to draw more than normal current. Too much current results in the generation of too much heat within the windings of the motor. For every 10°C above the maximum temperature rating of the motor, the expected life of the motor is reduced by 50%. This is sometimes referred to as the "half-life rule."

The Electrical Apparatus Service Association has a tremendous amount of technical literature available regarding electric motors and common failures. Check out its Web site: http://www.easa .com. Manufacturers of motors and motor controllers

FIGURE 7-1 Circuit breaker–type lock-off accessory. (*Delmar/Cengage Learning*)

also have an abundance of technical data available on their Web sites. Consult them often.

Bear in mind that as we discuss motor circuit design requirements found in *Article 430*, our calculations are based on *rated* voltage, *rated* frequency, *rated* current, *rated* horsepower, *rated* full-load amperes, and *rated* full-load speed. We will also discuss variances to *rated* values.

Motor circuits can be simple or complex depending on factors such as the type, size, and characteristics of the motor or motors, how each motor is to be operated, and the electrical requirements. Given this information, the electrician usually begins by locating the power source, then planning the circuit between the source and the motor. The first component in this circuit is usually the branch-circuit short-circuit and ground-fault protective device. See the discussion later in this chapter.

Disconnecting Means

The *Code* requires that all motors be provided with a means to disconnect the motor from its electrical supply. These requirements are found in *NEC Article 430, Part IX*.

NEC 430.109 gives the type of disconnecting means permitted. Included are motor circuit switches rated in horsepower, molded case circuit breakers, molded case switches, instantaneous-trip circuit breakers that are part of listed combination controllers, self-protected combination controllers, and manual motor controllers. For most applications, the motor circuit switch for a motor must be horsepower rated. The most common exceptions are

- for motors ⅛ horsepower or less where the branch-circuit overcurrent device can serve as the disconnecting means.

- for stationary motors of 2 horsepower or less, rated at 300 volts or less, for which a general-use switch is permitted.

- for stationary ac motors greater than 100 horsepower, for which a general-use disconnect switch, or an isolating switch is permitted when the switch is marked "Do Not Open Under Load."

- an attachment plug, which is permitted to serve as the disconnecting means if the attachment plug cap is horsepower-rated not less than that of the load. Refer to *NEC 430.109(F)*. The

horsepower rating of an attachment plug cap is considered to be the corresponding horsepower rating for 80% of the ampere rating of the attachment plug cap.

EXAMPLE ────────────────

A 2-wire, 125-volt, 20-ampere attachment plug cap:
$$20 \times 0.80 = 16 \text{ amperes}$$

────────────────────────

Checking *NEC Table 430.248* for a single-phase, 115-volt motor, we determine that a 1-horsepower motor has a full-load ampere rating of 16 amperes. Thus, this attachment plug cap is permitted to serve as the disconnecting means for this motor.

NEC 430.110(A) requires that the disconnecting means for motors rated at 600 volts or less must have an ampere rating of not less than 115% of the motor's full-load ampere rating.

The nameplate on all motor circuit switches, as well as the manufacturer's technical data, furnishes the horsepower rating, the voltage rating, and the ampere rating of the disconnect switch.

The *NEC* requires that a disconnecting means be provided for motors and motor controllers, *NEC 430.101*. This requirement is broken into two specific rules, and both rules must be met. Figures 7-2(A), 7-2(B), 7-2(C), and 7-2(D) illustrate the location of the disconnecting means in relation to the motor's controller, the motor, and the driven machinery.

Rule #1. See Figure 7-2(A). *NEC 430.102(A):* An individual disconnect must be in sight of the controller and must disconnect the controller. This is the basic rule. The *NEC* definition of "in sight" means that the controller must be visible and not more than 50 ft (15 m) from the disconnect.

Rule #2. See Figure 7-2(B). *NEC 430.102(B):* Disconnect must be in sight of motor location and driven machinery. If the disconnect as required in *430.102(A)* is in sight of the controller, the motor location, and driven machinery, then that disconnect meets the requirements of both *430.102(A)* and *430.102(B)*.

If, because of the nature of the installation, the disconnect is in sight of the controller, but not in sight of the motor location and the driven machinery, then, generally, another disconnect must be installed, Figure 7-2(C).

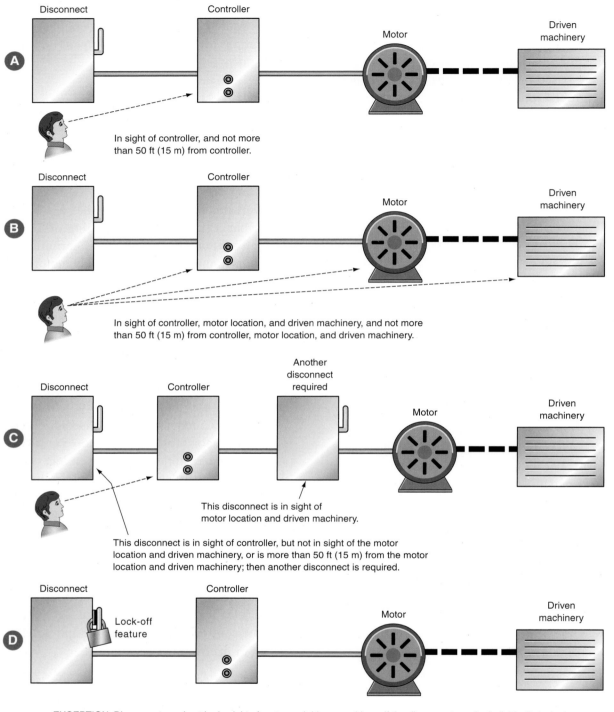

EXCEPTION: Disconnect need not be in sight of motor and driven machinery if the disconnect can be individually locked in the open (off) position. The locking provision must be of a permanent type installed on the switch or circuit breaker. The locking provision must remain in place with or without the lock installed.

Most disconnect switches, combination starters, and motor control centers have this "lock-off" feature.

To apply the EXCEPTION, *either* of two conditions must be met:

1. The disconnect location is impracticable . . . or might cause additional or increased hazards to persons or property.
2. It is an industrial installation where there is a *written* safety program in the workplace. . . and where conditions of maintenance and supervision ensure that only qualified persons will service the equipment.

FIGURE 7-2 Location of motor disconnect means. (*Delmar/Cengage Learning*)

As with many *NEC* requirements, there are exceptions. Figure 7-2(D) shows the exceptions to the disconnect requirements for motors, controllers, and driven machinery.

The reason the *Code* uses the term *motor location* instead of *motor* is that, in many instances, the motor is inside an enclosure and is out of sight until an access panel is removed. If the term *motor* were to be used, then it would be mandatory to install the disconnect inside of the enclosure. This is not always practical. The Commercial Building's rooftop air-conditioning units are examples.

Do Start–Stop and Lockout–Stop Buttons Qualify as a Disconnecting Means?

Start–stop or lockout–stop push buttons are not considered to be the required disconnecting means for a motor. These push buttons are connected into the control circuit wiring and control the operating (holding) coil. They do not disconnect the ungrounded (hot) supply conductors. Even if locked in the lockout–stop position, a ground fault or short circuit in the control circuit could result in the motor starting. Personal injury could happen. The word "off" means exactly that—"off," not "maybe off."

Air-Conditioning and Refrigerating Equipment Disconnect

Be careful about locating the disconnecting means for air-conditioning and refrigeration equipment!

NEC 440.14 requires that the disconnecting means be

- located within sight from the air-conditioning or refrigerating equipment.
- readily accessible.
- permitted to be installed on or within the air-conditioning or refrigerating equipment.
- located to allow access to panels that need to be accessed for servicing the equipment.
- mounted so as not to obscure the equipment nameplate(s).
- provided with a lock-off provision that *remains in place with or without the lock installed.* Snap-on covers for circuit breakers do not meet this requirement.

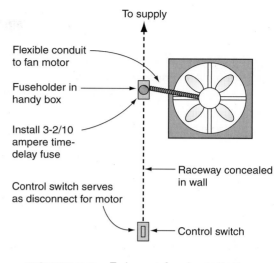

FIGURE 7-3   Exhaust fan installation. (*Delmar/Cengage Learning*)

Exhaust Fan

To turn the exhaust fan on and off, an ac general-use, single-pole snap (toggle) switch is installed on the wall below the fan. The electrician will install an outlet box approximately 1 ft from the ceiling (see the electrical plan), from which he will run a short length of flexible metal conduit to the junction box on the fan. The fan exhausts the air from the Bakery. The fan is installed through the wall, approximately 18 in. below the ceiling, Figure 7-3. Some occupancies will have a roof-mounted exhaust fan that is ducted to a supply grill on the ceiling of the facility or to hoods located above cooking equipment.

NEC 430.111 permits this switch to serve as both the controller and the required disconnecting means.

Important: As you study the following motor branch-circuit information, be sure to clearly understand that *430.6(A)(1)* tells us to select motor branch-circuit conductors, disconnecting means, controllers, and the branch-circuit short-circuit and ground-fault protection based on the current values found in Tables *430.247, 430.248, 430.249,* and *430.250* rather than on the motor nameplate.

NEC 430.6(A)(2) instructs us to use the actual motor nameplate current rating to determine the motor's overload protection.

Motor Branch-Circuit Conductors

Branch circuit conductors for a single motor used in a continuous-duty application shall not be smaller than 125% of the full-load current rating of the motor as found in the tables, *NEC 430.22.*

EXAMPLE

Determine the minimum size Type THHN/THWN conductors for a 7½ horsepower, 208-volt, 3-phase motor, service factor 1.15. The panelboard, circuit breakers, disconnect, and motor controller are all rated for 75°C. THHN/THWN conductors are rated 90°C in a dry location and 75°C in a wet location. The motor's leads are no problem because their temperature rating is considerably above that of the other equipment associated with the motor.

1. Referring to *NEC Table 430.250* to obtain the motor's full-load current rating, we find the full-load current rating is 24.2 amperes.
2. Multiply the motor's full-load current as found in *Tables 430.247, 430.248, 430.249,* or *430.250* by 1.25 to find the minimum conductor ampacity.

   24.2 amperes $\times$ 1.25 = 30.25 amperes

3. Check *NEC 110.14(C)(1)* to find limitations on terminations. For the size of the conductors used for this motor, the 75°C ampacity values found in *NEC Table 310.15(B)(16)* can be used. If it is unknown whether the panelboard and breakers are suitable for 75°C, the 60°C column for circuits 100 amperes or less, or 14 AWG through 1 AWG conductors, is required to be used.
4. *NEC Table 310.104(A)* shows that Type THHN conductors have a temperature rating of 90°C. *NEC Table 310.15(B)(16)* shows that a 75°C conductor:
   - 12 AWG Type THWN conductor has an allowable ampacity of 25 amperes.
   - 10 AWG Type THWN conductor has an allowable ampacity of 35 amperes.

The conclusion for this example is that at minimum 10 AWG THHN/THWN conductors are needed to meet the minimum conductor ampacity of 30.25 amperes for the motor.

When hooking up motors, *NEC 110.14(C)(1) (a)(4)* provides a little "wiggle room" for NEMA design B, C, or D motors. In this Section, permission is given to select the conductor ampacity using the 75°C column of *NEC Table 310.15(B)(16)*. Watch out when you try to apply *NEC 110.14(C)(1)(a)(4)*. Be sure that the temperature rating of the equipment on the other end of the conductor is marked 75°C. This may possibly allow the use of a smaller size conductor. Remember, a conductor has two ends. Conductor ampacity is governed by the temperature rating of the weakest link in the circuit!

Color Coding for Motor Branch-Circuit Conductors

Color coding was discussed in detail in Chapter 5. Color selection for ungrounded "hot" conductors can be just about any color except green, white, gray, or any color conductor (except green) that has three continuous white stripes. See *NEC 200.7(A)*. The choice is pretty much up to the electrician.

Over the years electricians have found the following color coding for motor branch circuits to be acceptable:

120-volt single-phase:
BLACK–WHITE

240-volt single-phase:
BLACK–BLACK
or BLACK–RED

208Y/120-volt 3-phase:
BLACK–BLACK–BLACK
or BLACK–RED–BLUE
(This is the typical color coding for services, feeders, and branch circuits.)

208-volt single-phase:
BLACK–BLACK
or BLACK–RED
or RED–BLUE
or BLUE–BLACK
(This is confusing when many single-phase motors are involved, so most electricians prefer merely to stick to BLACK–BLACK.)

480Y/277-volt 3-phase:
BLACK–BLACK–BLACK
or RED–RED–RED
or BROWN–ORANGE–YELLOW
(This is the typical color coding for services, feeders, and branch circuits.)

480-volt single-phase:
BLACK–BLACK

or RED–RED
or BROWN–ORANGE
or ORANGE–YELLOW
or YELLOW–BROWN
(This is confusing when many single-phase motors are involved, so most electricians prefer to stick merely to BLACK–BLACK.)

Motor Overload Protection

A motor might be considered to be "dumb." A motor will struggle to deliver its full horsepower. . . no matter how much load the motor is trying to handle, no matter what the supply voltage is, no matter how unbalanced the supply voltage is. The motor will keep trying until it commits suicide.

Motor burnout can be kept to a minimum if the motor has proper overload protection. If a properly sized overload device is selected, it will sense the increase in current and produce a trip, Table 7-2. Proper motor overload protection also protects the motor branch-circuit conductors from overheating. Overload relays are the most common form of motor overload protection. Overload relay types include thermal (eutectic alloy), fixed and interchangeable bimetallic, and electronic solid-state. Thermal overload devices are often referred to as *heaters, heater elements*, or *TOLs*. Solid-state devices offer additional features, such as they do not contain thermal units, are adjustable to cover a range of trip settings, provide phase-loss protection, provide unbalance current sensing that will trip in 3 seconds if the current in one phase is 25% greater than the average of all three phases, provide visual trip indication (some even have trip cause indication), and provide ground-fault protection.

Bimetallic devices work on the principle that different metals expand at different rates when heated.

When these two metals are bonded together and subjected to heat, the different rate of expansion causes them to bend/warp, making or breaking the contacts. Adjustable trip setting is accomplished by adjusting the space of the gap between the contacts. This is the same principle as found in many thermostats.

Some overload relays are manual reset; others are automatic reset. Obviously, automatic reset is used with caution to avoid personal injury should a motor that is not running suddenly start up again, much to the surprise of the individual working on the motor.

Figure 7-4 is a combination motor controller, complete with a lockout-type fusible disconnecting means, a 3-pole fuse block for the motor branch-circuit fuses, a relay for starting and stopping the motor, a step-down control circuit transformer, and in-the-cover start-stop buttons.

Figure 7-5 shows a manual resettable-type overload relay and a thermal overload "heater." When thermal overload heaters result in the relay tripping, a cooling-off period is needed before the relay can be reset.

Overload relays are available in Class 10, Class 20, and Class 30. Class 20 (standard trip) overload relays are the most common, are used for typical motor applications, and are generally under 200 amperes. Class 30 (slow trip) overload relays are used for motors having high inertia loads that are hard and/or slow to come up to speed as well as motors rated over 200 amperes. Class 10 (fast trip) overload relays are used on hermetic

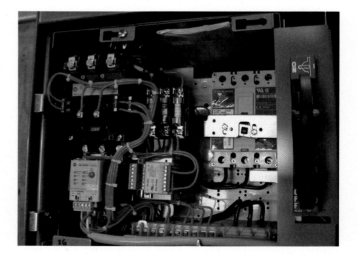

FIGURE 7-4 A fusible motor controller complete with disconnecting means, motor branch-circuit overcurrent protection, and motor overload protection. (*Courtesy of Schneider Electric North America/Square D Company*)

TABLE 7-2	
Overload protection.	

Motor Nameplate Rating	Overload Protection as a Percentage of Motor Nameplate Full-Load Current Rating
Service factor not less than 1.15	125%
Temperature rise not over (40°C)	125%
All other motors	115%

FIGURE 7-5 (A) is a typical manual resettable-type overload relay, part of a motor controller. Note the three positions where the thermal overload heater will be installed. (B) is a typical thermal overload "heater" of the type that would be installed in the relay. (*Courtesy of Schneider Electric North America/Square D Company*)

Table 3		
Motor Full-Load Current (Amp)		Thermal Unit Number
1 T.U.	3 T.U.	
0.29-0.31	0.28-0.29	B 0.44
0.32-0.36	0.30-0.33	B 0.51
0.37-0.39	0.34-0.36	B 0.57
0.40-0.47	0.37-0.44	B 0.63
0.48-0.56	0.45-0.52	B 0.71
0.57-0.63	0.53-0.59	B 0.81
0.64-0.69	0.60-0.64	B 0.92
0.70-0.77	0.65-0.71	B 1.03
0.78-0.86	0.72-0.80	B 1.16
0.87-0.97	0.81-0.90	B 1.30
0.98-1.12	0.91-1.03	B 1.45
1.13-1.24	1.04-1.14	B 1.67
1.25-1.39	1.15-1.27	B 1.88
1.40-1.57	1.28-1.44	B 2.10
1.58-1.78	1.45-1.63	B 2.40
1.79-1.96	1.64-1.79	B 2.65
1.97-2.20	1.80-2.01	B 3.00
2.21-2.41	2.02-2.19	B 3.30
2.42-2.75	2.20-2.52	B 3.70
2.76-3.25	2.53-2.95	B 4.15
3.26-3.50	2.96-3.17	B 4.85
3.51-3.87	3.18-3.50	B 5.50
3.88-4.13	3.51-3.73	B 6.25
4.14-4.69	3.74-4.22	B 6.90
4.70-5.20	4.23-4.68	B 7.70

Table 4		
Motor Full-Load Current (Amp)		Thermal Unit Number
1 T.U.	3 T.U.	
0.32-0.33	0.29-0.30	B 0.44
0.34-0.38	0.31-0.35	B 0.51
0.39-0.41	0.36-0.37	B 0.57
0.42-0.50	0.38-0.45	B 0.63
0.51-0.61	0.46-0.54	B 0.71
0.62-0.68	0.55-0.61	B 0.81
0.69-0.74	0.62-0.66	B 0.92
0.75-0.83	0.67-0.74	B 1.03
0.84-0.93	0.75-0.83	B 1.16
0.94-1.05	0.84-0.93	B 1.30
1.06-1.21	0.94-1.07	B 1.45
1.22-1.34	1.08-1.19	B 1.67
1.35-1.50	1.20-1.33	B 1.88
1.51-1.70	1.34-1.51	B 2.10
1.71-1.93	1.52-1.70	B 2.40
1.94-2.12	1.71-1.87	B 2.65
2.13-2.38	1.88-2.10	B 3.00
2.39-2.61	2.11-2.29	B 3.30
2.62-2.99	2.30-2.63	B 3.70
3.00-3.53	2.64-3.09	B 4.15
3.54-3.80	3.10-3.34	B 4.85
3.81-4.21	3.33-3.67	B 5.50
4.22-4.49	3.68-3.91	B 6.25
4.50-5.10	3.92-4.43	B 6.90
5.11-5.66	4.44-4.91	B 7.70

FIGURE 7-6 Sample tables for heaters for motor controllers. (*Delmar/Cengage Learning*)

motors, submersible pump motors, and other motors that do not have much locked-rotor capability.

The class indicates the trip time in seconds at a current rating of 600% of the overload relay's rating. For example, a Class 10 relay rated 5 amperes will trip in 10 seconds or less when seeing 30 amperes (5 × 6). A Class 20 relay rated 5 amperes will trip in 20 seconds or less when seeing 30 amperes. A Class 30 relay rated 5 amperes will trip in 30 seconds or less when seeing 30 amperes. The 6 times factor is based on typical locked-rotor (starting) current. Manufacturers of motor controls provide detailed time-current characteristic curves for overload relays. They also provide multipliers for derating the overload relay for installation in high- or low-temperature surroundings.

If you experience nuisance tripping of overload relay, and you're thinking about putting in a higher ampere rating overload relay—DON'T! If the motor hasn't started in 10 seconds, you want to have the motor overload devices take the motor off the line. Do not arbitrarily substitute a larger ampere rating overload relay. Instead, consider installing a higher class overload relay.

Motor controller manufacturers agree that when the motor and the motor controller are in the same location (temperature), use the recommended heater element according to their charts that matches the motor's full-load current to a specific heater element. See Figure 7-6 for sample heater element tables. When the controller is in a higher or lower ambient than the motor, manufacturers provide correction factor curves for selecting heater elements. Ambient compensated overload relays are also available. As always, check the motor and motor controller manufacturer's technical literature.

Insulation

Electric motors are available with different classes of insulation. Motors marked with the motor's insulation class are suitable for use in ambient temperatures shown in Table 7-3. Manufacturers of equipment select electric motors for their equipment based on their use.

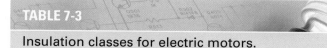

TABLE 7-3

Insulation classes for electric motors.

Insulation Class	Maximum Temperature for AC Motors with 1.00 Service Factor
A	105°C
B	130°C
C	155°C
D	180°C

Service Factor

Sizing motor overload protection depends on the service factor of the motor. The service factor is found on the motor nameplate.

Simply stated, a motor may be loaded continuously up to the horsepower obtained by multiplying the rated horsepower by the service factor. For example, a 10-horsepower motor marked with a service factor of 1.15 may be loaded to 11.5 horsepower. Consider service factor a margin of safety.

How Does Current Flow in the Windings of a Motor?

Single-phase motors require an overload device in one conductor, *Table 430.37*. The current that enters the motor on one conductor comes out of the motor on the other conductor—a simple series circuit. What goes in . . . comes out! When the overcurrent device opens, the flow of current to the motor is totally cut off.

Three-phase motors are a different story. We are talking about three phases, each 120° out of phase (symmetrical) with the other phases. Normal current in a 3-phase motor is shown in Figure 7-7. A motor

FIGURE 7-7   A view of the stator windings of a 3-phase motor that have not been damaged due to overloads, single-phasing, ground faults, or short-circuited windings. In this example, the motor has a normal full-load current draw of 10 amperes. Overload protection is sized at 125% of the motor's FLA. Time-delay fuses are sized at 175% of the motor's FLA for the motor's branch-circuit, short-circuit, and ground-fault protection. (*Courtesy of Electrical Apparatus Service Association, Inc.*)

subjected to a locked-rotor (stalled) condition is shown in Figure 7-8. An overloaded motor is shown in Figure 7-9.

Single Phasing

A 3-phase motor can be subjected to a situation referred to as "single phasing." Single phasing means one phase is open—somewhere. See Figure 7-10 and Figure 7-11. The current values shown in the figures are for a "pure" single phasing condition with one motor only in the circuit. Nothing can prevent single phasing. Single phasing can be caused by such things as a downed power line, the utility company's primary transformer fuse or cutout being open, an open connection, a bad termination, an open splice, a broken conductor, burned open contacts in a motor controller,

FIGURE 7-8 This photo clearly shows all stator windings in a motor burned out due to a locked-rotor (stalled) condition. Locked-rotor current depends on the code letter of the motor. The code letter is found on the nameplate of the motor. See *NEC Table 430.7(B)*. (*Courtesy of Electrical Apparatus Service Association, Inc.*)

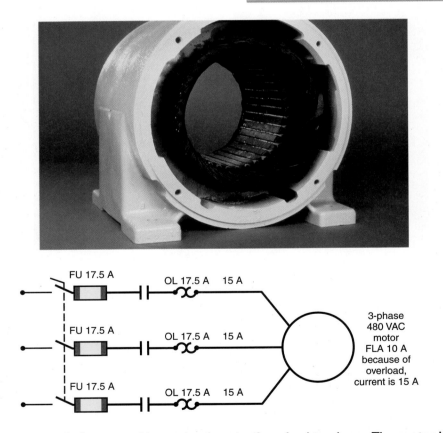

FIGURE 7-9 An overloaded motor with overload protection sized too large. The motor is rated 10 amperes full load. An overload causes the current in the windings to increase to 15 amperes. The overload devices are sized in violation of the *NEC* at 175% of the motor's full-load current rating instead of the required 125%. The time-delay fuses are sized at the correct 175% for the motor's branch-circuit, short-circuit, and ground-fault protection. Being sized too large, the overload devices cannot sense and respond to the 15-ampere current flow. The motor burns out. (*Courtesy of Electrical Apparatus Service Association, Inc.*)

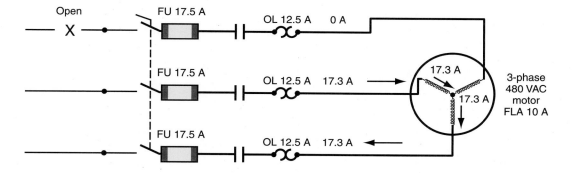

FIGURE 7-10   Example of secondary single phasing. The windings in the motor are wye-connected. When a secondary single phasing occurs, the current in one phase drops to zero amperes and the current in the other two phases increases to approximately 1.73 × 10 = 17.3 amperes. Two sets of phase windings are burned out because the motor did not have proper overload protection. Proper motor overload protection would have overload devices in all three phases sized at 125% of the motor's full-load current rating. When an "open" occurs somewhere in the branch circuit supplying the motor while the motor is running, the overload devices will sense the 17.3-ampere current flow and cause the relay to open the circuit, protecting the motor from burnout. (*Courtesy of Electrical Apparatus Service Association, Inc.*)

bad contacts in a switch or circuit breaker, an open winding in the supply transformer, or misalignment. Figure 7-12 shows a primary single-phase condition in a wye-delta connected transformation. Similar conditions occur in a delta-wye or delta-delta transformation.

Prior to 1971, the *NEC Table 430-37* permitted 3-phase motors to have overload protection in only two phases. At that time, when a primary and/or secondary single phasing occurred, there were numerous motor burnouts.

Why so many motor burnouts? Because the motor winding for the phase subjected to the damaging increase in current just might be the phase that did not have overload protection—a one-out-of-three chance of motor burnout! The 1971 *NEC* began to require overload protection in all three phases, *NEC Table 430-37*.

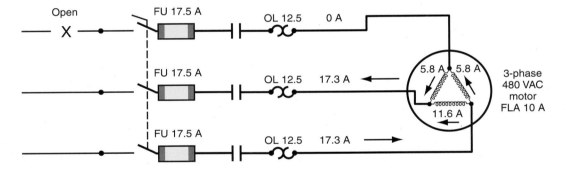

FIGURE 7-11   Example of secondary single phasing. The windings in the motor are delta connected. When a secondary single phasing occurs, the current in one phase conductor drops to zero amperes and the current in the other two phase conductors increases to approximately 1.73 × 10 = 17.3 amperes. Note that the current in one phase winding in the motor increases to 11.6 amperes and the current in the other two phase windings now in series have a current flow of 5.8 amperes. The phase winding with the 11.6 amperes burns out. Proper motor overload protection would have overload devices in all three phases sized at 125% of the motor's full-load current rating. When an "open" occurs somewhere in the branch circuit supplying the motor while the motor is running, properly sized overload devices will sense the 17.3-ampere current flow and cause the relay to open the circuit, taking the motor off line and protecting the motor from burnout. (*Courtesy of Electrical Apparatus Service Association, Inc.*)

Strange Things Happen

The current values in these diagrams are theoretical. In the real world, unexpected things can happen. A 3-phase motor will probably not start if one phase supplying the motor is open. However, if a large, lightly loaded 3-phase motor is running at the time of single phasing, and other smaller motors connected to the system are running, the large motor acts as a generator and will actually feed back into the system on the open phase. This does not create a true symmetrical 3-phase system (120° phase displacement), but it creates a distorted, asymmetrical 3-phase system. This generated phase might be called a "phantom phase." Smaller motors not fully loaded on the system might start or continue to run, leading to motor burnout unless the problem is corrected, Figure 7-13.

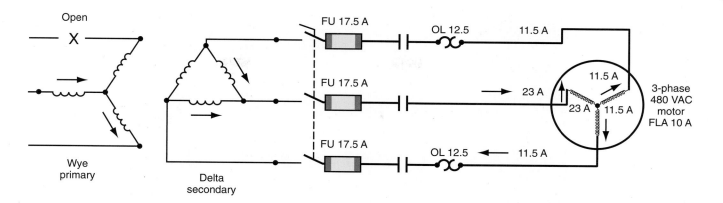

FIGURE 7-12   Example of primary single phasing on a wye-delta system. An "open" occurs in the primary while the motor is running. All three conductors are still supplying current to the motor. The motor has overload protection in only two phases as permitted prior to the 1972 *NEC*. One phase has no overload protection. The current in two phases increases to approximately 1.15 × 10 = 11.5 amperes. The current on one phase increases to approximately 2.3 × 10 = 23 amperes. The winding subjected to the 23 amperes burns out. Depending on which phase of the primary opened, there is a one-out-of-three chance of motor burnout. Had there been motor overload protection in all three phases, the overload devices would have sensed the increase in current and taken the motor off line. (*Delmar/Cengage Learning*)

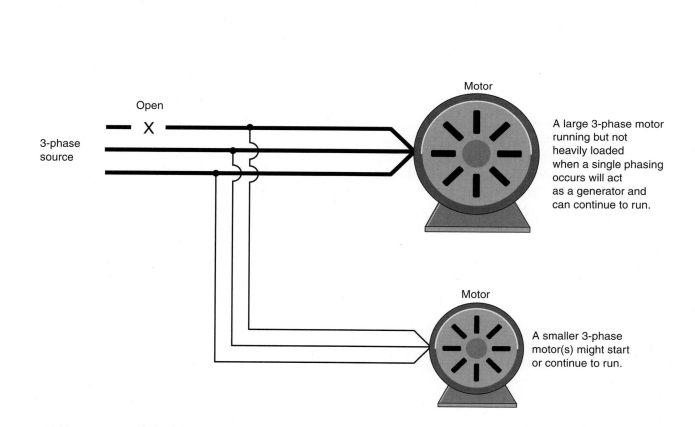

FIGURE 7-13   A 3-phase motor running at the time a single phasing occurs. If lightly loaded, this motor could continue to run. As it rotates, it will generate and feed back into the system a distorted third phase. Other motors on the line will "see" this distorted 3-phase supply and might start or continue to run. Motor failure can be expected—not a good situation. Properly sized motor overload protection is the key to keeping motor burnouts to a minimum. (*Delmar/Cengage Learning*)

FIGURE 7-14 Photo shows winding damage to a motor subjected to an unbalanced voltage condition. (*Courtesy of Electrical Apparatus Service Association, Inc.*)

Unbalanced Voltage

What really causes havoc to a motor is unbalanced voltage. Unbalanced voltage causes unbalanced currents in the motor's stator windings. This is the most common cause of motor failure. A slight amount of voltage unbalance causes a very high increase in temperature in the phase with the high current. Figure 7-14 shows motor winding damage due to an unbalanced voltage condition.

Motors might hum. NEMA recommends that for motors, the unbalanced voltage shall not exceed ±1%. Another way to recognize unbalanced voltage is that some lighting will dim while other lighting will brighten. Uninterruptible power supplies (UPS) systems might kick in and out.

Voltage Variation. Motors are affected when run at other than their nameplate rated voltage. NEMA rated motors are designed to operate at ±10% of their nameplate rated voltage. Good electrical design would be to keep the voltage levels at ±5% of the motor's nameplate rating. Table 7-4 shows changes that occur at 10% higher or 10% lower than nameplate rated voltage for a typical NEMA rated motor. The voltage readings between phases AB, BC, and CA are higher or lower than nameplate rated voltage rating, but the readings between phases are equal.

TABLE 7-4

Approximate changes in full-load current, starting current, and temperature rise for a typical general-purpose squirrel-cage induction motor when operated at overvoltage (110%) and undervoltage (90%) conditions.

Voltage Variation	Full-Load Current	Starting Current	Temperature Rise at Full-Load	Full-Load Speed
110% (10% over nameplate rated voltage)	7% decrease	10–12% increase	3–4°C (37–39°F) decrease	1% increase
90% (10% under nameplate rated voltage)	11% increase	10–12% decrease	6–7°C (43–45°F) increase	1½% decrease

Possible Causes of Unbalanced Voltage

- Large single-phase loads.
- Open connections at transformer.
- Open delta transformer connections.
- Unequal impedance of single-phase transformers in the same 3-phase bank.
- Capacitor banks unequal or open circuited.

Calculating Unbalanced Voltage

1. Take voltage readings. Let's say the voltage reading between AB is 218 volts, between BC is 200 volts, and between CA is 206 volts.

2. Add these voltages:

	218
	200
	206
Total	624

3. Find the average voltage:

$$\frac{624}{3} = 208 \text{ volts}$$

4. Subtract the average voltage from the voltage reading that results in the greatest difference. In this example:

$$218 - 208 = 10 \text{ volts}$$

5.

$$\frac{100 \times greatest\ voltage\ difference}{average\ voltage} = \frac{100 \times 10}{208}$$

$$= 4.8\% \text{ voltage unbalance}$$

6. Temperature rise in the winding with highest current:

$$2 \times (\text{percent voltage unbalance})^2$$

$$2 \times 4.8 \times 4.8 = 46\% \text{ temperature rise}$$

Fact: Single-phasing is an extreme example of voltage unbalance.

Fact: Ambient temperature means the temperature of the air where the motor or other equipment is installed.

Fact: Standard NEMA rated motors are designed to operate in a maximum ambient temperature of 40°C. This information is found on the nameplate of a motor. Motors with higher temperature limitations

TABLE 7-5	
Derating a motor when voltage unbalance exists.	
Percent Voltage Unbalance	**Derate Motor to**
1	98%
2	95%
3	88%
4	82%
5	75%

are available, depending on the class of insulation in the motor.

Fact: Temperature rise means the difference between motor winding temperature and the ambient temperature.

Fact: For every 10°C rise above the motor's rated temperature, the motor's insulation life is cut in half.

Fact: If it is determined that a voltage unbalance exists, but for some reason there is a need to run a motor, you can limp along by derating the motor according to the percentages shown in Table 7-5.

NEC Article 430, Part III addresses motor overload protection. This part discusses continuous and intermittent duty motors, motors that are larger than 1 horsepower, motors that are automatically started, and motors that are marked with a service factor that have integral (built-in) overload protection.

The exhaust fan in the Bakery draws 2.9 amperes, is less than 1 horsepower, and is nonautomatically started. According to *NEC 430.32(D)*, second sentence, *Any motor rated at 1 horsepower or less that is permanently installed shall be protected in accordance with NEC 430.32(B)*, which in part states that the overload protection shall be as shown in Table 7-2.

The same values indicated in Table 7-2 are also valid for continuous duty motors of more than 1 horsepower, *NEC 430.32(A)*.

Chapter 17, "Overcurrent Protection: Fuses and Circuit Breakers," covers in detail the subject of overcurrent protection using fuses and circuit breakers. Dual-element, time-delay fuses are an excellent choice for overload protection of motors. They can be sized close to the ampere rating of the motor.

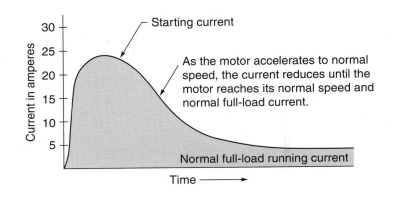

FIGURE 7-15 Typical motor starting current. (*Delmar/Cengage Learning*)

When the motor has integral overload protection or where the motor controller has thermal overloads, then dual-element, time-delay fuses are selected to provide backup overload protection to the thermal overloads.

The starting current of an electrical motor is very high for a short period of time as is illustrated in Figure 7-15. The current decreases to its rated value as the motor reaches full speed. Dual-element, time-delay fuses will carry five times their ampere rating for at least 10 seconds. This allows the motor to start and then provides good overload protection while it is running. For example, a 4-ampere, full-load rated motor could have a starting current as high as 24 amperes. The fuse must not open needlessly when it sees this inrush.

In this situation, a 15-ampere ordinary fuse or breaker might be required to allow the motor to start. This would provide the branch-circuit overcurrent protection, but it would not provide overload protection for the motor.

If the motor does not have built-in overload protection, sizing the branch-circuit overcurrent protection as stated would allow the motor to destroy itself if the motor were to be subjected to continuous overload or stalled conditions.

The solution is to install a dual-element, time-delay fuse that permits the motor to start, yet opens on an overload before the motor is damaged. To provide this overload protection for single-phase, 120- or 240-volt equipment, fuses and fuseholders as illustrated in Figure 7-16 are often used.

FIGURE 7-16 Type S plug fuse and adapter. (*Photo courtesy of Cooper Bussmann, Division of Cooper Industries*)

Type S fuses and the corresponding Type S adapter are inserted into the fuseholder, Figure 7-17. Once the adapter has been installed, it is extremely difficult to remove it. This makes it virtually

FIGURE 7-17 Fuseholders. (*Photo courtesy of Cooper Bussmann, Division of Cooper Industries*)

impossible to replace a properly sized Type S fuse with a fuse having an ampere rating too large for the specific load.

Table 7-6 compares the load responses of ordinary fuses and dual-element, time-delay fuses.

Table 7-7A shows how to size dual-element, time-delay fuses for motors that are marked with a service factor of not less than 1.15 and for motors marked with a temperature rise of not over 40°C.

These charts show one manufacturer's recommendations for the selection of dual-element, time-delay fuses for motor circuits. The Design B energy-efficient motors have a higher starting current than the older style motors. The nameplate should be checked to determine the design designation; then refer to *NEC Table 430.52* to determine proper branch-circuit, short-circuit, and ground-fault protection. Finally, check the protective device time-current

TABLE 7-6

Load response time for four types of fuses.

	(APPROXIMATE TIME IN SECONDS FOR FUSE TO BLOW)			
Load (amperes)	4-ampere Dual-element Fuse (time-delay)	4-ampere Ordinary Fuse (nontime-delay)	8-ampere Ordinary Fuse (nontime-delay)	15-ampere Ordinary Fuse (nontime-delay)
5	More than 300 sec	More than 300 sec	Won't blow	Won't blow
6	250 sec	5 sec	Won't blow	Won't blow
8	60 sec	1 sec	Won't blow	Won't blow
10	38 sec	Less than 1 sec	More than 300 sec	Won't blow
15	17 sec	Less than 1 sec	5 sec	Won't blow
20	9 sec	Less than 1 sec	Less than 1 sec	300 sec
25	5 sec	Less than 1 sec	Less than 1 sec	10 sec
30	2 sec	Less than 1 sec	Less than 1 sec	4 sec

TABLE 7-7(A)

Selection of dual-element fuses for motor overload protection.

A. MOTORS MARKED WITHOUT LESS THAN 1.5 SERVICE FACTOR OR TEMPERATURE RISE NOT OVER 40°C

Motor (40°C or 1.15 S.F.) Ampere Rating	Dual-Element Fuse Amp Rating Max. 125%	Motor (40°C or 1.15 S.F.) Ampere Rating	Dual-Element Fuse Amp Rating Max. 125%
1.00 to 1.11	1 ¼	20.0 to 23.9	25
1.12 to 1.27	1 ⁴⁄₁₀	24.0 to 27.9*	30
1.28 to 1.43	1 ⁶⁄₁₀	28.0 to 31.9	35
1.44 to 1.59	1 ⁸⁄₁₀	32.0 to 35.9	40
1.60 to 1.79	2	36.0 to 39.9	45
1.80 to 1.99	2 ¼	40.0 to 47.9	50
2.00 to 2.23	2 ½	48.0 to 55.9	60
2.24 to 2.55	2 ⁸⁄₁₀	56.0 to 63.9	70
2.56 to 2.79	3 ²⁄₁₀	64.0 to 71.9	80
2.80 to 3.19	3 ½	72.0 to 79.9	90
3.20 to 3.59	4	80.0 to 87.9*	100
3.60 to 3.99	4 ½	88.0 to 99.9	110
4.00 to 4.47	5	100 to 119	125
4.48 to 4.99	5 ⁶⁄₁₀	120 to 139	150
5.00 to 5.59	6 ¼	140 to 159	175
5.60 to 6.39	7	160 to 179*	200
6.40 to 7.19	8	180 to 199	225
7.20 to 7.99	9	200 to 239	250
8.00 to 9.59	10	240 to 279	300
9.60 to 11.9	12	280 to 319	350
12.0 to 13.9	15	320 to 359*	400
14.0 to 15.9	17 ½	360 to 399	450
16.0 to 19.9	20	400 to 480	500

***Note:** Disconnect switch must have an ampere rating at least 115% of motor ampere rating, *NEC 430.110(A).* Next larger size switch with fuse reducers may be required.

curve to determine whether it is capable of carrying the starting current until the motor reaches full speed.

Table 7-7B shows how to size dual-element, time-delay fuses for motors that have a service factor of less than 1.15 and for motors with a temperature rise of more than 40°C.

For abnormal installations, dual-element, time-delay fuses larger than those shown in Table 7-7A and Table 7-7B may be required. These larger-rated fuses (or circuit breakers) will provide short-circuit and ground-fault protection only for the motor branch circuit. *NEC 430.52* and *Table 430.52* show the maximum size fuses and breakers permitted. These percentages have already been discussed in this chapter.

Abnormal conditions might include

- situations where the motor is started, stopped, jogged, inched, plugged, or reversed frequently.
- high inertia loads such as large fans and centrifugal machines, extractors, separators, pulverizers, and so forth. Machines having large flywheels fall into this category.
- motors having a high *Code* letter and full-voltage start, as well as some older motors without *Code* letters.

TABLE 7-7(B)

All other motors.

LESS THAN 1.15 SERVICE FACTOR OR GREATER THAN 40°C RISE

All Other Motors—Ampere Rating	Dual-Element Fuse Amp Rating Max. 115%	All Other Motors—Ampere Rating	Dual-Element Fuse Amp Rating Max. 115%
1.00 to 1.08	1 1/8	17.4 to 20.0	20
1.09 to 1.21	1 1/4	21.8 to 25.0	25
1.22 to 1.39	1 4/10	26.1 to 30.0*	30
1.40 to 1.56	1 6/10	30.5 to 34.7	35
1.57 to 1.73	1 8/10	34.8 to 39.1	40
1.74 to 1.95	2	39.2 to 43.4	45
1.96 to 2.17	2 1/4	43.5 to 50.0	50
2.18 to 2.43	2 1/2	52.2 to 60.0	60
2.44 to 2.78	2 8/10	60.9 to 69.5	70
2.79 to 3.04	3 2/10	69.6 to 78.2	80
3.05 to 3.47	3 1/2	78.3 to 86.9	90
3.48 to 3.91	4	87.0 to 95.6*	100
3.92 to 4.34	4 1/2	95.7 to 108	110
4.35 to 4.86	5	109 to 125	125
4.87 to 5.43	5 6/10	131 to 150	150
5.44 to 6.08	6 1/4	153 to 173	175
6.09 to 6.95	7	174 to 195*	200
6.96 to 7.82	8	196 to 217	225
7.83 to 8.69	9	218 to 250	250
8.70 to 10.0	10	261 to 300	300
10.5 to 12.0	12	305 to 347	350
13.1 to 15.0	15	348 to 391*	400
15.3 to 17.3	17 1/2	392 to 434	450

*__Note:__ Disconnect switch must have an ampere rating at least 115% of motor ampere rating, _NEC 430.110(A)._ Next larger size switch with fuse reducers may be required.

Remember, the higher the _Code_ letter, the higher the starting current. For example, two motors have exactly the same full-load ampere rating. One motor is marked _Code_ letter J. The other is marked _Code_ letter D. The _Code_ letter J motor will draw more momentary starting inrush current than the _Code_ letter D motor. The _Code_ letter of a motor is found on the motor nameplate. _See NEC 430.7_ and _Table 430.7(B)._

Refer to the _NEC_ and to Bussmann's _Electrical Protection Handbook, Publication SPD,_ for more data relating to motor overload and motor branch-circuit protection, disconnecting means, controller size, conductor size, conduit size, and voltage drop.

Dual-element, time-delay fuses are designed to withstand motor-starting inrush currents and will not open needlessly. However, they will open if a sustained overload occurs. Dual-element, time-delay fuses of the Type S plug type, Figure 7-16, or of the cartridge type, as illustrated in Chapter 18 for motors, should be sized in the range of 115%, but not to exceed 125% of the motor full-load ampere rating.

For the exhaust fan motor that has a 2.9-ampere full-load current draw, select a dual-element, time-delay fuse in the range of

$$2.9 \times 1.15 = 3.34 \text{ amperes}$$
$$2.9 \times 1.25 = 3.62 \text{ amperes}$$

Referring to Table 7-7(A) and Table 7-7(B), find dual-element, time-delay fuses rated at 3½ amperes. If the motor has built-in or other overload protection, the 3½-ampere size will provide backup overload protection for the motor.

MOTOR BRANCH-CIRCUIT, SHORT-CIRCUIT, AND GROUND-FAULT PROTECTION

The basics of the typical motor installations will be presented, but for the not-so-common installations, refer to *NEC Article 430*.

- Motor branch-circuit conductors, controllers, and the motor must be provided with branch-circuit short-circuit and ground-fault overcurrent protection, *NEC 430.51*. Long-time overload protection is provided by the motor overload devices. This can be thought of as a "system approach."

- The branch-circuit short-circuit and ground-fault overcurrent protection must have sufficient time delay to permit the motor to be started, *NEC 430.52*.

- *NEC 430.52* and *Table 430.52* provide the maximum percentages of motor full-load current permitted for fuses and circuit breakers, for different types of motors.

- When applying the percentages listed in *NEC Table 430.52*, if the resulting ampere rating or setting does not correspond to the standard ampere ratings found in *NEC 240.6(A)*, we are permitted to round up to the next higher standard rating. See *NEC 430.52(C)(1), Exception 1*. See the column referred to as Step 1 in Table 7-8.

- If the size selected by rounding up to the next higher standard size will not allow the motor to start, such as might be the case on Design B (energy-efficient motors) motor circuits, then we are permitted to size the motor branch-circuit fuse or breaker even larger, *NEC 430.52(C)(1), Exception 2*. See the column referred to as Step 2 in Table 7-8.

- Always check the manufacturer's overload relay table in a motor controller to see whether the manufacturer has indicated a maximum size or type of overcurrent device, for example, "Maximum Size Fuse 25-ampere." Do not exceed this ampere rating, even though the percentage values listed in *NEC Table 430.52*

TABLE 7-8

Sizing of motor branch-circuit, short-circuit, and ground-fault protection.

	Step 1 Rating	Step 2 Rating
Nontime-delay fuses not over 600 amperes	300%	400%
Nontime-delay fuses over 600 amperes	300%	300%
Time-delay dual-element fuses not over 600 amperes	175%	225%
Fuses over 600 amperes	300%	300%
Inverse-time circuit breakers Motor FLA 100 amperes or less	250%	400%
Inverse-time circuit breakers Motor FLA over 100 amperes	250%	300%
Instant-trip circuit breakers for other than Design B energy-efficient motors	800%*	1300%**
Instant-trip circuit breakers for Design B energy-efficient motors	1100%*	1700%**

*Only permitted when part of a listed combination motor controller.

**Only permitted where necessary after doing an engineering evaluation to prove that the 800% and 1100% sizing will not perform satisfactorily.

result in a higher ampere rating. If the equipment is marked with a maximum fuse size only rather than overcurrent device, do not use a circuit breaker. To do so would be a violation of *NEC 110.3(B)*.

EXAMPLE

Determine the rating of dual-element, time-delay fuses for a Design B, 7½-horsepower, 208-volt, 3-phase motor, service factor 1.15. First refer to *NEC Table 430.250* to obtain the motor's full-load current rating, and then refer to *NEC Table 430.52*. The first column shows the type of motor, and the remaining columns show the maximum rating or setting for the branch-circuit, short-circuit, and ground-fault protection using different types of fuses and circuit breakers. Also refer to *NEC 430.52*.

$$24.2 \text{ amperes} \times 1.75 = 42.35 \text{ amperes}$$

NEC 240.6(A) lists 45-ampere as the next higher standard size.

If it is determined that 45-ampere, dual-element, time-delay fuses will not hold the starting current of the motor, we are permitted to increase the ampere rating of the fuses, but not to exceed 225% of the motor's full-load current draw.

$$24.2 \text{ amperes} \times 2.25 = 54.45 \text{ amperes}$$

The next lower standard OCPD size is rated 50-ampere.

For an easy-to-use computer version for doing motor circuit calculations, visit the Bussmann Web site at http://www.bussmann.com. Point to "Application Info" in the left column, and in the drop-down window click on "Software." In the following drop-down window click on "Motor Circuit Selection Guide."

MOTOR-STARTING CURRENTS/*CODE* LETTERS

Starting current/locked-rotor current is usually many times greater than normal full-load current.

Starting current is the instantaneous current draw when a motor is started. Locked-rotor current is the steady-state current draw with the motor at standstill. At the instant of start, locked-rotor current and starting current are the same. As a motor comes up to speed, the starting current draw drops rapidly until the motor stabilizes at its normal operating current. If the motor were locked in a standstill position, unable to start, the locked-rotor current would subside slightly as the motor's stator windings heat up. The higher resistance of the heated windings lowers the current a little.

Starting current of a motor is based on the type of motor and the *Code* letter found on the motor's nameplate. The higher the *Code* letter, the higher the starting current. This information is important when selecting branch-circuit, short-circuit, and ground-fault protection such as fuses and circuit breakers. Referring to the time-current characteristic curves for fuses and circuit breakers will give you an idea if they can handle the starting current without nuisance opening. Time-current characteristics curves are discussed in Chapter 17.

NEC Table 430.7(B) lists *Locked Rotor Indicating Code Letters* for electric motors. Code Letters are clearly marked on the motor nameplate. *NEC Table 430.7(B)* is based on kVA per horsepower. It is easy to calculate the range of the motor's starting current (locked rotor current) in amperes.

EXAMPLE

Calculate the starting current for a 240-volt, 3-phase, *Code* letter J, 7½-horsepower motor. The minimum starting current would be

$$I = \frac{kVA \times 1000}{E \times 1.73} = \frac{7.1 \times 7.5 \times 1000}{240 \times 1.73} = 128 \text{ amperes}$$

The maximum starting current would be

$$I = \frac{kVA \times 1000}{E \times 1.73} = \frac{7.99 \times 7.5 \times 1000}{240 \times 1.73} = 144 \text{ amperes}$$

Motor Design Designations

NEC 430.7, 430.52 and *Tables 430.52 and 430.251(B)* reference *design* types of motors. The *design* types of alternating-current, polyphase induction electric motors are defined in the National

TABLE 7-9			
NEMA design application.			
NEMA Design	**Starting Current**	**Locked Rotor Torque**	**Application**
A	High to Medium	Normal	A variation of Design B, having a higher starting current.
B	Low	High	For normal starting torque for fans, blowers, rotary pumps, unloaded compressors, some conveyors, metal cutting machine tools, misc. machinery. Very common for general-purpose across-the-line starting. Slight change of speed as load varies.
C	Low	High	For high inertia starts such as large centrifugal blowers, fly wheels, and crusher drums. Loaded starts, such as piston pumps, compressors, and conveyors, where rapid acceleration is needed. Slight change of speed as load varies.
D	Low	Very high	For very high inertia and loaded starts such as punch presses, shears and forming machine tools, cranes, hoists, elevators, oil well pumping jacks. Considerable change of speed as load varies.

Note: Torque is the turning force of the motor. Torque is generally classified by locked rotor (starting) torque . . . pull up torque . . . break down torque . . . full-load operating (rated) torque.

Electrical Manufacturers Association (NEMA) *Information Guide for General Purpose Industrial AC Small and Medium Squirrel-Cage Induction Motor Standard MG 1-2007.* To assign a design type to a motor, the speed, the torque, and the slip of the motor are taken into consideration. Most motors have about 5% slip and synchronous motors have none. "Slip" as used here refers to the fact the rotor is rotating at fewer revolutions than the rotating magnetic field. Do not confuse *Locked Rotor Indicating Code Letters* with NEMA *design letters.* Design letters are clearly marked on the nameplate of the motor. Different design types of motors have different applications as shown in Table 7-9.

TYPE 1 AND TYPE 2 COORDINATION

Because these terms might be found on the nameplate of a motor controller, familiarity with their meanings is necessary.

These terms are used by manufacturers of motor controllers based on the International Electrotechnical Commission (IEC) Standard 947–4-1. This is a European standard. Electrical equipment manufactured overseas to European standards is being shipped into the United States, and electrical equipment manufactured in this country to U.S. standards (NEMA and UL

Standard 508) is being shipped overseas. Conformance to one standard does not necessarily mean that the equipment conforms to the other standard.

In the United States, we prefer to use the term *protection* instead of *coordination* because coordination has a different meaning as discussed in Chapter 18, "Short-Circuit Calculations and Coordination of Overcurrent Protective Devices."

Type 1 short-circuit protection means that under short-circuit conditions, the contactor and starter shall not cause danger for persons working on or near the equipment. However, a certain amount of damage is acceptable to the controller, such as welding of the contacts or burning out the overload relays. Replacing the contacts of a controller does not constitute a hazard to personnel. Type 1 protection and UL 508A, *Industrial Control Panels* are similar in this respect. The steps necessary to get Type 1 protected controllers back on line after a short circuit or ground fault occurs are as follows:

1. Disconnect the power.

2. Locate and repair the fault.

3. Replace the contacts in the controller if necesary.

4. Replace the branch-circuit fuses or reset the circuit breaker.

5. Restore power.

Type 2 short-circuit protection means that under short-circuit conditions, the contactor and starter shall not cause danger for persons working on or near the equipment and, in addition, the contactor or starter shall be suitable for further use. No damage to the controller is acceptable. The steps necessary to get Type 2 protected controllers back on line after a short circuit or ground fault occurs are as follows:

1. Disconnect the power.
2. Locate and repair the fault.
3. Replace the branch-circuit fuses.
4. Restore power.

Type 2 protection requires extremely current-limiting overcurrent devices, such as Class J and Class CC fuses. Standard circuit breakers or older style fuses are typically not fast enough under short-circuit conditions to protect the motor starter from damage.

Although Type 2 protected controllers are marked by the manufacturer with the maximum size and type of fuse, the maximum size might be larger than that permitted for motor branch-circuit protection when applying the rules of *NEC 430.52* and *Table 430.52* for a particular motor installation. For example, the controller might be marked "Maximum Ampere Rating Class J Fuse: 8 amperes." Yet, the proper size Class J time-delay fuse for a particular motor circuit, applying the maximum values according to *NEC 430.52* and *Table 430.52,* might be 4½ amperes. In this example, the proper maximum size fuse is 4½ amperes.

EQUIPMENT INSTALLATION

There are a number of topics that must be addressed.

Industrial Control Panels

NEC Article 409 deals with industrial control panels operating at 600 volts or less. Industrial control panels are used in both industrial and commercial applications and might include a motor controller(s), overload relays, fused disconnect switches, circuit breakers, control devices, push-button stations, selector switches, timers, control relays, terminal blocks, pilot lights, and similar components. The equipment *and* the field wiring of industrial control panels must conform to *NEC Article 409* regarding conductor sizing, grounding, bonding, disconnecting means, rating or setting of the overcurrent protective device, construction, phase arrangement, wire space, wire bending, and marking. Many of the requirements in *Article 409* are mirror images of the requirements in *Article 430*.

For example, a motor controller shall be marked with the manufacturer's name, voltage, current or horsepower, and the short-circuit current rating, *NEC 430.8*.

An industrial control panel is required by *NEC 409.110* to be marked with:

- The manufacturer's name, trademark, or other descriptive marking identifying the manufacturer.

- The supply voltage, number of phases, frequency, and full-load current.

- When an industrial control panel is supplied by more than one power source such that more than one disconnecting means is required to disconnect all power within the control panel, it shall be marked to indicate that more than one disconnecting means is required to de-energize the equipment.

- Short-circuit current rating of the entire assembly. Unless otherwise marked, the short-circuit rating is the short-circuit rating of lowest rated device in the panel. The short-circuit current is the sum of the available fault current at the line-side terminals of the control panel plus motor contribution. Chapter 18 in this text covers various methods for calculating short-circuit current values.

- If intended for use as service equipment, it shall be so marked.

- A complete wiring diagram.

- The enclosure type.

When a "listed" industrial control panel is installed, it would conform to UL 508A. These are assembled in a production facility and bear an overall industrial control panel or industrial equipment listing mark. The electrician provides the field wiring for feeders and/or branch circuits to and from the control panel but does not do any other internal wiring. Any field wiring the electrician installs to or from the listed industrial control panel shall meet the requirements of *NEC Chapter 3*.

NEC 409.3 is helpful in that it lists other *Articles* in the *NEC* that also apply when connecting specific types of electrical equipment to an industrial control panel.

Related standards of interest are:

- UL 508, *Standard for Industrial Control Equipment*
- UL 508A, *Standard for Industrial Control Panels*
- NFPA 79, *Electrical Standard for Industrial Machinery*

Controller Enclosure Types

Appendix C of this text shows motor controller enclosure types.

Equipment and Lighting on a Branch Circuit

The exhaust fan in the Bakery is an example of an equipment load being connected on a branch circuit with lighting. Circuit 1 serves 1305 VA of lighting and the exhaust fan. This fan is a permanently connected motor-operated appliance. Permanently connected is often called hard wired, differentiating it from cord- and plug-connected.

NEC 210.23(A) permits, on 15- and 20-ampere branch circuits, the supply of lighting and equipment provided the hard-wired equipment does not exceed 50% of the circuit rating.

This particular exhaust fan has a $\frac{1}{10}$ horsepower, 120-volt motor with a full-load rating of 2.9 amperes. The load would be

$$1305 \text{ VA} + (120 \text{ V} \times 2.9 \text{ A} \times 1.25) \times\; = 1740 \text{ VA}$$

$$\frac{1740 \text{ VA}}{120 \text{ VA}} = 14.5 \text{ A}$$

In selecting the proper size conductor, remember what has been previously discussed. *NEC Table 310.15(B)(16)* requires the use of the 60°C

column. Accordingly, a 12 AWG Type THHN/THWN conductor has an allowable ampacity of 25 amperes, even though the allowable ampacity at the conductors 90°C rating is 30 amperes.

Unless otherwise specifically provided, the maximum overcurrent protection for a 12 AWG copper conductor is 20 amperes, *NEC 240.4(D)(4)*.

Considering Circuit 1 to be a "continuous load," the 80% maximum loading factor on the branch-circuit overcurrent device and branch-circuit conductors is

$$20 \times 0.80 = 16 \text{ amperes}$$

Therefore, 12 AWG Type THHN/THWN conductors protected by a 20-ampere circuit breaker meets the requirements of the *NEC*.

The fan motor is under the 50% limitation of *NEC 210.23(A)*.

Combination Load on Individual Branch Circuit

When a cord- and plug-connected appliance is supplied by an individual branch circuit, the load is limited to 80% of the circuit rating by *NEC 210.23(A)(1)*. An example of this is Circuit 7 serving the dough machine:

Motor load 792 volt-amperes
Heater load 2000 volt-amperes
Total load 2792 volt-amperes

Conversion to amperes

$$\frac{2792 \text{ VA}}{208 \times 1.732} = 7.8 \text{ A}$$

A 20-ampere-rated circuit, the minimum allowed, is assigned to supply this load.

Supplying a Specified Load

The bake oven installed in the Bakery is an electrically heated commercial-type bake oven (see Figure 7-22). The oven has a marked nameplate indicating a load of 16,000 volt-amperes, at 3-phase, 208-volt. This load includes all electrical heating elements, drive motors, timers, transformers, controls, operating coils, lights, and all other electrically powered apparatus. The nameplate also indicates that all necessary protective devices are installed.

The instructions furnished with the bake oven, and on the nameplate, specify that the supply conductors have a minimum ampacity of 56 amperes and that the terminations are rated for 75°C.

If the panelboard were to have terminations rated at 60°C, the conductor selection would be made in the 60°C column of *NEC Table 310.15(B)(16)*. A minimum size conductor 4 AWG having the lowest allowable ampacity higher than 56 is required. If the panelboard terminations are rated at 75°C, then the selection is made from the 75°C column and a 6 AWG having an allowable ampacity of 65 amperes would qualify. See *NEC 110.14(C)(1)*. A disconnecting means, without overcurrent protection, is installed at the appliance to allow maintenance and adjustments to be made to the internal electrical system. This disconnect is not required by the *NEC* if the branch-circuit panelboard is in sight of the appliance and the circuit breaker can be locked. The lock-off provision must be permanently installed on or at the switch or circuit breaker, *NEC 422.31(B)*.

Conductors Supplying Several Motors

▶*NEC 430.24* requires that the conductors supplying several motors, or a motor(s) and other load(s), shall have an ampacity not less than the sum of each of the following:

(1) 125 percent of the full-load current rating of the highest rated motor, as determined by 430.6(A)

(2) Sum of the full-load current ratings of all the other motors in the group, as determined by 430.6(A)

(3) 100 percent of the noncontinuous non-motor load

(4) 125 percent of the continuous non-motor load.*◀

For example, consider the Bakery's multimixers and dough divider, which are supplied by a single branch circuit and have full-load ratings of 7.48, 2.2, and 3.96 amperes as determined by referring to the equipment's nameplate and installation instructions. Therefore:

$$1.25 \times 7.48 = 9.35 \text{ amperes}$$

plus	2.20 amperes
plus	3.96 amperes
Total	15.51 amperes

*Reprinted with permission from NFPA 70-2011.

The branch-circuit conductors must have an ampacity of 15.51 amperes minimum. Specifications for this Commercial Building call for 12 AWG minimum. Checking *NEC Table 310.15(B)(16)*, a 12 AWG Type THHN/THWN copper conductor has an ampacity of 25 amperes, more than adequate to serve the three appliances. The ampacity of 25 amperes is from the 75°C column. When adjusting ampacities for more than three conductors in one raceway, or when correcting ampacities for high temperatures, we would begin the derating by using the 90°C column ampacity to determine the 90°C Type THHN/THWN conductor's ampacity in a dry location.

Several Motors and Other Loads on One Branch Circuit

This type of installation is not very common. This situation may occur on equipment that has multiple motors and other integral loads. In just about all encounters with this kind of equipment, it will be marked with the required conductor and overcurrent protection sizing.

This is covered in *NEC 430.53*. This kind of group installation does not have individual motor branch-circuit, short-circuit, and ground-fault protection. The individual motors do have overload protection. The individual motor controllers, circuit breakers of the inverse-time type, and overload devices must be listed for group installation. The label on the controller indicates the maximum rating of fuse or circuit breaker suitable for use with the particular overload devices in the controller(s).

The branch-circuit fuses or circuit breakers are sized using these steps:

1. Determine the fuse or circuit-breaker ampere rating or setting per *NEC 430.52* for the highest rated motor in the group.

2. To this, add the full-load current ratings for all other motors in the group.

3. To this, add the current ratings for the "other loads."

EXAMPLE ──────────────

To determine the maximum overcurrent protection using dual-element, time-delay fuses for a branch circuit serving three motors plus an additional load, the FLA rating of the motors are 27, 14, and 11 amperes. There is 20 amperes of "other load" connected to the branch circuit. First determine the maximum time-delay fuse for the largest motor:

$$27 \times 1.75 = 47.15 \text{ amperes}$$

The next higher standard ampere rating is 50 amperes; next add the remaining ampere ratings to this value:

$$50 + 14 + 11 + 20 = 95 \text{ amperes}$$

For these four loads the maximum ampere rating of dual-element, time-delay fuses is 95 amperes, or rounded up to the next higher standard rating of 100 amperes. The disconnect switch would be a 100-ampere rating.

It can be readily seen that the rating of the branch-circuit overcurrent device is a variable, depending on the type of overcurrent device used. For example, if time-delay fuses sized at 225% are chosen because of the electrician's familiarity with high inrush current during start-up:

$$27 \times 2.25 = 60.75 \text{ amperes (round down to 60)}$$
$$\text{So, } 60 + 14 + 11 + 20 = 105 \text{ amperes}$$

Again, using common sense with full knowledge of the variables in sizing the branch-circuit over-current protection, you would install 100-ampere time-delay fuses in a 100-ampere horsepower-rated disconnect switch.

Several Motors on One Feeder

NEC 430.24 and *430.62* set forth the requirements for this situation. The individual motors will have branch-circuit, short-circuit, and ground-fault protection sized according to *NEC 430.52*. The individual motors do have overload protection.

The feeder fuses or circuit breakers are sized using these steps:

1. Determine the fuse or circuit-breaker rating or setting from *NEC Table 430.52* for the highest rated motor in the group.

2. To this, add the full-load current ratings for all other motors in the group.

EXAMPLE

To determine the maximum overcurrent protection using dual-element, time-delay fuses for a feeder serving three motors, the FLA ratings of the motors are 27, 14, and 11 amperes.

First determine the maximum time-delay fuse for the largest motor:

$$27 \times 1.75 = 47.15 \text{ amperes}$$

The next higher standard ampere rating over-current device is 50 amperes.

Next add the remaining ampere ratings to this value:

$$50 + 14 + 11 = 75 \text{ amperes}$$

For these three motors, the maximum ampere rating of dual-element, time-delay fuses is 75 amperes.

The rule in *NEC 430.62(A)* says the process we have just completed provides the maximum size overcurrent device. Since we rounded up to the next larger overcurrent device rating as permitted in *NEC 430.52(C)(1) Exception No. 1*, the 75-ampere calculation above becomes the maximum rating permitted. In other words, we are not permitted to round-up twice. You could also install 70-ampere, dual-element, time-delay fuses if you are confident that they would have sufficient time delay to allow the motors to start. In either case, the size of the disconnect switch would be 100 amperes.

Sometimes there is confusion when one or more of the motors are protected with instantaneous-trip breakers, sized at 800% to 1700% of the motor's full-load ampere rating. This would result in a very large (and possibly unsafe) feeder overcurrent device. *NEC 430.62(A)*, *Exception No. 1* states that in such cases, make the feeder calculation as though the motors protected by the instant-trip breakers were protected by a circuit breaker or dual-element, time-delay fuses in accordance with *NEC 430.52*.

For an easy-to-use computer version for doing a group motor circuit calculation, visit the Bussmann Web site at http://www.bussmann.com. Go to "Electrical UL/CSA," then to "Technical & Application Information," then to "Software." Here you will find links to all of the Cooper

Bussmann software programs, including "Group Motor Protection Guide."

APPLIANCE DISCONNECTING MEANS

Except for the water heater, each of the three appliances in the Bakery is furnished with a 4-wire cord-and-plug (three phases plus equipment ground). This meets the requirements for the disconnecting of cord-and plug-connected appliances, *NEC 422.33*.

The disconnecting-means requirement for permanently connected appliances is covered in *NEC 422.31. NEC 422.32* requires that the disconnect means for motor-driven appliances be located within sight of the appliance or be capable of being locked in the off position. The lock-off provision must be permanently installed on or at the switch or circuit breaker, *NEC 422.31(B)*.

The disconnecting means shall have an ampere rating not less than 115% of the motor full-load current rating, *NEC 430.110*.

The disconnecting means for more than one motor shall not be less than 115% of the sum of the full-load current ratings of all of the motors supplied by the disconnecting means, *NEC 430.110(C)(2)*.

GROUNDING

WHY?   *NEC 250.4(A)(1)* and *(2)* explains *why* equipment must be grounded.

WHERE?   *NEC 250.110* explains locations *where* equipment fastened in place (fixed) or connected by permanent wiring methods must be grounded.

WHAT?   *NEC 250.112* explains *what* equipment that is fastened in place (fixed) or connected by permanent wiring methods must be grounded. *NEC 250.114* explains *what* equipment that is cord- and plug-connected must be grounded.

HOW?   *NEC 250.134* explains *how* to ground equipment that is fastened in place and is connected by permanent wiring methods. *NEC 250.138* explains *how* to ground equipment that is cord- and plug-connected.

SIZE?   *NEC Table 250.122* lists the size of an equipment grounding conductor based on the rating of the OCPD protecting the circuit.

OVERCURRENT PROTECTION

Overcurrent protection for appliances is covered in *NEC 422.11*. If the appliance is motor-driven, then *NEC Article 430, Part III* applies. In most cases, the overload protection is built into the appliance. The appliance meets Underwriters Laboratories standards if it bears a listing mark (label).

A complete discussion for individual motors and appliances can be found in the "Exhaust Fan" section earlier in this chapter.

THE BAKERY EQUIPMENT

A bakery can have many types and sizes of food-preparing equipment, such as blenders, choppers, cutters, disposers, mixers, dough dividers, grinders, molding and patting machines, peelers, slicers, wrappers, dishwashers, rack conveyors, water heaters, blower dryers, refrigerators, freezers, hot-food cabinets, and others, Figure 7-18 and Figure 7-19.

FIGURE 7-18   Cake mixer. (*Delmar/Cengage Learning*)

FIGURE 7-19 Dough divider.
(*Delmar/Cengage Learning*)

TABLE 7-10				
Appliance load data.				
Type of Appliance	**Voltage**	**Amperes**	**Phase**	**HP**
Multimixer	208	3.96	3	¾
Multimixer	208	7.48	3	1½
Dough divider	208	2.28	3	½

All appliances are connected with NEMA 15–20P on 4-wire cord.

outlets are supplied by one 3-phase, 208-volt branch circuit, Table 7-10.

The data for these food-preparation appliances has been taken from the nameplate and installation instruction manuals of the appliances.

A NEMA 15–20R receptacle outlet, Figure 7-20, is provided at each appliance location. These are part of the electrical contract.

Each of these appliances is purchased as a complete unit. After the electrician provides the proper receptacle outlets, the appliances are ready for use as soon as they are moved into place and plugged in.

All of this equipment is designed and manufactured by companies that specialize in food-preparation equipment. These manufacturers furnish specifications that clearly state such things as voltage, current, wattage, phase (single- or 3-phase), minimum required branch-circuit rating, minimum supply circuit-conductor ampacity, and maximum overcurrent protective device rating. If the appliance is furnished with a cord-and-plug arrangement, it will specify the NEMA size and type, plus any other technical data that is required in order to connect the appliance in a safe manner as required by the *Code*.

NEC Article 422, Part V sets forth the requirements for the marking of appliances.

The Mixers and Dough Dividers (Three Appliances Connected to One Circuit)

When more than one appliance is to be supplied by one circuit, a load calculation must be made. Note on the Bakery electrical plans that three receptacle

FIGURE 7-20 NEMA 15–20R receptacle and plate.
(*Delmar/Cengage Learning*)

The Doughnut Machine

An individual branch circuit provides power to the doughnut machine. This machine consists of (1) a 2000-watt heating element that heats the liquid used in frying and (2) a driving motor that has a full-load rating of 2.2 amperes.

As with most food-preparation equipment, the appliance is purchased as a complete, prewired unit. This particular appliance is equipped with a 4-wire cord to be plugged into a receptacle outlet of the proper configuration.

Figure 7-21 is the control-circuit diagram for the doughnut machine. The following components are listed in Figure 7-21:

S A manual switch used to start and stop the machine.

T1 A thermostat with its sensing element in the frying tank; this thermostat keeps the oil at the correct temperature.

T2 Another thermostat with its sensing element in the drying tank; this thermostat controls the driving motor.

A A 3-pole contactor controlling the heating element.

B A 3-pole motor controller operating the drive motor.

M A 3-phase motor.

OL Overload units that provide overload protection for the motor; note one thermal overload unit in each phase.

P Pilot light to indicate when power to the heating elements is on.

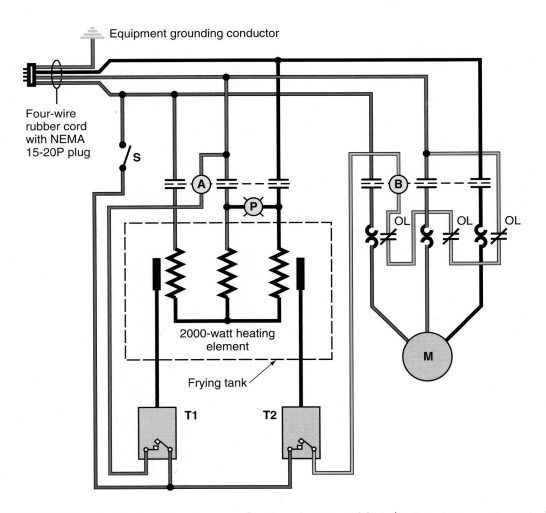

FIGURE 7-21   Control-wiring diagram for doughnut machine. (*Delmar/Cengage Learning*)

Because this appliance is supplied by an individual circuit, its current is limited to 80% of the branch-circuit rating according to *NEC 422.10(A)*. The branch circuit supplying the doughnut machine must have sufficient ampacity to meet the minimum load requirements as indicated on the appliance nameplate, or in accordance with *NEC 430.24*.

Doughnut Machine Load Data	
Heater load	2000 VA
Motor load	792 VA
25% motor load	198 VA
Total	2990 VA

The maximum continuous load permitted on a 20-ampere, 3-phase branch circuit is

16 amperes × 208 volts × 1.73 = 5760 VA

The load of 2990 volt-amperes is well within the 5760 volt-amperes permitted loading of the 20-ampere branch circuit.

The Bake Oven

As previously discussed, the bake oven installed in the Bakery is an electrically heated commercial-type bake oven, Figure 7-22.

According to the panel schedule, the bake oven is fed with a 60-ampere, 3-phase, 3-wire feeder consisting of three 6 AWG Type THHN/THWN copper

FIGURE 7-22　Bake oven. (*Delmar/Cengage Learning*)

conductors. The metal raceway is considered acceptable for the equipment grounding conductor.

A 3-pole, 60-ampere, 250-volt disconnect switch is mounted on the wall near the oven. From this disconnect switch, an appropriately sized raceway, often flexible metal conduit or liquidtight flexible metal conduit, if or Type MC cable is run to the control panel on the oven, Figure 7-23.

The bake oven is installed as a complete unit. All overcurrent protection is an integral part of the circuitry of the oven. The internal control circuit of the oven is 120 volts, which is supplied by an integral control transformer.

All of the previous discussion in this chapter relating to circuit ampacity, conductor sizing, grounding, etc., also applies to the bake oven, dishwasher, and food-waste disposer. Additional information relating to appliances is found in Chapter 3 and Chapter 21.

For the bake oven:

1. Ampere rating = 44.5 amperes

2. Minimum conductor ampacity = 44.5 × 1.25 = 55.6 amperes

3. Branch-circuit overcurrent protection = 44.5 × 1.25 = 55.6 amperes

It is permitted to install 55- or 60-ampere fuses in a 60-ampere disconnect switch.

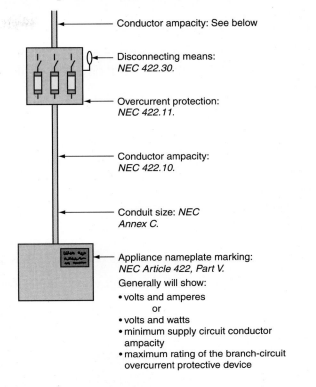

FIGURE 7-23   Motor branch circuit for appliance. (*Delmar/Cengage Learning*)

REVIEW

Refer to the *National Electrical Code* or the working drawings when necessary. Where applicable, responses should be written in complete sentences.

Cite the appropriate *NEC Article, Section,* or *Table* for each of the following:

1. Where does the *NEC* refer specifically to air-conditioning equipment?

   _____

2. Where does the *NEC* refer specifically to appliances? _____

3. Where does the *NEC* refer specifically to disconnecting means for a motor?

   _____

4. Where does the *NEC* refer specifically to grounding exposed noncurrent-carrying metal parts of motor circuits?

   _____

The following questions refer to cord- and plug-connected appliances.

5. In your own words, define an *appliance* and give examples.

_____

_____

6. A 20-ampere branch circuit shall not serve a single appliance with a rating greater than

_____

_____

7. An appliance is fastened in place and other cord- and plug-connected appliances are connected to a 20-ampere branch circuit. What is the maximum allowable load for the appliance that is fastened in place? _____

_____

_____

The following questions refer to motors and motor circuits.

8. Where shall the disconnect means for a 208-volt, 3-phase motor controller be located?

_____

_____

9. Where shall the disconnect means for a motor be located? _____

_____

10. When selecting conductors, from where is the motor full-load ampere rating to be taken? _____

_____

11. When selecting motor overload protective devices, from where is the full-load ampere rating taken? _____

_____

_____

12. A motor with a service factor of 1.25 shall be protected by an overload device set to trip at not more than _____ of the motor full-load current rating.

13. The full-load current rating of a 5-horsepower, 208-volt, 3-phase motor is 16.7 amperes. The minimum ampacity of the branch-circuit conductors supplying this motor would be not less than _____.

Questions 14, 15, and 16 are for a 5-horsepower, 3-phase motor that has a nameplate current rating of 16 amperes. Dual-element, time-delay fuses are to be used for the motor's branch-circuit short-circuit and ground-fault protection. Show your calculations.

14. The preferred rating of dual-element, time delay fuses (125% of motor full-load amperes) is _____ amperes.

15. If dual-element, time delay fuses with the preferred rating will not permit the starting of the motor, the dual-element, time delay fuse rating may be increased to a maximum (175% of motor full-load amperes) of _____ amperes.

16. If the motor is to be started and stopped repeatedly, possibly resulting in nuisance opening of the circuit, the dual-element, time delay fuse rating may be increased to an absolute maximum (225% of motor full-load amperes) of _____ amperes.

The following questions apply to alternating-current motors:

17. Define *in sight* as it applies to a motor and its disconnect switch.

_____

_____

_____

18. A motor has a full-load current rating of 74.8 amperes. What is the minimum allowable ampacity of the branch-circuit conductors?

_____

_____

19. A motor has a full-load current rating of 74.8 amperes and the terminations are rated for 75°C. What is the minimum conductor size and type?

_____

_____

20. Three 3-phase motors are supplied by a single set of conductors. The motors are rated 208 volts and 5, 10, and 15 horsepower. The minimum allowable ampacity of the conductors is _____ amperes.

21. A ½-horsepower, 115-volt motor (service factor 1.15) is to be installed, and a Type S fuseholder and switch is provided to control and protect the motor. The proper size fuse will have a rating of _____ amperes.

22. For a 25-horsepower, 480-volt, *Code* letter G motor, the minimum starting current will be _____ amperes, and the maximum starting current will be _____ amperes.

23. *Type 1* and *Type 2 Protection* are terms that might be found on the nameplate of a motor controller. Explain in your own words what these terms mean.

_____

_____

_____

_____

24. NEMA rated general-purpose motors can operate on voltages ±10% of their rated nameplate voltage. An unbalanced voltage situation is extremely damaging to a motor. Calculate the expected heat rise in a motor subjected to an unbalanced voltage situation where the voltages between phases AB = 465 volts, BC = 487 volts, and CA = 473 volts.

Add these voltages: _____

_____

_____

Total _____

Find the average voltage: Total ÷ 3 = _____ volts

Subtract the average voltage from the voltage reading that results in the greatest difference: _____ − _____ = _____ volts

$$\frac{100 \times \textit{greatest voltage difference}}{\textit{average voltage}} = \underline{\hspace{2cm}}\% \text{ voltage unbalance}$$

Temperature rise in the winding with highest current = 2 × (percent voltage unbalance)^2

_____ = _____%

temperature rise in the winding with highest current

Feeders

OBJECTIVES

After studying this chapter, you should be able to

- calculate the feeder loading.
- determine the minimum feeder overcurrent protective device rating.
- determine the minimum feeder conductor size.
- determine the appropriate correction and adjustment factors.
- calculate voltage drops.
- reduce the neutral conductor size as appropriate.
- determine the minimum raceway size.

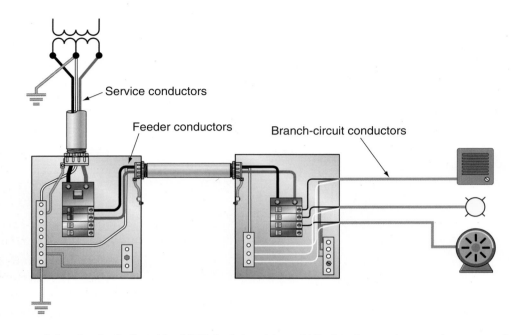

FIGURE 8-1 A feeder is defined in *NEC Article 100* as *"All circuit conductors between the service equipment, the source of a separately derived system, or other power supply source and the final branch-circuit overcurrent device."** (*Delmar/Cengage Learning*)

Feeders are the part of the electrical system that connects the branch-circuit panelboards to the electrical service equipment. See Figure 8-1. Conductors from on-site generators and transformers are also feeders, not service conductors. In the Commercial Building, a feeder is installed to each of the five occupancies, and one to the panelboard for the owner's circuits.

The feeder layout is shown in a riser diagram on working drawing E4. Specific requirements for feeders are given in *NEC Article 215* concerning installation requirements and in *NEC Article 220, Part II*, concerning calculated loads and demand factors.

Some of the information presented in this unit has been introduced previously. This redundancy is intentional because of its application to both feeders and branch circuits.

FEEDER REQUIREMENTS

Feeder Ampacity

The ampacity of a feeder must comply with the following requirements:

- A feeder must have an ampacity no smaller than the sum of the volt-amperes of the coincident

loads of the branch circuits supplied by the feeder as reduced by demand factors; see *NEC Article 220, Part II*.

- Demand factors are allowed in selected cases if it is unlikely that all the loads would be energized at the same time; see *NEC 220.60*.

- Only the larger of any noncoincident loads (loads unlikely to be operated simultaneously such as the air-conditioning and the heating) need be included; see *NEC 220.60*.

- Referring to *NEC 220.61*, a feeder neutral conductor may be reduced in size in certain situations. It may not be reduced in feeders consisting of 2-phase wires and a neutral conductor of a 3-phase, 4-wire, wye-connected system.

The neutral conductor of a 3-phase, 4-wire, wye-connected system may be reduced in size provided it remains of a size sufficient to

- carry the maximum unbalanced load.
- carry the nonlinear load.

Overcurrent Protection

The details of selecting overcurrent protective devices are covered in Chapters 17, 18, and 19.

*Reprinted with permission from NFPA 70-2011.

As previously stated, the rating of a branch-circuit is based on the rating of the circuit overcurrent protective device, *NEC 210.3*. The rating of the overcurrent device for a feeder shall not be less than the noncontinuous loading plus 125% of the continuous loading, *NEC 215.3*.

The basic requirement for overcurrent protection is that a conductor shall be protected in accordance with its ampacity, but several conditions exist where the rating of the overcurrent protective device (OCPD) can exceed the ampacity of the conductor being protected.

For example, if the ampacity of a feeder does not match a standard OCPD rating, then the next higher standard rating may be used, *NEC 240.4(B)*, provided the rating is 800-ampere or less. Feeder conductors with an ampacity of 130 amperes may be protected with a 150-ampere protective device. However, this "round-up" rule, as it is often referenced, does not allow the load to be greater than 130 amperes. The conductor must always be adequate for the load. See Table 8-1 for examples of the application of this rule.

Temperature Limitations

When selecting the components for a circuit, remember that no circuit is better than its terminations. *NEC 110.14(C)* ensures that the consideration of terminations is included in the selection of the circuit components.

Terminations, like other components in a circuit, have temperature ratings. The terminations on breakers, switches, or panelboards with a rating of 100-ampere or less may be rated for 60°C or 75°C. See Figure 8-2 for a graphic representation of these requirements. No terminals in equipment assemblies such as service equipment, panelboards, or motor controllers in listed 600-volt class of equipment are rated greater than 75°C. *NEC 110.14(C)(1)(a)* stipulates that for equipment rated 100 amperes or smaller, or marked for 14 AWG through 1 AWG conductors, the circuit ampacity shall not exceed the value given in the 60°C column of *NEC Table 310.15(B)(16)*. The section goes on to allow equipment marked with a 75°C temperature rating (for example, 60°C/75°C or 75°C), to be used at the 75°C allowable ampacity from *Table 310.15(B)(16)*. The reason for this is that the heat generated by the current (amperes) in a conductor is dependent on the conductor size. Conductors with a 90°C insulation system are permitted to be used for application of adjustment or correction factors as well as for the ampacity allowed in the 60°C or 75°C column of *NEC Table 310.15(B)(16)*.

The default temperature rating for equipment terminals for circuits rated over 100 amperes, or marked for conductors larger than number 1 AWG, is 75°C; see Figure 8-3. Conductors with higher temperature ratings, such as 90°C, are permitted to be used for these circuits so long as the allowable ampacity of the conductor is appropriate in the 75°C column of *NEC Table 310.15(B)(16)*.

The consequence of *NEC 110.14(C)(1)* is that it establishes a conductor size that is compatible with the temperature rating of the terminations. For example, the minimum conductor size for a noncontinuous load of 60 amperes would be a 4 AWG conductor unless the terminations are marked 75°C, in which case a 6 AWG would satisfy the requirements. The resistance of the 4 AWG is less than the 6 AWG; thus the heat generated would be less and, subsequently, the termination rating may be less.

TABLE 8-1

Application of round-up rule.

Load, Calculated or Connected	Minimum Conductor Ampacity	Typical Conductor Size	Conductor Ampacity	Maximum OCPD[1,2]
132 A	132 A	1/0 AWG	150	150 A
276 A	276 A	300 kcmil	285	300 A
360 A	360 A	500 kcmil	380	400 A
720 A	720 A	(2) 500 kcmil	760	800 A
920 A	920 A	(3) 400 kcmil	1005	1000 A

[1]It is permitted to round up to the next standard overcurrent device rating through 800 A.

[2]These values are not required to apply to motor circuits.

Conductor Selection

The process of selecting a conductor begins with identifying the wire material. An examination of

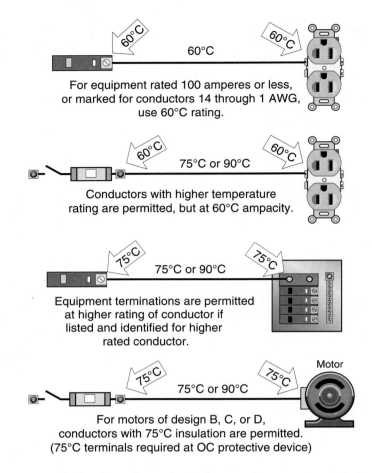

FIGURE 8-2 Temperature limitation of terminals for conductors sized 14 through 1 AWG or for overcurrent protective devices rated 100 amperes or lower. (*Delmar/Cengage Learning*)

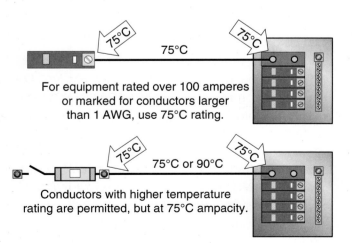

FIGURE 8-3 Temperature limitation of terminals for conductors sized larger than 1 AWG or for overcurrent protective devices rated greater that 100 amperes. (*Delmar/Cengage Learning*)

NEC Table 310.15(B)(16) reveals that only two classifications of wire are listed:

1. Copper

2. Aluminum or copper-clad aluminum

The table indicates that for the same allowable size and type, a copper wire will have the higher allowable ampacity. *NEC 110.5* stipulates that the wire will be assumed to be copper unless stated otherwise.

Aluminum wire is seldom used in branch circuits, but because of weight and cost factors, it is frequently selected for feeders and services. When installing aluminum conductors, the electrician must follow the directions and be especially careful to tighten all terminals to the recommended torque. It is also important to remember that listed connectors are required whenever it is necessary to join copper and aluminum conductors.

Some common problems associated with aluminum conductors when not properly connected may be summarized as follows:

- A corrosive action is set up when copper and aluminum wires come in contact with one another if moisture is present.

- The surface of an aluminum conductor oxidizes as soon as it is exposed to air. If this oxidized surface is not broken through, a poor connection results. When installing aluminum conductors, particularly in larger sizes, the electrician brushes an inhibitor onto the aluminum conductor and then scrapes the conductor with a stiff brush where the connection is to be made. The process of scraping the conductor breaks through the oxidation, and the inhibitor keeps the air from coming in contact with the conductor. Thus, further oxidation is prevented. Aluminum compression-type connectors usually have an inhibitor paste installed inside the connector.

- Aluminum wire expands and contracts to a greater degree than does copper for an equal load. This characteristic is another possible cause of a poor connection. Be certain to use connectors listed for making aluminum connections. Carefully follow manufacturer's installation instructions, including those for torque values. Crimp connections for aluminum conductors are usually longer than those for comparable copper conductors, resulting in greater contact surface of the conductor in the connector.

FEEDER COMPONENT SELECTION

Previously in this chapter, the information necessary for selecting feeder and branch-circuit components has been discussed in detail. Here, that information is first summarized and then presented in sequential steps. While studying these steps, it may be useful to review the detailed discussions and study the examples presented following the selection procedure.

Before the selection process can begin, it is necessary that the following circuit parameters be known:

- The continuous and noncontinuous loadings, in amperes, calculated in accordance with *NEC Article 220*.

- The type of conductors that are to be installed; see *NEC Table 310.104(A)*.

- The ambient temperature of the environment where the conductors are to be installed and the number of current-carrying conductors that will be in the cable or raceway; see *NEC 310.15(A)(3)*.

Step 1: OCPD Selection

- The OCPD must be rated not smaller than the sum of the noncontinuous load plus 125% of the continuous load; see *NEC 215.3*.

Step 2: Minimum Ungrounded Conductor Size Determination

- If the selected OCPD has a rating of 100 amperes or less, or if any of the terminations are marked 1 AWG or smaller, or are rated at 60°C, the minimum size conductor is determined by entering the 60°C column of *NEC Table 310.15(B)(16)* and selecting the conductor size that has an allowable ampacity not less than the sum of the noncontinuous load plus 125% of the continuous load.

- If the selected OCPD has a rating greater than 100-ampere, or if all the terminations are rated for 75°C, the minimum size conductor is determined by entering the 75°C column of *NEC Table 310.15(B)(16)* and selecting the conductor size that has an allowable ampacity equal to or greater than the sum of the noncontinuous load plus 125% of the continuous load.

- As stated above, the conductor must always be adequate for the load, and through 800 amperes the overcurrent device may be the next standard rating above the ampacity of the conductor; see *NEC Sections 215.2(A)(1), 215.3,* and *240.6(A)*.

Step 3: Conductor Type Determination

- If the selected OCPD has a rating of 100 amperes or less, the conductors to be installed may be selected from the 60°C, 75°C, or 90°C column of *NEC Table 310.15(B)(16)*. The standard insulation for the conductors used in this building is THHN/THWN copper.

- If the selected OCPD has a rating greater than 100-amperes, the conductors to be installed shall be selected from either the 75°C or 90°C column of *NEC Table 310.15(B)(16)*.

Step 4: Conductor Size Determination

To make the feeder conductor size determination: From *NEC Table 310.15(B)(16)* for the type of conductor selected in Step 3, choose a conductor that has an ampacity

- equal to or greater than the minimum calculated load (noncontinuous load plus 125% of the continuous load).

- equal to or greater than the connected load if it is greater than the calculated load (noncontinuous load plus 125% of the continuous load).

- that permits the use of the OCPD selected in Step 1.

Step 5: Neutral Conductor Size Determination

NEC 220.61 sets forth conditions for determining the minimum size of the neutral conductor. It provides the *Basic Calculation* in *(A)*, *Permitted Reductions* in *(B)*, and *Prohibited Reductions* in *(C)*.

Applying applicable and permitted factors to reduce the size of the neutral conductor can result in a reduction in wire and raceway cost.

To determine a reasonable minimum size, two types of loads must be considered—linear and nonlinear. Nonlinear loads are defined in *NEC Article 100*. In the *Informational Note* following the definition, the following items are listed: *Electronic equipment, electronic/electric discharge lighting, and adjustable speed drives.*

Linear Loads. In the past, most connected loads were linear, such as resistive heating and lighting, and motors. In a linear circuit, current changes in proportion to a voltage increase or decrease in that circuit. The voltage and current sine waves are sinusoidal.

Linear loads for 3-phase systems may be calculated by the following formula:

$$N = \sqrt{A^2 + B^2 + C^2 - AB - BC - AC}$$

where

N = Neutral conductor current
A = Phase A current = 30 amperes
B = Phase B current = 40 amperes
C = Phase C current = 50 amperes

$$N = \sqrt{30^2 + 40^2 + 50^2 - (30 \times 40) - (40 \times 50) - (30 \times 50)}$$

$$N = \sqrt{900 + 1600 + 2500 - 1200 - 2000 - 1500}$$

$$N = \sqrt{300} = 17.32 \text{ amperes}$$

Assuming phase loads of 30, 40, and 50 amperes, the neutral conductor load would be 17.32 amperes.

Additional calculation would indicate that if the 30-ampere load were disconnected, the neutral load would be 43.6 amperes; and if all but the 50-ampere load were disconnected, the neutral load would be 50 amperes. For the neutral conductor to comply with *NEC 220.61*, it must be sized for this maximum unbalanced condition.

Nonlinear Loads. Beginning in 1947 and continuing in each succeeding edition, the *NEC* has addressed the various problems of neutral conductor currents from nonlinear/harmonic sources.

The following quotes are from the 2011 *National Electrical Code*:

> *NEC Article 100, Nonlinear load A load where the wave shape of the steady-state current does not follow the wave shape of the applied voltage.*
>
> *(Informational Note): Electronic equipment, electronic/electric-discharge lighting, adjustable*

speed drive systems, and similar equipment may benonlinear loads.*

NEC 210.4(A), Informational Note, states that *A 3-phase, 4-wire, wye-connected power system used to supply power to non-linear loads may necessitate that the power system design allow for the possibility of high harmonic neutral-conductor currents.**

NEC 220.61(C) states that *There shall be no reduction of the neutral or grounded conductor capacity applied to the amount in 220.61(C)(1), or portion of the amount in 220.61(C)(2), from that determined by the basic calculation:*

(1) Any portion of a 3-wire circuit consisting of 2-phase wires and the neutral conductor of a 4-wire, 3-phase, wye-connected system.

*(2) That portion consisting of nonlinear loads supplied from a 4-wire, wye-connected, 3-phase system.**

*NEC 220.61, Informational Note 2: A 3-phase, 4-wire, wye-connected power system used to supply power to nonlinear loads may necessitate that the power system design allow for the possibility of high harmonic neutral conductor currents.**

NEC 310.15(B)(5)(c) states that *On a 4-wire, 3-phase wye circuit where the major portion of the load consists of nonlinear loads, harmonic currents are present in the neutral conductor; the neutral conductor shall therefore be considered a current-carrying conductor.**

NEC 400.5(B), fourth paragraph states that *On a 4-wire, 3-phase wye circuit where more than 50% of the load consists of nonlinear loads, there are harmonic currents present in the neutral conductor and the neutral conductor shall be considered to be a current-carrying conductor.**

NEC 450.3, Informational Note 2, states that *Nonlinear loads can increase heat in a transformer without operating its overcurrent protective device.**

NEC 450.9, Informational Note 2, states that *Additional losses may occur in some transformers where nonsinusoidal currents are present, resulting in increased heat in the transformer above its rating. See ANSI/IEEE C57.110-1993, Recommended Practice for Establishing Transformer Capability When Supplying Nonsinusoidal Load Currents, where transformers are utilized with nonlinear loads.**

Also refer to *NEC Annex D* in *Example D3(a)* for cautions relating to nonlinear loads.

Harmonics. The arrival of electronic equipment such as UPS systems, ac-to-dc converters (rectifiers), inverters (ac-to-dc to adjustable frequency ac), computer power supplies, programmable controllers, data processing, electronic ballasts, and similar equipment brought about problems in electrical systems on 3-phase systems. Unexplained events started to happen such as overheated neutrals, overheating and failure of transformers, overheated motors, hot busbars in switchboards, unexplained tripping of circuit breakers, incandescent lights blinking, fluorescent lamps flickering, malfunctioning computers, and hot lugs in switches and panelboards, even though the connected loads were found to be well within the conductor and equipment rating. All of the preceding electronic equipment is considered to be nonlinear, when the current in a given circuit does not increase or decrease in proportion to the voltage in that circuit. The resulting distorted voltage and current sine waves are nonsinusoidal. See Figure 8-4 for an illustration of nonsinusoidal sine waves. These problems can feed back into an electrical system and affect circuits elsewhere in the building, not just where the nonlinear loads are connected.

The root of the problem can be traced to electronic devices such as thyristors (silicon-controlled rectifiers [SCRs]) that can be switched on and off for durations that are extremely small fractions of a cycle. Another major culprit is switching-mode power supplies that switch at frequencies of 20,000 to more than 100,000 cycles per second. This rapid switching causes distortion of the sine wave. Harmonic frequencies are superimposed on top of the fundamental 60 Hz frequency, creating harmonic distortion.

Another culprit is IGBTs (insulated gate bipolar transistors) used in most variable frequency drives.

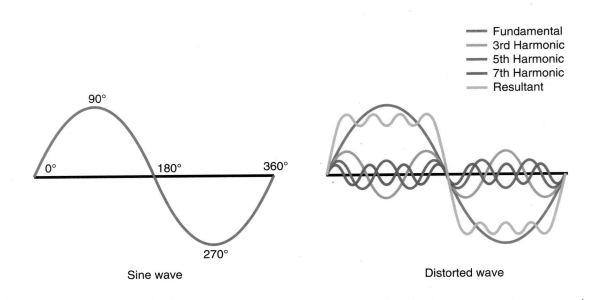

FIGURE 8-4 Distorted sine wave resulting from nonlinear loads. (*Delmar/Cengage Learning*)

Harmonics are multiples of a fundamental frequency. In the United States, 60 Hz (60 cycles per second) is standard or the fundamental. Other frequencies superimposed on the fundamental frequency can be measured, such as these:

Odd Harmonics		Even Harmonics	
3rd	180 cycles	2nd	120 cycles
5th	300 cycles	4th	240 cycles
7th	420 cycles	6th	360 cycles
9th	540 cycles	8th	480 cycles
etc.		etc.	

Currents of the 3rd harmonic and odd multiples of the 3rd harmonic (9th, 15th, 21st, etc.) add together in the common neutral conductor of a 3-phase system, instead of canceling each other. These odd multiples are referred to as triplens. For example, on an equally balanced 3-phase, 4-wire, wye-connected system in which each phase carries 100 amperes, the neutral conductor might be called on to carry 200 amperes or more. If this neutral conductor had been sized to carry 100 amperes, it would become severely overheated. This is illustrated in Figure 8-5.

Unusually high neutral current will cause overheating of the neutral conductor. There might also be excessive voltage drop between the neutral and the ground. Taking a voltage reading at a receptacle between the neutral conductor and ground with the

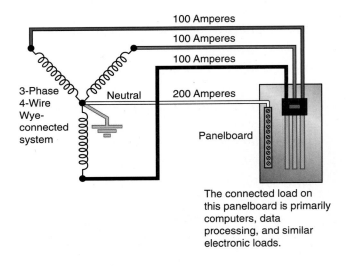

The connected load on this panelboard is primarily computers, data processing, and similar electronic loads.

FIGURE 8-5 Feeder neutral current resulting from nonlinear loads. (*Delmar/Cengage Learning*)

loads turned on and finding more than two volts present probably indicates that the neutral conductor is overloaded because of the connected nonlinear load.

Attempting to use a low-cost, average-reading, clamp-on meter will not give a true reading of the total current in the neutral conductor because the neutral current is made up of current values from many frequencies. The readings will be from 30% to 50% lower. To get accurate current readings where harmonics are involved, ammeters referred to as "true rms" must be used.

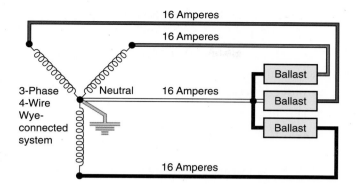
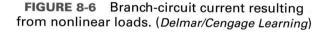

FIGURE 8-6 Branch-circuit current resulting from nonlinear loads. (*Delmar/Cengage Learning*)

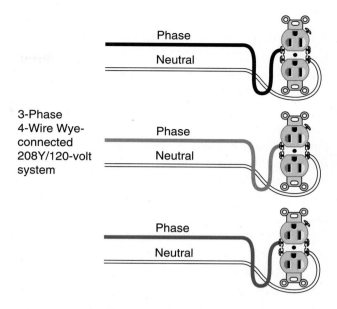

FIGURE 8-7 Recommended: separate neutrals serving nonlinear loads. (*Delmar/Cengage Learning*)

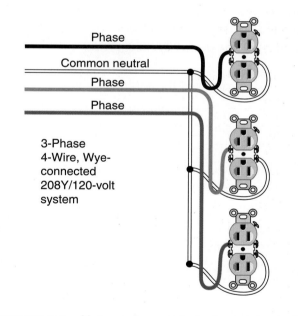

FIGURE 8-8 Not recommended: shared neutral serving nonlinear loads. (*Delmar/Cengage Learning*)

At least one major manufacturer of Type MC cable manufactures a cable that has an oversized neutral conductor. For cables in which the phase conductors are 12 AWG, the neutral conductor is an 8 AWG. For cables in which the phase conductors are 10 AWG, the neutral conductor is a 6 AWG. They also provide a Type MC cable that has a separate neutral conductor for each of the phase conductors. There is still much to be learned about the effects of nonlinear loads. Many experienced electrical consulting engineers are now specifying that 3-phase, 4-wire branch circuits supplying nonlinear loads have a separate neutral for each phase conductor instead of using a typical 3-phase 4-wire branch circuit using a common neutral; see Figure 8-6. They also specify that the neutral be sized double that of the phase conductor. The Computer and Business Equipment Manufacturers Association (CBEMA) offers the following recommendation:

Run a separate neutral to 120-volt outlet receptacles on each phase. Avoid using shared neutral conductors for single-phase 120-volt outlets on different phases. Where a shared neutral conductor for a 208Y/120 volt system must be used for multiple phases use a neutral conductor having at least 200% of the phase conductors.

What does this mean to the electrician? It simply means that if overheated neutral conductors, overheated transformers, overheated lugs in switches and panels, or other unexplained heating problems are encountered, they might be caused by nonlinear loads. Where a commercial or industrial building contains a high number of nonlinear loads, it is advisable to have the electrical system analyzed and redesigned by an electrical engineer specifically trained and qualified on the subject of nonlinear loads.

The effects of nonlinear loads, such as electric discharge lighting, electronic/data-processing equipment, and/or variable speed motors, are illustrated in Figure 8-5, Figure 8-6, Figure 8-7, and Figure 8-8.

In each case, the power is supplied from a 3-phase, 4-wire 208Y/120-volt system. Figure 8-5 illustrates the possible effect the load could have on a feeder. The neutral could be carrying 200% of the phase conductors. Figure 8-6 illustrates a branch circuit where the load is fluorescent lighting using core and coil ballast. If the ballast were electronic, the neutral current would be even greater. Figure 8-7 illustrates the recommended arrangement for serving three nonlinear loads from a 4-wire wye system. If the neutral is shared, as is illustrated in Figure 8-8, then the neutral should have at least 200% of the carrying capacity of the individual phase conductors. Using separate neutrals is the preferred arrangement.

It is clear from the information given that nonlinear loads can create neutral currents in excess of the phase currents in a 3-phase, 4-wire wye system. Considering what is known and unknown, it is prudent to take a conservative approach considering the minimal savings involved.

Three stipulations are clear:

1. The neutral must be able to carry the maximum unbalanced load.

2. On a 3-phase, 4-wire, wye system, no reduction is allowed for nonlinear loads.

3. In *NEC 220.61, Informational Note 2,* a warning is issued concerning the currents created by nonlinear loads.

Three conditions of failure may occur with a 3-phase, 4-wire wye system:

1. If all three phases are operable: In this, the normal situation, by definition the neutral is part of the feeder and must be sized for the same load and environmental requirements. The minimum size shall comply with *NEC 110.14(C),* and the adjustment and corrections must be applied as with the phase conductors.

2. If two phases are operable: In essence, this creates a situation similar to a feeder consisting of two phases of a 3-phase system, a condition where the *NEC* specifically prohibits a reduction in neutral size.

3. If only a single phase is operable (two phase conductors have opened for some reason): The circuit reverts to a single-phase circuit and the load characteristics are common to both ungrounded (hot) and grounded (neutral) conductor.

In this commercial building, the following strategy will be followed to determine the minimum neutral feeder conductor size:

- The load will be calculated according to *NEC 220.61* which results in the maximum unbalanced current on the feeder. Note that line-to-line connected loads do not add current to the neutral.

- The minimum size neutral conductor will be determined from *NEC 215.2(A)(2).* This results in a neutral conductor for the feeder that is sized not smaller than an equipment grounding conductor as required in *NEC 250.122.* The larger of the two conductors determined in the previous two activities will be specified.

Summary of Nonlinear Loads

Because of the complexity of calculating nonlinear loads, most consulting engineers will simply do load calculations per the *NEC*—double the size of the neutral.

You will also find that for installations where the nonlinear load is huge, most consulting engineers will specify K-rated transformers. A K-rated transformer can handle the heat generated by harmonic currents. A K-rated transformer is manufactured with heavier gauge wire for the windings and has double-size neutral terminals to accommodate the larger size neutral installed by the electrician.

Voltage Drop

NEC 215.2(A)(3), Informational Note 2, sets forth information on limiting the voltage drop on feeders. *NEC 210.19(A)(1), Informational Note 4,* provides similar information for branch circuits.

It is important to understand that an Informational Note is not a requirement. *NEC 90.5(C)* states that Informational Notes *are informational only and are not enforceable as requirements of this Code.*

These recommendations mean that there can be a voltage drop of 2% in the feeder and 3% in the branch circuit, or 3% in the feeder and 2% in the branch circuit. The bottom line is that the total voltage drop should not exceed 5%. The voltage at the service point to a building is governed by public utility regulations.

The utility records in its public service commission documents (utility regulatory commission), a standard nominal service voltage acceptable to the public service commission. The regulation requires that the voltage shall remain reasonably constant.

If you experience voltage drop problems, such as lights burning dim, lights "dipping" or flickering, or other unusual occurrences, check the voltage at the line-side terminals of the main service, using all the precautions of working on live equipment as required by *NFPA 70E*. If the voltage reading is low, the condition is probably the utility's problem.

If the voltage at the line-side terminals of the main service is satisfactory, but unacceptable voltage levels are found at equipment, the problem is somewhere in the premises wiring. The causes could be that the feeder and/or branch-circuit conductors are too small for the connected load, the feeder and/or the branch-circuit conductors are too long, there are bad (faulty) connections, and so on.

As an example, the branch circuit to a receptacle outlet is to be loaded to 1500 volt-amperes or 12.5 amperes, as illustrated in Figure 8-9. The distance from the panelboard to the receptacle outlet is 85 ft and the minimum wire size is 12 AWG. Three different methods will be demonstrated to estimate the voltage drop in this circuit.

The first method requires that the resistance per foot of the conductor be known. According to *NEC Chapter 9, Table 8*, the resistance of a 12 AWG uncoated copper conductor is 1.93 ohms per 1000 ft (300 m). Therefore, the resistance of the circuit is

$$R = \text{Unit resistance} \times \text{distance}$$

$$R = \left(\frac{1.93 \text{ ohm}}{1000 \text{ ft}}\right) \times (2 \times 85 \text{ ft}) = 0.3281 \text{ ohm}$$

$$R = \left(\frac{1.93 \text{ ohm}}{300 \text{ m}}\right) \times (2 \times 25.5 \text{ m}) = 0.3281 \text{ ohm}$$

The factor of two in the equation is required because both the phase and neutral conductors carry the current. The voltage drop is calculated using Ohm's Law:

$$VD = I \times R$$
$$= 12.5 \text{ amperes} \times 0.3281 \text{ ohms}$$
$$= 4.1 \text{ volts}$$

This is a voltage drop of

$$\frac{4.1}{120} = 0.0342 \text{ or } 3.42\%$$

This would limit the drop allowable for the feeder to 1.58% ($5 - 3.42 = 1.58$).

The second method requires that the circular mil area of the conductor be known. From *NEC Chapter 9, Table 8*, the kcmil area of a 12 AWG is 6530. The equation is

$$VD = \frac{K \times I \times L \times 2}{\text{kcmil}}$$

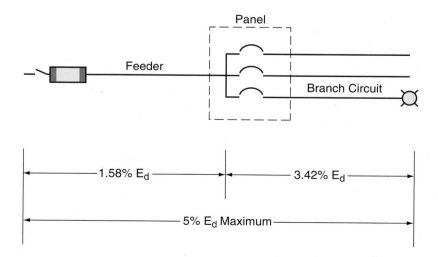

FIGURE 8-9   Voltage drop calculations. (*Delmar/Cengage Learning*)

Where K (K in the formula represents a constant for the resistance of the conductor material) for copper is 39.4 when the length is in meters and 12 when the length is in feet. For aluminum, K is 68 and 20. Therefore,

$$VD = \frac{12 \times 12.5 \text{ A} \times 85 \text{ ft} \times 2}{6530 \text{ kcmil}} = 3.9 \text{ volts}$$

$$VD = \frac{39.4 \times 12.5 \text{ A} \times 26 \text{ m} \times 2}{6530 \text{ kcmil}} = 3.9 \text{ volts}$$

Another method is one that considers the power factor, the type of raceway, and whether it is a single- or 3-phase circuit. Abbreviated Table 8-2(A) and Table 8-2(B) show values appropriate for use with the circuits in the Commercial Building. The values given are for installation in steel conduit. A complete table is provided in Appendix Table A-2 and Table A-3.

To calculate the voltage drop, multiply the current, the distance, and the proper factor from the table, and then move the decimal point six places to the left. The following is the calculation for the branch circuit to the show window receptacle outlets, assuming a power factor of 90%:

$$VD = \frac{12.5 \text{ A} \times 85 \text{ ft} \times 3659}{10^6} = 3.88 \text{ V}$$

$$VD = \frac{12.5 \text{ A} \times 25.9 \text{ m} \times 12,000}{10^6} = 3.88 \text{ V}$$

This is a voltage drop of

$$\frac{3.888 \text{ V}}{120 \text{ V}} = 0.032 \text{ or } 3.2\%$$

It is recommended that 10 AWG conductors be used.

Parallel Conductors

Often the load is such that parallel conductors should be considered. This usually occurs with loads over 400 amperes. In the Commercial Building, the service has three sets of conductors installed in parallel. The major reasons for this are the required size of the conductors and the raceways. Conductors larger than 500 kcmil are difficult to install, and often contractors do not have the equipment to thread and bend conduit in larger sizes.

TABLE 8-2A

Installation in steel conduit—meters.

Wire Size	12	10	8	6	4	3	2	1	1/0
90% pf, 1-phase	12,000	7260	4790	3070	2000	1643	1340	1105	860
90% pf, 3-phase	10,390	6288	4144	2662	1731	1422	1160	957	748
Wire Size	2/0	3/0	4/0	250	300	350	400	500	600
90% pf, 1-phase	744	613	515	465	410	370	344	308	282
90% pf, 3-phase	642	531	445	403	354	321	298	265	245

TABLE 8-2B

Installation in steel conduit—feet.

Wire Size	12	10	8	6	4	3	2	1	1/0
90% pf, 1-phase	3659	2214	1460	937	610	501	409	337	263
90% pf, 3-phase	3169	1918	1264	812	528	434	354	292	228
Wire Size	2/0	3/0	4/0	250	300	350	400	500	600
90% pf, 1-phase	227	187	157	142	125	113	105	94	86
90% pf, 3-phase	196	162	136	123	108	98	91	81	75

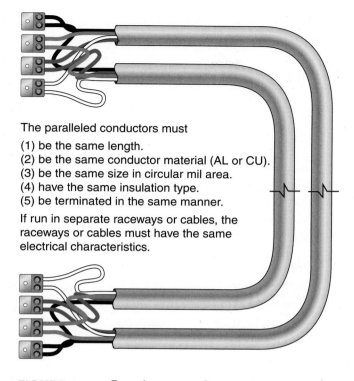

The paralleled conductors must

(1) be the same length.
(2) be the same conductor material (AL or CU).
(3) be the same size in circular mil area.
(4) have the same insulation type.
(5) be terminated in the same manner.

If run in separate raceways or cables, the raceways or cables must have the same electrical characteristics.

FIGURE 8-10 Requirements for parallel conductor installations. (*Delmar/Cengage Learning*)

NEC 310.10(H) sets forth the requirements for installing parallel conductors. These requirements are intended to ensure the conductors will share the current equally to prevent overloading individual conductors. The requirements are illustrated in Figure 8-10. The parallel conductors shall

- be 1/0 AWG or larger.

- be the same length.

- be of the same material (copper–copper or aluminum–aluminum).

- have the same insulation type.

- be the same size in circular mil area.

- be terminated in the same manner.

The raceway or cables shall have the same electrical characteristics. For example, for the raceways in a 3-parallel run installation, don't use Schedule 80 PVC for one raceway, aluminum for another, and steel for another.

It is critical that the conductors be the same length. A knowledgeable electrician will mark out the required length on a clean flat surface and cut all conductors to identical lengths. The following example will illustrate how a small difference in

length results in a significant difference in current in the conductor sets.

Problem

A 1600-ampere service consists of four conduits, each containing four 600-kcmil Type THHN/THWN copper conductors. The allowable ampacity of each conductor is 420 amperes. When the service is operating at 1600 amperes, what is the current in each of the four parallel sets of conductors if their lengths are 20 ft (6.09 m), 21 ft (6.4 m), 22 ft (6.71 m), and 23 ft (7.01 m)?

Solution. The total length of these conductors is 20 + 21 + 22 + 23 = 86 ft (26.21 m).

The current in each set is inversely proportional to the resistance, which is proportional to the conductor's length. The proportional resistance of each set is

$$\frac{20}{86} \times 1600 = 372.1 \text{ amperes}$$

$$\frac{21}{86} \times 1600 = 390.6 \text{ amperes}$$

$$\frac{22}{86} \times 1600 = 409.3 \text{ amperes}$$

$$\frac{23}{86} \times 1600 = 427.9 \text{ amperes}$$

Because current is inversely proportional to resistance, the currents would be 428 for the 20-ft conductors, 409 for the 21 ft conductors, 390 for the 22 ft conductors and 372 for the 23 ft conductors. With only 1-foot (0.3-m) difference in each set, the current exceeds the rating in one set of the conductors. An observation: Because the impact of conductor length differences is proportional to the total conductor lengths, a 1-foot (0.3-m) difference in individual conductor lengths for a 100-ft run would have a significantly reduced impact on current-carrying differences. Check it out!

What to do? There are a few choices:

- Reinstall conductors of the same length. However, this could prove to be very costly.

- Reduce the overcurrent protection to 1500 amperes. This can be accomplished by installing a nonstandard 1500-ampere fuse or by installing a 1600-ampere adjustable trip circuit breaker with the long-time setting at 1500 amperes.

This does not provide the intended 1600-ampere capability, but it does result in no conductor carrying current in excess of the conductor's allowable ampacity. This meets the intent of *NEC 310.15(A)(3)*. Check this out with your local electrical inspector before doing anything.

PANELBOARD WORKSHEET, SCHEDULE, AND LOAD CALCULATION

We developed a panelboard worksheet for the Drugstore in Chapter 4 where we discussed requirements for branch circuits. You may want to review that information before proceeding in this chapter.

Three tables (actually Excel spreadsheets) are developed for each tenant space: a panelboard worksheet, a panelboard schedule, and a load calculation table. All three work together. The branch circuits are entered in the panelboard worksheet, where compensation is made for continuous loads, for more than three current-carrying conductors in a raceway or cable and for operation at elevated ambient temperatures.

These loads are transferred to the panelboard schedule, where circuit numbers are identified, as well as the load of individual branch circuits. The spreadsheet totals the loads per phase to be certain no individual phase is overloaded. This is sometimes referred to as balancing the load.

Finally, the total loads are transferred to the load calculation spreadsheet, where the load on the ungrounded and grounded (neutral) conductors is determined. Keep in mind, there may be slight differences between the panelboard schedule and the load calculation spreadsheet. The loads in the panelboard schedule are the applicable connected or calculated loads. The load calculation spreadsheet will use the greater of the two. Also, in the load calculation spreadsheet, demand factors are applied where permitted and only the greater of noncoincident loads are used in the final calculation. The panelboard schedule shows loads for both coincident loads (loads that are likely to operate at the same time) as well noncoincident loads (loads that are not likely to operate at the same time). Heating and cooling loads are examples of noncoincident loads.

FEEDER AMPACITY DETERMINATION, DRUGSTORE

The Drugstore feeder load calculation form is used to calculate the minimum ampacity of the feeder supplying the panelboard. We have chosen to use a computer spreadsheet for this task as the formulas enable us to easily make the calculations. This form is shown in Table 8-3. The following is an outline illustrating the process of selecting the feeder components for the Drugstore. The other tenant occupancy calculation tables are found in Appendix A of this text.

Let's look at the structure of the form:

Column 1: The loads are listed here. We will review the contents in the column a little later.

Column 2, Count: This column is used to list the quantity of items to be calculated or connected to the panelboard.

Column 3, VA/Unit: Factors to be used to enter the volt-ampere loads or unit loads are entered in this column.

Column 4, NEC Load: The product (result) of multiplying the count by the volt-ampere or unit loads is entered in this column. Note that the loads in this column are considered calculated loads rather than connected loads. Calculated loads are actual loads rather than a unit load.

Column 5, Connected Load: This column contains the connected (actual) loads. As you can see, for the lighting loads, we use the greater of the calculated or connected loads.

Column 6, Continuous Loads: This column represents just what the name implies, loads that are expected to be in operation for 3 hours or more. A 125% factor is applied to the column total to comply with the rules for branch circuits in *NEC 210.19(A) (1)*, for feeders in *NEC 215.2(A)(1)* and in *NEC 230.42(A)* for services. This column can also be used to add 25% to the load of the largest motor to comply with *NEC 430.24*.

Column 7, Noncontinuous Loads: Those loads that are not required to be considered *continuous*.

Column 8, Neutral Load: The loads are entered in this column for line-to-neutral-connected loads. Note that neutral loads are not required to have a 125% factor applied as the loads do not connect to a circuit breaker terminal or terminal for a fusible switch.

TABLE 8-3

Drugstore feeder load calculation.

	Count[1]	VA/Unit[2]	NEC Load[3]	Connected Load[4]	Continuous Loads[5]	Non-continuous[6]	Neutral Load[7]
General Lighting:							
NEC 220.12	1395	3	4185				
Connected loads from worksheet				2630			
Totals:			4185	2630	4185		4185
Storage Area:							
NEC 220.12	858	0.25	215				
Connected load from worksheet				600			
Totals:			215	600		600	600
Other Loads:							
Receptacle outlets	23	180	4140			4140	4140
Copy machine	1	1250		1250		1250	1250
Receptacle for HACR	1	1500		1500		1500	1500
Sign outlet	1	1200	1200		1200		1200
Cash registers	2	240		480		480	480
Electric doors	2	360		720		720	720
Motors & Appliances	**Count**	**VA/Unit**					
Water heater	1	4500		4500	4500		
Unit htr (storage)	1	7500		7500	7500		
AC outdoor unit		9645		9645			
AC indoor unit[8]	1	14,760		14,760	14,760		
Exhaust fans	2	120		240		240	240
Totals Loads				32,145		8930	14,315
Minimum feeder load 125% of continuous loads plus 100% of noncontinuous:						49,111	14,315
Minimum ampacity:						136.4	38.6

Minimum 1/0 AWG ungrounded and 8 AWG Neutral for load. Minimum Neutral 6 AWG per *215.2(A)(1)*.

Minimum size EMT = (0.1855 × 3) + 0.0507 = 0.6072 sq. in. = 1-½ in.

[1]Count: Square footage of given area, the number of a given type load, or the ampere rating of the given type load.
[2]VA/Unit: Specific unit load values required by *NEC Article 220*, or actual load values for the given type load.
[3]*NEC* Load: Count × *NEC* VA/Unit. Also the "calculated load."
[4]Connected Load: Actual load connected to the electrical system.
[5]Continuous Loads: The maximum current is expected to continue for 3 hours or more. Column also used to apply 125% to the largest motor load.
[6]Noncontinuous: Loads that are not continuous.
[7]Neutral Load: The maximum unbalanced load calculated according to *NEC 220.61*.
[8]Because the heating and cooling loads are noncoincident, only the larger load is included.

Lighting Loads. We have separated the loads for the first floor (store area) and for the storage area in the basement. This allows us to more easily determine the *calculated* and *connected* loads. The calculated load is determined by multiplying the square footage (1395) by 3 volt-amperes per square ft (33 volt-amperes per square meter). This results in a calculated load of 4185 VA.

Next, we compare the calculated load with the actual connected load. To save space, we look at the panelboard worksheet for the connected loads. They total 2554 VA, so we use 4185 VA. This load is entered in the Continuous Load column. It is also entered in the Neutral Load column.

For the storage area, the connected load is larger than the calculated load so the connected load is entered in the Continuous Load column and in the Neutral Load column.

Other Loads. The loads that are not lighting loads are entered in the first column. The number of units is entered in the VA/Unit column. The results are entered in the appropriate column to facilitate the calculation. Note that these are actual calculated or connected loads. The 125% factor for continuous loads is applied just below the Total Loads row.

Motors and Appliances. Grouping motors allows us to enter 125% of the largest motor load as required by *NEC 430.24*. In our calculation you will see we are employing another provision of *NEC 220.60*, that of eliminating the smaller of the noncoincident load. The indoor unit heating load of 14,760 VA exceeds the outdoor unit load of 6856 so the air-conditioning load is not included in our calculation.

This also results in us disregarding the largest motor load from our calculation.

Total Calculations. The loads in the columns are added. The continuous load, at 125%, is added to the noncontinuous load. This results in

$$(30,185 \times 1.25) = 37,731 + 8760 = 46,491 \text{ VA}$$

The minimum ampacity of the feeder is determined as follows:

$$\frac{46,491}{208 \times 1.732} = \frac{46,491}{360} = 129.1 \text{ amperes}$$

$$\frac{13,905}{208 \times 1.732} = \frac{13,905}{360} = 38.6 \text{ amperes}$$

Another issue with the minimum size of the neutral is found in *NEC 215.2(A)(1)*. This section requires the neutral to be not smaller than an equipment grounding conductor. The minimum size of an equipment grounding conductor is found in *NEC Table 250.122* and is based on the rating of the overcurrent device on the supply side of the feeder. A 6 AWG conductor is the minimum size permitted for an overcurrent device rated from 100 to 200 amperes.

Overcurrent Protective Device Selection

The rating of the next larger overcurrent protective device that exceeds the minimum ampacity of the feeder conductors is 150 amperes. See *NEC 240.6(A)*. Because installing a 150 ampere feeder results in little spare capacity, we choose to install a 200 ampere feeder. This results in only a slight increase in cost and may actually be less expensive because 200-ampere equipment such as the service disconnecting means and panelboards are standard stock items at the electrical wholesale distributor.

As shown on the Drugstore feeder load calculation form, the minimum allowable ampacity of the feeder conductors must be not smaller than 1/0 AWG for the three ungrounded conductors and 6 AWG for the grounded or neutral conductor. The minimum size ungrounded conductor for the 200-ampere feeder overcurrent device is 3/0 AWG with THHN/THWN insulation.

Phase Conductor Selection

We need to consider the conditions of use where the feeder is installed and make any adjustments or corrections necessary. This includes the conductors being in a high ambient temperature or having more than three current conductors in a circular raceway or cable.

- If the ambient temperature where the conductors are installed is determined to be higher than 86°F (30°C), apply the correction factors from *NEC 310.15(B)(2)*.

- If the feeder conductors are installed in circular raceways on rooftops and are exposed to the direct rays of the sun, apply the temperature adjustments from *310.15(B)(3)(c)*.

- Determine whether the harmonic loads (fluorescent lighting, etc.) are a major portion of the loads. If so, the neutral will be considered as a current-carrying conductor. The adjustment factor for four conductors in *NEC Table 310.15(B)(3)(a)* is 0.8 and must be applied.

- If both installation conditions are present, apply both correction and adjustment factors.

- The conductor size selection must also meet two criteria. It must be at least the size of the minimum conductor size, and it must have an allowable ampacity that is equal to or greater than the minimum allowable ampacity. For the feeder to the Drugstore, a 1/0 AWG conductor satisfies the load and overcurrent protection requirements. As mentioned earlier, we have chosen to install 3/0 AWG ungrounded conductors with a 200-ampere overcurrent device.

- The conductor type is compliant with the construction specifications, and the allowable ampacity is taken from *NEC Table 310.15(B)(16)* for the conductor size.

- The ampacity is the allowable ampacity multiplied by the derating factor. This is checked against the minimum allowable ampacity, but if the steps have been followed correctly, it should never fail.

Voltage Drop. If the feeder length is significant, you should check the voltage drop on the feeder conductors. For our installation, the feeder length to the Drugstore panelboard is probably less than 30 ft (9.14 m) and is oversized, so voltage drop should not be a problem. If, after applying the actual length, the drop is in excess of 3%, serious consideration should be given to increasing the conductor size.

Neutral Conductor Size Determination

As is shown in the Drugstore feeder load calculation, the maximum neutral current is 14,315 volt-amperes. This is divided by 360 (120 volts times the square root of three, or 1.732) to give the current:

$$\frac{14,315}{120 \times 1.732} = \frac{14,315}{360} = 39.8 \text{ amperes}$$

Because all terminals in the service switch and distribution panelboard have terminals rated for 75°C and we are using a conductor with 75°C insulation (THHN/THWN), we can select a conductor from the 75°C column of *NEC Table 310.15(B)(16)*.

This conductor can be an 8 AWG that has an allowable ampacity of 50 amperes.

As mentioned earlier, the rule in *NEC 215.2(A)(1)* requires the grounded conductor in the feeder to be not smaller than required for an equipment grounding conductor. From *NEC Table 250.122*, we see the minimum size equipment grounding conductor would be 6 AWG. That is the conductor selected.

Raceway Size Determination

It has been concluded that the feeder to the Drugstore will consist of three 3/0 AWG and one 6 AWG. No equipment grounding conductor is necessary if a metal raceway such as EMT or rigid metal conduit is installed. The specification also permits the use of Schedule 80 rigid nonmetallic conduit, which would require the installation of an equipment grounding conductor.

From *NEC Chapter 9, Table 5*:

AWG	Count	Area in.²	Total	Area mm²	Total
3/0	3	0.2679	0.8037	172.8	518.4
6	1	0.0507	0.0507	32.71	32.71
Area Totals			0.8544		551.11

Consulting *NEC Chapter 9, Table 4,* for EMT and rigid metal conduit the smallest allowable size, for more than two conductors, is trade size 2 or metric designator 53.

Calculations for the Other Occupancies in the Commercial Building

In Appendix A of this text, you will find the panelboard worksheet, panelboard schedule, and load calculation form for the Bakery, Beauty Salon, Real Estate Office, Insurance Office, and Owner. The one-line diagram for the service and feeders also shows the rating of the overcurrent device and raceway size for all feeders.

REVIEW

Refer to the *National Electrical Code* or the working drawings when necessary. Where applicable, responses should be written in complete sentences.

1. Determine the conductor sizes for a feeder to a panelboard. It is a 120/240-volt, single-phase system. The OCPD has a rating of 100 amperes. The calculated load is 15,600 VA. All the loads are 120 volts.

2. Calculate the neutral current in a 120/240-volt, single-phase system when the current in phase A is 20 amperes and the current in phase B is 40 amperes. The load is resistive.

3. Calculate the neutral current in a 208Y/120-volt, 3-phase, 4-wire system when the current in phase A is 0, in phase B is 40, and in phase C is 60 amperes. The load is resistive.

4. Calculate the neutral current in a 208Y/120-volt, 3-phase, 4-wire system when the current in phase A is 20, in phase B is 40, and in phase C is 60 amperes. The load is resistive.

5. A balanced electronic ballast load connected to a 3-phase, 4-wire multiwire branch circuit can result in a neutral current of (zero times) (half) (two times) that of the current in the phase conductors. (Circle the correct answer.)

6. Harmonic currents associated with electronic equipment (add together) (cancel out) a shared common neutral on multiwire branch circuits and feeders. (Circle the correct answer.)

7. Check the correct statement:

   For circuits that supply computer/data-processing equipment, it is recommended that

   ☐ a separate neutral be installed for each hot-phase branch-circuit conductor.

   ☐ a common neutral be installed for each multiwire branch circuit.

8. Determine the feeder size and other information requested in the following table. The load data to be used is as follows:

Continuous	20,000 VA
Noncontinuous/receptacle	12,000 VA
Highest motor	8000 VA
Other motor	16,000 VA
Nonlinear	20,000 VA
Neutral	24,000 VA

A 3-phase, 4-wire feeder is to be installed. From a distribution panelboard, the feeder is installed underground in rigid PVC, Schedule 80, to an adjacent building where intermediate metal conduit (IMC) is used.

- The IMC is installed in a room where the ambient temperature will be 110°F (43°C).
- Type THHN/THWN conductors are used.
- The growth allowance will be 10%.
- The power factor will be maintained at 90% or higher.
- The conductor length is 180 ft.
- The voltage drop at maximum load shall not exceed 2%.

Load Summary, 208Y/120, 3 Phase, 4-Wire	Calculated	OCPD
Continuous load		
Noncontinuous + receptacle load		
Highest motor load		
Other motor load		
Growth		
Calculated load & OCPD load		
OCPD Selection	**Input**	**Output**
OCPD Load volt-amperes & amperes		
OCPD rating		
Minimum conductor size		
Minimum ampacity		

(*continued*)

Phase Conductor Selection	Input	Output
Ambient temperature & correction factor		
Current-carrying conductors & adjustment factor		
Derating factor		
Minimum allowable ampacity		
Conductor size		
Conductor type, allowable ampacity		
Ampacity		
Voltage drop, 0.9 pf, 3 ph, per 100 ft (30.48 m)		

Neutral Conductor Selection	Input	Output
OCPD load volt-amperes		
Balanced load volt-amperes		
Nonlinear load volt-amperes		
Total load volt-amperes & amperes		
Minimum neutral size		
Minimum allowable ampacity		
Neutral conductor type & size		
Allowable ampacity		

Raceway Size Determination	Input	Output
Feeder conductors size & total area		
Neutral conductor size & area		
Grounding conductor size & area		
Total conductor area		
Raceway type & minimum trade size		

9. A 1200-ampere service was installed consisting of three sets of 600-kcmil THHN/THWN copper conductors per phase. The electrical contractor was careful to cut the conductors the same length. When the utility crew made up the connections at the service heads, they cut the conductors to different lengths so as to make their connections simpler.

   The actual lengths of the service-entrance conductors in a given phase ended up being 20 ft (6.1 m), 22 ft (6.7 m), and 24 ft (7.3 m). The maximum ampacity of a 600-kcmil THHN/THWN copper conductor is 420 amperes using the 75°C column of *Table 310.15(B)(16)*. This is more than adequate for the calculated 1200 amperes when three conductors are run in parallel.

   Determine how the load of 1200 amperes would divide in each of the three paralleled conductors in a phase.

Special Systems

OBJECTIVES

After studying this chapter, you should be able to

- select and install a surface metal raceway.
- select and install multioutlet assemblies.
- calculate the loading allowance for multioutlet assemblies.
- select and install a floor outlet system.
- install wiring for a fire alarm system.

A number of electrical systems are found in almost every commercial building. Although these systems usually are a minor part of the total electrical work to be done, they are essential systems, and it is recommended that the electrician be familiar with the installation requirements of these special systems.

SURFACE METAL RACEWAYS

Surface metal raceways, either metal or nonmetallic, are generally installed as extensions to an existing electrical raceway system, and where it is impossible to conceal conduits, such as in desks, counters, cabinets, and modular partitions. The installation of surface metal raceways is governed by *NEC Article 386. NEC Article 388* governs the installation of surface nonmetallic raceways. The number and size of the conductors to be installed in surface raceways is limited by the design of the raceway. Catalog data from the raceway manufacturer will specify the permitted number and size of the conductors for specific raceways. Conductors to be installed in raceways may be spliced at junction boxes or within the raceway if the cover of the raceway is removable. See *NEC 386.56* and *388.56*. It should be noted that the combined size of the conductors, splices, and taps shall not fill more than 75% of the raceway at the point where the splices and taps occur.

Surface raceways are available in various sizes, Figure 9-1, and a wide variety of special fittings,

FIGURE 9-2 Surface metal raceway supports. (*Delmar/Cengage Learning*)

like the supports in Figure 9-2, makes it possible to use surface metal or nonmetallic raceways in almost any dry location. Two examples of the use of surface raceways are shown in Figure 9-3 and Figure 9-4.

MULTIOUTLET ASSEMBLIES

Multioutlet assembly is defined in *NEC Article 100* as *A type of surface, flush, or freestanding raceway designed to hold conductors and receptacles, assembled in the field or at the factory*. The installation requirements are specified in *NEC Article 380*. Multioutlet assemblies, as illustrated in Figure 9-5, are similar to surface raceways and are designed to hold both conductors and devices. These assemblies offer a high degree of flexibility to an installation and are particularly suited to heavy-use areas where many outlets are required or where there is a likelihood of changes in the installation requirements. The plans for the Insurance Office specify the use of a multioutlet assembly that will accommodate power, data, telecommunications, security, and audio/visual systems; see Figure 9-6. This installation will allow the tenant in the Insurance Office to revise and expand the office facilities as the need arises.

Receptacles for multioutlet assemblies are available with GFCI, isolated ground, and surge suppression features.

The covers are available in steel, aluminum, and PVC with vinyl laminates of different colors and wood veneers such as maple, cherry, mahogany, and oak.

Loading Allowance

The load allowance for a fixed multioutlet assembly is specified by *NEC 220.14(H)* as

- 180 volt-amperes for each 5 ft (1.5 m) of assembly, or fraction thereof, when normal loading conditions exist.

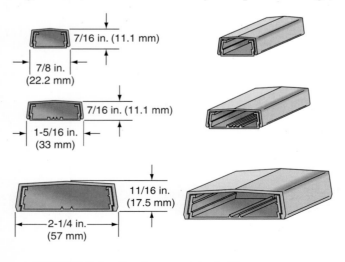

FIGURE 9-1 Surface nonmetallic raceways. (*Delmar/Cengage Learning*)

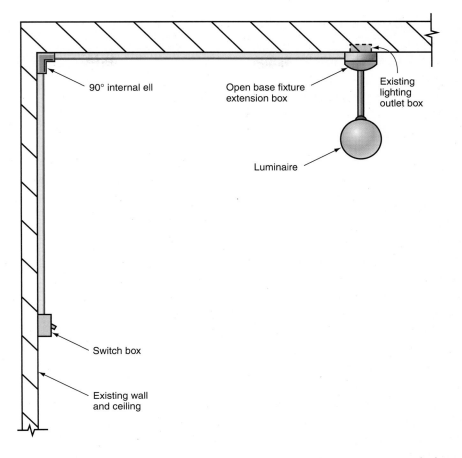

FIGURE 9-3 Use of surface metal raceway to install switch on existing lighting installation. (*Delmar/Cengage Learning*)

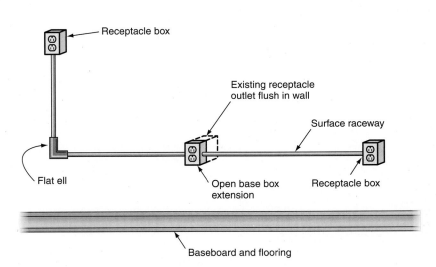

FIGURE 9-4 Use of surface metal raceway to install additional receptacle outlets. (*Delmar/Cengage Learning*)

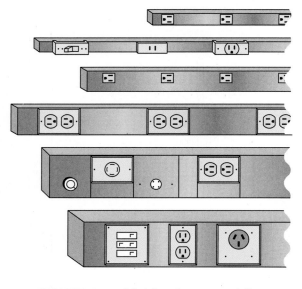

FIGURE 9-5 Multioutlet assemblies.
(*Delmar/Cengage Learning*)

- 180 volt-amperes for each 1 ft (300 mm) of assembly, or fraction thereof, when heavy loading conditions exist.

The usage in the Insurance Office is expected to be intermittent and made up of only small appliances (equipment); thus, it would qualify for the allowable minimum of 180 volt-amperes per 5 ft. Even though the contractor is required to install a duplex receptacle every 18 in. (450 mm) for a total of 56 receptacles, there is no requirement in the *NEC* that the number of receptacles be multiplied by 180 volt-amperes to determine the calculated load, which would be 10,080 volt-amperes. Five branch circuits are provided, which allows for future growth and for typical office equipment such as personal computers, copiers, scanners, and printers. These are considered noncontinuous loads. Additional circuits can be installed easily through the surface metal raceway.

When determining the feeder conductor size, the requirement for the receptacle load can be reduced by the application of the demand factors given in *NEC Table 220.44*. If we had been required to consider all the receptacles at 180 volt-amperes, the total receptacle loading in the Insurance Office would be 12,780 volt-amperes. As provided in *NEC Table 220.44*, the first 10,000 volt-amperes is included at 100%. The remaining load is included at 50%. The calculated load on the feeder due to receptacles would have been 11,390 amperes.

Receptacle Wiring

The plans indicate that the receptacles to be mounted in the multioutlet assembly must be spaced 18 in. (450 mm) apart. The receptacles may be connected in either of the arrangements shown in Figure 9-7. That is, all of the receptacles on a phase can be connected in a continuous row, or the receptacles can be connected on alternate phases.

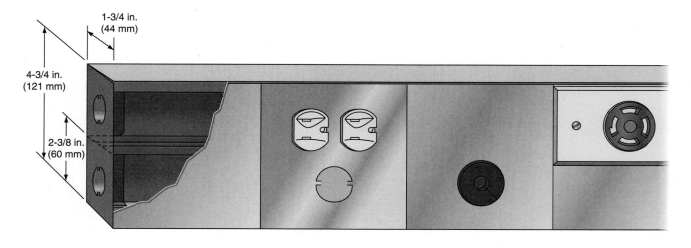

1-3/4 in.
(44 mm)

4-3/4 in.
(121 mm)

2-3/8 in.
(60 mm)

FIGURE 9-6 Multioutlet assemblies are available for data, telecommunications, security, and audio/visual systems. (*Delmar/Cengage Learning*)

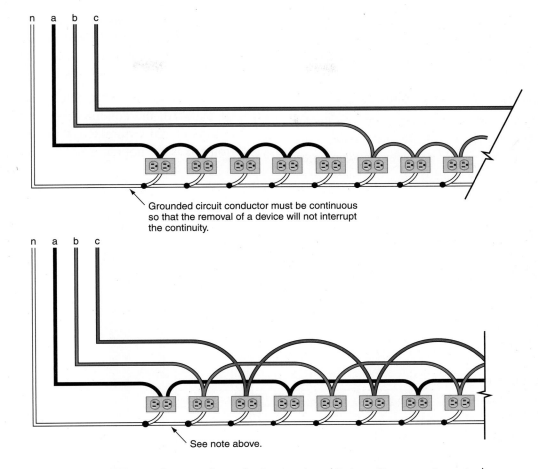

Grounded circuit conductor must be continuous so that the removal of a device will not interrupt the continuity.

See note above.

FIGURE 9-7 Connection of receptacles. (*Delmar/Cengage Learning*)

COMMUNICATIONS SYSTEMS

The installation of the telephone system in the commercial building will consist of two separate installations. The *electrical contractor* will install an empty conduit system according to the specifications for the commercial building and in the locations indicated on the working drawings. In addition, the installation will meet the rules, regulations, and requirements of the communications company that will serve the building. When the conduit system is complete, the electrical contractor, telephone installing company, or telephone company will install all cables and equipment for a complete telephone system.

Power Requirements

An allowance of 600 volt-amperes is made for the installation of the telephone equipment to provide the power required to operate the special switching equipment for the telephones used by the tenants. Because of the importance of the telephone as a means of communication, the receptacle outlets for this equipment are connected to the optional standby power system.

The Communications System

Communication/telephonic requirements vary widely according to the type of business and the extent of the communication convenience desired. Details on communication systems are given in *NEC Article 800*.

Because it is difficult to install the system wiring after the building construction is completed, it is often a requirement that conduits be installed from each of the occupancies to a designated place or area where they are accessible to the installers of the system. The materials used in the installation and the method of installing the raceway lines are the same as those used for light and power wiring with the following modifications:

- Because of small openings and limited space, junction boxes are used rather than the standard conduit fittings, such as ells and tees.

- Even though multipair conductors are no longer used extensively in telephone installations, the size of the conduit should be trade size ¾ or larger.

- The number of bends and offsets is kept to a minimum; when possible, bends and offsets are made using greater minimum radii or larger sweeps (the allowable minimum radius is 5 in. [127 mm]).

- When basement communication wiring is to be exposed, the conduits are dropped into the basement and terminated with a bushing (junction boxes are not required). No more than 2 in.

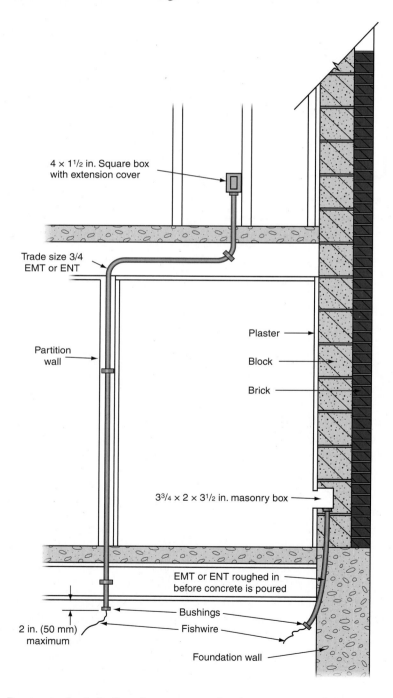

FIGURE 9-8 Raceway installation for communications system. (*Delmar/Cengage Learning*)

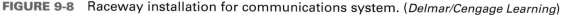

(50 mm) of conduit should project beyond the joists or ceiling level in the basement.

- The inside conduit drops need not be grounded unless they can become energized; see *NEC 250.86 Exception No. 2.*

- The service conduit carrying communications cables from the exterior of a building to the interior must be permanently and effectively grounded; see *NEC 250.104(B).*

Installing the Communications System Raceways

The plans for the commercial building indicate that each of the occupancies requires access to a communication system. Access is achieved by installing a trade size ¾ EMT or ENT to the basement from each occupancy, Figure 9-8. As previously indicated, the Insurance Office uses a multioutlet assembly as a distribution means. In the other areas, the distribution is the responsibility of the occupant. All the cables and equipment will be installed by others.

FLOOR OUTLETS

In an area the size of the Insurance Office, it may be necessary to place equipment and desks where wall outlets are not available. Floor outlets are installed to provide the necessary electrical supply to such equipment. Two methods can be used to provide floor outlets: (1) installing underfloor raceway or (2) installing floor boxes.

Underfloor Raceway

The installation requirements for underfloor raceway are set forth in *NEC Article 390.* It is common for this type of raceway to be installed to provide both power and communication outlets in a dual-duct system similar to the one shown in Figure 9-9. The junction box is constructed so that the power and communications

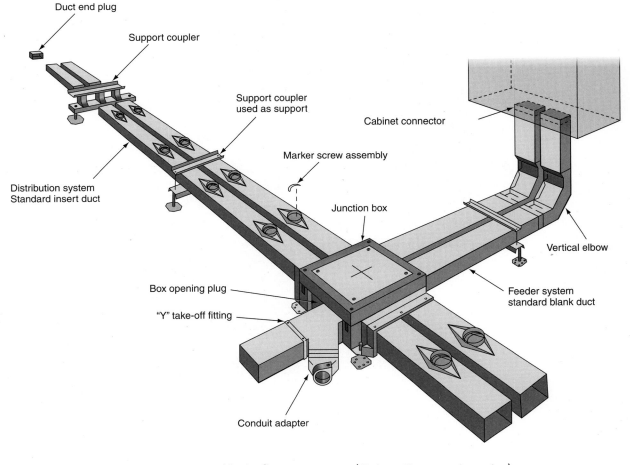

FIGURE 9-9 Underfloor raceway. (*Delmar/Cengage Learning*)

systems are always separated from each other. Service fittings are available for the outlets, Figure 9-10.

Floor Boxes

Many types of floor boxes are available, including metal and nonmetallic, poured in place, and poke through. The exact type to be used is most often included in the specifications by the electrical designer. Floor boxes, either metallic or nonmetallic, can be wired using any suitable raceway such as rigid metal conduit, rigid nonmetallic conduit, electrical metallic tubing, or electrical nonmetallic tubing. These boxes, particularly those of the poke-through type, are wired with a cable wiring system such as Type MC. Some boxes must be installed to the correct height by adjusting the leveling screws before the concrete is poured, Figure 9-11. One

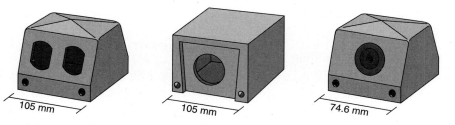

For high-potential service

Dimensions: 4-1/8 in. (105 mm) long; 4-1/8 in. (105 mm) wide; 2-15/16 in. (74.6 mm) high.

FIGURE 9-10   Service fittings. (*Delmar/Cengage Learning*)

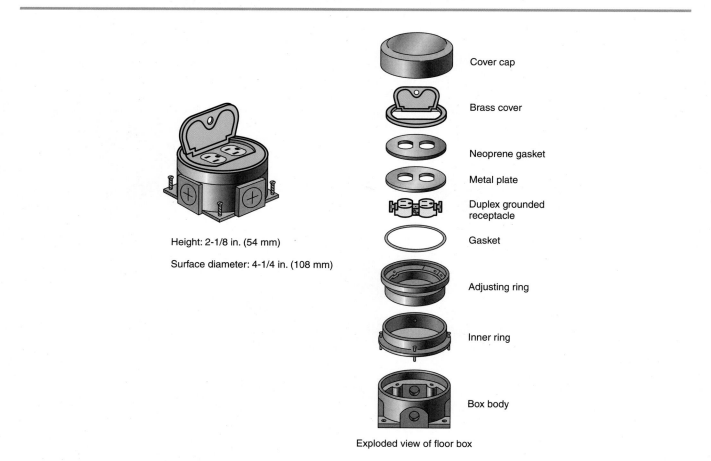

Height: 2-1/8 in. (54 mm)

Surface diameter: 4-1/4 in. (108 mm)

Cover cap

Brass cover

Neoprene gasket

Metal plate

Duplex grounded receptacle

Gasket

Adjusting ring

Inner ring

Box body

Exploded view of floor box

FIGURE 9-11   A floor box with leveling screws. (*Delmar/Cengage Learning*)

nonmetallic box can be installed without adjustment and cut to the desired height after the concrete is poured, Figure 9-12 and Figure 9-13.

The floor plans show the installation of combination floor receptacles with provision for 120-volt power, telecommunications and data for Internet access.

FIRE ALARM SYSTEM

A fire alarm system with battery backup is specified for our commercial building. These systems are commonly required by the local building code. The system is considered a power-limited fire alarm system and thus will be required to comply with *NEC Article 760, Part III*.

1 Level the floor box using a nonmetallic slotted screw (Max. ht. adjustment 1¾ in.)

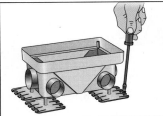

2 Secure mounting pads to form, rebar, etc.

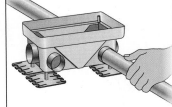

3 Make connection into outlets by solvent cementing conduit into proper outlet in box.

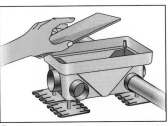

4 Slip on throwaway plastic cap and seal unused outlets with solvent cemented plug.

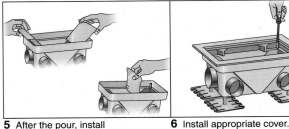

5 After the pour, install appropriate plates/barrier.

6 Install appropriate cover. Secure with screws provided.

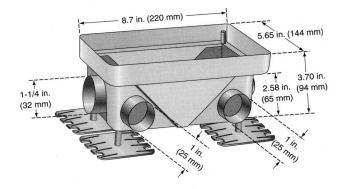

FIGURE 9-12 Installation of multifunction nonmetallic floor box. (*Delmar/Cengage Learning*)

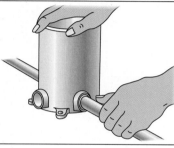

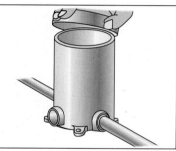

1 Fasten the box to the form or set on level surface.

2 Make connections into outlets by solvent cementing conduit into the proper outlet in box.

3 Slip on temporary plastic cover and seal unused outlets with solvent cemented plugs.

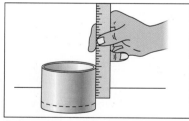

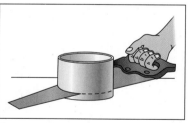

4 Remove temporary plastic cover and determine thickness of flooring to be used. Scribe a line around box at this distance from the floor.

5 Using a handsaw, cut off box at scribed line.

6 Install leveling ring to underside of cover, so that four circular posts extend through slots in outer ring.

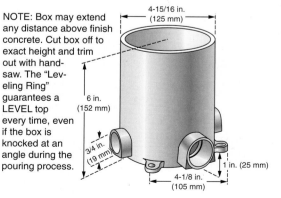

7 Apply PVC cement to leveling ring and to top inside edge of box.

8 Press outer ring cover assembly into floor box for perfect flush fit.

NOTE: Box may extend any distance above finish concrete. Cut box off to exact height and trim out with hand-saw. The "Leveling Ring" guarantees a LEVEL top every time, even if the box is knocked at an angle during the pouring process.

4-15/16 in. (125 mm)

6 in. (152 mm)

3/4 in. (19 mm)

1 in. (25 mm)

4-1/8 in. (105 mm)

FIGURE 9-13 Installation and trim-out of nonmetallic floor box. (*Delmar/Cengage Learning*)

FIGURE 9-14 Fire alarm control panel.
(*Photo courtesy Honeywell Notifier*)

A control panel is installed in the owner's room in the basement, see Figure 9-14. An annunciator is installed inside the back door near the stairways. Smoke detectors are located as required by the manufacturer for all three levels of the building. Pull stations are located near the elevator and near the exit doors and stairways on all floors. Strobe-type annunciators are located as required by the manufacturer on all three floors. The specifications in Appendix A provides specific requirements for the fire alarm system.

These fire alarm systems formerly were required to be wired in one continuous loop with no "tee-taps." In other words, there could be no branches from the detector or annunciator loop. The concern was that the fire alarm control panel could not monitor the branches from the detection or annunciator wiring conductors. Now, each detector, pull station, and annunciator is addressable, see Figure 9-15. In other words, each device has a unique electronic address, and its connection to the system, as well as functionality, can be monitored by the control panel. As a result, wiring the fire alarm system is greatly simplified.

The cable used for the fire alarm system is typically identified as "FPL." Cables are also permitted to be installed in raceways. Other types of cables are permitted to be substituted as shown in *NEC 760.154(D) Cable Substitution Hierarchy* and *Cable Substitutions.*

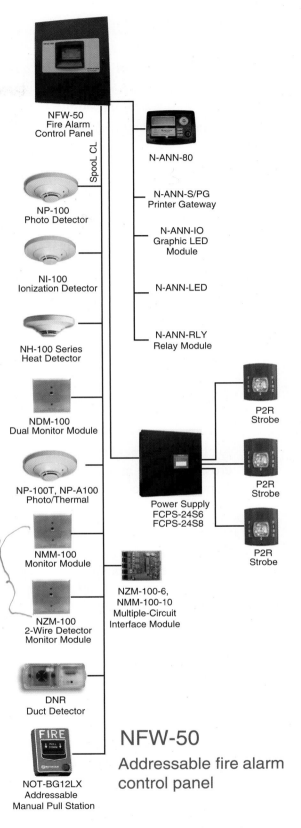

NFW-50
Fire Alarm
Control Panel

SpooL CL

N-ANN-80

N-ANN-S/PG
Printer Gateway

N-ANN-IO
Graphic LED
Module

N-ANN-LED

N-ANN-RLY
Relay Module

NP-100
Photo Detector

NI-100
Ionization Detector

NH-100 Series
Heat Detector

NDM-100
Dual Monitor Module

P2R
Strobe

NP-100T, NP-A100
Photo/Thermal

P2R
Strobe

Power Supply
FCPS-24S6
FCPS-24S8

NMM-100
Monitor Module

P2R
Strobe

NZM-100
2-Wire Detector
Monitor Module

NZM-100-6,
NMM-100-10
Multiple-Circuit
Interface Module

DNR
Duct Detector

NFW-50

Addressable fire alarm
control panel

NOT-BG12LX
Addressable
Manual Pull Station

Relays

FIGURE 9-15 Typical equipment for fire alarm system. (*Photo courtesy Honeywell Notifier*)

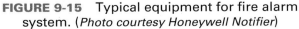

REVIEW

Refer to the *National Electrical Code* or the working drawings when necessary. Where applicable, responses should be written in complete sentences.

The responses to Questions 1–3 are to cite the applicable *NEC* references.

1. Surface raceways may be extended through dry partitions if_____

   _____

   _____

2. Where power and communications circuits are to be installed in a combination raceway, the different types of circuits must be installed in_____

   _____

   _____

3. If it is necessary to make splices in a multioutlet assembly, the conductors plus the splices shall not fill more than _____% of the cross-sectional area at the point where the splices are made._____

   _____

   _____

Respond to Questions 4 and 5 by showing required calculations and citing *NEC Sections*.

4. Using a maximum conductor fill of 40% for a surface metal raceway that measures ½ in. × ¾ in., the maximum allowable cross-sectional area fill is _____ sq. in.

5. Each compartment of the multioutlet assembly installed in the Insurance Office has a cross-sectional area of 3.7 sq. in. (2390 sq. mm). The maximum number of 10 AWG, Type THWN conductors that may be installed in a compartment of the multioutlet assembly is _____

   _____.

6. Several computer circuits are to be installed in a building where nonmetallic cable and metal boxes are being used. Specify the number of conductors required for a cable serving a computer outlet similar to those in the Insurance Office. Note any special actions that must be taken to use that cable, and cite the *NEC* references.

   _____

   _____

   _____

7. Typical fire-alarm cable is marked with the letters _____ .

Working Drawings— Upper Level

OBJECTIVES

After studying this chapter, you should be able to

- tabulate materials required to install an electrical rough-in.

- select the components to install wiring for an electric water heater.

- discuss the advantages/disadvantages between single- and 3-phase supply systems.

Before examining the three offices on the upper level of the Commercial Building, the working drawings, the panelboard schedules, and the panelboard worksheets should be reviewed. The panelboard schedules are located on blueprint sheet E-4 and worksheets are located in the Appendix.

INSURANCE OFFICE

Several special wiring systems are shown in the Insurance Office, most of which have been discussed in earlier chapters.

- Combination floor receptacles are installed in the center of the office area.

- A special raceway is installed in the two exterior walls. This raceway provides for both electrical power and communications.

- Special circuits for personal computers are installed to the computer room and special luminaires are installed in that room.

- A reception area is provided with a combination lighting system to provide a different "look" for the clients.

Computer Room Circuits

A room in the Insurance Office is especially designed to be a computer room, but it is really not an information technology equipment room as described in *NEC 645.4*. The computer room in the Insurance Office is a room with some computers in it and plenty of receptacles, but the special requirements for an information technology equipment room are not applicable.

The following actions have been taken to comply with good practice:

- A special surge protection receptacle is specified.

- A special grounding system is provided.

- A separate neutral conductor is provided with each of the receptacle circuits.

- A luminaire lens that lessens the glare on the computer screens has been specified.

The special receptacles are discussed in Chapter 5, "Switches and Receptacles." The grounding terminal of each of these receptacles is connected to an insulated equipment grounding conductor and not to the box. These insulated equipment grounding conductors are installed in the EMT or Type MC cable with the circuit conductors. The equipment grounding conductors are connected to a special isolated grounding terminal in the panelboard; see *NEC 250.146(D)*. A separate, insulated-grounding conductor is installed with the feeder circuit and is connected to the neutral conductor at the service. This grounding system reduces, if not eliminates, the electromagnetic interference that often is present in the conventional grounding system.

The surge protection feature of the receptacle provides protection from lightning and other severe electrical surges that may occur in the electrical system. These surges are often severe enough to cause a loss of data in the computer memory and, in some cases, damage to the computer. These surge protectors are available in separate units, which can be used to provide protection on any type of sensitive equipment, such as computers and printers. Three separate neutral conductors are installed to reduce the possibility of a conductor overheating from nonlinear currents. There are six current-carrying conductors in the raceway or cable, three neutral conductors, and three phase conductors; thus their ampacity must be adjusted as required by *NEC 310.15(B)(3)(a)*.

Material Take-off

A journeyman electrician should be able to look at an electrical drawing and prepare a list of the materials required to install the wiring system. A great deal of time can be lost if the proper materials are not on the job when needed. It is essential that the electrician prepare in advance for the installation so that the correct variety of material is available in sufficient quantities to complete the job.

Experience is by far the best teacher for learning to tabulate materials. Guidance can be given in preparing for the experience:

- Be certain there is a clear understanding of how the system is to be installed. Decide from where home runs (raceway or cable directly to panelboard) will be made, and establish the sequence of connecting the outlet boxes. The electrician has considerable freedom in making these decisions, even when a scheme is shown on the working drawings.

- Take special note of long runs of raceway or cable and determine whether voltage drop is a problem. This can mean an increase in conductor and/or raceway size.

- Be organized and meticulous. Have several sheets of paper (ruled preferred) and a sharp pencil with a good eraser. Start with the home run, and complete that system before starting another.

BEAUTY SALON

Of special interest in the Beauty Salon are the connection to the water heater, a small laundry unit, and the lighting for the stations where customers are given special attention. A reduced image is presented in Figure 10-1. For a larger image refer to working drawing E3.

Water Heater Circuits

Water heaters are available for commercial occupancies in many sizes and voltages, and for both single and 3-phase supply. The size of water heater as well as the voltage and phase ratings are shown for each tenant and for the owner in the water heater schedule included in Appendix A of this text.

A beauty salon (Figure 10-1) uses a large amount of hot water. To accommodate this need, the specifications, electrical drawings, and the panelboard schedule indicate that a separate circuit is to supply an electric water heater for the Beauty Salon. The water heater is not furnished by the electrical contractor, but the circuit is to be installed and connected by the electrical contractor, Figure 10-2. To save room in the public space on the second floor, the water heater is located in the storage space in the basement. Because the circuit breaker for the water heater is located in the panelboard on the second floor and is out of sight from the water heater, a disconnecting means at the water heater is required by *NEC 422.31(B)*.

The water heater is specified to be of the 120-gallon (450-L) size and rated for 24,000 watts at 208 volts 3-phase because 3-phase power is readily available. Three-phase circuits result in a smaller circuit than if a 1-phase circuit were installed. In most cases, the rating of the overcurrent protective device will also be indicated in the panelboard schedule. If not, the following method can be used to determine the correct circuit rating. The water heater is connected to an individual branch circuit; see *NEC Article 100*.

NEC 422.13 requires that the branch circuit for a storage-type electric water heater with a capacity of 120 gallons (450 L) or less is considered to be a continuous load. This means that the conductor and overcurrent protection sizing is subject to the 125% factor as found in *NEC 210.19(A)(1), 210.20(A), 215.2(A)(1), 215.3*, and *230.42*.

Load current:

$$\frac{24,000}{208 \times 1.732} = 66.67 \text{ Amperes}$$

Minimum branch-circuit rating:

$$66.67 \text{ A} \times 1.25 = 83.3 \text{ A}$$

NEC 422.11(E)(3) establishes the maximum OCPD rating as 150% of the current or the next higher standard OCPD rating.

Maximum OCPD rating:

$$66.67 \text{ A} \times 1.50 = 100 \text{ A}$$

The next higher standard OCPD rating is permitted but not applicable in our example.

The minimum size conductor, if all terminals are rated 75°C or higher, is 4 AWG as the allowable ampacity in *NEC Table 310.15(B)(16)* is 85 amperes. The next standard overcurrent device in *NEC 240.6(A)* is 90 amperes. This is the overcurrent device given in the electric heater heater schedule in Appendix A. If the terminal on either end of the branch circuit conductors is rated 60°C, a 3 AWG conductor is required. Finally, if the maximum rating of overcurrent device of 100 amperes is selected, a 3 AWG conductor with THHN/THWN insulation is required.

Washer-Dryer Combination

One of the uses of the hot water is to supply a washer-dryer combination used to launder the many towels and other items commonly used in a beauty salon. The unit is rated for 4000 VA at 208Y/120 volts single-phase.

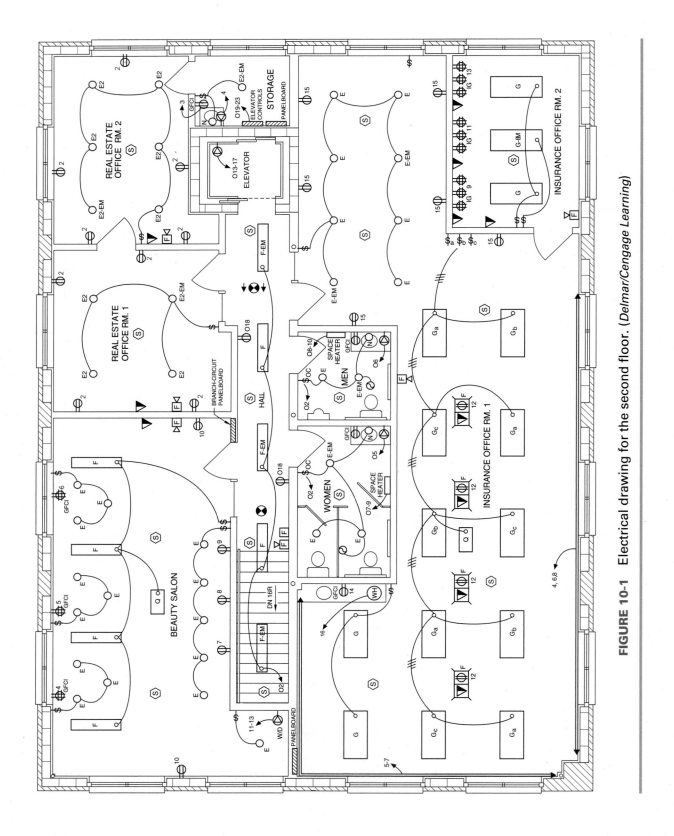

FIGURE 10-1 Electrical drawing for the second floor. (*Delmar/Cengage Learning*)

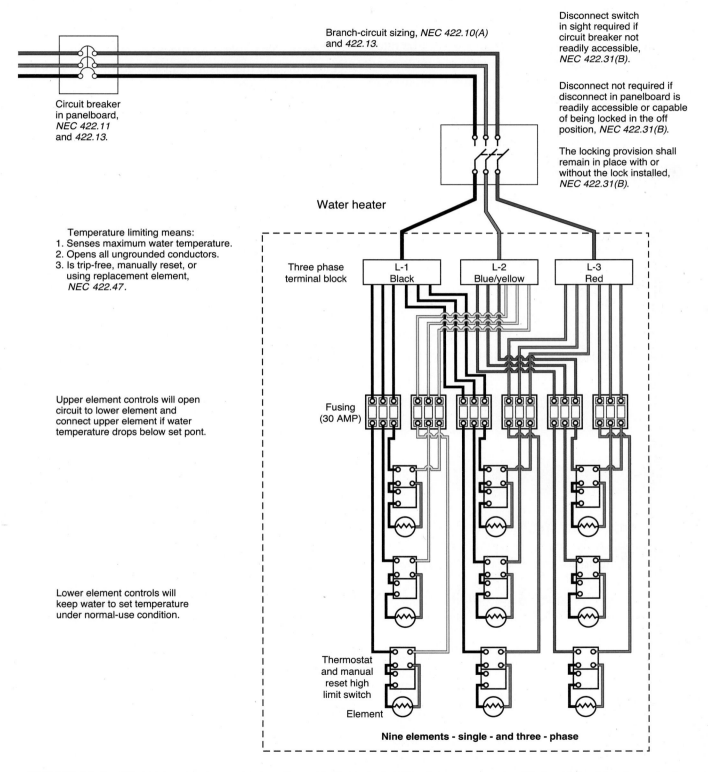

Branch-circuit sizing, *NEC 422.10(A)*
and *422.13.*

Disconnect switch
in sight required if
circuit breaker not
readily accessible,
NEC 422.31(B).

Disconnect not required if
disconnect in panelboard is
readily accessible or capable
of being locked in the off
position, *NEC 422.31(B).*

The locking provision shall
remain in place with or
without the lock installed,
NEC 422.31(B).

Circuit breaker
in panelboard,
NEC 422.11
and *422.13.*

Water heater

Temperature limiting means:
1. Senses maximum water temperature.
2. Opens all ungrounded conductors.
3. Is trip-free, manually reset, or
   using replacement element,
   NEC 422.47.

Three phase
terminal block

L-1
Black

L-2
Blue/yellow

L-3
Red

Upper element controls will open
circuit to lower element and
connect upper element if water
temperature drops below set pont.

Fusing
(30 AMP)

Lower element controls will
keep water to set temperature
under normal-use condition.

Thermostat
and manual
reset high
limit switch

Element

Nine elements - single - and three - phase

FIGURE 10-2   Water heater branch circuit and internal electrical components. (*Delmar/Cengage Learning*)

- The branch circuit supplying this unit must have a rating of at least 125% of the appliance load, *NEC 422.10(A)*.

- The branch-circuit ampacity must be at least 20 amperes (4000 VA/208 V = 19.23 A).

- The overcurrent protective device must be rated at least 25 amperes (load in amperes multiplied by 1.25 and then raised to the next standard size overcurrent device).

- The conductor ampacity must be at least 21 amperes to allow the use of an overcurrent device with a rating of 25-ampere, *NEC 240.6(A)*.

- The use of a 25-ampere OCPD requires the installation of a size 10 AWG conductor or larger; see *NEC 240.4(D)*.

REAL ESTATE OFFICE

The Real Estate Office has a smaller storage space behind the elevator. The distribution panelboard is located in this space as is a small counter and sink and an undercounter water heater. See the electric water heater schedule in Appendix A for the branch circuit requirements.

The control cabinet for the elevator is also located in this room. The elevator specified for this building is very efficient and can be referred to as "Green." The operating mechanism is located on the elevator, and there is no elevator equipment room as would be required for a typical hydraulic elevator. The elevator is referred to as "elevator-equipment-room-less!"

TOILET ROOMS

The toilet rooms are intended to be accessible by the handicapped. Each has an undercounter water heater for hand washing. Each room is provided with an electrical wall insert space heater. In addition, each room is equipped with an occupancy sensor to control the lighting and exhaust fan. All of these loads are supplied by the Owner's panelboard as the toilet rooms serve all tenants on the second floor in addition to any of the public who may visit the premises.

REVIEW

Refer to the *National Electrical Code* or the working drawings when necessary. Where applicable, responses should be written in complete sentences.

Questions 1–3 refer to the water heater circuit to be installed in the Beauty Salon.

1. Is it permissible to install a receptacle for the water heater in the salon? What section of the *Code* permits or prohibits this installation?

_____

_____

2. The water heater has nine elements. Under what conditions are these elements heating?

_____

_____

3. If the water heater is not marked with the maximum overcurrent protection rating, the maximum overcurrent protection is to be sized not to exceed _____% of the water heater's ampere rating; see *NEC 422.11(E)*.

The following questions refer to the branch-circuit panelboard in the Real Estate Office. The panelboard schedule is shown on drawing E4 and the panelboard summary in Appendix A.

4. How many single-pole branch circuits can be added to the panelboard? _____

   _____

   _____

5. How many general-purpose branch circuits are to be installed? _____

   _____

6. If an appliance that will operate for long periods of time is marked with the rating of the overcurrent device, is it permissible to install a device with the next higher rating?

   _____

Refer to the working drawings for the Insurance Office for Question 7.

7. Tabulate the following materials:

   a. Luminaires installed in the ceiling of the Insurance Office Room 1?

      Style/count _____

      _____

      Manufacturers' catalog number(s) _____

   b. Luminaires installed in the ceiling of the Insurance Office Reception Room?

      Style/count _____

      _____

      Manufacturers' catalog number(s) _____

      _____

   c. Luminaires installed in the ceiling of the Insurance Office Room 2?

      Style/count _____

      _____

      Manufacturers' catalog number(s) _____

   d. Receptacles in the Insurance Office Room 1, Insurance Office Room 2, and Rception Room?

      Type/count _____

      _____

      Ampere/voltage rating _____

e. Wall switches in the Insurance Office Room 1, Insurance Office Room 2, and Reception Room?

Type/count _____

_____

_____

Ampere/voltage rating _____

f. There are nine receptacles along one wall of the Insurance Office Room 2 (Computer Room). Explain what they are.

_____

_____

_____

_____

R509
1K
D301 IN414 B
D302
IN414 B
501
936
R513

CHAPTER

11

Special Circuits
(Owner's Circuits)

OBJECTIVES

After studying this chapter, you should be able to

- describe typical connection schemes for photocells and timers.

- understand requirements for wiring an elevator.

- describe the connections necessary to control a sump pump.

The occupants of the Commercial Building are each responsible for the electric energy used within their areas. It is the owner's responsibility to provide power for the elevator; illumination for the public areas such as hallways, stairways, and public toilet rooms; and heat for the common areas of the building. The circuits supplying those areas and the devices for which the building is responsible will be called the owner's circuits.

PANELBOARD WORKSHEET, PANELBOARD SCHEDULE, AND LOAD CALCULATION FORM

The three documents that are used to determine branch circuit layout, panelboard circuit layout, and the load calculation form for the owner's circuits are printed in the Appendix. These circuits serve:

- the elevator equipment that has the greatest electrical demand.

- heating and lighting for the public restrooms, hallway, and stairways as well as the exit signs for the basement, rear entrance hallway, and the second floor hallway.

- loads that are to be supplied by the optional standby power system.

LIGHTING CIRCUITS

Those lighting circuits that are considered to be the owner's responsibility can be divided into six groups according to the method used to control the circuit:

- Continuous operation
- Manual control
- Automatic control
- Timed control
- Photocell control
- Exit signs in a portion of the building

Continuous Operation

The luminaires installed for the stairway to the second floor and hallways are in continuous operation when the building is occupied because those spaces have no windows. The power to these luminaires is supplied directly from the owner's panelboard.

Manual Control

A conventional manual lighting control system has been selected to operate second-floor corridor lighting, stairway lighting, and lighting in the basement hallways.

Automatic Control

The sump pump is an example of equipment that is operated by an automatic control system. No human intervention is necessary—the control system starts the pump when the liquid level in the sump rises above the maximum allowable level, and then it is pumped down to the minimum level.

The toilet rooms have occupancy sensors installed for the lighting and exhaust fans. These are also considered automatic control systems.

Timed Control

Lighting can be controlled by timers, also referred to as time switches or timer switches. Electronic and mechanical timers are available with many different features such as these:

- Same day/same time on/off
- Same day/different time on/off
- Different day(s) same time on/off
- Different day(s)/different times on/off
- Self-adjustment for seasonal changes
- Options to skip a day(s)
- Automatic adjustment for daylight saving time
- Battery backup to save settings in the event of a major power outage
- Multiple on/off time settings

Typical ratings found on the nameplate of a timer are voltage, amperes, resistive or inductive loads (motor/transformer loads), horsepower, locked rotor amperes, and/or tungsten loads. Switching options are single-pole, single-throw; single-pole, double throw; and double-pole, single throw.

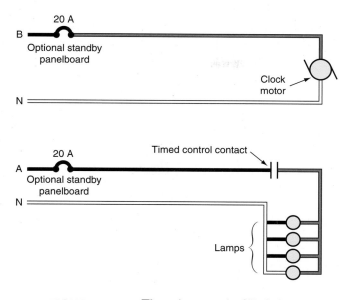

FIGURE 11-1 Timed control of lighting.
(*Delmar/Cengage Learning*)

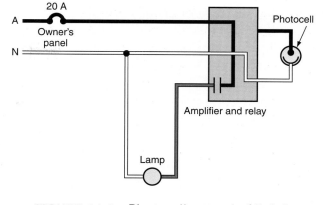

FIGURE 11-2 Photocell control of lighting.
(*Delmar/Cengage Learning*)

Figure 11-2 shows the typical connection for a photocell.

Exit Signs

Exit signs are installed to indicate the pathway to exit the building. The location of the exit signs is controlled by the building code that is adopted by the authority having jurisdiction.

The exit signs specified for the Commercial Building have a battery backup, and they are maintained fully charged by being connected to a 120-volt branch circuit. They are required to be connected so power is maintained to the signs at all times. If connected to the lighting circuits, they must be connected ahead of any local switches.

An astronomic feature is extremely desirable. It provides sunset ON and sunrise OFF that change automatically day by day as the sunrise and sunset times change every day. No need to keep resetting the times—set once and forget it.

The most common timer is to control 1 circuit, but the more elaborate timers are available to control 2, 4, 8, and 16 circuits.

Figure 11-1 shows a typical connection of a timer.

Refer to the Specifications in the Appendix of this text for more information.

Photocell Control

Outdoor luminaires can be controlled by individual photocells. Luminaires of the type specified for the Commercial Building are often available with or without an individual photocell. Multiple luminaires can also be connected to have all of them controlled by one strategically placed photocell. Another option is to order one luminaire with the photocell and connect the others to be supplied from the first luminaire. If you choose this last option, be certain to verify that the photocell has a rating for the load of all luminaires supplied. The photocell control device consists of a light-sensitive cell and an amplifier that increases the signal until it is sufficient to operate a relay that controls the light. A circuit is connected to the photocell to provide power for the amplifier and relay.

SUMP PUMP CONTROL

The sump pump is used to remove water entering the building because of sewage line backups, water main breakage, minor flooding due to natural causes, or plumbing system damage within the building. Because the sump pump is a critical item, it is connected to the optional standby power panel. The pump motor is protected by a manual motor starter, Figure 11-3, and is controlled by a float switch, Figure 11-4. When the water in the sump rises, a float is lifted, which mechanically completes the circuit to the motor and starts the pump, Figure 11-5. When the water level falls, the pump shuts off.

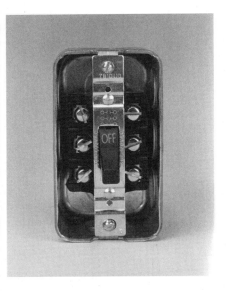

FIGURE 11-3 Manual motor controller for sump pump. (*Delmar/Cengage Learning*)

FIGURE 11-4 Float switch. (*Delmar/Cengage Learning*)

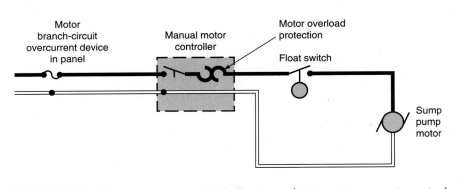

FIGURE 11-5 Sump pump control diagram. (*Delmar/Cengage Learning*)

Another type of sump pump that is available commercially has a start-stop operation due to water pressure against a neoprene gasket. This gasket, in turn, pushes against a small integral switch within the submersible pump. This type of sump pump does not use a float.

WATER HEATER AND SPACE HEATING

The owner is responsible for supplying the hot water for the public restrooms on the second floor. As shown on the electrical drawing and on the panelboard schedule for the owner's circuits, a water heater is installed in each toilet room for this purpose. These water heaters are considered point-of-use and are cord-and-plug connected to a receptacle installed under the counter. A dedicated 120-volt, 20-ampere branch circuit supplies each water heater. Here's an excellent opportunity to install a multiwire branch circuit consisting of two ungrounded (hot) conductors on opposite phases and a shared neutral. Keep in mind a 2-pole circuit breaker or two single-pole breakers with a tie handle are required for a multiwire branch circuit, *NEC 210.4(B)*.

Space heating is installed in the restrooms in the form of wall-mounted, electric forced air heaters. Several options are available for wiring the heaters. Either a single, 30-ampere branch circuit can be installed to supply both heaters or an individual 15-ampere branch circuit can be installed to each heater. (Note that the specifications for the building require a minimum 20-ampere branch circuit and 12 AWG conductors.) The following table summarizes the options:

	Heater Rating (Watts)	Continuous Load Factor	Adjusted Rating	Minimum Circuit Ampacity	Conductors (2 in EMT)	Overcurrent Device	14 AWG 90°C Ampacity	Conductors (4 in EMT)
Men's Toilet Room	1500	125%	1875	9.0	14 AWG	15 Amperes	25	20
Women's Toilet Room	2000	125%	2500	12.0	14 AWG	15 Amperes	25	20

Note: Installing 4, 14 AWG current-carrying conductors in one raceway or Type MC cable is permitted as 80% of the 90°C allowable ampacity of 25 amperes exceeds the minimum circuit ampacity and rating of the overcurrent protective device.

As shown in the above table, the 14 AWG conductors are permitted with four conductors in an EMT or Type MC cable. The ampacity of the branch circuit conductors is required to be adjusted in accordance with *NEC Table 310.15(B)(3)(a)* if four branch circuit conductors are installed in the same EMT or Type MC cable. However, we are permitted to use the 90°C allowable ampacity of the conductor for derating purposes.

$$25 \times .8 = 20 \text{ amperes}$$

Another option for installing the branch circuit for the heaters is to install one 25- or 30-ampere branch circuit with 10 AWG conductors. This is permitted by *NEC 424.3(A)*.

$$1500 \text{ VA} + 2000 \text{ VA} = 3500 \text{ VA} \times 1.25$$
$$= 4375 \text{ VA} \div 208 = 21.0 \text{ amperes}$$

⎍⎍⎍ ELEVATOR WIRING

An electric elevator is specified to serve both floors and the basement. This allows persons with disabilities to access the office spaces on the second floor as well as the building tenants to easily take freight and other commodities to and from their office space and their storage space in the basement.

The specifications require a high-efficiency elevator known as a "machine-room-less" elevator. As indicated by the title, all the elevator lifting mechanism is located in the elevator shaft either on the elevator or on the wall. Hydraulic pumps, pistons and controls are not used in this design. As a result, it is "environmentally friendly" as there is no hydraulic oil to leak or be disposed of. The elevator uses a much smaller motor with a solid-state electronic controller that gives it the soft-start

capability. The new elevator design is indeed "Going Green!"

The feeder for the elevator is supplied from the owner's panelboard that is in turn supplied from the automatic transfer switch, so electric power will be available during an outage from the electric utility.

The wiring for electric elevators is covered in *NEC Article 620*. Generally, the electrical contractor is responsible for providing and installing the "ungrounded main power supply" disconnecting means required in *NEC 620.51*. The electrical contractor is also responsible for installing lighting and receptacle outlets in the elevator pit. All other wiring in the elevator shaft as well as wiring of the elevator car and all controls are the responsibility of the elevator contractor.

The disconnecting means for the elevator is required to be an enclosed, externally operable fused motor circuit switch or circuit breaker capable of being locked in the open position; see *NEC 620.51(A)*. The disconnecting means is required to be readily accessible to qualified persons. The location of the disconnecting means is to be on the wall near the elevator control panel in the Real Estate Office storage room.

A separate branch circuit is required for the lighting outlet and receptacle outlet in the hoistway (elevator) pit. The switch for the luminaire is required to be readily accessible from the pit access door and the luminaire is not permitted to be connected on the load side of a GFCI device; see *NEC 620.24*. The receptacle in these spaces is required to be a 15- or 20-ampere, 125-volt receptacle of the GFCI type. Although GFCI protection of the branch circuit is not prohibited, it is intended that a person servicing the elevator could reset a GFCI in the area work is being performed; see *NEC 620.85*.

Another separate branch circuit is required for lighting, exhaust fan, and other services in the elevator car; see *NEC 620.22*. This branch circuit is supplied from the panelboard for these services, as indicated below. An additional branch circuit is required for heating and air-conditioning if these are provided for the elevator.

A definition of "separate branch circuit" is not found in *NEC Articles 100 or 620*. No doubt, the committee that is responsible for *Article 620* intends that it be a single branch circuit that serves no loads or outlets other than the luminaire and receptacle outlet. The switch for the luminaire is required to be readily accessible from the pit access door, and the luminaire is not permitted to be connected on the load side of a GFCI device; see *NEC 620.23*. The receptacle(s) in these spaces is required to have GFCI protection, *NEC 620.85*.

An additional feeder is installed to the elevator equipment location for loads related to the elevator such as for elevator car lighting and heating and cooling. These loads are determined to be 9600 volt-amperes. We calculate the minimum feeder ampacity as

$$\frac{9600}{208 \times 1.732} = 26.6 \text{ amperes}$$

A minimum 30 ampere, 208-volt, 3-phase feeder is installed for these loads.

OPTIONAL ELECTRIC BOILER

In some areas, the owner may choose to install an electrically heated boiler to supply heat to all areas of the Commercial Building. The boiler would have a full-load rating of 200 kW or as calculated to compensate for the heat loss of the building. As purchased, the boiler is completely wired except for the external control wiring, Figure 11-6. A heat

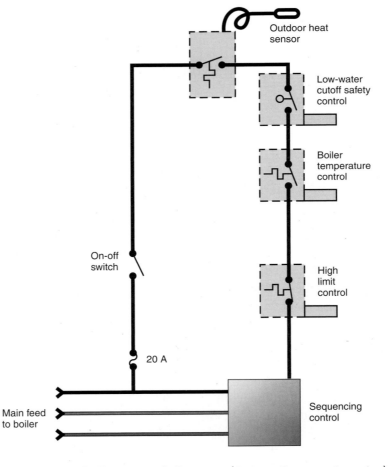

FIGURE 11-6 Boiler control diagram. (*Delmar/Cengage Learning*)

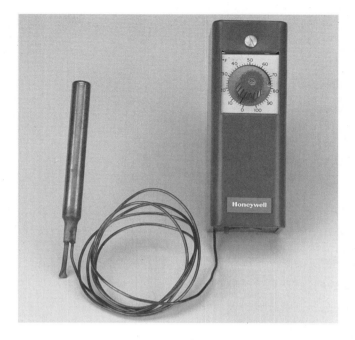

FIGURE 11-7 Heat sensor with remote mounting bulb. (*Delmar/Cengage Learning*)

sensor plus a remote bulb, Figure 11-7, is mounted so that the bulb is on the outside of the building. The bulb should be mounted where it will not receive direct sunlight and be spaced at least ½ in. from the brick wall. If these mounting instructions are not followed, the bulb will give inaccurate readings. The heat sensor is adjusted so that it closes when the outside temperature falls below a set point, usually 65°F (18°C). The sensor remains closed until the temperature increases to a value above the differential setting, approximately 70°F (21°C). When the outside temperature causes the heat sensor to close, the boiler automatically maintains a constant water temperature and is always ready to deliver heat to any zone in the building that requires heat.

A typical heating control circuit is shown in Figure 11-8.

Figure 11-9 is a schematic drawing of a typical connection scheme for the boiler heating elements.

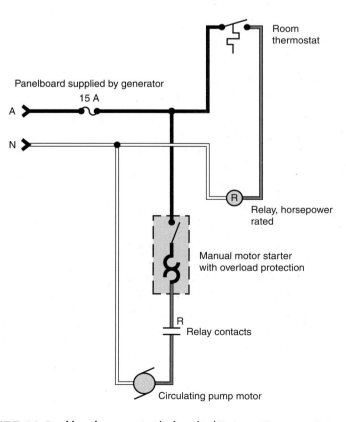

FIGURE 11-8 Heating control circuit. (*Delmar/Cengage Learning*)

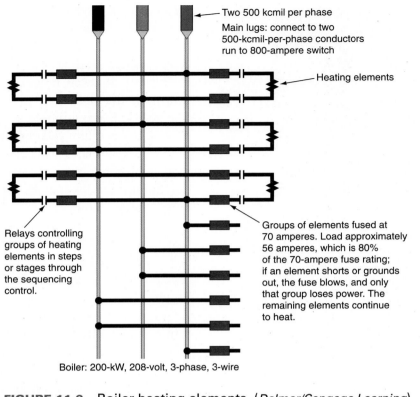

Two 500 kcmil per phase
Main lugs: connect to two
500-kcmil-per-phase conductors
run to 800-ampere switch

Heating elements

Relays controlling groups of heating elements in steps or stages through the sequencing control.

Groups of elements fused at 70 amperes. Load approximately 56 amperes, which is 80% of the 70-ampere fuse rating; if an element shorts or grounds out, the fuse blows, and only that group loses power. The remaining elements continue to heat.

Boiler: 200-kW, 208-volt, 3-phase, 3-wire

FIGURE 11-9 Boiler heating elements. (*Delmar/Cengage Learning*)

The following points summarize the wiring requirements for a boiler:

- A boiler is a fixed electric heating load and is considered to be a continuous load.

- A disconnecting means shall be installed to disconnect all ungrounded supply conductors.

- The disconnecting means shall be rated not less than 125% of the boiler load.

- The disconnecting means shall be in sight of the boiler or be capable of being locked in the open position. The lock-off provision shall remain in place with or without a lock installed, *424.19*.

- The branch-circuit conductors shall be sized at not less than 125% of the boiler load because it is considered to be a continuous load.

- The overcurrent protective devices are to be sized at 125% of the boiler load.

NEC Article 424 establishes the requirements for the installation of fixed electric space-heating equipment. *Part VII* in particular covers electric resistance-type boilers.

The building's occupants will each have a heating thermostat located within their particular area. The thermostat operates a relay, Figure 11-10, which controls a circulating pump, Figure 11-11, in the hot-water piping system serving the area. The circuit to the circulating pump will be supplied from the optional standby power panel so that the water will continue to circulate if the utility power fails, Figure 11-12. In subfreezing weather, the continuous circulation prevents the freezing of the boiler water; such freezing could damage the boiler and the piping system. The water in the piping typically contains an antifreeze solution to protect the boiler and piping in the event of electrical system failure.

FIGURE 11-10 Relay in NEMA 1 enclosure. (*Delmar/Cengage Learning*)

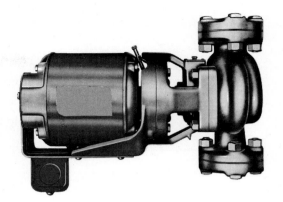

FIGURE 11-11 Water-circulating pump. (*Delmar/Cengage Learning*)

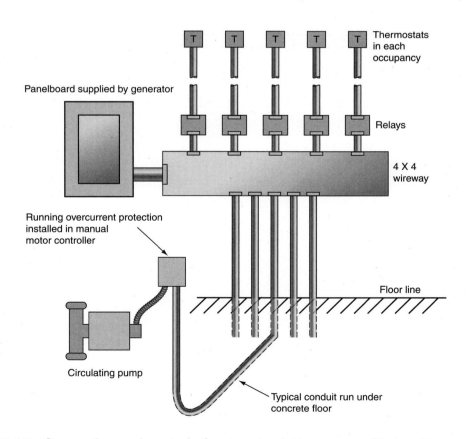

FIGURE 11-12 Connection and control of water-circulating pumps. (*Delmar/Cengage Learning*)

REVIEW

Refer to the *National Electrical Code* or the working drawings when necessary. Where applicable, responses should be written in complete sentences.

The first four questions refer to the installation of the electrical supply to the boiler.

1. What is the proper name of the electrical supply to the boiler?

    _____

    _____

2. Describe, in detail, the electrical supply to the boiler, indicating options available and any special precautions.

    _____

    _____

    _____

    _____

    _____

    _____

3. Show the calculations to determine the minimum size Type THHN conductors and the rating of the overcurrent protective device (fuses) for the circuit supplying the boiler. Use two conductors per phase in two raceways. Also determine the size of the equipment grounding conductors to be installed in each raceway. The expected maximum temperature in the room where the boiler is located is 100°F (37.8°C). The boiler electrical panel and the distribution panel are rated 75°C. All conductors are copper.

4. Verify, or disprove, the correctness of the sizing of the boiler branch circuit.

The following three questions refer to the installation of the various items of equipment in the equipment room.

5. What is the proper rating of the overload protection to be installed for the sump pump?

_____

_____

_____

_____

_____

_____

6. Detail the installation of the outside heat sensor.

_____

_____

_____

_____

_____

7. The Specifications for the Commercial Building call for the electrician to furnish, install, and connect a timer to control all outdoor Wal-Pak® luminaires and the two ceiling lumininaires in the front entry. Refer to the Specifications in Appendix A in the back of this text for specific requirements.

   There are many different ways the wiring of these luminaires can be accomplished. It is up to the electrician to decide how to run the branch circuit to the luminaires.

   Draw a sketch of how you would lay out the wiring of the Wal-Pak luminaires and the two ceiling luminaires in the front entry. Use separate 8½ × 11 paper for your drawing(s) if needed.

Panelboard Selection and Installation

OBJECTIVES

After studying this chapter, you should be able to

- identify the criteria for selecting a panelboard.
- correctly place and number circuits in a panelboard.
- compute the correct feeder size for a panelboard.
- determine the correct overcurrent protection for a panelboard.
- prepare a panelboard directory.

The installation of panelboards is always a part of new building construction, such as the Commercial Building. After a building is occupied, it is common for the electrical needs to exceed the capacity of the installed system. In this case, the electrician is often expected not only to install a new panelboard, but also to select a panelboard that will satisfy the needs of the occupant.

PANELBOARDS

Separate feeders are to be run from the main service equipment to each of the areas of the Commercial Building. Each feeder will terminate in a panelboard, which generally will be installed in the area to be served. Figure 12-1 shows two types of typical panelboards. The first panel has a main breaker plus many branch-circuit breakers. The second panel has branch-circuit breakers only and is referred to as a main-lug-only panelboard or MLO.

A panelboard is defined by the *NEC* as *A single panel or group of panel units designed for assembly in the form of a single panel, including buses and automatic overcurrent devices, and equipped with or without switches for the control of light, heat, or power circuits; designed to be placed in a cabinet or cutout box placed in or against a wall, partition, or other support; and accessible only from the front.**

As indicated in the definition, panelboards are the bussing and overcurrent devices that are mounted in enclosures such as cabinets. The installation of cabinets is covered in *NEC Article 312*. This provides overlapping requirements as the general rules on construction and installation of panelboards are in *NEC Article 408*.

For commercial installations, three different components are often selected; the cabinet, the panelboard, and the cover for either flush or surface installations. For residential applications, "enclosed panelboards" are often installed. Enclosed panelboards include the cabinet, panelboard interior, and cover, all in one carton.

Panelboards shall be dead front, *NEC 408.38*. Thus, no current-carrying parts are exposed to a person operating the devices.

*Reprinted with permission from NFPA 70-2011.

FIGURE 12-1   Panelboards. (*Delmar/Cengage Learning*)

A panelboard is accessible from the front and is usually wall mounted, whereas a switchboard may be accessible from the rear as well and is often floor mounted.

In the past, panelboards were defined two ways: (1) as lighting and appliance panelboards and (2) as power panelboards. The *2008 NEC* removed the differentiation.

The maximum number of overcurrent devices permitted in a panelboard is now governed by the design, rating, and listing of the panelboard, *NEC 408.36* and *NEC 408.54*. In the past, a lighting and appliance panelboard was limited to 42 overcurrent devices. This is no longer true. The design, rating, and listing of the panelboard are now the governing factors.

NEC 408.36(A) stipulates that if the panelboard contains snap switches with a rating of 30 amperes or less, the panelboard overcurrent protection shall not exceed 200 amperes.

Snap switches are switches that are installed in a fusible panelboard. They are additional to the overcurrent devices and are to be rated for the load they are to control. Older-style fusible panelboards had fuseholders and snap switches to turn the power for a given branch circuit on and off. Today, a circuit breaker provides both overcurrent protection and the disconnecting means.

Panelboards are used on single-phase, 3-wire, and 3-phase, 4-wire electrical systems to provide individual branch circuit protection for equipment of all sizes and loads, for lighting and receptacle outlets and to provide branch circuits for the connection of appliances such as the equipment in the Bakery.

Panelboard Construction

In general, panelboards are constructed so the main feed busbars run vertically for nearly the height of the panelboard. The buses to the branch-circuit protective devices are connected to the alternate main buses as shown in Figures 12-2(A) and 12-2(B). The busing arrangement shown in these figures is typical.

Always check the wiring diagram furnished with each panelboard to verify the busing arrangement. In an arrangement of this type, the connections directly across from each other are on the same phase, and the adjacent connections on each side

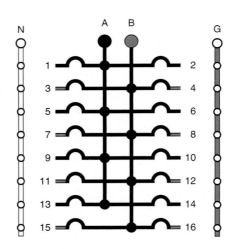

FIGURE 12-2A   Single-phase, 3-wire panelboard. (*Delmar/Cengage Learning*)

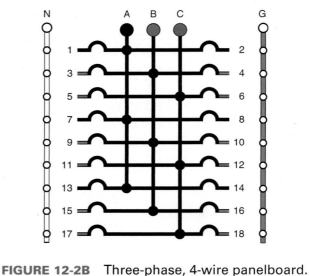

FIGURE 12-2B   Three-phase, 4-wire panelboard. (*Delmar/Cengage Learning*)

are on different phases. As a result, multiple protective devices can be installed to serve the 208-volt equipment. The numbering sequence shown in Figures 12-2 and 12-3 is the system used for most panelboards. Figures 12-4 and 12-5 show the phase arrangement requirements of *NEC 408.3(E)* and *408.3(F)*.

Numbering of Circuits

The number of overcurrent devices in a panelboard is determined by the needs of the area being served. Using the owner's panelboard as an example, there are nine single-pole circuits,

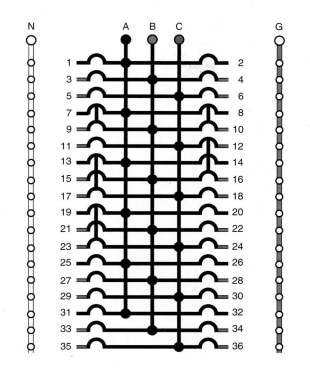

FIGURE 12-3 Panelboard circuit numbering scheme. (*Delmar/Cengage Learning*)

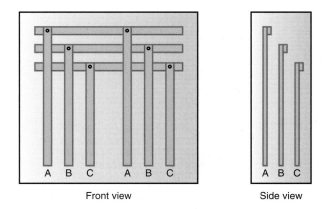

Top view

Front view   Side view

FIGURE 12-4 Phase arrangement requirements for switchboards and panelboards. (*Delmar/Cengage Learning*)

two 2-pole circuits, and three 3-pole circuits. See Figure 12–3 for a panelboard circuit numbering scheme example. This shows a total of 36 poles. When using a 3-phase supply, the incremental

number is six (a pole for each of the three phases on both sides of the panelboard). Panelboards with space for 42 poles are often installed as this provides considerable flexibility for growth.

Note that the numbering of the overcurrent devices continues with odd numbers on the left and even numbers on the right irrespective of whether 1-pole, 2-pole, or 3-pole overcurrent devices are installed or created with handle ties.

Neutral and equipment grounding conductor terminal bars are installed as needed or desired.

Electrical distributors often standardize on sizes and ratings of panelboards. So a 42-circuit panelboard with a 200-ampere rating may be less expensive than one designed for 30 or 36 overcurrent devices.

Panelboard Ratings

Panelboards are available in various ratings including 100-, 200-, 225-, 400-, 600-, and greater amperes. These ratings are determined by the current-carrying capacity of the internal main bus. After the feeder capacity is determined, a panelboard rating is selected in compliance with *NEC 408.30*, which has a rating not less than the load calculated in accordance with *NEC Article 220*.

Panelboard Overcurrent Protection

Figures 12-5, 12-6, 12-7, and 12-8 are variations for providing overcurrent protection for panelboards.

Two extremely important requirements must be followed regarding overcurrent protection for panelboards:

- Overcurrent protection for a panelboard shall not exceed the rating of the panelboard, *NEC 408.30*.

- Overcurrent protection for a panelboard is permitted to be located in the panelboard or at any point on the supply side of the panelboard, *NEC 408.36*.

Figure 12-5 shows a panelboard being supplied from a 120/240, 3-phase, 4-wire transformer. The delta connected system with one transformer grounded will create what is commonly referred to as a "high-leg," "red leg," or "wild-leg" system.

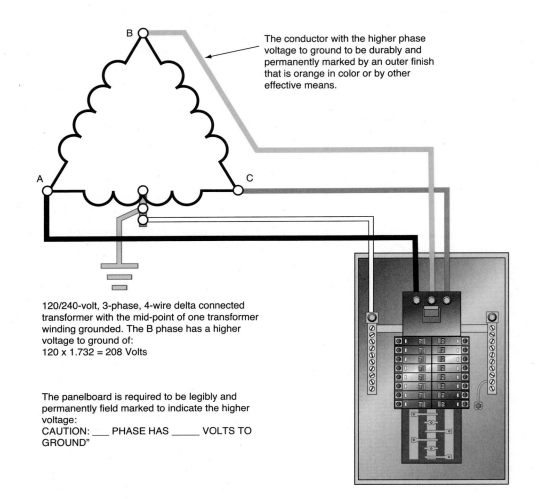

The conductor with the higher phase voltage to ground to be durably and permanently marked by an outer finish that is orange in color or by other effective means.

120/240-volt, 3-phase, 4-wire delta connected transformer with the mid-point of one transformer winding grounded. The B phase has a higher voltage to ground of:
120 x 1.732 = 208 Volts

The panelboard is required to be legibly and permanently field marked to indicate the higher voltage:
CAUTION: _____ PHASE HAS _____ VOLTS TO GROUND"

FIGURE 12-5 Panelboard supplied by 4-wire, delta-connected system. (*Delmar/Cengage Learning*)

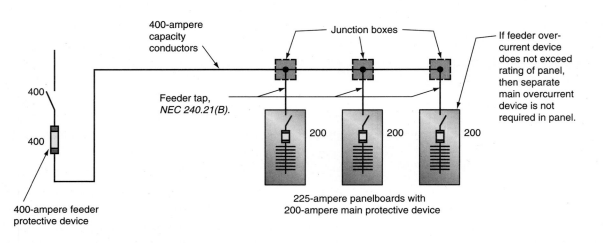

400-ampere capacity conductors

Junction boxes

If feeder over-current device does not exceed rating of panel, then separate main overcurrent device is not required in panel.

400

Feeder tap, *NEC 240.21(B).*

400

200   200   200

400-ampere feeder protective device

225-ampere panelboards with 200-ampere main protective device

FIGURE 12-6 Panelboards with main overcurrent protection in each panel. (*Delmar/Cengage Learning*)

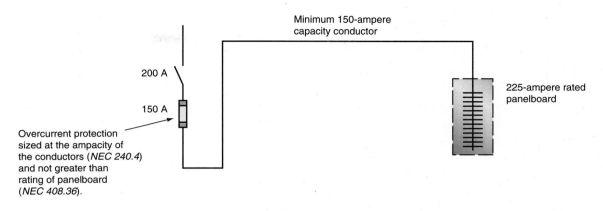

FIGURE 12-7   Panelboard without main, *NEC 408.30* and *408.36*. (*Delmar/Cengage Learning*)

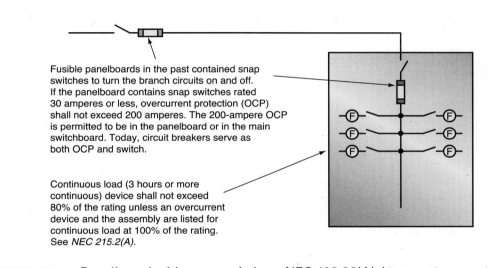

FIGURE 12-8   Panelboard with snap switches, *NEC 408.36(A)*. (*Delmar/Cengage Learning*)

The *NEC* refers to it as the conductor with a *higher voltage to ground.*

The conductor is generally required to be identified with an orange insulation or marking or other effective means. See *NEC 110.15* and *230.56.* Also note that *NEC 408.36(B)* requires that panelboards supplied through a transformer have overcurrent protection on the secondary side of the transformer. An exception allows the provisions of protecting the panelboard by the overcurrent device on the transformer primary through the ratio of the transformer according to *NEC 240.21(C)(1).*

Figure 12-6 shows a feeder supplying more than one panelboard. The feeder overcurrent protection is sized for the ampacity of the feeder conductor.

Individual overcurrent protection is provided in each panelboard. In some instances where the feeder overcurrent protection rating does not exceed the lowest rating of any of the panelboards, separate overcurrent protection in each panelboard is not required. Feeder taps meeting the definition in *NEC 240.2* are shown in the figure. Overcurrent protection must be provided for the conductors in compliance with *NEC 240.21(B).*

In Figure 12-7, a main overcurrent device is not required because the panelboard is protected by the feeder protective device.

Figure 12-8 shows a panelboard that contains snap switches, in which case the overcurrent protection shall not exceed 200 amperes, *NEC 408.36(A).*

Panelboard Cabinet Sizing

If there are special requirements for using the wiring gutters in a panelboard, the physical size of the cabinet shall be sufficient to accommodate those requirements. Often a feeder serves two or more panelboards, in which case either a junction box or wireway is installed or the feeder connects to the first panelboard and then is extended to the others. In some cases, double lugs are installed for the phase conductors and neutral, if used. Other designs have feed-through lugs with the supply-side feeder conductors connected to the line terminals of the first panelboard and conductors of the same size connected from the load terminals of the first panelboard to the line terminals of the second panelboard. Both methods require the installation of additional feeder conductors. The cabinet must have sufficient gutter width for bending these conductors and for the required splices if splices are made. These requirements are illustrated in Figure 12-9 and are given in *NEC 312.6*, *312.7*, *312.8*, and *312.9*.

Panelboard Directory

One of the final actions in installing an electrical circuit or system is to update or create the panelboard directory. A holder will be located on the inside of the panelboard door. An examination of Figure 12-1 reveals that the size and shape of these holders will vary. The single-column schedule shown on the left has the advantage that the numbers appear in numerical order (1, 2, . . . *x*). The advantage of the holder on the right is that the schedule mimics the layout of the overcurrent devices and their numbering (odd numbers on the left, even on the right).

A circuit directory is required by *NEC 110.22* and *408.4*. There is the option of creating a directory in your choice of size and shape and securing it to the inside of the panelboard door. If this is done, a sheet of clear plastic covering the directory will lengthen its life. Securing the plastic sheet only at the top will permit removal to make modifications.

Circuit Directory or Circuit Identification

NEC 408.4 is very clear as to its requirements for preparing a circuit directory. *Every circuit and circuit modification shall be legibly identified*

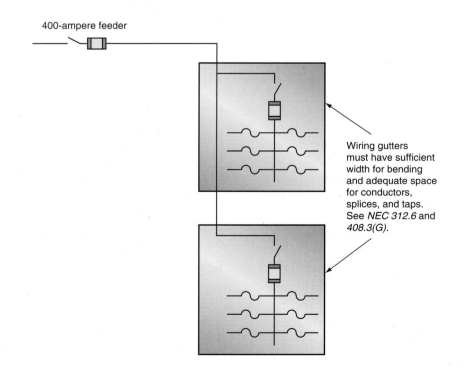

400-ampere feeder

Wiring gutters must have sufficient width for bending and adequate space for conductors, splices, and taps. See *NEC 312.6* and *408.3(G)*.

FIGURE 12-9 Panelboards with adequate space for wiring. (*Delmar/Cengage Learning*)

TABLE 12-1			
Drugstore panelboard directory.			
Cir	**Load Name**	**Load Name**	**Cir**
1	Sales area lighting	Pharmacy luminaires	2
3	Lights-work, toilet, fans	Storage lighting	4
5	Receptacles Main Fl N	Receptacles Main Fl S	6
7	Receptacles Cash registers, floor, toilet and storage	Receptacle for AC	8
9	Receptacles Basement	Receptacle Copier	10
11	Electric doors	Sign outlet	12
13			14
15	Water heater	Air conditioner, outside unit	16
17			18
19			20
21	Air conditioner, indoor unit	7.5 kW unit heater basement	22
23			24
25	Spare	Spare	26
27	Spare	Spare	28
29	Spare	Spare	30
31			32
33			34
35			36
37			38
39			40
41			42

as to its clear, evident, and specific purpose or use. The identification shall include sufficient detail to allow each circuit to be distinguished from all others. Spare positions that contain unused overcurrent devices or switches shall be described accordingly. The identification shall be included in a circuit directory that is located on the face or inside of the panel door in the case of a panelboard, and located at each switch on a switchboard. No circuit shall be described in a manner that depends on transient conditions of occupancy. *

Table 12-1 exhibits a directory prepared for the Drugstore. All directories for the Commercial Building can be easily extracted from the panelboard schedules. It is common for panelboard schedules to appear in the specifications or on the working drawings. The directory gives the "load/area" in a wording that would be clear for "nonelectrical" people to decipher and identify the circuit number for luminaires, receptacles, appliances, or equipment.

Identification of Branch Circuits and Feeders

NEC 210.5(C) requires that where more than one system voltage is used for premise wiring, all panelboards and similar branch-circuit distribution equipment shall be permanently marked to inform persons working on the equipment of the system voltage(s) of how the ungrounded phase conductors are color coded. This requirement includes panelboards, motor controllers, switchboards, and similar electrical equipment.

---

*Reprinted with permission from NFPA 70-2011.

TABLE 12-2

Typical identification of conductors by system.

3-Ph. System Voltage	Phase A	Phase B	Phase C	Neutral
208Y/120 V	Black	Red	Blue	White
480Y/277 V	Brown	Orange	Yellow	Gray

Examples of typical labels that could be attached to electrical panelboards to identify loads served and the color coding of conductors where different system voltages are used, such as 480Y/277 volts and 208Y/120 volts. This is required by *NEC 210.5(C)* for branch circuits, and in *NEC 215.12(C)* for feeders.

Ungrounded conductors and the grounded conductor for branch-circuit wiring are identified at all splice points, terminations, and connections throughout the installation by having established a color coding for branch circuits. This was discussed in detail in Chapter 5 and should be reviewed to make sure you fully understand the concept of conductor identification.

It is obvious that the reasoning for identifying the system voltage(s) and phases is safety. Anyone working on an installation having different voltage systems can readily understand the importance of knowing precisely what the system voltages are and the hazards involved.

This identification requirement applies only to installations that have more than one nominal voltage system. It does not apply to an installation having one nominal voltage system. However, there is nothing stopping you from attaching phase and conductor identification labels to panelboards and distribution equipment even if only one system voltage is present, as in this Commercial Building, which is served by a 208Y/120-volt system.

A similar requirement is found in *NEC 215.12(C)* for feeders. Note that for both branch circuit and feeder conductor identification, the method used for identification is permitted to be documented rather than posted. This means the identification method can be included in information readily available to maintenance staff and to others who might work on the systems.

Most large commercial buildings use two voltages: 480Y/277 volts and 208Y/120 volts. The 480Y/277-volt system generally feeds motor loads, heating, and air-conditioning; and 277-volt serves the lighting loads. The 208Y/120-volt system feeds 120-volt lighting loads, receptacles, and other loads

requiring a 120-volt connection. To establish the 208Y/120-volt system, a 480-volt circuit delivers power to a dry-type transformer to step down the voltage from 480 volts to 208Y/120 volts.

Conductors for 480Y/277-volt systems are generally *brown*, *orange*, and *yellow* for the ungrounded conductors and *gray* for the grounded conductor. Conductors for 208Y/120-volt systems are usually *black*, *red*, and *blue* for the ungrounded conductors and *white* for the grounded conductor.

Table 12-2 is an example of how a panelboard might be marked. On the identifying label, you could use the actual word "Phase" or the electrical symbol Ø. In addition to identifying the ungrounded phase conductors and the grounded conductor of an electrical system(s), some electrical inspectors insist on also color coding switch legs and travelers. Electrical distributors carry labels in a multitude of designs. Labels for dot-matrix and ink-jet printers as well as simple hand-written labels are available.

Identify the Source

►All switchboards and panelboards supplied by a feeder in other than one- and two-family dwellings are required to be marked as to where the power supply originates.◄ See *NEC 408.4(B)*.

The marking can be combined with other markings such as on the panelboard circuit directory. Others may attach a phenolic label to the panelboard cover.

Close Unused Openings!

It seems obvious that there should be no unused openings in panelboards and switches. Sometimes a circuit breaker "knockout" in the cover of a panelboard or a conduit knockout is inadvertently removed, leaving an opening for someone or something to enter the panelboard and come in contact with live parts. Very hazardous! Unused openings, other than those intended for the operation of equipment, those intended for mounting purposes, or those permitted as part of the design for listed equipment, shall be closed to afford protection substantially equivalent to the wall of the equipment. See *NEC 408.7* and *NEC 110.12(A)*.

WORKING SPACE AROUND ELECTRICAL EQUIPMENT

Personnel working on or needing access to electrical equipment shall have enough room and adequate lighting to safely move around and work on the equipment. This will enable them to perform the required function such as examining, adjusting, servicing, or maintaining the equipment. Many serious injuries and deaths have occurred as a result of arc blasts and electrical shock. Should this occur, the personnel must have enough space to get away safely and quickly from the faulted electrical equipment.

NEC 110.26 covers the minimum working space, access, headroom, and lighting requirements for electrical equipment such as switchboards, panelboards, and motor control centers operating at 600 volts or less that are likely to be examined, adjusted, serviced, or maintained while energized. Although common sense as well as the OSHA rules and *NFPA 70E* limit working on energized electrical equipment, sometimes doing so is necessary.

Working Space Considerations

When discussing working space, the major concern is ACCESS TO and ESCAPE FROM the working space.

NEC 110.26 addresses *working space about electrical equipment*, not generally the room itself. Do not confuse the term *working space* with *room*. The concern for safety requires adequate working space around the equipment, which may or may not be the entire room.

The diagrams on the following pages show the required working space in front of electrical equipment. If access is needed on the sides and/or rear of the electrical equipment, the required space shall be provided on the sides and/or rear of the equipment.

Working space requirements are classified by conditions and voltages. As the conditions get more hazardous (for voltages of 150-volts to ground or greater), the depth of the working space increases. The depth of the working space remains at 3 ft for systems 150 volts to ground or lower. In these conditions, metal water or steam pipes; electrical conduits; electrical equipment; metal ducts; all metal grounded surfaces or objects; and all concrete, brick, tile, or similar conductive building materials are considered as "grounded." Refer to *NEC Table 110.26(A)(1)*.

Unless otherwise indicated, equipment likely to be worked on while energized in the following figures is indicated by the following conditions:

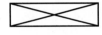

Condition 1—Exposed live parts on one side of the working space and no live or grounded parts on the other side of the working space, or exposed live parts on both sides of the working space that are effectively guarded by insulating materials. See Figure 12-10.

Condition 2—Exposed live parts on one side of the working space and grounded parts on the

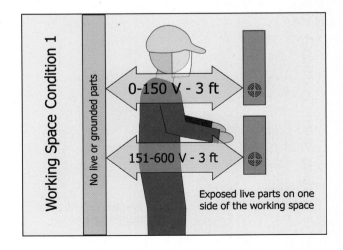

FIGURE 12-10 Working space for Condition 1. (*Delmar/Cengage Learning*)

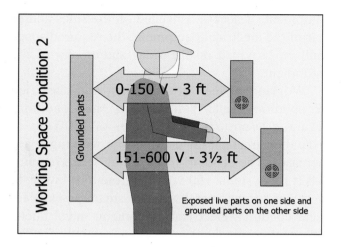

FIGURE 12-11 Working space for Condition 2. (*Delmar/Cengage Learning*)

other side of the working space. Concrete, brick, or tile walls shall be considered as grounded. See Figure 12-11.

Condition 3—Exposed live parts on both sides of the working space. See Figure 12-12.

Different Voltages to Ground . . . Different Working Space Requirements

For system voltages of 0 to 150 volts to ground, the minimum depth of working space is 3 ft (914 mm). An example of this is the 208Y/120-volt,

wye-connected system in the Commercial Building discussed in this text, where the voltage to ground is 120 volts. See Figure 12-10.

For system voltages of 151 to 600 volts to ground, the minimum working space requirement varies from 3 ft to 4 ft (914 mm to 1.2 m), depending upon the Condition. An example of this would be a 480/277-volt, wye-connected system in which the voltage to ground is 277 volts. See Figures 12-11 and Figure 12-12.

The Commercial Building in this text does not have a "high-voltage" service. For systems over 600 volts, the requirements for working clearances, working spaces, guarding, accessibility, and so on, are found in *NEC Article 110, Part III*.

Working Space "Depth" [*NEC 110.26(A)(1)*]

Working space is that space required to provide safe access to and egress from the live parts of electrical equipment. Generally, this is in front of the equipment, but it could be on the sides or back, depending upon the design of the equipment. If there is a need to access live parts from the sides and/or rear of the equipment, minimum working space shall be provided for those needs as illustrated in Figure 12-13.

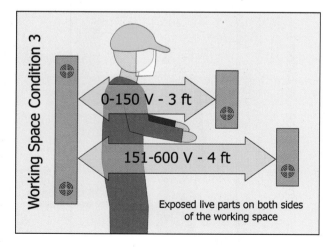

FIGURE 12-12 Working space for Condition 3. (*Delmar/Cengage Learning*)

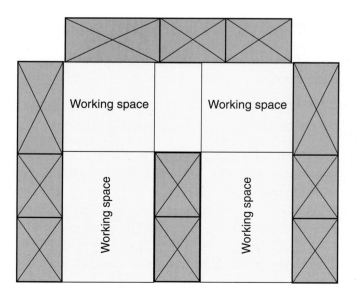

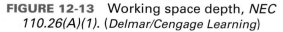

FIGURE 12-13 Working space depth, *NEC 110.26(A)(1)*. (*Delmar/Cengage Learning*)

All working spaces shall allow for the opening of equipment doors and hinged panels to at least a 90° position as is shown in Figure 12-14. This is usually not a problem for panelboards, motor control centers, and switchboards that have covers that are bolted onto the equipment. Some industrial control equipment is mounted in cabinets that have wide doors for providing access.

Working spaces shall be kept clear and not be used for storage as is illustrated in Figure 12-15.

Working space distances are measured from exposed live parts, or from the enclosure "dead front" of panelboards, motor control centers, and switchboards. When this enclosed equipment is "opened," live parts will be exposed; see Figure 12-16.

Working Space "Width"

For equipment with a width equal to or less than 30 in. (762 mm), as shown in Figure 12-17, provide a working space with a width of not less than 30 in. (762 mm). For equipment with a width greater than 30 in. (762 mm), as shown in Figure 12-18, provide a working space width of not less than the width of the equipment. This working space width will enable a person to stand in front of the equipment and not be in contact with a grounded object.

Working Space "Height" [NEC 110.26(A)(3)]

▶Working space height shall not be less than 6.5 ft (2.0 m). For equipment more than 6.5 ft (2.0 m) high, the working space height shall be at least as high as the equipment.◀ This working space height will enable most people to stand in front of the equipment. See Figure 12-19 and *NEC 110.26(E)*.

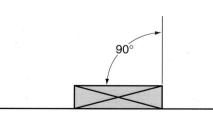

FIGURE 12-14 Hinged door and panels, *NEC 110.26(A)(2)*. (*Delmar/Cengage Learning*)

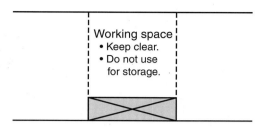

FIGURE 12-15 Clear working spaces, *NEC 110.26(B)*. (*Delmar/Cengage Learning*)

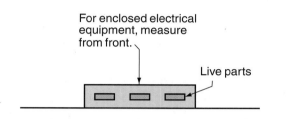

FIGURE 12-16 Measured from live parts, *NEC 110.26(A)(1)*. (*Delmar/Cengage Learning*)

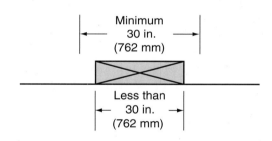

FIGURE 12-17 Equipment 30 in. (762 mm) or less wide. (*Delmar/Cengage Learning*)

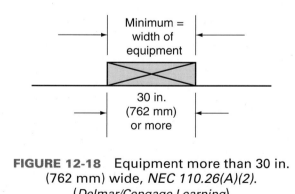

FIGURE 12-18 Equipment more than 30 in. (762 mm) wide, *NEC 110.26(A)(2)*. (*Delmar/Cengage Learning*)

A new exception has been added to this section that allows meters installed in meter sockets to encroach on the working space. It reads, ▶"*Exception No. 2: Meters that are installed in meter sockets shall be permitted to extend beyond the other equipment. The meter socket shall be required to follow the rules of this section."*◄ This new exception has the effect of permitting meter sockets to encroach up to 6 in. (150 mm) into the working space plus the depth of the meter itself. Meters will have different dimensions depending on the type of meter and the manufacturer.

Entrance to and Egress from Working Space

Access to and *egress from* refer to the actual working space requirements about electrical equipment, not to the room itself. However, in instances where the electrical equipment is located in a small room, the working space could, in fact, be the entire room.

For electrical equipment not considered large, at least one entrance of sufficient area required to provide access to and egress from the working space. A door is permitted but not required. Dimensions of this entrance is not given in the NEC. See Figure 12-20.

For electrical equipment rated 1200 amperes (see Figure 12-21) or more and greater than 6 ft (1.8 m) wide, one entrance to and egress from the working space is required at each end of the working space. The access to and egress from this space shall have a minimum width of 2 ft (610 mm) and a minimum height of 6.5 ft (2.0 m). Note that this rule does not specify a door. It is simply the dimensions required for the point of access.

When there is an unobstructed, continuous way of egress travel from the working space, one means of egress is permitted, as illustrated in Figure 12-22.

Reprinted with permission from NPFA 70-2011

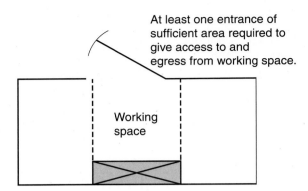

At least one entrance of sufficient area required to give access to and egress from working space.

Working space

FIGURE 12-20 Access to working space, *NEC 110.26(C)(1)*. (*Delmar/Cengage Learning*)

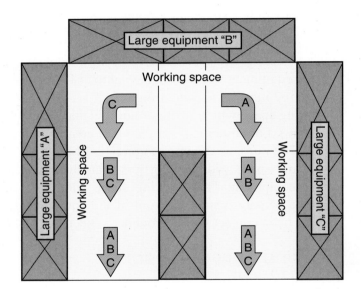

Large equipment "B"

Working space

Large equipment "A"

Working space

Large equipment "C"

Working space

FIGURE 12-21 Space for large electrical equipment, *NEC 110.26(C)(2)*. (*Delmar/Cengage Learning*)

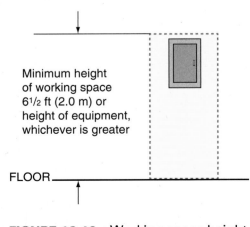

Minimum height of working space 6½ ft (2.0 m) or height of equipment, whichever is greater

FLOOR

FIGURE 12-19 Working space height, *NEC 110.26(A)(3)*. (*Delmar/Cengage Learning*)

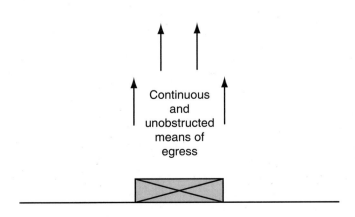

Continuous and unobstructed means of egress

FIGURE 12-22 Unobstructed exit travel, *NEC 110.26(C)(2)(a)*. (*Delmar/Cengage Learning*)

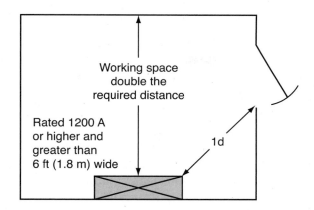

FIGURE 12-23 Single exit, *NEC 110.26(C)(2)(b)*. (*Delmar/Cengage Learning*)

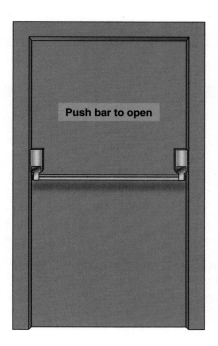

FIGURE 12-24 A typical personnel door that opens easily by pushing on the panic bar. (*Delmar/Cengage Learning*)

If the minimum working space requirements listed in *NEC Table 110.26(A)(1)* are doubled (2d), only one entrance to the working space is required, as shown in Figure 12-23.

Large Equipment Access and Egress

Equipment rated 1200 amperes or more, as shown in Figures 12-20 and 12-22, requires that personnel doors for access and egress open in the direction of egress, and that they *be equipped with panic bars, pressure plates, or other devices that are normally latched but open under simple pressure*. See *NEC 110.26(C) (3)*. This requirement is for the safety of personnel attempting to get out of the working area in the event of a catastrophic arc-fault. An arcing fault can result in a deafening explosion that destroys the ear drums, severe burns from the arc-blast, blinding from the brightness of the arc flash, and a tremendous amount of resulting smoke. When an arc-fault occurs, everything possible must be done to make it as easy as possible to get out of the area—fast! Figure 12-24 is a drawing of a typical personnel door with panic hardware.

Dedicated Electrical Space [NEC 110.26(E)]

A "zone" equal to the width and depth of the electrical equipment and extending from the floor to a height of 6 ft (1.8 m) above the equipment or to the structural ceiling, whichever is lower, shall be dedicated to the electrical installation (see Figure 12-25). This zone shall be kept clear of all piping, ducts, or

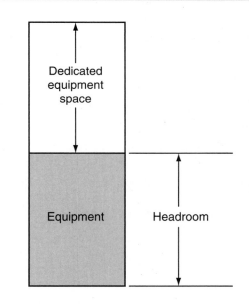

FIGURE 12-25 Dedicated space and headroom, *NEC 110.26(A)(3) and (E)*. (*Delmar/Cengage Learning*)

other equipment that is not related to the electrical installation. See *NEC 110.26(E)(1)(a)*.

Above the dedicated space, foreign systems are permitted, but only if adequate protection is provided so water leaks, condensation, breaks, ruptures, and so on, will not damage the electrical equipment. See *NEC 110.26(E)(1)(b)*.

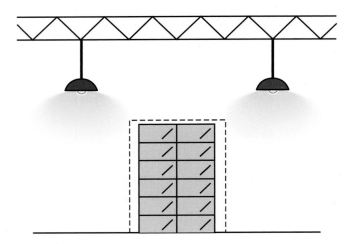

FIGURE 12-26 Illumination, *NEC 110.26(D)*.
(*Delmar/Cengage Learning*)

The minimum headroom (shown in Figure 12-25) for the working space around electrical equipment is 6.5 ft (2.0 m), or the height of the electrical equipment, whichever height is greater.

Illumination *[NEC 110.26(D)]*

For safety reasons, *NEC 110.26(D)* requires that illumination be provided for all working spaces around electrical equipment. This could be the general illumination in the room where the equipment is located, or it might be dedicated illumination (see Figure 12-26). Control for this illumination shall not be by automatic systems only; there shall be a manual means to turn on the illumination. In the Commercial Building, the service equipment is located outside where illumination is not required. The panelboards located inside the building are in normally illuminated spaces so additional dedicated illumination is not required.

SUMMARY

- The working space width shall be not less than 30 in. (762 mm) for each panelboard.
- The working space depth shall be not less than 3 ft (914 mm).
- The working space height shall be not less than 6.5 ft (2.0 m).
- A dedicated equipment space, not less than the depth and width of the panelboard, shall extend from the panelboard to the structural ceiling.
- The illumination in the space shall have a manual control.

REVIEW

Refer to the *National Electrical Code* or the working drawings when necessary. Where applicable, responses should be written in complete sentences.

The *NEC* refers to switchboards and panelboards. For Questions 1 and 2, write the distinguishing characteristics of each. By observing these characteristics, a person would know which type of board was being viewed.

1. Switchboard:

   _____

   _____

   _____

2. Panelboard:

   _____

   _____

   _____

Refer to the Drugstore panelboard directory (Table 12-1) and the Drugstore load calculation form in Appendix A for the information necessary to answer Questions 3–7.

3. The panelboard must be rated for at least _____ ampere.

4. The panelboard has a total of _____ poles.

5. There is space for the addition of _____ 3-pole breaker(s).

6. There is space for the addition of _____ 2-pole breaker(s).

7. There is space for the addition of _____ single-pole breaker(s).

8. How much working space must be provided in front of a 480/277-volt dead front switchboard according to *Table 110.26(A)(1)*? The switchboard faces a concrete block wall.　　　　3 ft (914 mm)　　　　3½ ft (1.0 m)　　　　4 ft (1.2 m)

   Circle the correct answer.

9. A wall-mounted panelboard has a number of electrical conduits running out of the top and bottom of the cabinet. A sheet metal worker started to run a cold-air return duct through the wall approximately 8 ft (2.5 m) high directly above the panel. The electrician stated that this is not permitted, and that this space is only to be used for electrical equipment. He cited *NEC* _____ of the *National Electrical Code* to support his argument. Is the electrician correct? Yes or No. (Circle your answer.)

10. For the service equipment located outside the Commercial Building, what rule describes the installation relative to working space? Refer to *NEC 110.26*.

11. *NEC 408.4* is quite demanding in how to prepare the directories in panelboards. On the jobs you work on, do you (circle your choice)

    a. use a pencil to mark in the circuit directory?

    b. have the circuit directory typed up to be neat and easy to read?

12. You are filling out the directory in a panelboard. Circle the best way to indicate one of the circuits that runs to a computer room:

    a. Computer room

    b. Receptacles in computer room

    c. Receptacle for computer no. 3.

    d. Receptacles in middle of north wall

    e. Tom Jones' office

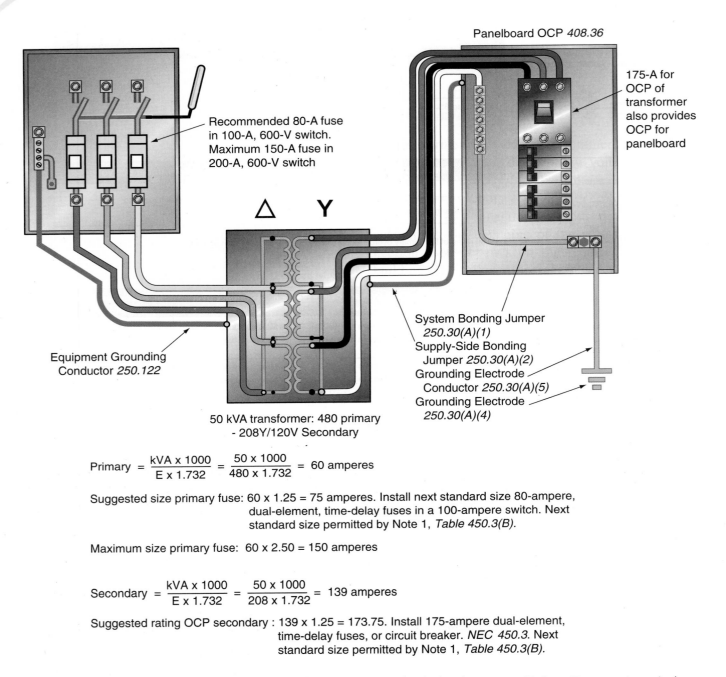

Panelboard OCP *408.36*

Recommended 80-A fuse in 100-A, 600-V switch. Maximum 150-A fuse in 200-A, 600-V switch

175-A for OCP of transformer also provides OCP for panelboard

Δ   Y

System Bonding Jumper *250.30(A)(1)*
Supply-Side Bonding Jumper *250.30(A)(2)*
Grounding Electrode Conductor *250.30(A)(5)*
Grounding Electrode *250.30(A)(4)*

Equipment Grounding Conductor *250.122*

50 kVA transformer: 480 primary - 208Y/120V Secondary

$$\text{Primary} = \frac{\text{kVA} \times 1000}{\text{E} \times 1.732} = \frac{50 \times 1000}{480 \times 1.732} = 60 \text{ amperes}$$

Suggested size primary fuse: 60 x 1.25 = 75 amperes. Install next standard size 80-ampere, dual-element, time-delay fuses in a 100-ampere switch. Next standard size permitted by Note 1, *Table 450.3(B)*.

Maximum size primary fuse: 60 x 2.50 = 150 amperes

$$\text{Secondary} = \frac{\text{kVA} \times 1000}{\text{E} \times 1.732} = \frac{50 \times 1000}{208 \times 1.732} = 139 \text{ amperes}$$

Suggested rating OCP secondary : 139 x 1.25 = 173.75. Install 175-ampere dual-element, time-delay fuses, or circuit breaker. *NEC 450.3*. Next standard size permitted by Note 1, *Table 450.3(B)*.

FIGURE 13-3   Diagram of dry-type transformer separately derived system. (*Delmar/Cengage Learning*)

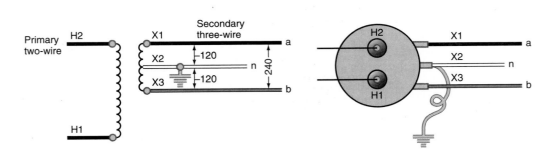

FIGURE 13-4   Single-phase transformer connection. (*Delmar/Cengage Learning*)

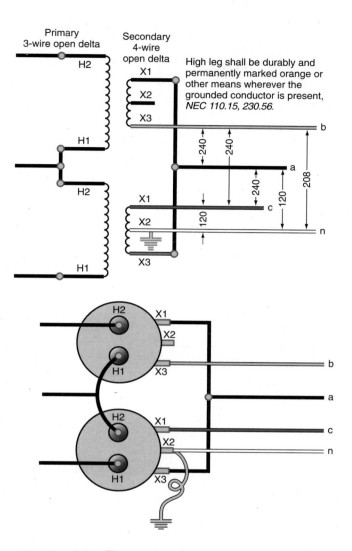

Primary
3-wire open delta

Secondary
4-wire
open delta

High leg shall be durably and permanently marked orange or other means wherever the grounded conductor is present, *NEC 110.15, 230.56.*

FIGURE 13-5 Three-phase open delta connection. (*Delmar/Cengage Learning*)

Open Delta System

This connection scheme has the advantage of being able to provide either 3-phase or 3-phase and single-phase power using only two transformers. It is usually installed where there is a strong probability that the power requirement will increase, at which time a third transformer can be added. The open delta connection is illustrated in Figure 13-5. Another advantage of a 3-phase delta-connected system with three transformers is, if one transformer fails in service, the transformer bank can operate at reduced capacity until a replacement transformer can be installed.

In an open delta transformer bank, 86.6% of the capacity of the transformers is available. For example, if each transformer in Figure 13-5 has a 100-kilovolt-ampere rating, then the capacity of the bank is

$$100 + 100 = 200 \text{ kVA}$$
$$200 \text{ kVA} \times 0.866 = 173 \text{ kVA}$$

Another way of determining the capacity of an open delta bank is to use 57.7% of the capacity of a full delta bank. Thus, the capacity of three 100-kilovolt-ampere transformers connected in full delta is 300 kilovolt-amperes. Two 100-kilovolt-ampere transformers connected in open delta have a capacity of

$$300 \text{ kVA} \times 0.577 = 173 \text{ kVA}$$

When an open delta transformer bank is to serve 3-phase power loads only, the center tap is not connected.

Four-Wire Delta System

This connection provides both 3-phase and single-phase power from an open delta, Figure 13-5, or a closed delta, Figure 13-6. One transformer is midpoint center tapped. This conductor is grounded and becomes the neutral conductor to a single-phase, 3-wire system.

The voltage between phases "A" and "C" is 240 volts. The voltage from "A" to "N" and from "C" to "N" is 120 volts.

The voltage between "B" and "N" is 208 volts. The "B" phase is called the high leg and cannot be used for lighting purposes. The high leg is identified by using orange-colored conductors or by effectively identifying it as the "B" phase conductor whenever it is present in a box or cabinet with the neutral conductor of the system. See *NEC 110.15, 230.56,* and *408.3(E).*

NEC 408.3(F) requires that all switchboards and panelboards where the high leg is present shall be legibly and permanently marked as follows:

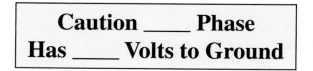

Caution ____ Phase Has ____ Volts to Ground

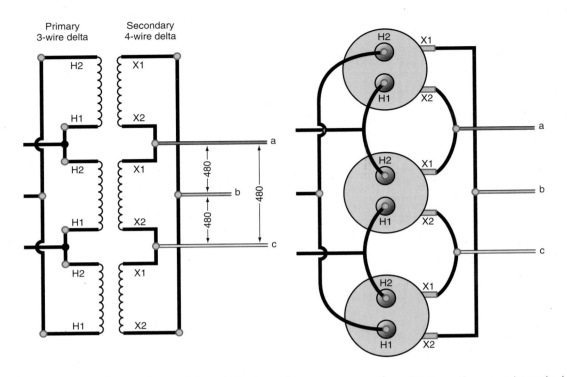

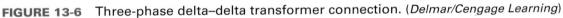

FIGURE 13-6 Three-phase delta–delta transformer connection. (*Delmar/Cengage Learning*)

Three-Wire Delta System

The secondary of this connection, illustrated in Figure 13-6, provides only 3-phase power. One phase of the system may be grounded, in which case it is often referred to as a "corner-grounded delta." The power delivered and the voltages measured between phases remains unchanged. Overcurrent devices are not generally permitted to be installed in the grounded phase. See Chapter 17, Figure 17-22.

Three-Phase, 4-Wire Wye System

The most commonly used system for modern commercial buildings is the 3-phase, 4-wire wye, Figure 13-7. This system has the advantage of being able to provide 3-phase power and permitting lighting to be connected between any of the three phases and the neutral conductor. Typical voltages available with this type of system are 208Y/120, 460Y/265, and 480Y/277. In each case, the transformer connections are the same.

Notes to All Connection Diagrams

All 3-phase connections are shown with the primary connected in a delta configuration. This is most common as only three conductors are required for the primary. Generally, there is no advantage to using a wye connection for the primary. For other connections, it is recommended that qualified engineering assistance be obtained.

UTILITY SUPPLY

The regulations governing the method of bringing the electric power into a building or structure are established by the local utility company. These regulations vary considerably between utility companies. Several of the more common methods of installing the service entrance are shown in Figures 13-8, 13-9, 13-10, and 13-11.

Pad-Mounted Transformers

Liquid-insulated transformers as well as dry-type transformers are used for this type of installation.

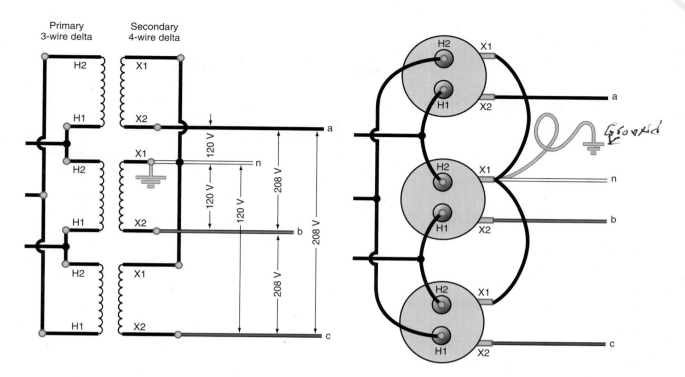

FIGURE 13-7   Three-phase delta–wye transformer connection. (*Delmar/Cengage Learning*)

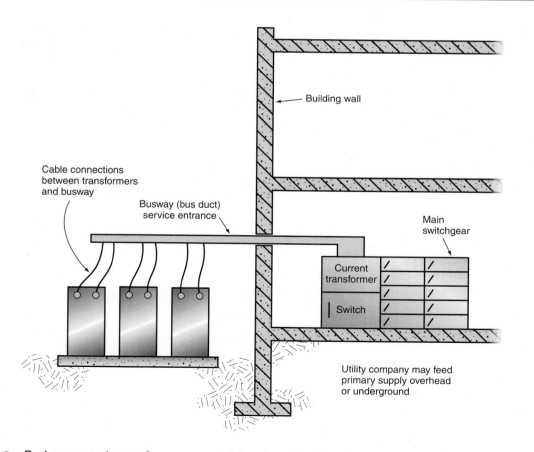

FIGURE 13-8   Pad-mounted transformers supplying bus duct service entrance. (*Delmar/Cengage Learning*)

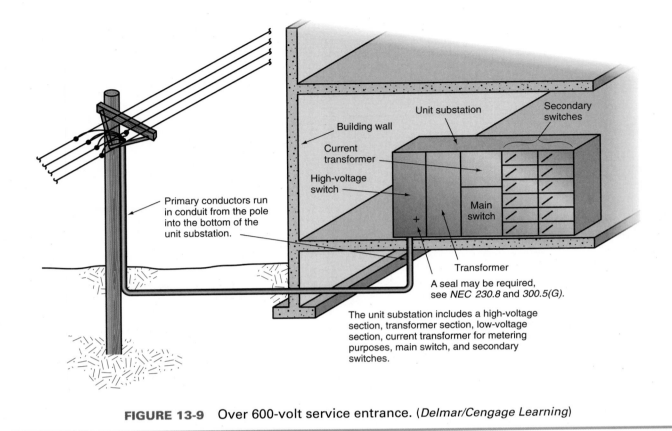

Primary conductors run in conduit from the pole into the bottom of the unit substation.

Building wall

Current transformer

High-voltage switch

Unit substation

Secondary switches

Main switch

Transformer

A seal may be required, see *NEC 230.8* and *300.5(G)*.

The unit substation includes a high-voltage section, transformer section, low-voltage section, current transformer for metering purposes, main switch, and secondary switches.

FIGURE 13-9 Over 600-volt service entrance. (*Delmar/Cengage Learning*)

These transformers may be fed by an underground or overhead service, Figure 13-8. The secondary normally enters the building through a bus duct or large cable.

Unit Substation

For this installation shown in Figure 13-9, the primary runs directly to the unit substation where all of the necessary equipment is located. The owner of the building typically furnishes and installs unit substation equipment. This equipment is often installed where the utility or owner's electrical distribution is rated at the primary voltage. Unit substation equipment is often installed for industrial plants where the electrical energy is distributed by the owner at the primary voltage level.

Pad-Mounted Enclosure

Figure 13-10 illustrates an attractive arrangement in which the transformer and metering equipment

(typically current-transformers) are enclosed in a weatherproof cabinet. This type of installation (also known as a *transclosure*) is particularly adapted to smaller service-entrance requirements.

Underground Vault

This type of service is used when available space is an important factor and an attractive site is desired. The metering may be at the utility pole or in the building, Figure 13-11.

Except when it has been legally rejected or amended or is specifically not covered, all wiring shall conform to the requirements of the *NEC*, NFPA 70. As stated in *NEC 90.2(B)(5)*, installations under the exclusive control of electric utilities are exempt from the *NEC* and are governed by the requirements of the *National Electrical Safety Code*. These regulations are considerably different from the *NEC* and are intended to govern the installation of electrical equipment installed by an electric utility for the purpose of generating electricity,

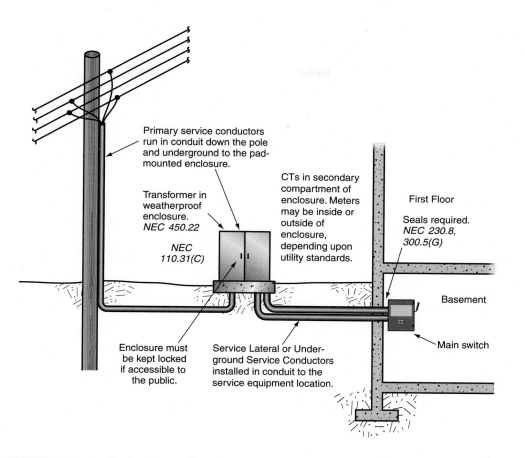

Primary service conductors run in conduit down the pole and underground to the pad-mounted enclosure.

CTs in secondary compartment of enclosure. Meters may be inside or outside of enclosure, depending upon utility standards.

Transformer in weatherproof enclosure. *NEC 450.22*

NEC 110.31(C)

First Floor

Seals required. *NEC 230.8, 300.5(G)*

Basement

Main switch

Enclosure must be kept locked if accessible to the public.

Service Lateral or Underground Service Conductors installed in conduit to the service equipment location.

FIGURE 13-10   Pad-mounted enclosed transformer with underground conductors to service. (*Delmar/Cengage Learning*)

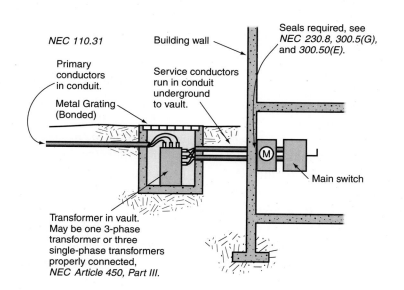

NEC 110.31

Primary conductors in conduit.

Metal Grating (Bonded)

Building wall

Service conductors run in conduit underground to vault.

Seals required, see *NEC 230.8, 300.5(G), and 300.50(E).*

Main switch

Transformer in vault. May be one 3-phase transformer or three single-phase transformers properly connected, *NEC Article 450, Part III.*

FIGURE 13-11   Transformer in underground vault supplying underground service entrance. (*Delmar/Cengage Learning*)

metering, distributing, transforming, and so on, as they function as a utility.

- ▶ *NEC 90.2(B)(5)* clearly states what is not covered by the *NEC* regarding installations under the exclusive control of an electric utility where such installations

  - consist of service drops or service laterals and associated metering; or

  - are on property owned or leased by the electric utility for the purpose of communications, metering, generation, control, transformation, transmission, or distribution of electric energy; or

  - are located in legally established easements or rights-of-way; or

  - are located by other written agreements either designated by or recognized by public service commissions, utility commissions, or other regulatory agencies having jurisdiction for such installations. These written agreements shall be limited to installations for the purpose of communications, metering, generation, control, transformation, transmission, or distribution of electric energy where legally established easements or rights-of-way cannot be obtained. These installations shall be limited to Federal Lands, Native American reservations through the U.S. Department of the Interior Bureau of Indian Affairs, military bases, lands controlled by port authorities and state agencies and departments, and lands owned by railroads. ◀

⎍⎍⎍ METERING

Electricians working on commercial installations seldom make metering connections for other than line- and load-side connections in direct metering equipment, such as plug-in meter sockets. Electricians also install current transformer cabinets, including mounting the current transformers, selecting and bolting lugs on the CTs, and connecting line-voltage conductors. The electric utility typically does all the wiring from the CTs to the meter socket. However, electricians should be familiar with the following two basic methods of metering.

Primary High-Voltage Metering

When a commercial building is occupied by a single tenant, the utility company may elect to meter the primary or high-voltage side of the transformer. To accomplish this, potential transformers and current transformers are installed on the primary high-voltage lines, and the leads are brought to the meter.

Two meters are provided with 15-minute demand attachments, one of which registers kilowatt (kW) and the other, kilovolt-ampere reactive (kVAR) values. These demand attachments will indicate the maximum usage of electrical energy for a 15-minute period during the interval between the readings made by the utility company. The rates charged by the utility company for electrical energy are based on the maximum demand and the power factor as determined from the two meters. A high demand or a low power factor will result in higher rates.

Low-Voltage Metering

For loads of 200 amperes or less, the service conductors from the transformer are run directly to the meter socket line terminals; see Figure 13-12. Conductors from the load side of the meter are run to the supply terminals of the service disconnecting means.

Low-voltage metering of loads greater than 200 amperes is accomplished in various ways. Some electric utilities will require the use of 320-ampere meter sockets for installations with calculated or connected loads greater than 200-amperes and that are within the rating of the equipment. These systems are typically single phase. Other electric utilities require the use of current transformers for metering service loads greater than 200 amperes. If all calculated loads are within the rating of the equipment, meter/main breaker equipment is often installed. This equipment is available in various ampere ratings for the bussing and the individual meter/main breakers. For installations with not more than six meters/mains, a main breaker or fusible switch on the supply side is not required. If more than six service disconnects are required, a main breaker or fusible switch is installed on the supply side.

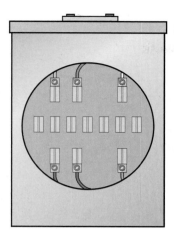 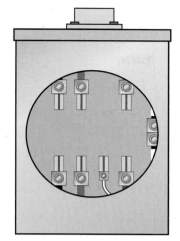 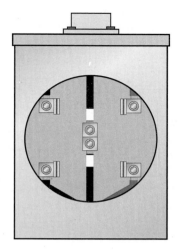

13-terminal, instrument-rated meter socket typically used for 208Y/120 Volt, 3-Phase, 200 to 800-ampere services

7-terminal meter socket typically used for 208Y/120 Volt, 3-Phase, 200-ampere services

4-terminal meter socket typically used for 120/240 Volt, 1-Phase, 200-ampere services

FIGURE 13-12   Meter sockets. Verify type and operating characteristics with service electric utility. (*Delmar/Cengage Learning*)

Other metering designs are used such as up to six meters in individual enclosures ahead of up to six enclosed circuit breakers or fusible switches. This is often the most economical, although the labor to install all the components must be considered and compared to that of installing one unit. This is the method selected for the Commercial Building discussed in this text. Another advantage of this method is that the individual occupancies determine the rating of service disconnecting means and metering. The service can be easily tailored to the load.

IMPORTANT DEFINITIONS

Several significant changes were made in the 2011 *NEC* to the definitions of components of the electrical service. These changes were made to add clarity to where the rules of the *National Electrical Safety Code (NESC)* (apply to the utility system) and the *NEC* begin and end regarding the installation of the service; see Figure 13-13 for a graphic representation. Additional changes were made to the requirements in *NEC Article 230* to incorporate the changes made to the definitions. The significantly changed definitions are as follows (note that words added are underlined and words deleted are crossed out so you can see the changes made):

▶**Service Drop.** *The overhead* <u>service</u> *conductors* <u>between the utility electric supply system and the service point</u> ~~from the last pole or other aerial support to and including the splices, if any, connecting to the service-entrance conductors at the building or other structure.~~ ◀ These conductors are installed by the electric utility according to the NESC and their policies and tariffs.

▶**Service-Entrance Conductors, Overhead System.** *The service conductors between the terminals of the service equipment and a point usually outside the building, clear of building walls, where joined by tap or splice to the service drop* <u>or overhead service conductors.</u> ◀ These conductors are installed by the electrical contractor according to the NEC.

Service-Entrance Conductors, Underground System. *The service conductors between the terminals of the service equipment and the point of connection to the service lateral* ~~or underground service conductors.~~

Informational Note: Where service equipment is located outside the building walls, there may be no

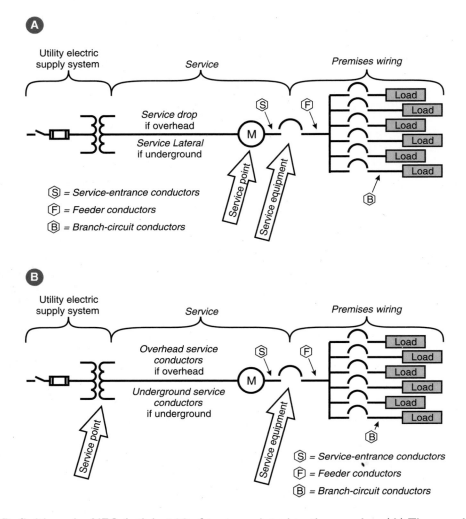

FIGURE 13-13 Definitions in *NEC Article 100* of terms related to the service. (A) The service point is at the building. (B) The service point is at the transformer. (*Delmar/Cengage Learning*)

service-entrance conductors or they may be entirely outside the building.

These conductors are installed by the electrical contractor according to the NEC.

▶**Service Lateral.** *The underground* ~~service con-~~ *ductors between the* utility electric supply system and the service point ~~street main, including any risers at a pole or other structure or from transform-ers, and the first point of connection to the service-entrance conductors in a terminal box or meter or other enclosure, inside or outside the building wall. Where there is no terminal box, meter, or other enclosure, the point of connection is considered to be the point of entrance of the service conductors into the building.~~◀ These conductors are installed

by the electric utility according to the NESC and their policies and tariffs.

Service Point

What code governs an electrical service instal-lation? It depends. These are some questions that need to be answered: Who is responsible for what part of the installation? Who owns and maintains that part of the installation? The *NEC* provides a definition of "service point" that is important in making these determinations. The service point is defined in *NEC Article 100* as *The point of connection between the facilities of the serving utility and the premises wiring. An Informational*

Note has been added to explain this definition and reads, ▶*The service point can be described as the point of demarcation between where the serving utility ends and the premises wiring begins. The serving utility generally specifies the location of the service point based on their conditions of service.*◀

The *NEC* covers all premises wiring beyond the *service point*. Up to the *service point*, the electric utility follows the *NESC* and its applicable public service commission regulations. The service point varies by electric utility based on their tariffs, policies, and procedures. Their practices are almost always governed by a state government agency such as a utilities regulatory commission.

For discussion purposes, in Figure 13-8, the *service point* is the connection to the transformers where the cables from the busway to the transformer have been installed by the electrical contractor. In Figure 13-9, depending on utility policies, the *service point* may be where the service lateral connects to the utility lines at the top of the pole or at the service equipment. Be careful here as *NEC 230.70(A)(1)* requires the location of the service disconnecting means to be *"nearest the point of entrance of the service conductors."* The electrical inspector may require the conduit in the crawl space or basement inside the building to be encased in concrete or to otherwise comply with *230.6* to be considered outside the building. In Figure 13-10, the *service point* is either where the service-lateral conductors connect to the utility lines at the top of the pole or at the transformer. For many services, the utility installs, owns, and maintains pad-mounted transformers that are installed outdoors. The underground conduit, the conduit riser up the pole, and the primary conductors are either installed and are maintained by the building owner or the electric utility, depending on utility policies and agreements. In Figure 13-11, the *service point* is at the secondary terminals of the transformer in the vault. The underground raceway and the primary service conductors have been installed and are maintained by the electric utility. The secondary conductors and corresponding raceways have been installed by the electrical contractor and are owned by the building owner.

SERVICE-ENTRANCE EQUIPMENT

When the transformer is installed at a location away from the building, the service-entrance equipment consists of the service-entrance conductors, the main switch or switches, the metering equipment, and the secondary distribution switches, Figure 13-14. The Commercial Building shown in the plans is equipped in this manner and will be used as an example for the following paragraphs.

The Service

The service for the Commercial Building is similar to the service shown in Figure 13-14. A pad-mounted, 3-phase transformer is located outside the building, and PVC conduit running underground serves as the service raceway. Note than many inspection jurisdictions permit direct buried cable to be used for the service lateral. PVC conduit is installed at the transformer and service equipment locations to protect the conductors from physical damage. *NEC 300.3* requires that single conductors specified in *Table 310.104(A)* are only installed as part of a recognized wiring method of Chapter 3.

The service-entrance equipment is located outside the Commercial Building. This equipment and the service and feeder raceways are shown in Figure 13-14.

Underground Service Conductors

The underground service conductors from the transformer are routed to a termination box that is installed on the outside wall of the building. Previously, these *underground service conductors* were referred to as *service lateral conductors*. They are required to be sized for the calculated load according to *NEC Article 220* by *NEC 230.31(A)*. The calculated load on the service is 850 amperes, as shown in the service load calculation form.

Rating of Underground Service Conductors and Termination Box

The load calculation for the service determines the minimum size of the underground service

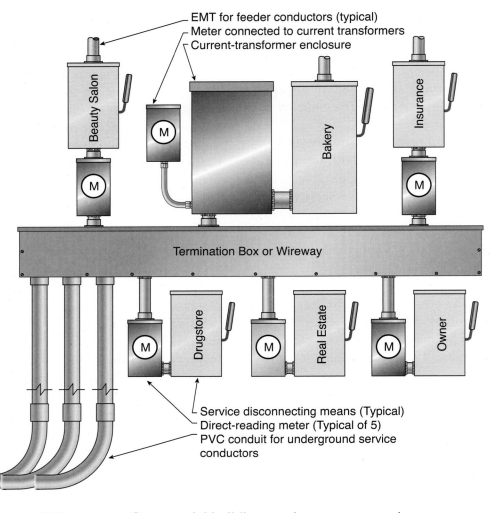

EMT for feeder conductors (typical)
Meter connected to current transformers
Current-transformer enclosure

Beauty Salon

Bakery

Insurance

Termination Box or Wireway

Drugstore

Real Estate

Owner

Service disconnecting means (Typical)
Direct-reading meter (Typical of 5)
PVC conduit for underground service conductors

FIGURE 13-14 Commercial building service-entrance equipment, pictorial view. (*Delmar/Cengage Learning*)

conductors and the rating of the termination box. The minimum rating of the individual service disconnecting means was determined by the load calculation for the feeder. As mentioned previously, a panelboard worksheet and load calculation form are provided in Appendix A of this text. Panelboard schedules are shown on blueprint sheet E-4. Table 13-1 shows the load calculation for the service.

Note that very few demand factors can be applied for the Commercial Building load calculation. These demand factors that are applied include the following:

- The branch circuit for the electric heat is required to be sized at 125% of the load by *NEC 424.3(B)* but the feeder is permitted to be sized at 100% of the load by *NEC 220.51*.

- Receptacle loads determined from *NEC 220.14(H)* for multioutlet assemblies or at 180 volt-amperes in accordance with *NEC 220.14(1)* are permitted to be subject to the demand factors of *NEC Table 220.44*. The first 10 kVA is calculated at 100% and the remainder at 50%.

- A demand factor of 65% is applied to both the feeder for the Bakery kitchen equipment and to the load calculation for the service. This is permitted in *NEC Table 220.56* as explained in *NEC 220.56*.

Other loads are calculated at 125% of the continuous loads and 100% of the noncontinuous load.

The footnotes to the calculation table explain the application of loads in the various columns.

TABLE 13-1

Service load calculation.

	Count[1]	VA/Unit[2]	*NEC* Load[3]	Connected Load[4]	Continuous Loads[5]	Noncontinuous[6]	Neutral Load[7]
General Lighting:							
NEC 220.12	5160	3	15,480				
Connected loads from Worksheets				4887			
Totals:			15,480	4887	15,480	0	15,480
Storage Area:							
NEC 220.12	2580	0.25	645				
Connected load from worksheet				3285			
Totals:			645	3285	0	3285	3285
Other Loads:							
Receptacle outlets	96	180	17,280	0	0	10,000	10,000
7280 VA @ 50%	0	0	3640	0	0	3640	3640
Other receptacle outlets	23	9780	2400	21,900	6000	18,300	24,300
Motors & Appliances:	Count[1]	VA/Unit					
Kitchen equipment[9]	6	71,098	0	71,098	36,992	9222	0
Exhaust fans	4	708	0	708	0	708	708
HVAC equipment[8]	12	1,18,640	0	109,532	54,656	44,760	0
Washer/dryer (BS)	1	4000	0	0	0	4000	832
Water heaters	7	39,300	0	40,800	40,800	0	6300
Elevator	1	9000	0	9000	0	9000	0
Elevator equipment	1	9600	0	9600	0	9600	0
Sump pump	1	1127	0	1127	0	1127	1127
				Total Loads	153,928	113,642	65,672
Minimum feeder load 125% of continuous loads plus 100% of noncontinuous:						306,051	65,672
					Minimum ampacity:	850.1	182.4

Minimum (3) 350 kcmil ungrounded and (3) 1/0 AWG neutral in parallel.

Minimum size Schedule 40 PVC = (0.5242 × 3) + 0.1825 = 1.7551 sq. in. = 2½ in.

[1] Count: Square footage of given area, the number of a given type load, or the ampere rating of the given type load.

[2] VA/Unit: Specific unit load values required by *NEC Article 220,* or actual load values for the given type load.

[3] *NEC* Load: Count × *NEC* VA/Unit. Also the "calculated load."

[4] Connected Load: Actual load connected to the electrical system.

[5] Continuous Loads: The maximum current is expected to continue for 3 hours or more. Column also used to apply 125% to the largest motor load.

[6] Noncontinuous: Loads that are not continuous.

[7] Neutral Load: The maximum unbalanced load calculated according to *NEC 220.61.*

[8] Because the heating and cooling loads are noncoincident, only the larger load is included.

[9] A 65% factor is applied to commercial cooking equipment as allowed by *220.56.*

The ungrounded conductors and termination box must have an allowable ampacity or rating of not less than 850 amperes. The grounded (neutral) conductors must have an ampacity of not less than 182 amperes.

Because the service conductors are installed underground and on the outside of the building, they are in a wet location. They also are subject to the ambient temperatures where the building is located. The conductors must have an allowable ampacity of not less than 850 amperes after application of any correction or adjustment factors. For our purposes, we will assume the building is located where the summer ambient temperature is from 96°F to 104°F (36°C to 40°C). It is anticipated that between 4 and 5 ft (1.2 to 1.5 m) of underground service conductors are installed above ground and are exposed to the higher ambient temperature. As a result, we are unable to apply the 10% exception to *NEC 310.15(A)(2)*. (We discussed this exception previously in detail.)

We are anticipating that the neutral conductor is not required to be counted as a current-carrying conductor under the terms of *NEC 310.15(B)(5)* as the major portion of the load is not nonlinear loads. Of the 182 amperes (65,672 VA) of neutral load shown in the service load calculation form, a maximum of 17,221 VA are lighting loads. If all lighting loads are nonlinear, they represent approximately 26% of the neutral load. This is determined as follows:

$$\frac{17,221}{65,672} = 26.2\%$$

The selection of the underground service conductors is demonstrated in Table 13-2.

As can be seen, the three 350 kcmil conductors with THHN/THWN insulation connected in parallel are not large enough after application of the correction factor for elevated temperature.

The three 300 kcmil conductors with XHHW-2 insulation connected in parallel are adequate because the correction factor is applied to their 90°C allowable ampacity. The 75°C ampacity of these conductors of 855 amperes (3 × 285) exceeds the 850-ampere calculated load. The three 350 kcmil conductors connected in parallel are obviously adequate as their allowable ampacity after application of the correction factor for the high ambient temperature is 955 amperes. We are choosing to install the three 350 kcmil conductors with XHHW-2 insulation that are connected in parallel as they provide adequate spare capacity.

Termination Box at Service

As shown in Figures 13-14 and 13-15, a termination box is installed at the service equipment to facilitate connections between the underground service conductors and the service-entrance conductors that supply each of the service switches. The term "termination box" is often thought of as a short-run busway or a tap box. This "busway" is referred to in the UL *White Book* as a "Termination Box" and has a Guide Card category of XCKT.

In Figure 13-14, the termination box appears as though it might be a wireway. However, the termination box contains copper busbars, as shown in Figure 13-15. This termination box is designed to make it easy to make connections of the service-entrance conductors that supply each service disconnecting means.

TABLE 13-2					
Selection of underground service conductors.					
Conductor Insulation	Size	Ampacity in Wet Location	Correction Factor	Corrected Ampacity Each	Ampacity for Three Conductors in Parallel
THHN/THWN	350 kcmil	310	0.88	272.8	818.4
XHHW-2	300 kcmil	320	0.91	291.2	873.6
XHHW-2	350 kcmil	350	0.91	318.5	955.5

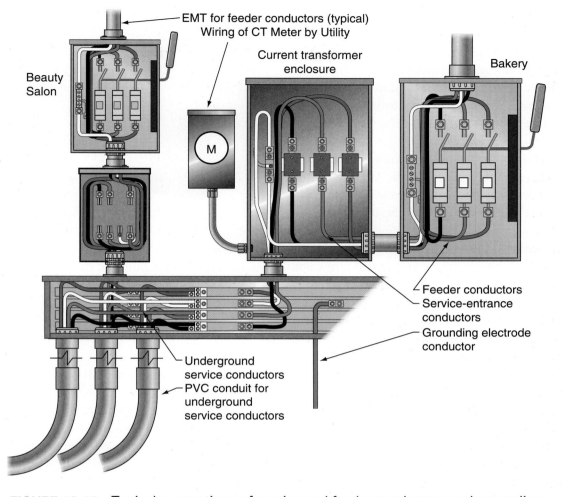

Beauty
Salon

EMT for feeder conductors (typical)
Wiring of CT Meter by Utility

Current transformer
enclosure

Bakery

M

Feeder conductors
Service-entrance
conductors
Grounding electrode
conductor

Underground
service conductors
PVC conduit for
underground
service conductors

FIGURE 13-15 Typical connections of service and feeder conductors and grounding
electrode conductor. (*Delmar/Cengage Learning*)

Information in the *White Book* states, "This category covers termination boxes rated 600 V or less that (1) consist of lengths of busbars, terminal strips, or terminal blocks with provision for wire connectors to accommodate incoming or outgoing conductors or both, and (2) are intended to be used in accordance with ANSI/NFPA 70, *National Electrical Code*. Termination boxes have a rating in amperes based on the size of the bus located within the termination box.

"Termination boxes do not contain switching devices, overcurrent protective devices, or any control components (see RELATED PRODUCTS).

"This category also covers termination bases to be field installed in termination boxes, and termination boxes in which termination bases are to be field installed.

"USE AND INSTALLATION

"Termination boxes rated and marked for use on the line side of service equipment may also be used on the load side of service equipment. Termination boxes not marked for use on the line side of service equipment and rated 100 A or less are only for use on the load side of service equipment.

"Termination boxes may have knockouts or openings for the connection of cable fittings, conduit or electrical metallic tubing. They may also have openings for connection with openings in other equipment, such as meter sockets, panelboards, switch or circuit breaker enclosures, wireways, raceways and the like.

"Termination boxes are intended for use with copper conductors unless marked to indicate which terminals are suitable for use with aluminum conductors. Such marking is independent of any marking on terminal connectors and is on a wiring diagram or other readily visible location.

"Termination boxes intended for use with field-installed wire connectors are marked stating which pressure terminal connectors, component terminal assemblies or termination bases are to be used.

"Factory-installed field wiring connectors requiring the use of a special tool (such as crimp connectors) are provided with instructions concerning the proper tool to be used for the termination of conductors.

"Termination boxes are marked with their short-circuit current ratings in rms symmetrical amps and with the words 'Short-Circuit Current Rating.'

"Termination boxes are marked with an enclosure type as described in Electrical Equipment for Use in Ordinary Locations (AALZ). Termination boxes marked with an enclosure Type designation of Type 3, 3S, 4, 4X, 6 or 6P may additionally marked 'Raintight.' A termination box marked Type 3R may additionally be marked 'Rainproof.'

"Termination boxes suitable for use on the line side of service equipment are marked 'Suitable for Use on the Line Side of Service Equipment,' or equivalent."

The electrical contractor is responsible to select a terminal box (resembles a short-run busway) that has the correct rating for the calculated load current for the service. The terminal box must also be rated for its location on the supply side of the service and for the wet location on the outside of the building.

Installation of Wireways

Figures 13-14 and 13-15 show the installation of a termination box to facilitate connections of the service disconnecting means. A wireway could just as easily have been installed for this purpose. Obviously, if a wireway is installed rather than the termination box, connections of the conductors will need to be made with split bolts or tap connectors. One of the challenges of using split bolts for these connections is that all of the parallel underground service conductors must be connected to the service-entrance conductors at the same point. Split bolt connectors are typically permitted to connect no more

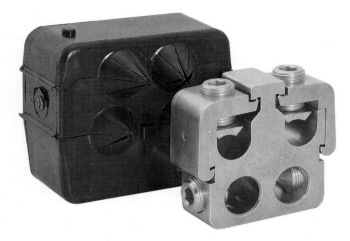

FIGURE 13-16 A pressure wire connector with insulator that is suitable for connecting service-entrance conductors to the parallel underground service conductors. (*Photo courtesy ILSCO*)

than two conductors. So, a connector designed for the purpose of connecting the parallel underground service conductors and the service-entrance conductors together at the same point must be used. See Figure 13-16 for a photo of a connector specifically designed for this purpose. Connectors are available in several configurations for parallel and tap connections including 3 parallel and one or two tap connections. Another bonus of using this connector is that it is available with a snap-on insulator, so the laborious task of taping up the connection is unnecessary.

Uses permitted for metal wireways are covered in *NEC 376.10*. Uses not permitted are covered in *NEC 376.12*. *NEC 376.22* provides an incentive for using wireways as for the purposes shown in Figure 13-14. Although the cross-sectional area is limited to 20% fill, so long as the number of current-carrying conductors in the wireway does not exceed 30, an adjustment (derating) for containing more than three current-carrying conductor is not required.

Table 13-3 shows the correct method for sizing a wireway, if used, for the service installation.

As can be seen in Table 13-3, we end up installing two wireways connected by a nipple. We chose to install two wireways so the application of an adjustment factor due to having more than 30 current-carrying conductors in the wireway is not required. Note that the rule in *NEC 376.22(B)* exempts the requirement for applying an adjustment factor so long as the number

TABLE 13-3

Sizing wireways for service.

Section 1	Size	Insulation	Number	In.² Area	Area
Underground Service Conductors	350	XHHW-2	9	0.5166	4.6494
	1/0	XHHW-2	3	0.1825	0.5475
Beauty Salon	3/0	THHN/THWN	3	0.2679	0.8037
	2	THHN/THWN	1	0.1158	0.1158
Bakery	350	THHN/THWN	3	0.5242	1.5726
	1/0	THHN/THWN	1	0.1855	0.1855
Drugstore	1/0	THHN/THWN	3	0.1855	0.5565
	4	THHN/THWN	1	0.0842	0.0842
			24	Total In.² Area	8.5152
			Minimum Square-inch Area of Wireway		**42.6**

Section 2	Size	Insulation	Number	In.² Area	Area
Underground Service Conductors	350	XHHW-2	9	0.5166	4.6494
	1/0	XHHW-2	3	0.1825	0.5475
Real Estate	3	THHN/THWN	3	0.0973	0.2919
	8	THHN/THWN	1	0.0366	0.0366
Insurance	1/0	THHN/THWN	3	0.1855	0.5565
	4	THHN/THWN	1	0.0842	0.0842
Owner	1/0	THHN/THWN	3	0.1855	0.5565
	4	THHN/THWN	1	0.0842	0.0842
			24	Total In.² Area	6.8068
			Minimum Square-inch Area of Wireway		**34.0**

of current-carrying conductors in the wireway does not exceed 30. There would be 36 current-carrying conductors if we install one long wireway and count all the current-carrying conductors. We are counting the neutral conductors as current-carrying according to *NEC 310.15(B)(5)*. You may be able to convince the authority having jurisdiction that the majority of the load on the neutral conductors is linear and the neutral conductors should be excluded. Counting 36 conductors as current-carrying forces us out of the provisions of *NEC 376.22(B)* and into the ampacity adjustment requirements of *NEC Table 310.15(B)(3)(a)*. We would have to apply an adjustment factor of 40%. In other words, a conductor with an allowable ampacity of 420 amperes has a reduced ampacity of 420 × .4 = 168 amperes. It seems more practical to install two sections of wireway to avoid the penalty.

As shown in Table 13-3, calculate the area of the conductors installed in the wireway. Then, multiply this conductor area by 5 to get the minimum area of a wireway. You will want to select the next larger size unless a wireway is being custom built for the job. The minimum square-inch wireway in standard dimensions is 8 in. × 8 in. = 64 in.² (20.3 cm × 20.3 cm = 412.9 cm²).

A caution on installing wireways: Verify that the wire-bending requirements of *NEC 312.6(A)* are satisfied if the conductors are deflected in the wireway. In our example, the largest conductor to be installed is a 350 kcmil copper underground service conductor. Even though the conductors are being installed in parallel for efficiency, and three conductors per terminal would normally be required, *NEC 378.23(A)* allows one wire per terminal in *NEC Table 312.6(A)* to be applied. By reference to *NEC Table 312.6(A)*, we see the minimum height of the wireway where the 350 kcmil conductors are being deflected is 5 in. (203 mm). Obviously, an 8 in. (20.3 mm) wireway satisfies the wire-bending space rules.

Service Disconnect Location

NEC 230.70(A)(1) requires that *The service disconnecting means shall be installed at a readily accessible location either outside of a building or structure or inside nearest the point of entrance of the service conductors.* * Service conductors do not have overload protection until they reach the main service disconnecting means overcurrent protection.

The only overcurrent protection for those service conductors is the transformer primary fuses that are sized by the utility to protect its transformer—not the secondary service-entrance conductors. If service conductors are installed inside a building or structure, they must be kept as short as possible because they are not provided with overcurrent protection.

There is a service switch and meter for each of the panelboards. One service disconnecting means is shown for each tenant and one for the owner's equipment.

Figure 13-17 provides a one-line diagram of the service equipment. For each occupant, there is

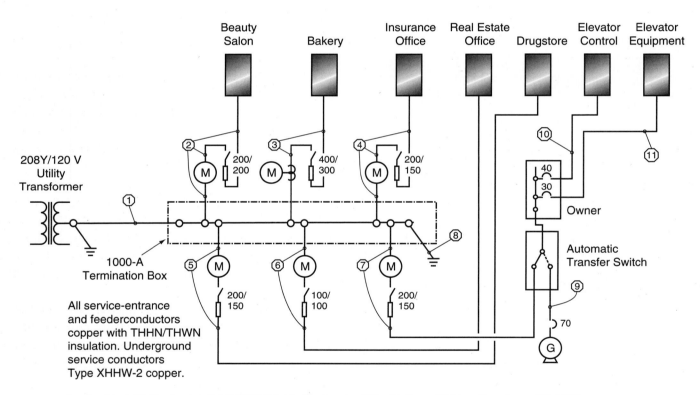

1 - Service: (3) Trade size 2-1/2 PVC (Metric Designator 63) with three 350 kcmil and one 1/0 AWG
2 - Beauty Salon: Trade size 2 EMT (Metric Designator 53) with three 3/0 AWG and one 2 AWG
3 - Bakery: Trade size 2-1/2 EMT (Metric Designator 63) with three 300 kcmil and one 4 AWG
4 - Insurance Office: Trade size 1-1/2 EMT (Metric Designator 41) with three 1/0 AWG and one 4 AWG
5 - Drugstore: Trade size 1-1/2 EMT (Metric Designator 41) with three 1/0 AWG and one 4 AWG
6 - Real Estate Office: Trade size 1 EMT (Metric Designator 27) with three 3 AWG and one 8 AWG
7 - Owner: Trade size 1-1/2 EMT (Metric Designator 41) with three 1/0 AWG and one 4 AWG
8 - Grounding electrode conductor 2/0 AWG to metal water pipe with 4 AWG jumper to concrete-encased grounding electrode.
9 - Feeder to transfer equipment: Trade size 1-1/4 EMT (Metric Designator 35) with four 4 AWG
10 - Feeder to elevator control panel: Trade size 3/4 EMT (Metric Designator 21) with three 8 AWG
11 - Feeder to elevator equipment panelboard: Trade size 3/4 EMT (Metric Designator 21) with four 10 AWG

FIGURE 13-17 One-line diagram for the service for the Commercial Building. (*Delmar/Cengage Learning*)

a meter, an overcurrent protection device (a set of fuses), a service/feeder disconnect switch, and terminals for connection of the feeder. Another set is provided for the owner. As many as six disconnects may be installed at one location and be in compliance with *NEC 230.71(A)*.

It is possible that the sum of the feeder overcurrent devices is greater than the rating of a main service overcurrent device if one were installed. This is because each feeder is calculated for the load on that feeder. The service-entrance conductors and overcurrent devices are calculated according to the requirements for service-entrance calculations, which may permit applying a demand factor.

Adding the feeder fuse sizes in the Commercial Building will yield a sum of 1050 amperes. Since the calculated load is 850 amperes, if a main breaker were installed, a 1000-ampere frame size with an adjustable trip set at 900 amperes would be permitted. If a fusible switch were installed, a 1,200 ampere switch with 900 ampere fuses would be permitted. Keep in mind, we have oversized several of the feeders to allow for spare capacity, as shown in Table 13-4.

In Chapter 8, "Feeders," the process for sizing the feeder to the Drugstore panelboard was explained in detail. A panelboard worksheet, panelboard schedule, and load calculation have been developed for each occupancy. A load calculation and schedule has been developed for the service. All

documents are located in Appendix A or on blueprint sheet E-4.

Noncoincident load is provided for in the minimum branch circuit capacity entry in the nameplate information for the HVAC equipment as well as in the load calculation for each feeder. The circulating fan motor (evaporator motor) in the equipment operates for both heating and cooling functions. The electric heating coils operate only when the thermostat calls for heat. Similarly, the compressor runs only when the thermostat calls for cooling. Only the greater of the heating or cooling load, plus the circulating fan motor is used in the feeder and service load calculation.

Ampere Rating of Service Disconnects

NEC 230.79 tells us that *The service disconnect shall have a rating not less than the calculated load to be carried, determined in accordance with Parts III, IV, or V of Article 220.* * These calculations are shown in this chapter, tenant by tenant.

NEC 230.71 permits not more than six disconnects to serve as the building service disconnecting means. The combined ampere ratings of all of the switches or circuit breakers (not more than six) shall be not less than the ampere rating of a single disconnecting means as calculated in *NEC 230.79*.

The Commercial Building has six service disconnects. This can be cost effective. One service disconnect for the owner's loads and five for the tenant's power needs. See Figures 13-14 and 13-17.

With a single main disconnect, shutting that main disconnect turns the power off to the building's service-entrance equipment. Power up to and including the line side of the single main disconnect is still on. Power on the load side of the single disconnect has been "killed." One motion of the hand shuts all the power off.

Without a single main disconnect, it will take two, three, four, five, or six motions of the hand to shut off the power. The *NEC* permits up to six disconnects to serve as the service disconnecting means for each service permitted by *NEC 230.2*.

TABLE 13-4

Spare capacity in electrical system.

Occupancy	Calculated Load	Installed Feeder	Spare Capacity
Beauty Salon	171.5 A	200	28.5 A
Bakery	241.3 A	300	58.7 A
Insurance Office	117.2 A	150	32.8 A
Drugstore	136.4 A	150	13.6 A
Real Estate Office	59.6 A	100	40.4 A
Owner	122.1 A	150	27.9 A
Service	850 A	Service conductors 930 A	80 A

Service-Entrance Conductor Size

A set of service-entrance conductors is installed from the termination box to each service disconnecting means. The minimum size of the service-entrance conductors is determined by performing a load calculation in accordance with *NEC Article 220*. The load calculation form for each occupancy can be found in Appendix A of this text.

On reviewing the load calculation form for any of the occupancies, you will find the service disconnecting means and overcurrent protection is sized not smaller than 125% of the continuous load and 100% of the noncontinuous loads. In addition, the minimum size of the neutral or grounded conductor is determined.

Finally, the minimum size is determined for the raceway that encloses the service-entrance conductors and feeders.

The following table shows the switch rating, the rating of the fuse in the switch, and the sizes of the ungrounded (hot) and grounded (neutral) as well as the size of the EMT to be installed to enclose the conductors.

Sizing the Neutral Conductors

Several factors come into play when sizing the neutral conductors for the service as well as for the feeders. Requirements include these:

- *NEC 215.2(A)(1)* requires that the grounded conductor be not smaller than an equipment grounding conductor determined from *NEC Table 250.122*. We made adjustments in the size of some neutral conductors as shown in the load calculation forms. All are included in Appendix A.

- *NEC 220.61* provides procedures for calculating the feeder or service neutral load. We used these requirements in calculating the neutral load for each feeder and service disconnecting means. These calculations are included in Appendix A.

- *NEC 250.24(C)(1)* requires that the grounded conductor, that is, the neutral conductor, in a service shall not be smaller than the required grounding electrode conductor. Grounding electrode conductor sizes are given in *NEC Table 250.66*. Table 13-5 shows compliance with this rule.

- In *NEC 250.24(C)(2)* the grounded conductor is not permitted to be smaller than 1/0 AWG when part of a parallel set if the conductors are installed in separate raceways. This rule applies to the underground service conductors as they are installed in parallel but does not apply to any of the service disconnecting means as none of these have parallel service conductors.

Optional Service for Electric Boiler

The operation of the boiler is discussed in Chapter 11, "Special Circuits (Owner's Circuits)" and is shown as an option for heating the building. The supply would be 3-phase, and no neutral

TABLE 13-5

Sizing service conductors and raceways.

Service Equipment Enclosure: Surface				Voltage: 208Y120 3 Ph. 4-W Location: West End of Building				Min. AIC Rating: 32 kAIC Fed From: Util. Transformer		
Cir	Load Name	Switch/ Fuse	Service Entrance Conductors/ Feeders	Raceway	Raceway	Service-Entrance Conductors/ Feeders	Switch/ Fuse	Load Name	Cir
1	Beauty Salon	200/200	3/0, 2	2	2½	350, 1/0	400/300	Bakery	2
3	Insurance Office	200/150	1/0, 4	1½	1½	1/0, 4	200/150	Drugstore	4
5	Real Estate Office	100/100	3, 8	1	1½	1/0, 4	200/150	Owner	6

Underground Service Conductors: (3) sets of 350 kcmil ungrounded conductors and 1/0 AWG neutral conductors in 3 trade size 2½ (metric designator 63) PVC conduits

TABLE 13-6

Optional Commercial Building boiler branch-circuit calculation.

Load Summary, 208Y 3-Phase, 4-Wire	Connected Load	Calculated Load
Continuous load	200,000	250,000
OCPD Selection		
OCPD selection amperes	556	695
OCPD ampere rating		700
Minimum conductor sets and size	2	500 kcmil CU
Allowable ampacity	380 × 2	760
Raceway Size Determination		
Circuit conductors size and area	500 kcmil = 0.7073 in.² × 3 =	2.1219 in.² (1369 mm²)
Equipment grounding conductor size and area	1/0 AWG	0.1855 in.² (119.7 mm²)
Total conductor area		2.3074 in.² (1489 mm²)
Minimum raceway type and trade size	RMC EMT	Trade Size 3 Trade Size 2½

is required. An equipment grounding conductor would be installed in the feeder raceway because of the necessity of installing a short section of flexible metal conduit to reduce vibration transmissions. The sizing of this equipment grounding conductor is discussed later in this chapter. The branch-circuit conductors to the boiler are required to have an ampacity of not less than 125% of the load to be in compliance with *NEC 424.82. NEC 220.5(B)* simplifies calculations when it states:

Fractions of an Ampere. ▶*Calculations shall be permitted to be rounded to the nearest whole ampere, with decimal fractions smaller than 0.5 dropped.*◀

As shown in Table 13-4 and in the specifications, the electric boiler has a rating of 200 kilowatts at 208 volts, 3-phase. In amperes, this equates to

$$I = \frac{kW \times 1000}{E \times 1.73} = \frac{200 \times 1000}{208 \times 1.73} = 556 \text{ ampres}$$

Apply the 1.25 multiplier as required by *NEC 424.82:*

$$556 \times 1.25 = 695 \text{ amperes}$$

Therefore, the conductors and the overcurrent protective device for the boiler would be required to have an ampacity of not less than 695 amperes. See Table 13-6.

To derate, correct, and adjust the ampacity of the conductor to be used, *NEC 110.14(C)* permits the use of the allowable ampacity in the column that indicates the temperature rating of the type of conductor to be used. In this case, THHN/THWN conductors with a 90°C rating are being used. *NEC Table 310.15(B)(16)* indicates that a 500-kcmil copper THHN/THWN conductor has an allowable ampacity of 430 amperes.

Two conductors are run per phase. The total allowable ampacity for two of these conductors in parallel, in separate raceways, is

$$430 \times 2 = 860 \text{ amperes}$$

Because the boiler room would be a relatively hot location, to be safe and conservative, an arbitrary correction factor of 0.91 found in *NEC Table 310.15(B)(3)(a)* is used:

$$860 \times 0.91 = 783 \text{ amperes}$$

Therefore, two 500-kcmil copper THHN/THWN conductors in parallel would be more than adequate for the boiler load requirement of 695 amperes.

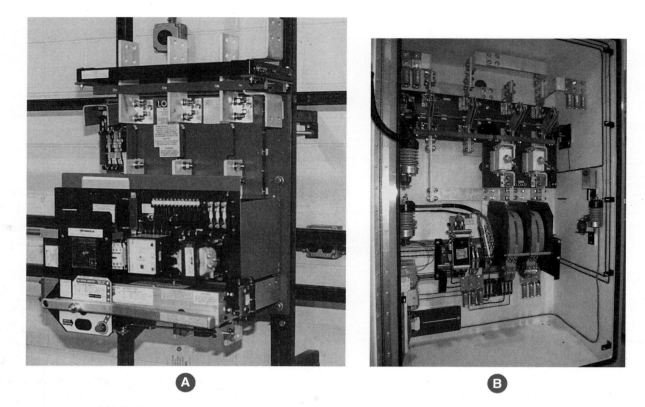

FIGURE 13-18 (A) Bolted pressure contact switch for use with high-capacity fuses. (*Courtesy of Schneider Electric Square D*) (B) A bolted pressure contact switch with shunt trip operator, integral ground-fault protection, auxiliary contacts, and key interlock. (*Courtesy of Eaton Corporation*)

When paralleling conductors, all of the phase conductors are required to be identical in length, type, size, material, and terminations. It is customary for electricians to lay all the conductors, in this case six, on a floor where a perfect match in lengths can be ensured. See *NEC 310.10(H)* and Chapter 8.

The disconnecting means for the boiler would likely be an 800-ampere bolted pressure switch [illustrated in Figure 13-18(A) and (B)]. They are designed and listed for operation at 100% of rating. The fuses in this switch are 700-ampere Class L fuses.

Fixed electric heating is considered to be a continuous load and is subject to the 125% multiplier as indicated in *NEC 210.19(A)(1), 210.20(A), 215.2(A) (1), 215.3,* and *230.42.* By *Code* definition, a *continuous load* is a load that is expected to continue for 3 hours or more.

For 600 kcmil with a feeder neutral, the balanced load is subtracted, and the nonlinear load is added.

GROUNDING/BONDING

The *NEC* covers grounding and bonding in several articles, but the primary coverage is in *Article 250.* Before we get too far into the subject of grounding and bonding, an understanding of some special terms is needed.

Typical commercial electrical systems are grounded systems. Within the system, some things are *grounded* . . . some things are *bonded*. The terms *grounding* and *bonding* are used throughout the *NEC*. Bonding and grounding requirements overlap somewhat, but they all have a common interest—that of preventing electrical shocks and electrical fires.

There is a distinct difference between grounding and bonding. Each serves a different purpose. However, in most cases equipment that is grounded is also bonded by being connected together. Let's take a look at these two very important terms. As you already know, the red triangles ▶◀ indicate a change in the 2011 edition of the *NEC* from the previous 2008 edition.

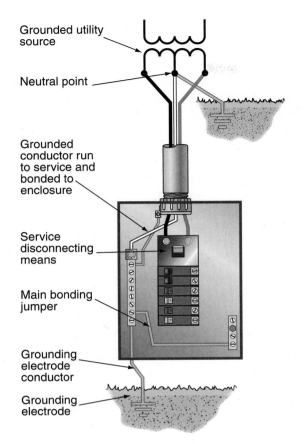

Grounded utility source

Neutral point

Grounded conductor run to service and bonded to enclosure

Service disconnecting means

Main bonding jumper

Grounding electrode conductor

Grounding electrode

FIGURE 13-19 Grounding of an electrical service connects the electrical system to earth. This might be connecting the grounding electrode conductor to an underground metal water pipe, to a concrete-encased grounding electrode (Ufer ground), to rebars, or to a ground rod(s). The grounding electrode conductor must be sized according to *NEC Table 250.66.* Equipment grounding conductors are sized according to *NEC Table 250.122.* (*Delmar/Cengage Learning*)

What Does Grounded (Grounding) Mean?

Grounded (grounding) is defined in *NEC Article 100* as *Connected (connecting) to ground or to a conductive body that extends the ground connection.** See Figure 13-19.

Grounding

The following definitions related to grounding are found in *NEC Articles 100, 250.2,* and *250.4(A).*

Ground. *The earth.** Throughout this country, and for that matter, throughout the world, the earth is the common reference point for all electrical systems. The earth is not permitted to serve as an effective ground-fault current path, *NEC 250.4(A)(5).* Simply stated, this means that trying to establish a good ground on a piece of electrical equipment such as a lamp post with a ground rod by itself will not meet *Code.*

Grounded Conductor. *A system or circuit conductor that is intentionally grounded.** In most wiring, this is the white conductor. It might also be gray. It is most often referred to as the neutral conductor or simply the neutral.

►**Grounding Conductor, Equipment (EGC).** *The conductive path(s) installed to connect normally non-current-carrying metal parts of equipment together and to the system grounded conductor or to the grounding electrode conductor or both.**◄ In commercial wiring, the equipment grounding conductor might be a bare or green insulated conductor, the metal jacket of Type MC cable, the metal jacket plus the bonding strip in armored cable, or metal raceways. Different acceptable EGCs are listed in *NEC 250.118.* Note a significant concept is included here. The equipment grounding conductor *connects normally non-current-carrying metal parts of equipment together.* As such, the equipment grounding conductor serves to bond conductive equipment together, a bonding function. Here's where you can't separate grounding and bonding!

Ground Fault. *An unintentional, electrically conducting connection between an ungrounded conductor of an electrical circuit and the normally non-current-carrying conductors, metallic enclosures, metallic raceways, metallic equipment, or earth.** Ground faults occur when an ungrounded "hot" conductor comes in contact with a grounded surface or grounded equipment. This could be a result of insulation failure or an ungrounded conductor connection coming loose.

Ground-Fault Current Path. *An electrically conductive path from the point of a ground fault on a wiring system through normally non-current-carrying conductors, equipment, or the earth to the electrical supply source.** This is the path that the flow of ground-fault current will take. Whenever fault current flows, it must flow through a complete circuit, and that fault current must return to its source. What goes out must come back! What goes up must come down! The current might return through connectors, couplings, bonding jumpers, grounding conductors, ground clamps, and other components that make up the ground-fault return path.

Ground-Fault Current Path, Effective. As defined in *NEC 250.2, An intentionally constructed, low-impedance electrically conductive path designed and intended to carry current under ground-fault conditions from the point of a ground fault on a wiring system to the electrical supply source and that facilitates the operation of the overcurrent protective device or ground-fault detectors on high-impedance grounded systems.* If and when a ground fault occurs, we want the overcurrent device ahead of the ground fault to open as fast as possible to clear the fault. To accomplish this, the integrity of the ground-fault current path must be unquestionable.

Where Does Fault Current Travel?

We sometimes think that electricity follows the path of least resistance. That is not totally correct! Electricity follows all available paths. As stated above, ground-fault current might return through connectors, couplings, bonding jumpers, grounding conductors, ground clamps, and any other components that make up the ground-fault return path. Some of the fault current might flow on sheet metal ductwork and on metal water or metal gas piping. That is why the *NEC* is so strict about keeping the impedance of the ground-fault current path as low as possible. We want the ground-fault current to return on the electrical system, not on other nonelectrical parts in the building. Think about it this way: the lower the impedance, the higher the current flow. The higher the current flow, the faster the overcurrent device will clear the fault.

Grounding, Electrical System. *Electrical systems that are grounded shall be connected to earth in a manner that will limit the voltage imposed by lightning, line surges, or unintentional contact with higher voltage lines and that will stabilize the voltage to earth during normal operation.* NEC 250.4(A)(1)

Grounding of Electrical Equipment. *Normally non-current-carrying conductive materials enclosing electrical conductors or equipment, or forming part of such equipment, shall be connected to earth so as to limit the voltage to ground on these materials.* NEC 250.4(A)(2)

Grounding Electrode. *A conducting object through which a direct connection to earth is established.* In commercial wiring, examples of a grounding electrode might be an underground metal water piping supply to the building, a ground rod(s), or a concrete-encased conductor (Ufer ground).

Grounding Electrode Conductor. *A conductor used to connect the system grounded conductor or the equipment to a grounding electrode or to a point on the grounding electrode system.*

Ungrounded. *Not connected to ground or a conductive body that extends the ground connection.*

Now Let's See How All of the Above Comes Together. One benefit of grounding electrical systems, equipment, and conductive materials enclosing the preceding is to facilitate (i.e., enable) the immediate response of overcurrent protective devices to ground faults. This action removes the hazard to people and animals that would otherwise exist with equipment that is energized at a dangerous voltage.

As stated earlier, electrical systems are *connected to earth in a manner that will limit the voltage imposed by lightning, line surges, or unintentional contact with higher-voltage lines and that will stabilize the voltage to earth during normal operation, NEC 250.4(A)(1).*

Electrical equipment and conductive materials enclosing electrical equipment are *connected to earth so as to limit the voltage to ground on these materials, NEC 250.4(A)(2).*

The system/circuit grounded conductor and the equipment/materials grounding conductor are to be joined at only one location on a property, and that is at the main service equipment or at a separately derived system. Beyond the main service equipment or source of a separately derived system, the system grounded neutral conductor and the equipment grounding conductor are kept separated. In feeder

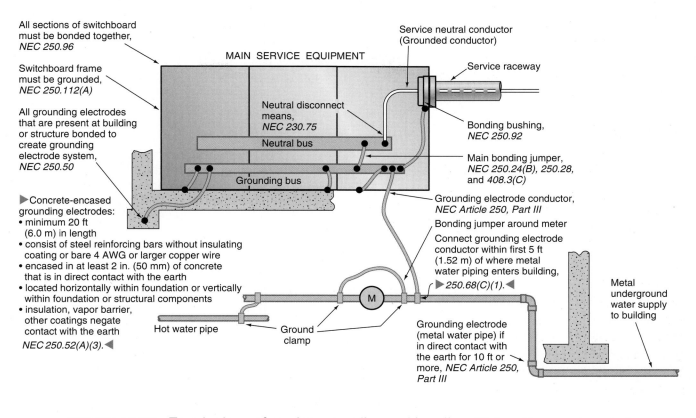

All sections of switchboard
must be bonded together,
NEC 250.96

Switchboard frame
must be grounded,
NEC 250.112(A)

All grounding electrodes
that are present at building
or structure bonded to
create grounding
electrode system,
NEC 250.50

▶Concrete-encased
grounding electrodes:
• minimum 20 ft
  (6.0 m) in length
• consist of steel reinforcing bars without insulating
  coating or bare 4 AWG or larger copper wire
• encased in at least 2 in. (50 mm) of concrete
  that is in direct contact with the earth
• located horizontally within foundation or vertically
  within foundation or structural components
• insulation, vapor barrier,
  other coatings negate
  contact with the earth
NEC 250.52(A)(3).◀

MAIN SERVICE EQUIPMENT

Neutral disconnect
means,
NEC 230.75

Neutral bus

Grounding bus

Service neutral conductor
(Grounded conductor)

Service raceway

Bonding bushing,
NEC 250.92

Main bonding jumper,
NEC 250.24(B), 250.28,
and *408.3(C)*

Grounding electrode conductor,
NEC Article 250, Part III

Bonding jumper around meter

Connect grounding electrode
conductor within first 5 ft
(1.52 m) of where metal
water piping enters building,
▶*250.68(C)(1).*◀

Hot water pipe

Ground
clamp

Grounding electrode
(metal water pipe) if
in direct contact with
the earth for 10 ft or
more, *NEC Article 250,
Part III*

Metal
underground
water supply
to building

M

FIGURE 13-20   Terminology of service grounding and bonding. (*Delmar/Cengage Learning*)

panelboards, the neutral conductor IS NOT connected to the panel enclosure. An exception to this is in the case of separately derived systems, *NEC 250.30(A)* under the conditions provided there. The Commercial Building in the text does not have a separately derived system. Separately derived systems are found in buildings where the electrical service is 480Y/277 volts or higher and step-down transformers are used to get 208Y/120 volts.

Failure to maintain the separation of neutral and equipment grounding conductors will result in extraneous currents from the electrical conductors, that is, through the metal gas pipes, metal water pipes, metal ducts, and so on. Stated another way, connecting the grounded neutral conductor and the equipment grounding conductor together anywhere beyond the main service panel sets up multiple (parallel) paths for the return neutral current to flow on. These currents increase the level of electromagnetic radiation, which many have alleged has a harmful effect on the human body.

It is important that the ground connections and the grounding electrode system be properly

installed. To achieve the best possible grounded system, the electrician must use the recommended procedures and equipment when making the installation. Figure 13-20 illustrates some of the terminology used in *NEC Article 250.*

As you study several figures in this chapter and other similar figures in this text, you will note that locknuts and bushings are quite often used to secure raceways to equipment. To achieve the positive bonding required in *NEC 250.92(B)* for service equipment, these bushings will be bonding bushings, grounding bushings, or they might be dual-purpose insulated bonding bushings. It is important to note that *NEC 300.4(G)* requires that these bushings shall be of the insulating type where the raceways containing insulated circuit conductors 4 AWG or larger enter a cabinet, box enclosure, or raceway. Similar insulating bushing requirements are found in *NEC 312.6(C)*, *314.17(D)*, *354.46*, and *362.46*. Insulating sleeves that slide into the bushing are also permitted. Some EMT connectors have insulated throats.

If the grounding and bonding are installed in accordance with *NEC Article 250*, there will be a

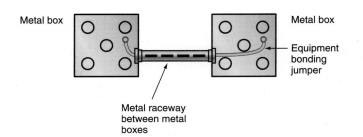

FIGURE 13-21   Two methods for bonding the metal parts together. One method is using the metal raceway. The second method is to install a separate equipment grounding conductor between the two metal boxes, as would be required if a nonmetallic raceway had been used. An equipment bonding jumper must be sized large enough to handle any fault current that it may be called on to carry. (*Delmar/Cengage Learning*)

good current path should a ground fault occur. The reason for this is that the lower the impedance of the grounding path, the greater the ground-fault current. This increased ground-fault current causes the overcurrent device protecting the circuit to respond faster. For example, compare the fault current at different impedance values.

$$I = \frac{E}{Z} = \frac{277}{1} = 277 \text{ amperes}$$

At a lower impedance:

$$I = \frac{E}{Z} = \frac{277}{0.1} = 2{,}770 \text{ amperes}$$

With still lower impedance:

$$I = \frac{E}{Z} = \frac{277}{0.01} = 27{,}700 \text{ amperes}$$

You can compare opening times for typical fuses and circuit breakers by using the time-current characteristic curves, Figures 17-25, 17-26, and 17-31 in Chapter 17.

The amount of ground-fault damage to electrical equipment is related to (1) the response time of the overcurrent device and (2) the amount of current. One common term used to relate the time and current to the ground-fault damage is *ampere squared seconds* (I^2t):

amperes $\times$ amperes $\times$ time in seconds = I^2t

It can be seen in the equation for I^2t that when the current (I) and time (t) in seconds are kept to a minimum, a low value of I^2t results. Lower values mean that less ground-fault damage will occur. The same I^2t theory can be applied to conductors that are carrying current greatly in excess of their ampacity under short-circuit conditions.

Chapters 17, 18, and 19 provide detailed coverage of overcurrent protective devices, fuses, and circuit breakers.

Bonding

What Does Bonded (Bonding) Mean? Bonded (bonding) is defined in *NEC Article 100* as: *Connected to establish electrical continuity and conductivity.** See Figure 13-21.

Which shows two metal boxes bonded together with the metal raceway installed between the two boxes.

▶**Bonding Conductor or Jumper:** *A reliable conductor to ensure the required electrical conductivity between metal parts required to be electrically connected.**◀ This definition was expanded in the 2011 *NEC* to include bonding conductors as these conductors are often 20 ft (6 m) or longer as used in *NEC Chapter 8*. Figure 13-21 shows two metal boxes bonded together by a conductor installed in the raceway. This bonding jumper would be installed if the raceway were nonmetallic.

Bonding Jumper, Equipment: *The connection between two or more portions of the equipment grounding conductor.** A conductor installed in the raceway between the two boxes in Figure 13-21 would be classified as an equipment bonding jumper.

Bonding Jumper, Main: *The connection between the grounded circuit conductor and the equipment*

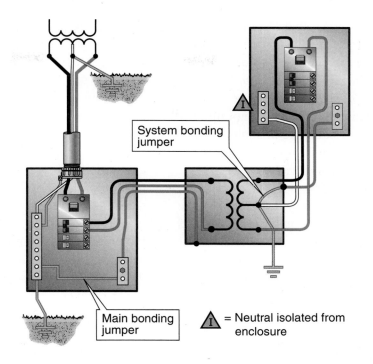

FIGURE 13-22 Main bonding jumper installed in service. System bonding jumper installed in separately derived system. Both provide a path for ground-fault current to return to source. (*Delmar/Cengage Learning*)

*grounding conductor at the service.** A main bonding jumper is clearly shown in Figures 13-20 and 13-22.

Figure 13-22 illustrates a main bonding jumper that is installed at the service equipment and a system bonding jumper that is installed for separately derived systems. As can be seen, these bonding jumpers perform identical functions: providing a low-impedance path for ground-fault currents to return to their source.

►**Bonding Jumper, Supply-Side:** *A conductor installed on the supply side of a service or within a service equipment enclosure(s), or for a separately derived system, that ensures the required electrical conductivity between metal parts required to be electrically connected.**◄ This definition was added to the 2011 *NEC*. It describes a conductor used to bond metallic parts of services and separately derived systems together on the supply side of an overcurrent device. Because no overcurrent device is located at this position, sizing of the Supply Side Bonding Jumper is selected from *NEC Table 250.66.* Specific reference is made in *NEC 250.30* for separately derived systems and in *NEC 250.102(C)* for services.

Bonding of Electrically Conductive Materials and Other Equipment: *Normally non-current-carrying electrically conductive materials that are likely to become energized shall be connected together and to the electrical supply source in a manner that establishes an effective ground-fault current path.** NEC 250.(A)(4)*

Bonding of Electrical Equipment: *Normally non-current-carrying conductive materials enclosing electrical conductors or equipment, or forming part of such equipment, shall be connected together and to the electrical supply source in a manner that establishes an effective ground-fault current path* 250.4(A)(3)*

Why Bond? *NEC 250.4(A)(3)* states that *Normally non-current-carrying conductive materials enclosing electrical conductors or equipment, or forming part of such equipment, shall be connected together and to the electrical supply source in a manner that establishes an effective ground-fault current path.** This refers to the metal raceways, metal cables, boxes, panelboards, nipples, transformers, and any other electrical materials and equipment related to the electrical installation.

NEC 250.4(A)(4) states that *Normally non-current-carrying electrically conductive materials that are likely to become energized shall be connected together and to the electrical supply source in a manner that establishes an effective ground-fault current path.** This refers to items not directly associated with the electrical system, such as metal frames of buildings, metal gas piping, metal water piping, and similar metallic materials.

After proper grounding has been taken care of, it is also the responsibility of the electrician to make sure that all conductive materials are properly bonded together. This includes all metallic piping, sheet metal ducts, metal framing, metal partitions, metal siding, and all other items that could come in contact with electrical system wiring or circuits and with a person or animal. An electrician makes sure that all electrical raceways, panelboards, and other electrical equipment are properly grounded and bonded in conformance to the *NEC*.

The issue of grounding and bonding ductwork and metal siding is another problem. For the most part, these building components become grounded and bonded to the electrical system by chance. These components are installed by other trades.

Ideally, except in special cases where isolation is necessary, bonding of all conductive materials should be the goal.

Grounding Electrode System

The wording in *NEC 250.50* is that all of the following items *if present* shall be bonded together to form the grounding electrode system:

- Metal underground water piping
- Metal frame of a building
- Concrete-encased electrode (rebars or 4 AWG wire)
- Rod and pipe electrodes
- Other listed electrodes such as chemical rods
- Plate electrodes
- Ground ring

The term *if present* is often misunderstood, particularly when it comes to concrete-encased electrodes (rebars). It is perfectly clear that in most building or structure construction, rebars are present at some point during the time of construction. This makes it necessary for the electrician and the individual installing the rebars to work together, making sure that a rebar is brought out of the foundation in an accessible and dry location so that it can be connected to a grounding electrode conductor. An option is for the electrician to connect a properly sized grounding electrode conductor or bonding conductor to the reinforcing steel bar before the concrete is poured. A ground clamp that is listed for connection to the rebar, connection of the copper grounding electrode conductor, and suitability for concrete encasement is required.

Figures 13-17, 13-20, 13-22, and 13-26 show typical grounding and bonding at a main electrical service. Equipment grounding and bonding beyond the main electrical service is accomplished through metal raceways and/or separate equipment grounding conductors in the raceways or cables.

The interconnection of the equipment/material makes it highly unlikely that any of the items could become isolated, for even if one connection failed, the integrity of the grounding would be maintained through other connections. The value of this concept is illustrated in Figure 13-23 and described in the following scenario:

1. A live wire contacts the gas pipe. The bonding jumper A is not installed originally.
2. The gas pipe is now energized at 120 volts.
3. The insulating joint in the gas pipe installed at the gas meter results in no current flow as the circuit is open.
4. The 20-ampere fuse or circuit breaker will not open but the gas pipe remains energized.
5. If a person touches the live gas pipe and the water pipe at the same time, a current passes through the person's body. If the body's resistance is 1,100 ohms, the current passing through the body is:

$$I = \frac{E}{R} = \frac{120}{1,100} = 0.109 \text{ amperes}$$

THIS VALUE OF CURRENT PASSING THROUGH A HUMAN BODY CAN CAUSE DEATH.

6. The fuse or circuit breaker does not open.
7. If proper bonding had been achieved, bonding jumper A would have kept the voltage

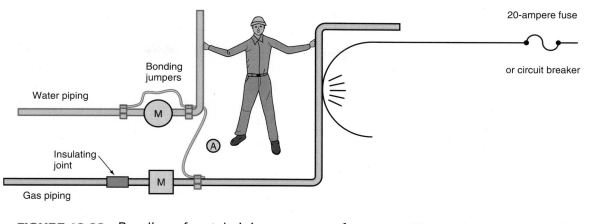

FIGURE 13-23   Bonding of metal piping ensures safe contact. (*Delmar/Cengage Learning*)

difference between the water pipe and the gas pipe at virtually zero, and the fuse would have opened. If 10 ft (3 m) of 4 AWG copper wire were used as the jumper, then the resistance of the jumper would be 0.00308 ohm per *NEC* Chapter 9, Table 8. The current is

$$\frac{120}{0.00308 \text{ ohms}} = 38,961 \text{ amperes}$$

In an actual system, the impedance of all of the parts of the circuit would be much higher. Thus, a much lower current would result. The value of the current, however, would be enough to cause the fuse or circuit breaker to open quickly.

These advantages of proper bonding and grounding:

- The potential voltage differentials between the different parts of the system are kept at a minimum, thereby reducing shock hazard.

- Impedance of a ground path is kept at a minimum, which results in higher current flow in the event of a ground fault. The lower the impedance, the greater is the current flow, and the faster the overcurrent device opens.

Grounding Systems and Electrodes

All alternating-current systems of the type installed in the Commercial Building are required to be grounded as it is a 208/120-volt wye-connected system; see *NEC 250.20(B)*. The system can be grounded so the maximum voltage to ground is 150 volts or less.

In Figures 13-20, 13-21, 13-23, and 13-27, the main service equipment, the service raceways, the neutral bus, and the equipment grounding bus have been bonded together. The hot and cold water pipes have been bonded together in Figures 13-21 and 13-27.

For discussion purposes of the Commercial Building, at least 20 ft (6 m) of 3/0 AWG bare copper conductor is installed in the footing to serve as a concrete-encased grounding electrode. It also serves as the additional supplemental electrode that is required by *NEC 250.52(A)(3)* for the water pipe grounding electrode. The minimum size permitted is 4 AWG copper. This conductor, buried in the concrete footing, is supplemental by the water pipe electrode but is also recognized as a grounding electrode in *NEC 250.52(A)(3)*. *NEC 250.53(D)(2)* requires the supplemental electrode to be bonded to the grounding electrode conductor, the grounded service-entrance conductor, the grounded service raceway, or the grounded service enclosure. As mentioned earlier, most commercial buildings of the size and type in this text will have reinforcing steel bars in the concrete. If present, these bars must be connected and become a part of the grounding electrode system.

▶For commercial or industrial installations, the supplemental electrode may be bonded to the interior metal water piping more than 5 ft (1.52 m) inside the building only where the entire length of water piping that will serve as the conductor is exposed and will be under qualified maintenance conditions. See *NEC 250.68(C)(1), Exception.*◀

Had a 3/0 AWG bare copper conductor not been selected for installation in the concrete footing, the supplemental grounding electrode could have been

- the metal frame of the building where effectively grounded (not applicable in our Commercial Building).
- at least 20 ft (6 m) of steel-reinforcing bars (rebars) 1/2-inch (12.7 mm) minimum diameter, encased in concrete at least 2 in. (50 mm) thick, in direct contact with the earth, near the bottom of the foundation or footing.
- at least 20 ft (6 m) of bare copper wire, encircling the building, minimum size 2 AWG, buried directly in the earth at least 2.5 ft (750 mm) deep.
- ground rods.
- ground plates.

▶Two ground rods or plates are required unless it can be proven by testing that the resistance of a single rod or plate does not exceed 25 ohms. Multiple rods shall be installed at least 6 ft (1.8 m) apart; see *NEC 250.53(A)(2)*.◀

All of these factors are discussed in *NEC Article 250, Parts III, V,* and *VI.*

Many *Code* interpretations are made by local electrical inspectors regarding the grounding electrode systems concept. Therefore, the local *Code* authority should be consulted. For instance, some electrical inspectors may not require the bonding jumper between the hot and cold water pipes, as shown in Figure 13-24. They may determine that an adequate bond is provided through the water heater itself of either type, electric

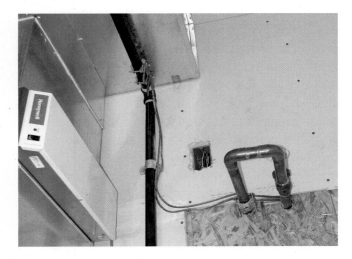

FIGURE 13-24 Bonding jumper between metal water and gas pipes. (*Delmar/Cengage Learning*)

or gas or by connection to metal mixing valves at some showers, bathtubs, or washing machines. Other electrical inspectors may require that the hot and cold water pipes be bonded together because some water heaters contain insulating fittings that are intended to reduce corrosion inside the tank caused by electrolysis. According to their reasoning, even though the originally installed water heater contains no fittings of this type, such fittings may be included in a future replacement heater, thereby necessitating a bond between the cold and hot water pipes. The local electrical inspector should be consulted for the proper requirements. However, bonding the pipes together, as shown in Figure 13-24, is the recommended procedure.

Grounding Requirements

When grounding service-entrance equipment, the following *Code* rules shall be observed:

- The electrical system shall be grounded when maximum voltage to ground on the ungrounded (hot) conductors does not exceed 150 volts; see *NEC 250.20(B)(1)*.
- The electrical system shall be grounded when the neutral conductor is used as a circuit conductor (for example, a 480/277-volt, wye-connected, 3-phase, 4-wire system), *NEC 250.20(B)(2)*. This is similar to the connection of the system shown in Figure 13-7.
- The electrical system shall be grounded where the midpoint of one transformer is used to establish the grounded neutral circuit conductor, as on a 120/240-volt, 3-phase, 4-wire delta system; see *NEC 250.20(B)(3)* and refer to Figure 13-5.
- The earth shall not be considered as an effective ground-fault current path; see *NEC 250.4(A)(5)*.
- *NEC 250.4(A)(5)* requires that an effective ground-fault current path be established for all electrical equipment, wiring, and other electrically conductive material likely to become energized. This path shall be
  - low impedance and
  - capable of safely carrying the maximum ground-fault current likely to be imposed on it;
  - the earth is not considered to be an effective ground-fault current path.

  Details of this subject are discussed in Chapters 17, 18, and 19.

- All grounding schemes shall be installed so that no objectionable currents will exist in the grounding conductors and other grounding paths; see *NEC 250.6*. This requirement is somewhat unclear as the *NEC* does not define objectionable current.

- The grounding electrode conductor shall be connected at an accessible location from the load end of the service drop or service lateral to the termination point of the neutral or grounded conductor in the service disconnecting means; see *NEC 250.24(A)(1)*.

- The identified neutral conductor is the conductor that shall be grounded; see *NEC 250.26*.

- Tie (bond) grounding electrodes together to create the grounding electrode system; see *NEC 250.50*.

- Do not use interior metal water piping to serve as a means of interconnecting steel framing members, concrete-encased electrodes, and the ground ring for the purposes of establishing the grounding electrode system as required by *NEC 250.104*. Should any PVC nonmetallic water piping be interposed in the metal piping runs, the continuity of the bonding-grounding system is broken.

  If metal water pipe is used for bonding the grounding electrode conductor, the bonding conductors associated with the metal framing members, concrete-encased electrodes, and the ground ring, the connection must generally be made within the first 5 ft (1.5 m) of metal water piping after it enters the building. The first 5 ft (1.5 m) may include a water meter. See *NEC 250.68(C)(1)*.

  The only exception to this is for industrial, commercial, and institutional buildings, if the entire length of the water piping being used as the conductor to interconnect the various electrodes is exposed and where only qualified people will be doing maintenance on the installation; see *NEC 250.68(C)(1), Exception*.

- The grounding electrode conductor is to be sized as given in *NEC 250.66* and *Table 250.66*.

- The metal frame of the building is required to be used as a grounding electrode if it qualifies as a grounding electrode as indicated in *NEC 250.52(A)(2)*.

- The hot and cold water metal piping system shall be bonded to the service-equipment enclosure, to the grounded conductor at the service, and to the grounding electrode conductor, *NEC 250.104*.

- *NEC 250.64(C)* states that grounding electrode conductors generally shall not be spliced. The grounding electrode conductor may be spliced by either exothermic welding or irreversible compression-type connectors that are listed for that purpose. Bolted connections of busbars being used as a grounding electrode conductor are also permitted. ▶Bolted, riveted, or welded connections of structural metal frames of buildings or structures are permitted as are threaded, welded, brazed, soldered or bolted-flange connections of metal water piping.◀

- The grounding electrode conductor shall be connected to the metal underground water pipe when 10 ft (3 m) long or longer and it is in direct contact with the earth. The 10 ft (3 m) includes the metal well casing bonded to the water pipe; see *NEC 250.52(A)(1)*.

- In addition to grounding the service equipment to the underground water pipe, one or more additional electrodes are required, such as a bare conductor in the footing, a grounding ring, rod or pipe electrodes, or plate electrodes. All of these items shall be bonded if they are available on the premises. See *NEC 250.50* and *250.104*.

- *NEC 250.52(B)* prohibits using a metal underground gas piping system as the grounding electrode. However, where metal gas piping comes into a building, and if the gas piping is likely to become energized, it shall be bonded to the service equipment, the grounded conductor at the service, the grounding electrode conductor, or to one or more of the accepted grounding electrodes used; see *NEC 250.104(B)*.

- The grounding electrode conductor shall be copper, aluminum, or copper-clad aluminum, *NEC 250.62*. If used outside, aluminum and copper-clad aluminum are not permitted to be terminated within 18 in. (450 mm) of the earth; see *NEC 250.64(A)*.

- The grounding electrode conductor may be solid or stranded, uninsulated, covered, or bare, and shall not generally be spliced; see *NEC 250.62* and *250.64(C)*.

- Bonding is required around all insulating joints or sections of the metal piping system that might be disconnected, *NEC 250.68(B)*.

- The connection to the grounding electrode shall generally be accessible, *NEC 250.68(A)*. A connection to a concrete-encased, driven, or buried grounding electrode does not have to be accessible.

- Make sure that the length of bonding conductor is long enough so if the equipment is removed, the bonding will remain in place, *NEC 250.68(B)*. The bonding jumper around the water meter is a good example where this requirement applies.

- The grounding electrode conductor shall be tightly connected by using proper lugs, connectors, clamps, or other approved means, *NEC 250.70*. Typical ground clamps are shown in Figure 13-25.

Sizing the Grounding Electrode Conductors

The grounding electrode conductor connects the grounding electrode to the equipment grounding conductor and to the system grounded (neutral) conductor. In the Commercial Building, this means that the grounding electrode conductor connects the underground metal water pipe and concrete encased electrode to the grounding bus (equipment grounding conductor) and to the neutral bus [system grounded (neutral) conductor]. See Figure 13-21.

FIGURE 13-25 Typical ground clamps. (A) For ground rods; (B) for connection to concrete-encased grounding electrodes or smaller pipes; (C) for connecting conduit containing grounding electrode conductor and grounding electrode conductor to metal pipe. (*Courtesy of Thomas and Betts*)

TABLE 13-7

Table 250.66 Grounding Electrode Conductor for Alternating-Current Systems

Size of Largest Ungrounded Service-Entrance Conductor or Equivalent Area for Parallel Conductors[a] (AWG/kcmil)		Size of Grounding Electrode Conductor (AWG/kcmil)	
Copper	Aluminum or Copper-Clad Aluminum	Copper	Aluminum or Copper-Clad Aluminum[b]
2 or smaller	1/0 or smaller	8	6
1 or 1/0	2/0 or 3/0	6	4
2/0 or 3/0	4/0 or 250	4	2
Over 3/0 through 350	Over 250 through 500	2	1/0
Over 350 through 600	Over 500 through 900	1/0	3/0
Over 600 through 1100	Over 900 through 1750	2/0	4/0
Over 1100	Over 1750	3/0	250

Notes:
1. Where multiple sets of service-entrance conductors are used as permitted in 230.40, Exception No. 2, the equivalent size of the largest service-entrance conductor shall be determined by the largest sum of the areas of the corresponding conductors of each set.
2. Where there are no service-entrance conductors, the grounding electrode conductor size shall be determined by the equivalent size of the largest service-entrance conductor required for the load to be served.

[a]This table also applies to the derived conductors of separately derived ac systems.
[b]See installation restrictions in 250.64(A).

Reprinted with permission from NFPA 70-2011.

NEC Table 250.66 (Table 13-7) is referred to when selecting grounding electrode conductors for services where there is no overcurrent protection ahead of the service-entrance conductors other than the utility company's primary or secondary overcurrent protection.

In *NEC Table 250.66*, the service conductor sizes are given in the wire size and not the ampacity values. To size a grounding electrode conductor for a service with three parallel 350-kcmil service conductors, it is necessary to total the wire size and select a grounding electrode conductor for an equivalent 1050-kcmil service conductor. In this case, the grounding electrode conductor is

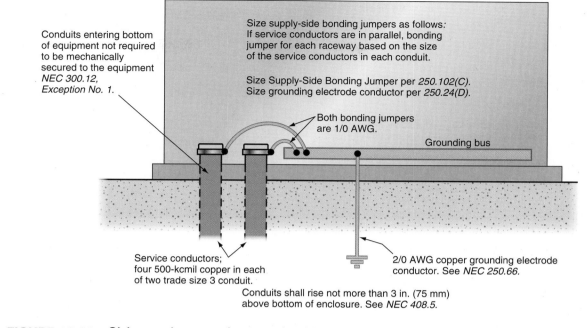

Conduits entering bottom of equipment not required to be mechanically secured to the equipment *NEC 300.12, Exception No. 1.*

Size supply-side bonding jumpers as follows: If service conductors are in parallel, bonding jumper for each raceway based on the size of the service conductors in each conduit.

Size Supply-Side Bonding Jumper per *250.102(C).* Size grounding electrode conductor per *250.24(D).*

Both bonding jumpers are 1/0 AWG.

Grounding bus

Service conductors; four 500-kcmil copper in each of two trade size 3 conduit.

2/0 AWG copper grounding electrode conductor. See *NEC 250.66.*

Conduits shall rise not more than 3 in. (75 mm) above bottom of enclosure. See *NEC 408.5.*

FIGURE 13-26   Sizing and connecting supply-side bonding jumpers to conduits supplying open bottom switchboard. (*Delmar/Cengage Learning*)

2/0 AWG copper (see Figure 13-17). Figure 13-26 shows another example.

Sizing the Main Bonding Jumper

A main bonding jumper, Figure 13-15, connects the neutral busbar to the service equipment enclosure in each of the six service disconnecting means. The jumper is sized according to *NEC 250.28(D).* As can be seen, *250.28* sends us to *Table 250.66* for sizing. For each of the six service disconnecting means, it is expected that listed equipment will be installed. The manufacturer of the listed service switch is required to provide an appropriately sized main bonding jumper for each switch. The obvious benefit to the installer is a calculation of the main bonding jumper is not required.

If a service is installed with a single main overcurrent device or a single enclosure with six disconnecting means, the size of the main bonding jumper can be determined as follows. The underground service conductors to the Commercial Building consist of three 350-kcmil conductors in parallel. This adds up to a total conductor area of 1050 kcmil. By reference to *NEC Table 250.66*, a main bonding jumper not smaller than 2/0 AWG copper is required.

Sizing the Grounded Service Conductors

The neutral or grounded conductor for each of the six disconnecting means is shown in the feeder load calculation form for the occupancy. The same size grounded conductor as shown for the feeder is installed from the termination box to the service disconnecting means. This would seem to be adequate. However, there are other *Code* sections that must be checked.

Referring to *NEC 250.24(C)*, the requirement is set forth that the minimum size for the neutral conductor is determined by the requirement for a grounding electrode conductor. A grounded conductor sized as though it is a grounding electrode conductor is considered large enough to carry fault current back to the utility transformer. The neutral conductor is required to be routed with the phase conductors and shall be sized no smaller than the required grounding electrode conductor, *NEC Table 250.66.* Rules for sizing the grounded conductor for parallel sets of service-entrance conductors is found in *NEC 250.24(C)(1)* and *(C)(2).* By reference to *NEC Table 250.66,* the 350-kcmil underground service conductors in each service raceway requires that the neutral conductor be no smaller than 2 AWG. However, because three conductors are installed

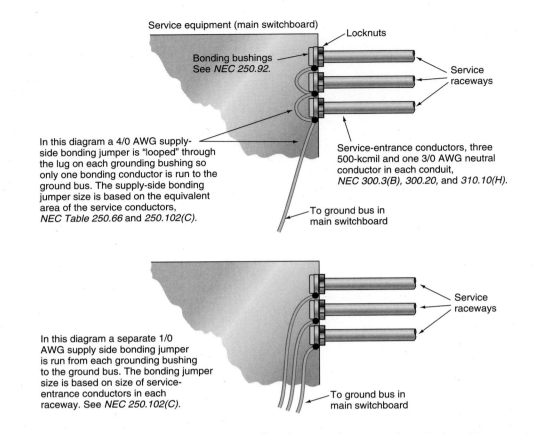

Service equipment (main switchboard)

Locknuts

Bonding bushings
See *NEC 250.92*.

Service raceways

In this diagram a 4/0 AWG supply-side bonding jumper is "looped" through the lug on each grounding bushing so only one bonding conductor is run to the ground bus. The supply-side bonding jumper size is based on the equivalent area of the service conductors, *NEC Table 250.66* and *250.102(C)*.

Service-entrance conductors, three 500-kcmil and one 3/0 AWG neutral conductor in each conduit, *NEC 300.3(B)*, *300.20*, and *310.10(H)*.

To ground bus in main switchboard

Service raceways

In this diagram a separate 1/0 AWG supply side bonding jumper is run from each grounding bushing to the ground bus. The bonding jumper size is based on size of service-entrance conductors in each raceway. See *NEC 250.102(C)*.

To ground bus in main switchboard

FIGURE 13-27 Sizing of the supply-side bonding jumper for a service. (*Delmar/Cengage Learning*)

in parallel or connected together at each end, the grounded conductor in each raceway is required to be not smaller than 1/0 AWG; see *NEC 250.24(C)(1)* and *310.10(H)*. Although the calculated neutral conductor load may allow a smaller conductor, a 1/0 AWG is the minimum size because this conductor would carry high-level fault currents if a ground fault occurs, and 1/0 AWG is the smallest size that is generally permitted to be installed in parallel.

Sizing Bonding Jumpers on the Supply Side of the Service

Figure 13-26 illustrates the appropriate sizing of supply-side bonding jumpers. These bonding jumpers are installed to ensure a low-impedance and adequate path for ground-fault current to flow should a ground fault occur on the line or supply side of the service overcurrent protection.

Since the point of bonding is ahead of an overcurrent device, the sizing requirements of *250.102(C)* apply. This rule once again points us

to *Table 250.66* for sizing the supply-side bonding jumper. Since individual bonding jumpers are shown in Figure 13-26 and in the lower illustration in Figure 13-27, we use the first column of *Table 250.66* and find the row with the 500 kcmil conductors. By moving to the next column to the right we find the minimum size supply-side bonding jumper to be 1/0 AWG.

Sizing Bonding Jumper on the Supply Side of the Service for More Than One Conduit

Figure 13-27 illustrates the two methods of bonding raceways that contain service-entrance conductors. The upper method shows using a single supply-side bonding jumper to bond the 3 metal raceways that contain service-entrance conductors. As required by *NEC 250.102(C)(2)*, the size of the ungrounded service conductors are added together to create the size of a single conductor. This size is used in *Table 250.66* to determine the minimum size of a single supply-side bonding jumper that is looped or daisy-chained from bonding fitting to bonding fitting. The minimum size is determined as illustrated in the following steps.

Step 1: Determine the total area of the ungrounded conductors.

$$500 \text{ kcmil} \times 3 = 1500 \text{ kcmil}$$

Step 2: Since the size is larger than 1100 kcmil copper, the supply-side bonding jumper must not be smaller than 12½% of the area of the ungrounded conductors.

$$1500 \text{ kcmil} \times 1.25 = 187.5 \text{ kcmil}$$

Step 3: Since 187.5 kcmil is not a standard size, refer to *NEC Table 8 of Chapter 9*. In the third column, find that 187.5 is between 167,500 for a 3/0 AWG and 211,600 for a 4/0 AWG conductor.

Step 4: Determine the minimum size supply-side bonding jumper for the installation is one 4/0 AWG conductor.

The lower illustration in Figure 13-27 is identical to the solution in Figure 13-26. If an individual supply-side bonding jumper is installed to each metal conduit, simply refer to *Table 250.66* for the minimum size.

Remember that *NEC 250.4(A)(5)*, *250.90*, and *250.96* require that the grounding and bonding conductors be capable of carrying any fault current that they might be called on to carry under fault conditions. This subject is covered in Chapters 17, 18, and 19.

Other Grounding and Bonding Requirements for Service Equipment

Several requirements related to grounding and bonding of service equipment are stated in Figure 13-28, including these:

- Placing a bonding jumper around concentric or eccentric knockouts in service equipment enclosures
- Sizing supply-side bonding jumpers
- Solder lugs not permitted for connections
- Locating the main bonding jumper
- Sizing and locating the grounding electrode conductor
- Using one continuous length for grounding electrode conductor
- Bonding metal enclosure for grounding electrode conductor
- Rules on connecting to water pipe grounding electrode
- Bonding around water meters and the like
- Attachment of the grounding electrode conductor to the grounding electrode

Other Grounding and Bonding Requirements for Service Equipment

Nonconductive paint, enamel, or similar coating shall be removed at contact points when sections of electrical equipment are bolted together to ensure that the sections are effectively bonded. Locknuts that have sharp edges on the tangs and are turned against the painted surface generally will bite through paints and enamels, thus making the removal of the paint unnecessary. See *NEC 250.96(A)* and refer to Figure 13-29.

Sizing Equipment Grounding Conductors

NEC Table 250.122 is referred to when selecting equipment grounding conductors when there is overcurrent protection ahead of the conductor supplying the equipment. For example, if a 700-ampere overcurrent device is located on the supply side of a feeder or branch circuit, *Table 250.122* requires a 1 AWG equipment grounding conductor be installed. An option is to use one of the appropriate wiring methods included in *NEC 250.118*. If this is done, the *NEC* does not require the installation of an equipment grounding conductor of the wire type.

NEC Table 250.122 (Table 13-8) is based on the current setting or rating, in amperes, of the overcurrent device installed ahead of the equipment (other than service-entrance equipment) being supplied.

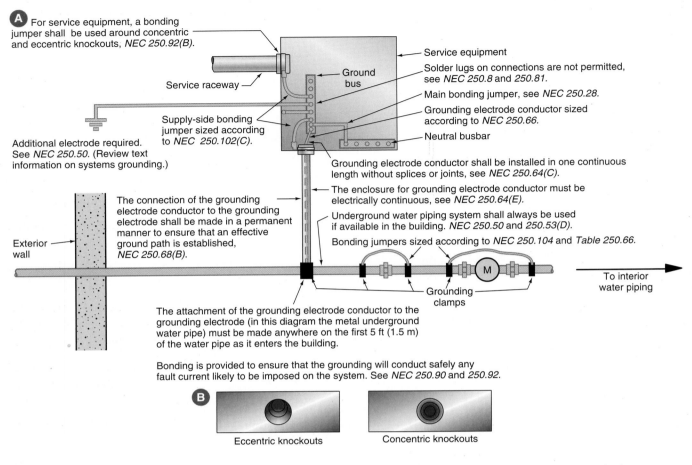

A For service equipment, a bonding jumper shall be used around concentric and eccentric knockouts, *NEC 250.92(B).*

Service raceway

Additional electrode required. See *NEC 250.50.* (Review text information on systems grounding.)

Exterior wall

The connection of the grounding electrode conductor to the grounding electrode shall be made in a permanent manner to ensure that an effective ground path is established, *NEC 250.68(B).*

Ground bus

Supply-side bonding jumper sized according to *NEC 250.102(C).*

Service equipment

Solder lugs on connections are not permitted, see *NEC 250.8* and *250.81.*

Main bonding jumper, see *NEC 250.28.*

Grounding electrode conductor sized according to *NEC 250.66.*

Neutral busbar

Grounding electrode conductor shall be installed in one continuous length without splices or joints, see *NEC 250.64(C).*

The enclosure for grounding electrode conductor must be electrically continuous, see *NEC 250.64(E).*

Underground water piping system shall always be used if available in the building. *NEC 250.50* and *250.53(D).*

Bonding jumpers sized according to *NEC 250.104* and *Table 250.66.*

M

To interior water piping

Grounding clamps

The attachment of the grounding electrode conductor to the grounding electrode (in this diagram the metal underground water pipe) must be made anywhere on the first 5 ft (1.5 m) of the water pipe as it enters the building.

Bonding is provided to ensure that the grounding will conduct safely any fault current likely to be imposed on the system. See *NEC 250.90* and *250.92.*

B

Eccentric knockouts

Concentric knockouts

FIGURE 13-28 Grounding and bonding of service equipment. (*Delmar/Cengage Learning*)

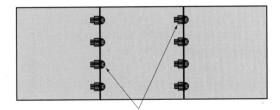

Nonconductive paint, enamel, or similar coating must be removed at contact points.

FIGURE 13-29 Removing paint or other nonconductive coatings at connection points. (*Delmar/Cengage Learning*)

As can be seen in Figure 13-30, a full size equipment grounding conductor is required in each conduit or cable when they are installed in parallel. If the 700-ampere overcurrent device is located at the origin of the feeder or branch circuit, a 1/0 AWG copper equipment grounding conductor is required in each conduit or cable. See *NEC 250.122(F).*

GROUND FAULT PROTECTION FOR EQUIPMENT

The *NEC* requires the use of ground-fault protection for equipment (GFPE) devices on services that meet the conditions outlined in *NEC 230.95.* Similar requirements for feeders are found in *NEC 215.10.* Thus, ground-fault protection devices are required to be installed

- on solidly grounded wye services or feeders above 150 volts to ground, but not greater than 600 volts between phases on disconnecting means rated at 1000 amperes or more (for example, on 480Y/277-volt systems).

- to operate at 1200 amperes of ground-fault current or less.

- so that the maximum time of opening the service switch or circuit breaker does not exceed

TABLE 13-8

Table 250.122 Minimum Size Equipment Grounding
Conductors for Grounding Raceway and Equipment

Rating or Setting of Automatic Overcurrent Device in Circuit Ahead of Equipment, Conduit, etc., Not Exceeding (Amperes)	Size (AWG or kcmil)	
	Copper	Aluminum or Copper-Clad Aluminum*
15	14	12
20	12	10
60	10	8
100	8	6
200	6	4
300	4	2
400	3	1
500	2	1/0
600	1	2/0
800	1/0	3/0
1000	2/0	4/0
1200	3/0	250
1600	4/0	350
2000	250	400
2500	350	600
3000	400	600
4000	500	750
5000	700	1200
6000	800	1200

Note: Where necessary to comply with 250.4(A)(5) or (B)(4), the equipment grounding conductor shall be sized larger than given in this table.
*See installation restrictions in 250.120.

Reprinted with permission from NFPA 70-2011.

1 second for ground-fault currents of 3000 amperes or more.

- to limit damage to equipment and conductors on the load side of the service or feeder disconnecting means. GFPE will not protect against damage caused by faults occurring on the line side of the service or feeder disconnect.

These ground-fault protection requirements do not apply on services or feeders for a continuous industrial process where a nonorderly shutdown will introduce additional or increased hazards, *NEC 215.10, Exception No. 1; 230.95, Exception 1;* and *240.12.*

When a fuse-switch combination serves as the service disconnect, the fuses shall have adequate interrupting capacity to interrupt the available fault current *(NEC 110.9)* and shall be capable of opening any fault current that exceeds the interrupting rating

of the switch during any time when the ground-fault protective system will not cause the switch to open; see *NEC 230.95(B).*

Ground-fault protection of equipment is not required on

- delta-connected, 3-phase systems.
- ungrounded wye-connected, 3-phase systems.
- single-phase systems.
- 120/240-volt, single-phase systems.
- 208Y/120-volt, 3-phase, 4-wire systems.
- systems greater than 600 volts; for example, 4160Y/2400 volts.
- service or feeder disconnecting means rated at less than 1000 amperes.
- systems in which the service is subdivided; for example, a 1600-ampere service may be divided between two 800-ampere switches.

The time of operation of the device as well as the ampere setting of the GFPE device must be considered carefully to ensure that the continuity of the electrical service is maintained. The time of operation of the device includes

1. the sensing of a ground fault by the GFPE monitor.
2. the monitor signaling the disconnect switch to open.
3. the actual opening of the contacts of the disconnect device (either a switch or a circuit breaker).

The total time of operation may result in a time lapse of several cycles or more (Chapters 17 and 18).

GFPE circuit protective devices were developed to overcome a major problem in circuit protection: the low-value phase-to-ground arcing fault, Figure 13-31. The amount of current in an arcing phase-to-ground fault can be low when compared to the rating or setting of the overcurrent device. For example, an arcing fault can generate a current of 600 amperes. A main breaker rated at 1600 amperes will allow this current without tripping, because the 600-ampere current appears to be just another load current. The operation of the GFPE device assumes that under normal conditions the total instantaneous current in all of the conductors of a circuit will

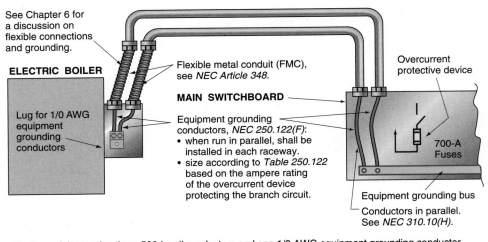

Each conduit contains three 500-kcmil conductors and one 1/0 AWG equipment grounding conductor, which is solidly connected to the equipment grounding bus in the switchboard and to the terminal box on the boiler. The equipment grounding conductor may be insulated or bare. See *NEC 250.118*. All conductors in the commercial building are copper.

FIGURE 13-30 Sizing equipment grounding conductor based on ampere rating of overcurrent protection. (*Delmar/Cengage Learning*)

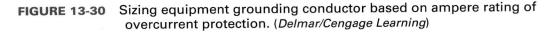

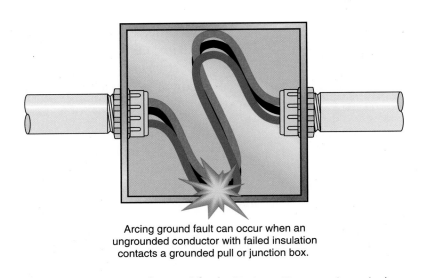

Arcing ground fault can occur when an ungrounded conductor with failed insulation contacts a grounded pull or junction box.

FIGURE 13-31 Ground fault. (*Delmar/Cengage Learning*)

exactly balance, Figure 13-32. Thus, if a current coil is installed so that all of the circuit conductors run through it, the normal current measured by the coil will be zero. If a ground fault occurs, some current will return through the grounding system, and an unbalance will result in the conductors. This unbalance is then detected by the GFPE device, Figure 13-33.

The purpose of ground-fault protection devices is to sense and protect equipment against *low-level*

ground faults. GFPE monitors do not sense phase-to-phase faults, 3-phase faults, or phase-to-neutral faults. These monitors are designed to sense phase-to-ground faults only.

High-magnitude ground-fault currents can cause destructive damage even though a GFPE is installed. The amount of arcing damage depends on

1. how much current flows.

2. the length of time that the current exists.

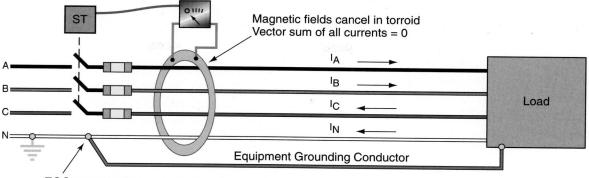

FIGURE 13-32 Ground-fault protection of equipment in normal conditions. (*Delmar/Cengage Learning*)

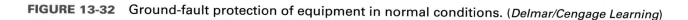

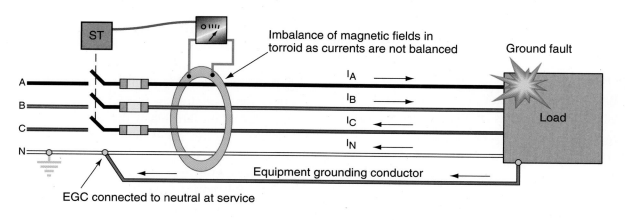

FIGURE 13-33 Ground-fault protection of equipment in abnormal conditions. (*Delmar/Cengage Learning*)

For example, if a GFPE device is set for a ground fault of 500 amperes and the time setting is six cycles, then the device will need six cycles to signal the switch or circuit breaker to open the circuit whether the ground fault is slightly more than 500 amperes or as large as 20,000 amperes. The six cycles needed to signal the circuit breaker plus the operation time of the switch or breaker may be long enough to result in damage to the switchgear.

The damaging effects of high-magnitude ground faults, phase-to-phase faults, 3-phase faults, and phase-to-neutral faults can be reduced substantially by the use of current-limiting overcurrent devices, *NEC 240.2*. These devices reduce both the peak let-through current and the time of opening once the current is sensed. For example, a ground fault of 20,000 amperes will open a current-limiting fuse in less than one-half cycle. In addition, the peak let-through current is reduced to a value much less than 20,000 amperes (see Chapters 17 and 18).

Ground-fault protection for equipment is not to be confused with ground-fault protection for personnel (GFCI). In commercial buildings, GFCI protection is required for receptacles in bathrooms, in kitchens, on rooftops, outdoors, where the receptacle will be within 6 ft (1.8 m) from the outside edge of any sink, ▶indoor wet locations, for locker rooms with adjacent showering facilities and garages, service bays, and similar areas. This is found in *NEC 210.8(B)*.◀

GFCI protection can be attained using GFCI circuit breakers or GFCI receptacles. The choice of which to use is generally based on economics—in other words, labor and material.

The GFPE is connected to the normal fused switch or circuit breaker, which serves as the circuit protective device. For feeders, the GFPE requirement is found in *NEC 215.10.* For services, the GFPE requirement is found in *NEC 230.95.* GFPE is required on services and feeders of solidly grounded systems of 1000 amperes or more where the voltage to ground is more than 150 volts but not more than 600 volts. The GFPE device is adjusted so that it will signal the protective device to open under ground-fault conditions. The maximum setting of the GFPE is 1200 amperes.

In the plans for the Commercial Building, the service voltage is 208Y/120 volts. The voltage to ground on this system is 120 volts. This value is not large enough to sustain an electrical arc. Therefore, it is not required (according to the *NEC*) that ground-fault protection of the service disconnecting means be installed for the Commercial Building. The electrician can follow a number of procedures to minimize the possibility of an arcing fault. Examples of these procedures follow:

- Be sure that conductor insulation is not damaged when the conductors are pulled into raceway.
- Be sure that the electrical installation is properly grounded and bonded.
- Tighten locknuts and bushings.
- Tighten all electrical connections.
- Tightly connect bonding jumpers around concentric and/or eccentric knockouts.
- Be sure that conduit couplings and other fittings are installed properly.
- Check insulators for minute cracks.
- Install insulating bushings on all raceways.
- Insulate all bare busbars in switchboards when possible.
- Be sure that conductors do not rest on bolts or other sharp metal edges.
- Be sure that electrical equipment does not become damp or wet either during or after construction.
- Be sure that all overcurrent devices have an adequate interrupting capacity.
- Do not work on hot panels.
- Be careful when using fish tape because the loose end can become tangled with the electrical panelboard.

- Be careful when working with live parts; do not drop tools or other metal objects on top of such parts.
- Avoid large services; for example, it is usually preferable to install two 800-ampere service disconnecting means rather than to install one 1600-ampere service disconnecting means.

SAFETY IN THE WORKPLACE

Although we covered much of this information in Chapter 1 of this text, a reminder about electrical safety can't hurt!

Many injuries have occurred by individuals working electrical equipment "hot." A fault, whether line-to-line, line-to-line-to-line, or line-to-ground, can develop tremendous energy. In such conditions, splattering of melted copper and steel, the heat of the arc blast, the blinding light of the arc, and electric shock are hazards. The temperatures of an electrical arc, such as might occur on a 480/277-volt solidly grounded, wye-connected system, are hotter than the surface of the sun. The enormous pressures of an "arc blast" can blow a person clear across the room. These awesome pressures will vent (discharge) through openings, such as the open cover of a panel, just where the person is standing.

Federal laws are in place to protect workers from injury while on the job. When working on or near electrical equipment, certain safety practices must be followed.

In the Occupational Safety and Health Act (OSHA), Sections 1910.331 through 1910.360 are devoted entirely to safety-related work practices. Proper training in work practices, safety procedures, and other personnel safety requirements are described. Such things as turning off the power, locking the switch off, or properly tagging the switch are discussed in the OSHA Standards.

The fundamental rule is **NEVER WORK ON EQUIPMENT WITH THE POWER TURNED ON!**

In the few cases where the equipment absolutely must be left on, proper training regarding electrical installations, proper training in established safety practices, proper training in first aid, properly insulated tools, safety glasses, hard hats, protective insulating shields (rubber blankets to cover live parts), rubber gloves, nonconductive and nonflammable clothing, are required.

Proper training is required to enable a person to become qualified. The *NEC* defines a "qualified person" in *Article 100* as *One who has skills and knowledge related to the construction and operation of the electrical equipment and installations and has received safety training to recognize and avoid the hazards involved.**

The National Fire Protection Association Standard NFPA 70B, *Electrical Equipment Maintenance*, discusses safety issues and safety procedures and mirrors the OSHA regulations in that **NO ELECTRICAL EQUIPMENT SHOULD BE WORKED ON WHILE IT IS ENERGIZED.**

The National Electrical Manufacturers Association (NEMA) in Publication PB 2.1–2002, "General Instructions for Proper Handling, Installation, Operation, and Maintenance of Deadfront Distribution Switchboards Rated 600 Volts or Less," repeats the safety rules just discussed. This standard states that "The installation, operation, and maintenance of switchboards should be conducted only by qualified personnel." This standard discusses the **"Lock-Out, Tag-Out"** procedures. It clearly states that:

WARNING: HAZARDOUS VOLTAGES IN ELECTRICAL EQUIPMENT CAN CAUSE SEVERE PERSONAL INJURY OR DEATH. UNLESS OTHERWISE SPECIFIED, INSPECTION AND PREVENTATIVE MAINTENANCE SHOULD ONLY BE PERFORMED ON SWITCHBOARDS, AND EQUIPMENT TO WHICH POWER HAS BEEN TURNED OFF, DISCONNECTED AND ELECTRICALLY ISOLATED SO THAT NO ACCIDENTAL CONTACT CAN BE MADE WITH ENERGIZED PARTS.

There have been lawsuits where serious injuries have occurred on equipment that had ground-fault protection of equipment (GFPE). The claims were that the ground-fault protection for equipment also provides protection for personnel. **This is not true!** Throughout the *NEC*, references are made to ***Ground Fault Protection For Personnel (GFCI)*** and ***Ground Fault Protection For Equipment (GFPE).***

Many texts have been written about the hazards of electricity. All of these texts say

TURN THE POWER OFF! LOCK IT OFF! TEST IT OFF! ASSUME IT IS ENERGIZED UNTIL LOCK/OUT TAG/OUT PROCEDURES ARE COMPLETED.

IT'S THE LAW!

NEC 110.16 has a requirement for placing a notice on certain equipment to warn qualified persons of the serious hazard of arc flash events. This section states:

▶*110.16. Arc-Flash Hazard Warning. Electrical equipment, such as switchboards, panelboards, industrial control panels, meter socket enclosures, and motor control centers, that are in other than dwelling units, and are likely to require examination, adjustment, servicing, or maintenance while energized shall be field marked to warn qualified persons of potential electric arc flash hazards. The marking shall be located so as to be clearly visible to qualified persons before examination, adjustment, servicing, or maintenance of the equipment.**◀

The *Informational Notes* to *110.16* reference NFPA 70E *Standard for Electrical Safety in the Workplace*, for *assistance in determining severity of potential exposure, planning safe work practices, and selecting personal protective equipment*, and to ANSI Z535.4, *Product Safety Signs and Labels, for guidelines for the design of safety signs and labels for application to products.**

Flash protection is a very complicated subject, and will be discussed over and over again as to how the electrical contractor and electrician will comply.

If you want to learn more, visit Bussmann's Web site at http://www.bussmann.com. There you will find an easy-to-use computer program for making arc-flash and fault-current calculations.

---

*Reprinted with permission from NFPA 70-2011.

REVIEW

Refer to the *National Electrical Code* or the working drawings when necessary. Where applicable, responses should be written in complete sentences.

A 300-kilovolt-ampere transformer bank consisting of three 100-kilovolt-ampere transformers has a 3-phase, 480-volt delta primary and a 3-phase, 208Y/120 secondary.

For Questions 1 through 4 show calculations and/or source of information.

1. What is the full-load current in the primary?

2. What is the full-load current in the secondary?

3. What is the proper rating of the fuses to install in the secondary?

4. What is the kilovolt-ampere of the bank if one of the transformers is removed?

5. Sketch the secondary connections, and indicate where a 120-volt load could be connected.

6. A cabinet with panelboard, shown on the right in the following figure, is added to an existing panelboard installation. Two knockouts, one near the top of the panelboard and the other near the bottom, are cut in the adjoining sides of the cabinets. List any parts to be checked or added and indicate on the drawing the proper way of extending the phase and neutral conductors to the new panelboard.

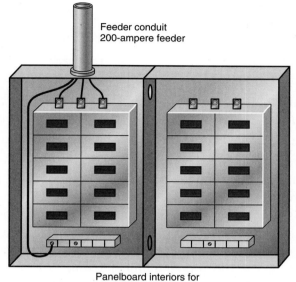

Feeder conduit
200-ampere feeder

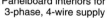

Panelboard interiors for
3-phase, 4-wire supply

The following questions refer to grounding requirements. Give the source of your responses.

7. What is the proper size copper grounding electrode conductor for a 100-ampere service that consists of 3/0 AWG phase conductors? _____

_____

_____

8. What is the minimum size copper grounding electrode conductor for a service with three sets of parallel 350-kcmil conductors (three conductors per phase)? _____

_____

_____

9. If they are present, what items shall be bonded together to form a grounding electrode system? _____

_____

_____

10. The engineering calculations for an 800-ampere service-entrance call for two 500-kcmil copper conductors per phase connected in parallel. The neutral conductor calculations show that the neutral conductors need only be 3 AWG copper conductors. The riser diagram shows two rigid metal conduits, each containing three 500-kcmil phase conductors and one 1/0 AWG neutral conductor. Why is the neutral conductor sized as a 1/0 AWG, or is this a mistake?

_____

_____

_____

_____

_____

11. Electric utility installations and equipment under the exclusive control of the utility do not have to conform to the *National Electrical Code*. However, they do have to conform to the _____.

12. You are an electrical apprentice. The electrical foreman on the job tells you to connect some equipment grounding conductors to the ground bus that runs along the front bottom of a dead-front switchboard. The system is 480Y/277-volt solidly grounded wye. The building is still under construction, and a number of other trades are working in the building. The switchboard has already been energized, and it is totally enclosed except for two access covers near the front bottom where the ground bus runs. The ground bus already has the proper number of and size of lugs for the equipment grounding conductors. You ask permission to turn off the main power so that you can work on connecting the equipment grounding conductors to the ground bus without having to worry about accidentally touching the phase buses with your hands or with a screwdriver or wrench. The foreman says, "No." What would you do?

_____

_____

_____

_____

_____

_____

13. Electrical explosions directly into the face of a person working on or near energized electrical equipment can and will cause serious injury or death. OSHA laws and NFPA 70E have strict safety requirements for working on electrical equipment. The basic rules are to turn off the power before working on the equipment and to wear the proper type of nonflammable clothing. To further provide safety for the worker, the *National Electrical Code* has added a new requirement pertaining to **Flash Protection**. This requirement is found in _____ of the *NEC*.

Lamps and Ballasts for Lighting

OBJECTIVES

After studying this chapter, you should be able to

- understand the technical terms associated with lamps and ballasts.

- identify the lamps scheduled to be used in the Commercial Building.

- understand the basics of incandescent, halogen, fluorescent, LED, and HID lamps.

- understand the basics of electronic and magnetic ballasts.

- understand the practical application of lamps used in the Commercial Building.

- understand more about energy savings for lamps and ballasts.

- identify lamp types according to certain characteristics and letter designations.

- be aware of the hazards of disposing lamps and ballasts.

Electrical Wiring—Commercial introduces some of the concerns relating to lighting, such concerns as ballasts, lamps, and the *NEC* requirements for the lighting branch-circuit, feeder, and service-entrance calculations.

As you study this chapter, don't get hung up on the terms *lamp* and *light bulb*. The *National Electrical Code* and manufacturers' literature use the terms *lamp* and *lampholder*. We still seem to feel comfortable using the term *light bulb*. Those of us in the electrical arena have seen the *National Electrical Code* switch from using the term *lighting fixture* to *luminaire*. We understand either term.

For most construction projects, the electrical contractor is required to purchase and install lamps in the luminaires. For large installations of fluorescent luminaires, major manufacturers will ship their luminaires with the lamps already installed.

Thomas Edison provided the talent and perseverance that led to the development of the incandescent lamp in 1879 and the fluorescent lamp in 1896. Peter Cooper Hewitt produced the mercury lamp in 1901.

The first practical use of a light-emitting diode (LED) occurred in 1962 when they were used as indicating lamps. Over the years, the technology of LEDs has increased phenomenally. LEDs are now a very viable source of illumination. Manufacturers of luminaires are coming out with new LED luminaires at a rate almost impossible to keep up with.

All four of these lamp types have been refined and greatly improved since they were first developed. The *2011* edition of the *NEC* added many references to light-emitting diodes (LEDs). Whereas fluorescent lamps require a ballast, an LED requires a "driver." The heat generated in the LED and the driver presents a major thermal problem. The heat is dissipated using heat sinks. All of this is covered in UL Standard 8750 entitled *Standard for Light Emitting Diode (LED) Light Sources for Use in Lighting Products*. An electrician in the field needs to be assured that the luminaire is listed by UL or another qualified electrical product testing laboratory.

NEC Article 410 contains the provisions for the wiring and installation of luminaires, lampholders, and lamps.

As you study the Plans for the Commercial Building, you will note that high-efficiency energy-saving fluorescent lighting and LED lighting is used. Very few incandescent lamps are used. Most of the energy used by incandescent lamps is heat—not light! Today, the buzzword is GO GREEN! The lighting design for the Commercial Building meets the GO GREEN concept. The *NEC* refers to lamps as either incandescent or electric discharge lamps. In the industry, an incandescent lamp is also referred to as a filament lamp because the light is produced by a heated wire filament. For example, the term *tungsten-filament* is found in the *NEC* in a number of places.

Electric discharge lamps include a variety of types, but all require a ballast. The most common types of electric discharge lamps are fluorescent, mercury, metal halide, high-pressure sodium, and low-pressure sodium. As previously mentioned, LEDs require a "driver."

Mercury, metal halide, and high-pressure sodium lamps are also classified as high-intensity discharge (HID) lamps. The *NEC* considers these and LEDs to be nonlinear loads, making them subject to the possibility of harmonic currents that would flow in the neutral conductor of the branch circuit, feeder, and service-entrance conductors and equipment. Electronic equipment such as computers, printers, scanners, and copiers are nonlinear loads. You may have noticed that some distribution transformers have a "K" rating. These are used where supplying loads that are significantly nonlinear loads.

LIGHTING TERMINOLOGY

Candela (cd)

The luminous intensity of a source, when expressed in candelas, is the candlepower (cp) rating of the source.

Lumen (lm)

Lumen (lm) is the amount of light received in a unit of time on a unit area at a unit distance from a point source of one candela, Figure 14-1. The surface area of a sphere is 12.57 times the radius. When the measurement is in customary units, the unit area is 1 square foot and the unit distance is 1 foot; thus a 1-candela source produces 1 lumen on each

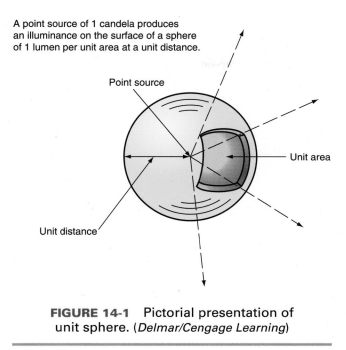

A point source of 1 candela produces an illuminance on the surface of a sphere of 1 lumen per unit area at a unit distance.

Point source

Unit area

Unit distance

FIGURE 14-1 Pictorial presentation of unit sphere. (*Delmar/Cengage Learning*)

square foot (1 footcandle) on the sphere for a total of 12.57 lumens. If the units are in SI, then the unit area is 1 square meter, and the unit distance is 1 meter; thus a 1-candela source produces 1 lumen on each square meter (1 lux) for a total of 12.57 lumen.

Illuminance

The measure of illuminance on a surface is the lumen per unit area expressed in footcandles (fc) or lux (lx). The recommended illuminance levels vary greatly, depending on the task to be performed and the ambient lighting conditions. For example, although 5 footcandles (fc), or 54 lux (lx), is often accepted as adequate illumination for a dance hall, 200 fc (2152 lx) may be necessary on a drafting table for detailed work.

LUMENS PER WATT (lm/W)

Lumens per watt is very important when comparing different light sources for the purpose of energy consumption for the amount of light created. To put this term into perspective, you might think of lumens per watt as miles per gallon. Now *that* you can understand!

Lumens per watt is a measure of the effectiveness (efficacy) of a light source in producing light from electrical energy. Checking lumens per watt enables the comparison of the energy effectiveness of various lamps.

A 100-watt incandescent lamp producing 1670 lumens has an efficacy of 16.7 lumens per watt. A typical 32-watt T8 fluorescent lamp might have an efficacy of approximately 95 lumens per watt. An energy-efficient 32-watt fluorescent lamp might actually consume only 25 watts and might have an efficacy of 100 lumens per watt. Energy-saving lamps in combination with high-efficiency energy savings ballast create a strong incentive for conformance to the "green" concept. Every lamp manufacturers' literature provides all of the benefits and features of their particular lamp. Check their Web sites for this type of information.

As a technical note, efficiency is when the input and output have the same units of measure; an efficacy is when the units are different. A motor has an efficiency of watts in–watts out. A lamp has an efficacy of watts in–light out.

Kelvin (K)

The Kelvin (sometimes incorrectly called degree Kelvin) is measured from absolute zero; it is equivalent to a degree value in the Celsius scale plus 273.16. The color temperature of lamps is given in Kelvin. The lower the number, the warmer appearing is the light (more red/orange/yellow content). See Figure 14-2. The higher the number, the cooler appearing is the light (more blue content). 2000–3000K is considered to be "warm." Greater than 4000K is considered to be "cool." Different manufacturers have different names, for example, white, warm white, warm white deluxe, warm white plus, cool white, cool white plus, cool white deluxe, natural white, daylight. It all boils down to the Kelvin rating of the lamp.

The fluorescent lamps specified for the commercial building are 3500 K. This is a good all-around choice.

Color Rendering Index (CRI)

This value is often given for lamps so the user can have an idea of the color rendering probability.

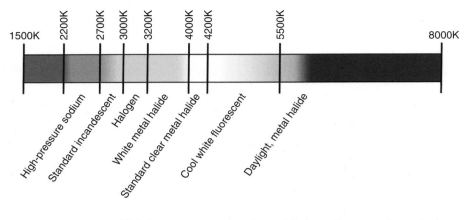

FIGURE 14-2 Kelvin temperature chart. (*Delmar/Cengage Learning*)

It is the ability of the lamp to portray objects as normally as possible. The numbering system goes from 0 to 100. The higher the number, the better and more natural is the color appearance of the object. A low number will cause colors to appear washed out. The CRI uses filament light as a base for 100 and the warm white fluorescent for 50.

The CRI can be used only to compare lamps that have the same color temperature. It is our recommendation that this system be avoided. The only sure way to determine whether a lamp will provide good color rendition is to see the objects in the light produced by that lamp.

Lumen Maintenance

This term refers to how a lamp maintains its light output over its lifetime. This varies by lamp type. For comparison, after 50% of a lamp's rated life, a T5 fluorescent lamp may still have 94% of its initial rated light output, compared to 88% for a T8 lamp and 73% for a T12 lamp. All of this information is available from the lamp/ballast manufacturer.

Ambient Temperature

Be careful of this one. The term *ambient temperature* refers to the surrounding temperature. Fluorescent lamps might not start in cold temperatures (outdoors, unheated garages and warehouses, refrigerated areas). If they do start, they may flicker, flutter, or give off a significantly reduced light output.

Lamp manufacturers provide light output data based on specific ideal temperatures, such as 77°F (25°C). They have tables showing expected light output at different ambient temperatures.

Jacketed lamps and cold weather ballast are available. Typical minimum starting temperatures might be 50°F (10°C), 20°F (6.66°C), or 0°F (−17.77°C). These ballasts have a higher open-circuit voltage and are marked with the minimum temperature at which they will operate properly.

In high ambient temperatures, a Class P ballast might cycle on and off repeatedly until the high-temperature condition is corrected.

INCANDESCENT LAMPS

The incandescent lamp has the lowest efficacy (lumens per watt) of the types listed in Table 14-1. Incandescent lamps are losing their popularity, giving way to compact fluorescent lamps (CFLs) and the up-and-coming LED lighting, all of this brought about by energy savings laws.

A halogen lamp is a unique type of incandescent lamp. Its tungsten filament is sealed in a small transparent quartz capsule filled with an inert gas and a small amount of a halogen (an electronegative element) such as iodine or bromine. Think of it as a "lamp within a lamp." See Figure 14-3. A halogen lamp operates its filament at a higher temperature. This results in brighter whiter light, longer life than a conventional gas-filled lamp of the same wattage

TABLE 14-1

Characteristics of electric lamps.*

	Filament (incandescent)	Fluorescent	Mercury (banned in United States in 2008)	Metal Halide	HPS	LPS
Lumens per watt	6–23	25–100	30–65	65–120	75–140	130–180
Wattage range	40–1500	4–215	40–1000	175–1500	35–1000	18–180
Life (hours)	750–8000	9000 to 20,000	16,000–24,000	5000–20,000	20,000–24,000	18,000
Color temperature	2400–3100	2700–7500	3000–6000	3000–5000	2100	1700
Color Rendition Index	90–100	50–110	25–55	60–85	20–25	0
Potential for good color rendition	high	highest	fair	good	**fair**	fair
Lamp cost	low	moderate	moderate	high	high	moderate
Operational cost	high	good	moderate	Moderate. Can use pulse-start (PS) ballasts. Energy-efficient, cool, long lasting.	low	low

*The values given above are typical. These values can be used to compare lamp characteristics. Values vary greatly by manufacturer. Consult manufacturer's technical specification data for actual values.

	LED	Compact Fluorescent (CFL)
Lumens per watt	29 to >100 (True lumens per watt should include the wattage of the lamp and the driver.)	Varies considerably. Can exceed 100.
Wattage range	7–21. Higher wattages with clusters. Depends on mfg.	14–34. Depends on mfg.
Life (hours)	40,000 hours. Longer life with fewer starts.	16,000/20,000 hours. Longer life with fewer starts.
Color temperature	3100–4200	≤3000–≥5000
Color Rendition Index	85	82
Potential for good color rendition	Very good. Improving as fast moving technology advances.	Excellent
Lamp cost	Moderate. Luminaires must be suitable for use with LEDs	Moderate to low
Operational cost	Very low considering lumens per watt.	Very low considering lumens per watt.

rating, and a higher efficacy (10–30 lm/W). Halogen lamps operate hotter than a conventional lamp. Because of their higher lumens per watt, they are generally acceptable as a replacement for an incandescent lamp where incandescent lamps are being legislated out.

The phase-out of incandescent lamps will occur over a period of time. Between 2012 and 2014, the phase-out will begin with the 100-watt lamp in 2012 and end in 2014 with the 40-watt. Federal law requires that light bulbs become 25–35% more efficient by 2014. By the year 2020, the Feds are

FIGURE 14-3 Halogen lamp. Note the filament sealed inside of an inner capsule. (*Delmar/Cengage Learning*)

looking for a 70% greater efficiency over today's light bulbs.

For all practical purposes, this will pretty much ban the sale of incandescent light bulbs, other than specialty lamps such as appliance lamps, colored lamps, and 3-way lamps.

Federal energy legislation that became effective in 1994 removed several of the more commonly used lamps from the marketplace. No longer can the standard reflector lamps like the R40, or standard PAR lamps like the PAR38 be manufactured. The newer types of lamps in these categories now have lower wattage ratings. The use of halogen and krypton-filled elliptical reflector (ER), and low-voltage lamps is encouraged. The ER lamp is designed primarily for use as a replacement and is not recommended for new installations. Typical

of the recommended substitutions are a 50-watt PAR30 halogen lamp with a 65° spread to replace the 75-watt R30 reflector lamp, and a 60- or 90-watt PAR halogen flood with a 30° spread to replace the 150-watt PAR38. A full list of substitutions is available at major hardware stores and electrical distributors.

Construction

The light-producing element in the incandescent lamp is a tungsten wire called the filament, Figure 14-4. This filament is supported in a glass envelope (lamp). The air is evacuated from the lamp and is replaced with an inert gas, such as argon or halogen. The filament is connected to the base by the lead-in wires. The base of the incandescent lamp supports the lamp and provides the connection

FIGURE 14-4 Typical A-19 incandescent lamp (light bulb). (*Delmar/Cengage Learning*)

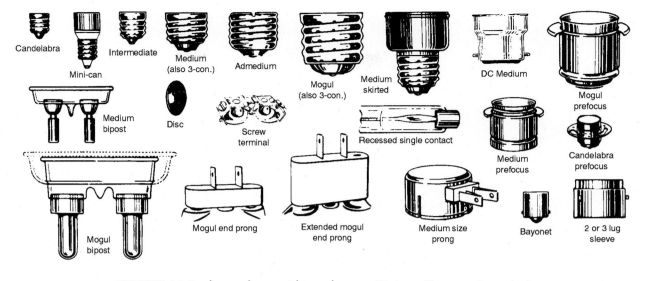

FIGURE 14-5 Incandescent lamp bases. (*Delmar/Cengage Learning*)

means to the power source. The lamp base may be any one of the base styles shown in Figure 14-5.

Characteristics

Incandescent lamps are classified according to the following characteristics:

Voltage Rating. Incandescent lamps are available with many different voltage ratings. When installing lamps, the electrician should be sure that a lamp with the correct rating is selected because a small difference between the rating and the actual voltage has a great effect on the lamp life and lumen output, Figure 14-6.

Wattage. Lamps are usually selected according to their wattage rating. This rating is an indication of the consumption of electrical energy but is not a true measure of light output. For example, at the rated voltage, a 60-watt lamp produces 840 lumens and a 300-watt lamp produces 6000 lumens; therefore, one 300-watt lamp produces more light than seven 60-watt lamps.

Shapes. Figure 14-7 illustrates the common lamp configurations and their letter designations.

Size. Lamp size is usually indicated in eighths of an inch and is the diameter of the lamp at the widest place. Thus, the lamp designation A19 means that the lamp has an arbitrary shape and is $^{19}/_8$ in. or $2^3/_8$ in. in diameter, Figure 14-8.

A T8 fluorescent lamp has a diameter of eight-eighths of an inch, or 1 in. A T12 fluorescent lamp has a diameter of twelve-eighths of an inch, or $1^1/_2$ in. An R30 incandescent lamp has a diameter of thirty-eighths of an inch, or $3^3/_4$ in.

Operation. The light-producing filament in an incandescent lamp is a resistance element that is heated to a high temperature by an electric current. This filament is usually made of tungsten, which has a melting point of 6170°F (3410°C), making it excellent for use as a filament in a light bulb. At this temperature, the tungsten filament produces light with an efficacy of 53 lumens per watt. However, to increase the life of the lamp, the operating temperature is lowered, which also means a lower efficacy. For example, if a 500-watt lamp filament is heated to a temperature of 3000 K, the resulting efficacy is 21 lumens per watt.

Catalog Designations. Catalog designations for incandescent lamps usually consist of the lamp wattage followed by the shape and ending with the diameter and other special designations as appropriate. Here are some common examples:

60A19	60 watt, arbitrary shape, $^{19}/_8$ in. (2.375 in.) diameter
65PAR38/FL	65-watt parabolic reflector flood, $^{38}/_8$ in. (4.75 in.) diameter.
65PAR38/SP	65-watt parabolic reflector spot, $^{38}/_8$ in. (4.75 in.) diameter.

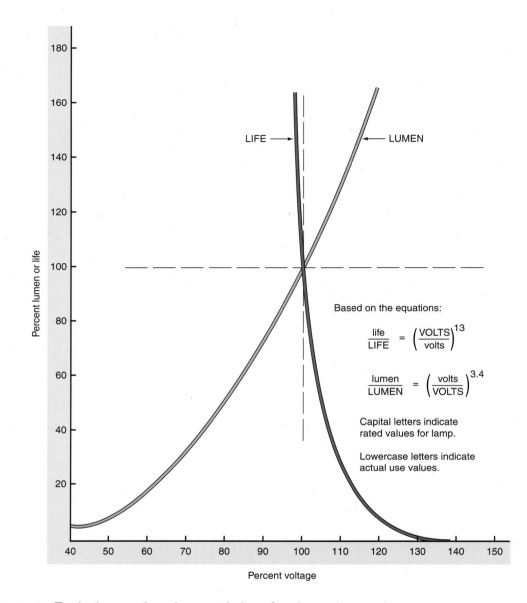

Based on the equations:

$$\frac{\text{life}}{\text{LIFE}} = \left(\frac{\text{VOLTS}}{\text{volts}}\right)^{13}$$

$$\frac{\text{lumen}}{\text{LUMEN}} = \left(\frac{\text{volts}}{\text{VOLTS}}\right)^{3.4}$$

Capital letters indicate rated values for lamp.

Lowercase letters indicate actual use values.

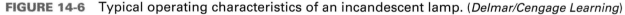

FIGURE 14-6 Typical operating characteristics of an incandescent lamp. (*Delmar/Cengage Learning*)

LOW-VOLTAGE INCANDESCENT LAMPS

In recent years, low-voltage (usually 12-volt) incandescent lamps have become very popular for accent lighting. Many of these lamps are tungsten lamps. They have a small source size, as shown in Figure 14-9. This feature allows precise control of the light beam. A popular size is the MR16, which has a multifaceted reflector diameter of just 2 in. These lamps provide a whiter light than regular incandescent lamps. Common wattage ratings are 20, 37, 50, and 65 watts. When dimming tungsten halogen lamps, a special dimmer is required because of the transformer that is installed to reduce the voltage. The dimmer is installed in the line voltage circuit supplying the transformer. Dimmed lamps will darken if they are not occasionally operated at full voltage. Dimming also changes the color temperature of the lamp.

Recently, LED lamps have become available to use as a retrofit for the conventional MR16 lamp. They consume considerably less power than the incandescent type—just 2.8 watts. LED lamps cannot be dimmed. Caution: LEDs do not work on a low-voltage system where an electronic transformer is used to reduce the voltage to 12 volts. They do work on magnetic transformers.

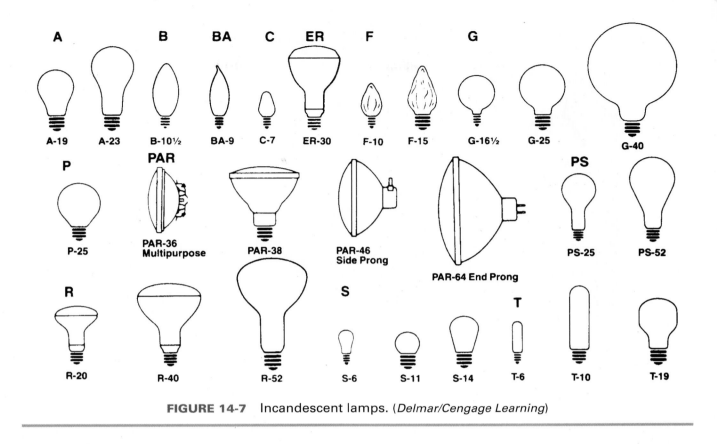

FIGURE 14-7 Incandescent lamps. (*Delmar/Cengage Learning*)

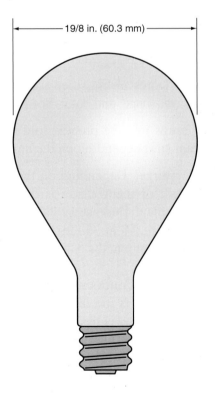

‹— 19/8 in. (60.3 mm) —›

FIGURE 14-8 An A-19 lamp.
(*Delmar/Cengage Learning*)

The requirements for low-voltage lighting are covered in *Article 411* of the *NEC*.

A word of caution regarding low-voltage lighting. Although considered safer than conventional 120-volt and 277-volt lighting regarding electric shock hazard, one must be aware of the current draw of a low-voltage system. Conductors have to be sized properly.

EXAMPLE

A 480-watt load at 120 volts draws:

$$I = \frac{W}{E} = \frac{480}{120} = 4 \text{ amperes}$$

Checking *Table 310.15(B)(16)*, a 4-ampere load is easily handled by a 14 AWG conductor.

A 480-watt load at 12 volts draws:

$$I = \frac{W}{E} = \frac{480}{12} = 40 \text{ amperes}$$

Checking *Table 310.15(B)(16)*, a 40-ampere load requires an 8 AWG conductor.

FIGURE 14-9 Assortment of MR (multifaceted reflector) halogen lamps. (*Delmar/Cengage Learning*)

Current draw of such a large magnitude also results in significant voltage drop. The length of the branch circuit wiring is limited.

UL requires that low-voltage lighting assemblies be clearly marked with the correct size and length of conductors to use.

Catalog Designations

Catalog designations for low-voltage incandescent lamps are similar to other incandescent lamps except for some special cases, such as:

MR16   mirrored reflector, 2 in. diameter

FLUORESCENT LAMPS

Luminaires using fluorescent lamps are considered to be electric-discharge lighting by the *NEC*. Fluorescent lighting has the advantages of energy savings because of their high lumens per watts (high efficacy), and long life.

Construction

A fluorescent lamp consists of a glass tube with an electrode and a base at each end, Figure 14-10.

The inside of the tube is coated with a phosphor (a fluorescing material), the air is evacuated, and an inert gas plus a small quantity of mercury is released into the tube. The base styles for fluorescent lamps are shown in Figure 14-11.

Characteristics

Fluorescent lamps are classified according to type, length or wattage, shape, and color. See Figure 14-12.

Type. The lamps may be preheat (old type), rapid start, or instant start depending on the ballast circuit.

Length or Wattage. Depending on the lamp type, either the length or the wattage is designated. To name a few, typical fluorescent lamp lengths are 18 in., 24 in., 36 in., 48 in., 60 in., 72 in., and 96 in. There are many other lengths. Check manufacturers' literature.

Wattage can be confusing. A lamp might be marked F32T8. This is the lamp that has replaced the older F40T12. This is assumed to be a 32-watt lamp. However, some energy-saving lamps with this marking actually consume 25, 28, or 30 watts. The only sure way of knowing the true energy consumption of a luminaire is to check the ballast label for total input wattage or total input volt-amperes.

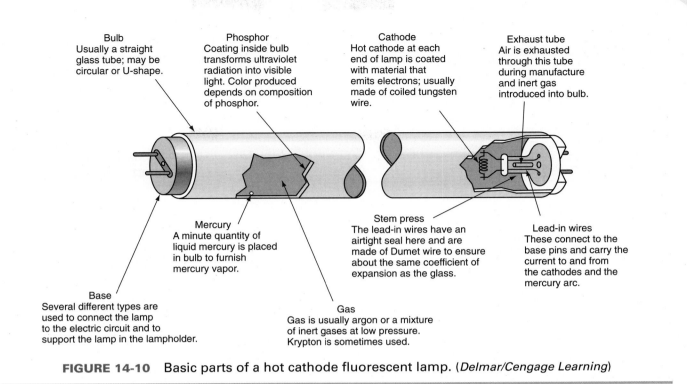

Bulb
Usually a straight glass tube; may be circular or U-shape.

Phosphor
Coating inside bulb transforms ultraviolet radiation into visible light. Color produced depends on composition of phosphor.

Cathode
Hot cathode at each end of lamp is coated with material that emits electrons; usually made of coiled tungsten wire.

Exhaust tube
Air is exhausted through this tube during manufacture and inert gas introduced into bulb.

Mercury
A minute quantity of liquid mercury is placed in bulb to furnish mercury vapor.

Stem press
The lead-in wires have an airtight seal here and are made of Dumet wire to ensure about the same coefficient of expansion as the glass.

Lead-in wires
These connect to the base pins and carry the current to and from the cathodes and the mercury arc.

Base
Several different types are used to connect the lamp to the electric circuit and to support the lamp in the lampholder.

Gas
Gas is usually argon or a mixture of inert gases at low pressure. Krypton is sometimes used.

FIGURE 14-10   Basic parts of a hot cathode fluorescent lamp. (*Delmar/Cengage Learning*)

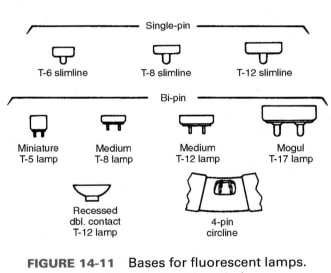

Single-pin

T-6 slimline   T-8 slimline   T-12 slimline

Bi-pin

Miniature T-5 lamp   Medium T-8 lamp   Medium T-12 lamp   Mogul T-17 lamp

Recessed dbl. contact T-12 lamp   4-pin circline

FIGURE 14-11   Bases for fluorescent lamps. (*Delmar/Cengage Learning*)

Today, most commercial lighting installations use T8 lamps as opposed to the older style T12 lamps because of the increased lumens per watt and longer life. Matching energy-efficient ballasts with the energy-efficient lamps results in significantly less energy consumption.

Shapes. The fluorescent lamp usually has a straight, tubular shape. Exceptions are the T-9, 4-pin circline lamp, which forms a complete circle (these are available in outside diameters of 6½ in., 8 in.,

12 in., and 16 in.) and the U-shaped lamp, which is a T8 or T12 lamp having a 180° bend in the center to fit a 2-ft (610-mm) long luminaire, and the PL lamp, which has two parallel tubes with a short connecting bridge at the ends opposite the base. Several other shapes are covered in the section on compact fluorescent lamps.

Catalog Designation and Color

An earlier reference was made to Federal energy legislation that impacted the manufacturing of fluorescent lamps and ballasts. The popular cool white and warm white lamps that have been used for decades are still available. Today's fluorescent lamps are available with more color rendition choices, considerably less energy consumption, and longer life.

As previously mentioned, fluorescent lamps are designed with a Kelvin color temperature rating. This rating is part of the lamp's description. The higher the K value, the whiter (cooler) the light output.

A fluorescent lamp with a 30 in its nomenclature indicates it produces light with a color temperature of 3000 K, which would provide good rendition

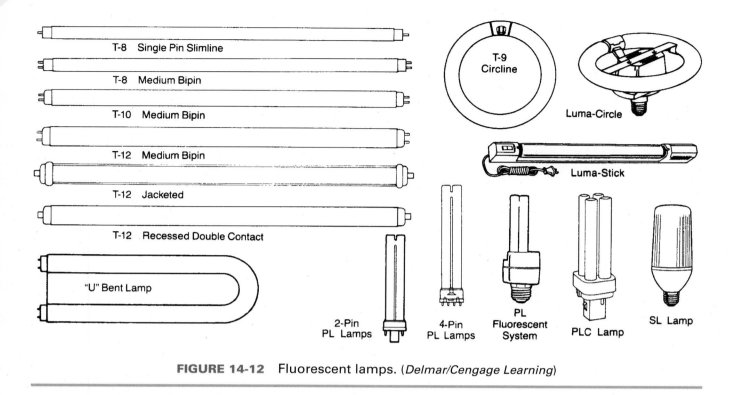

FIGURE 14-12 Fluorescent lamps. (*Delmar/Cengage Learning*)

to warm colors. Warm ratings bring out the red in objects. Lamps with a rating of 41 (4100 K), 50 (5000 K), or 65 (6500 K) would provide good rendition to cool colors.

Here are common examples of fluorescent lamp designations:

F34T12/41U/RS	This is a T12, U-shaped, rapid-start, 34-watt lamp with a color temperature of 4100 K.
F32T8/SPX/35/IS	This is a straight T8, instant-start, 32-watt, deluxe specification grade lamp with a color temperature of 3500 K.

Lamp manufacturers have many ways to further identify their products. Not all designations are an industry standard. For example, one manufacturer uses the letters XL, meaning extra-long life; SP, meaning specification series with good color rendition; and SPX, meaning deluxe specification grade with true and natural color rendition.

Consulting engineers, manufacturers, electrical contractors, and electrical distributors who specialize in lighting have computer programs to do the calculations and lighting layout. These programs include such things as lumen output, wattage, volt-amperes, light photometrics, wall-ceiling-floor reflection factors, candlepower, number of luminaires needed, number of lamps (2-3-4) in the selected luminaires, and other installation requirements. And, maybe more importantly, they know the requirements of federal, state, and local energy laws! Installing luminaires that do not meet these requirements could prove very costly!

Toxicity Characteristic Leaching Procedure (TCLP)

What is TCLP and who performs the TCLP?

This is all about toxic waste! There are many businesses, in almost every state, that can perform these analyses. Look in the yellow pages under "Laboratories—Analytical".

This procedure will be listed in the manufacturers' literature. The TCLP rating means that the lamp has passed Environmental Protection Agency (EPA) tests regarding the toxicity of the lamp. They test the toxicity of just about everything that might be disposed of. This means lower disposal costs when the lamp reaches the end of its life.

Disposing of fluorescent lamps, because they may contain mercury, is a problem. Contact your state or local governmental agency to learn more about how to dispose of fluorescent lamps. The subject of hazardous waste disposal is covered later on in this chapter.

Operation

If a substance is exposed to such rays as ultraviolet and X-rays and emits light as a result, then the substance is said to be fluorescing. The inside of the fluorescent lamp is coated with a phosphor material, which serves as the light-emitting substance. When sufficient voltage is applied to the lamp electrodes, electrons are released. Some of these electrons travel between the electrodes to establish an electric discharge or arc through the mercury vapor in the lamp. As the electrons strike the mercury atoms, radiation is emitted by the atoms. This radiation is converted into visible light by the phosphor coating on the tube, Figure 14-13.

As the mercury atoms are ionized, the resistance of the gas is lowered. The resulting increase in current ionizes more atoms. If allowed to continue, this process will cause the lamp to destroy itself. As a result, the arc current must be limited. The standard method of limiting the arc current is to connect a reactance (ballast) in series with the lamp.

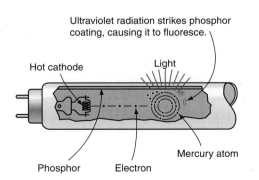

Ultraviolet radiation strikes phosphor coating, causing it to fluoresce.

Hot cathode

Light

Phosphor   Electron

Mercury atom

FIGURE 14-13 How light is produced in a hot cathode fluorescent lamp. (*Delmar/Cengage Learning*)

RETROFITTING EXISTING INSTALLATIONS

Retrofitting an existing installation that has luminaires using the older F40T12 lamps is becoming the "Way to Go Green." A couple of options:

1. Old magnetic ballasts are replaced with energy savings ballasts . . . continuing to use the old T12 lamps.

2. Old magnetic ballasts are replaced with energy savings ballasts, and the old style F40T12 lamps are replaced with energy savings T8 lamps.

Disposing of the old ballasts and lamps must be done properly. This subject is covered later on in this chapter.

Ballasts and Ballast Circuits

Without a ballast, if an arc within a fluorescent lamp is established, it will "run away." It will destroy itself. The ballast is needed to "choke" the flow of current.

As you will have already noted, the Commercial Building is served by a 208Y/120-volt, 3-phase, 4-wire system. Many larger commercial buildings are served by a 480Y/277-volt system. If a 480-volt, 3-phase system is installed, lighting is usually served by 277-volt branch circuits. Step-down transformers are installed to serve the 120-volt loads.

The trend today is to use luminaires that are equipped with electronic ballasts rated 120/277 volts. The same ballast can be connected to either a 120-volt branch circuit or a 277-volt branch circuit or any voltage in between. Manufacturer and distributors prefer this type. Inventory is significantly reduced. They do not have to stock two types of luminaires, and they do not have to stock two types of ballasts. Electricians like it because they need not be concerned with improperly hooking up a luminaire to a voltage not matching the ballast rating. A win–win situation for all concerned.

Today, the focus is on energy conservation. This has resulted in a major change in the design of ballasts. Many types of magnetic ballasts were "outlawed" as of April 1, 2005. The Department of Energy (DOE) mandated that magnetic ballasts may

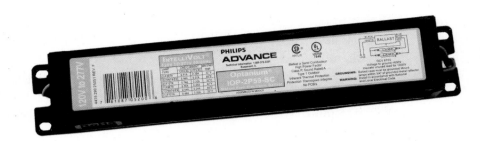

FIGURE 14-14 An electronic Class P ballast. (*Photo courtesy of Philips/Advance*)

not be sold through electrical distributors after July 1, 2010.

The most popular type of ballast today is the electronic ballast. Figure 14-14 is a photo of a Class P electronic fluorescent ballast. This particular ballast can be connected to either a 120-volt source or a 277-volt source. One ballast serves both voltages. Today's electronic ballasts have a power factor of 0.98 (98%) to 0.99 (99%), are quieter and extremely more efficient, and weigh considerably less than older core-and-coil ballasts. Fluorescent lighting is considered to be an electric discharge type of load. As such, this also increases the nonlinear load component of a multiwire branch circuit. Care must be exercised when sizing branch-circuit and feeder conductors. This is of major concern to electrical installations. Nonlinear loads can cause havoc (seriously overload), particularly in a neutral conductor of a multiwire branch circuit.

The *NEC* addresses nonlinear loads. The definition of a nonlinear load is

Nonlinear Load. *A load where the wave shape of the steady-state current does not follow the wave shape of the applied voltage.*

Informational Note: Electronic equipment, electronic/electric-discharge lighting, (both inductive and electronic ballasts, and light-emitting diode (LED) drivers for lighting), adjustable-speed drive systems, and similar equipment may be nonlinear loads.

Other places in the *NEC* that reference nonlinear loads are: *210.4, 220.61(C)(2), 310.15(B) (5), 400.5(A)(5), 450.3, 450.9, 520.2, Table 520.44 (notes)*, and in the examples in *Annex D*.

The older magnetic ballasts consisted of an assembly of a core and coil, a capacitor, and a thermal protector installed in a metal case. When the assembled parts are placed in the case, it is filled with a potting compound to improve the heat dissipation and reduce ballast noise.

Ballasts are required for high-intensity discharge lamps.

Ballasts serve two basic functions:

1. to provide the proper voltage for starting.
2. to control the current during operation.

The installation requirements for ballasts are covered in *NEC Article 410, Part XIII*.

Preheat Circuit. The first fluorescent lamps developed were of the preheat type and required a starter in the circuit. This type of lamp is now obsolete and is seldom found except on older fluorescent luminaires. The starter serves as a switch and closes the circuit until the cathodes are hot enough. The starter then opens the circuit, and the lamp lights. The cathode temperature is maintained by the heat of the arc after the starter opens.

Note in Figure 14-15 that after the starter opens, the ballast is in series with the lamp and acts as a choke to limit the current through the lamp.

Trigger Start. This type of ballast provides the preheating of the lamps cathodes without using a starter. The cathodes are heated continuously. Lamps start almost instantly, similar to rapid-start.

You might recognize a common type of preheat circuits: the trigger-start circuit. The low-wattage fluorescent lamps such as desk lamps used the trigger-start concept without the use of a starter. The ON/OFF switch served as the starter. To turn on the desk lamp, you pushed down on the switch for a few seconds, allowing the lamps cathodes to get warm. Releasing the switch resulted in the trigger-start ballast giving a "kick" to start the lamp.

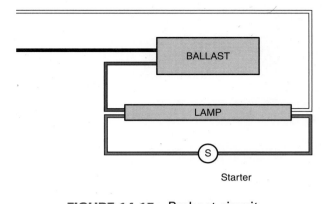

FIGURE 14-15 Preheat circuit.
(*Delmar/Cengage Learning*)

Preheat ballasts and preheat lamps have pretty much been replaced by rapid-start ballasts and lamps.

Rapid-Start Circuit. See Figure 14-16. In the rapid-start circuit, the cathodes are heated by a low voltage continuously by a separate winding in the ballast. At the same time, a starting voltage lower than that used in instant-start ballasts provides the "kick" to start the lamp. The result is almost instantaneous starting. You will notice a slight delay in starting. This type of fluorescent lamp requires the installation of a continuous grounded metal strip within an inch of the lamp. The metal wiring channel or the reflector of the luminaire can serve as this grounded strip. The standard rapid-start circuit operates with a lamp current of 430 mA. Two variations of the basic circuit are available; the high-output (HO) circuit operates with a lamp current of 800 mA, and the very high-output (VHO) circuit has 1500 mA

of current. HO and VHO ballasts and lamps provide a greater concentration of light, thus reducing the required number of luminaires. With the more common 430 mA ballasts, lamps will not start at temperatures below 50°F (10°C). Luminaires with 800 mA or 1500 ma ballasts are often specified for installation in areas having cold temperatures, such as outdoors in colder climates or in refrigerated rooms.

Instant-Start Circuit. The lamp cathodes in the instant-start circuit are not preheated. Sufficient voltage is applied across the cathodes to create an instantaneous arc, Figure 14-17. As in the preheat circuit, the cathodes are heated during lamp operation by the arc. The instant-start lamps require single-pin bases, Figure 14-18, and are generally called *slimline lamps*. Bipin base fluorescent lamps are available, such as the 40-watt F40T12/CW/IS lamp. For this style of lamp, the pins are shorted together so that the lamp will not operate if it is mistakenly installed in a rapid-start circuit.

Programmed Start (PS). This type of ballast applies power to the cathodes prior to lamp ignition. As soon as the lamp has ignited, the power that heated the cathodes is removed or reduced. The lamp cathodes are not hit by brute force voltage. Lamp life is extended by this type of "soft start." This type of ballast is preferred where there are expected to be lots of on/off cycles, such as luminaires controlled by occupancy sensors.

Dimming Circuit. The light output of a fluorescent lamp can be adjusted by maintaining a constant voltage on the cathodes and controlling the current passing through the lamp. Rapid-start lamps are used for dimming applications. Dimming ballasts

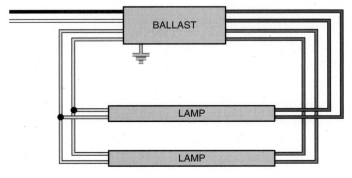

FIGURE 14-16 Rapid-start circuit.
(*Delmar/Cengage Learning*)

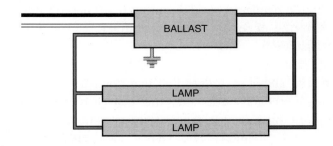

FIGURE 14-17 Instant-start circuit.
(*Delmar/Cengage Learning*)

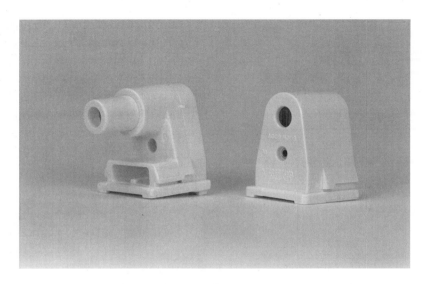

FIGURE 14-18 Single-pin base for an instant-start fluorescent lamp. (*Delmar/Cengage Learning*)

are less efficient than conventional ballasts when operating in a dimming mode because the ballast is still providing full voltage to the electrodes. Light output in a dimming mode can be as low as 1% of full output.

The manufacturer of the ballast should be consulted about the installation instructions for dimming circuits. They offer dimming ballasts in their product line. There is a potential of savings up to 65% in energy costs using dimming ballasts.

A major ballast manufacturer offers computer-controlled dimming for the more demanding installations such as theaters, lobbies, offices, conference rooms, and classrooms. Features are the ability to sense occupancy, measure and analyze the amount of light entering windows, and respond with dimming to keep lighting levels in that particular area to the predetermined desired level. The more light entering through the window(s), the less light is needed from the luminaires. These systems need the expertise of the manufacturer's lighting consultants.

This same manufacturer offers a simpler version of the above, a sensor that can control simultaneously up to 20 luminaires that have a specific type of dimming ballast. The sensor is mounted in one of the luminaires. This sensor monitors reflected light from a surface (table, desktop) and dims the lamps to the desired level. Desired levels are easily adjusted by the turn of a knob on the sensor.

Flashing Circuit. The burning life of a fluorescent lamp is greatly reduced if the lamp is turned on and off frequently. Special ballasts are available that maintain a constant voltage on the cathodes and interrupt the arc current to provide flashing. You will need to consult the ballast manufacturer when you are confronted with a flashing issue.

High-Frequency Circuit. Fluorescent lamps operate more efficiently at frequencies above 60 hertz. The gain in efficacy varies according to the lamp size and type. However, the gain in efficacy and the lower ballast cost generally are offset by the initial cost and maintenance of the equipment necessary to generate the higher frequency. You will need to consult the ballast manufacturer when you are confronted with a high-frequency issue.

Direct-Current Circuit. Fluorescent lamps can be operated on a dc power system if the proper ballasts are used. A ballast for this type of system contains a current-limiting resistor that provides an inductive kick to start the lamp. You will need to consult the ballast manufacturer when you are confronted with a direct-current issue.

Caution: The wiring diagrams in Figures 14-15, 14-16, and 14-17 are typical. Always follow the wiring diagram found in the instructions and/or on the label of ballasts.

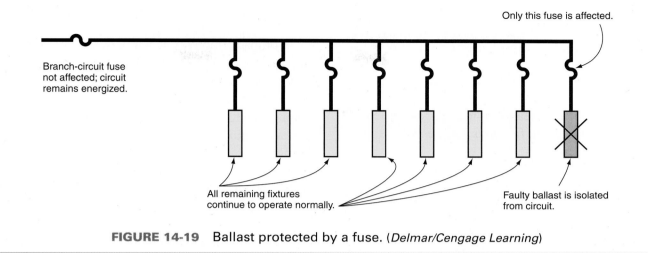

Only this fuse is affected.

Branch-circuit fuse not affected; circuit remains energized.

All remaining fixtures continue to operate normally.

Faulty ballast is isolated from circuit.

FIGURE 14-19   Ballast protected by a fuse. (*Delmar/Cengage Learning*)

Class P Ballast. Some time ago, the National Fire Protection Association reported that the second most frequent cause of electrical fires in the United States was the overheating of fluorescent ballasts. To lessen this hazard, Underwriters Laboratories established a standard for a thermally protected ballast, which is designated as a Class P ballast. This type of ballast has an internal protective device that is sensitive to the ballast temperature. This device opens the circuit to the ballast if the average case temperature exceeds 194°F (90°C) when operated in a 77°F (25°C) ambient temperature. The requirements for the thermal protection of ballasts are established in *NEC 410.130(E)(1)*. After the ballast cools, the protective device is automatically reset. As a result, a fluorescent lamp with a Class P ballast is subject to intermittent off–on cycling when the ballast overheats.

It is possible for the internal thermal protector of a Class P ballast to fail. The failure can be in the welded-shut mode or it can be in the open mode. Because the welded-shut mode can result in overheating and possible fire, some consulting engineers recommended that the ballast be protected with in-line fuses, as indicated in Figure 14-19. These fuseholders can be located inside the luminaire or mounted through the enclosure of the luminaire for easy access from the outside.

External fuses can be added for each ballast, so a faulty ballast can be isolated to prevent the shutdown of the entire circuit because of a single failure. See Figure 14-20. This keeps the other luminaires

on, whereas the faulty ballast is taken off the circuit. The ballast manufacturer normally provides information on the fuse type and its ampere rating. The specifications for the Commercial Building require that all ballasts shall be individually fused; the fuse size and type are selected according to the ballast manufacturer's recommendations.

Sound Rating. All magnetic ballasts emit a hum that is caused by magnetic vibrations in the ballast core. The industry developed a noise rating numbering system. Ballasts are given a sound rating (from A to D) to indicate the severity of the hum. The quietest ballast has a rating of A. The loudest ballast has a D rating.

Older magnetic ballasts were very noisy and had a sound rating of D. Energy-efficient magnetic ballasts are considerably less noisy and have a sound rating of A. All electronic ballasts are A rated.

Because all new luminaire installations have electronic ballasts, the problem of choosing a sound rated ballast is gone.

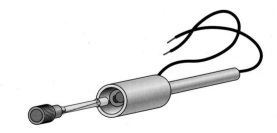

FIGURE 14-20   In-line fuseholder for ballast protection. (*Delmar/Cengage Learning*)

For those of you still involved with the older magnetic ballasts, a guide to the application of sound rated ballasts follows:

A: Private offices, libraries, and so on.
B: Offices, residential
C: Large office areas, commercial, stores, etc.
D: Manufacturing facilities, offices where large equipment is used.

The need for a quiet ballast is determined by the ambient noise level of the location where the ballast is to be installed. For example, the additional cost of the A ballast is justified when it is to be installed in a doctor's waiting room. In the Bakery work area, however, a ballast with a C rating is acceptable; in a factory, the noise of an F ballast probably will not be noticed.

Power Factor. The ballast limits the current through the lamp by providing a coil with a high reactance in series with the lamp. An inductive circuit of this type has an uncorrected power factor of from 40% to 60%; however, the power factor can be corrected to within 5% of unity by the addition of a capacitor. In any installation where there are to be a large number of ballasts, it is advisable to install ballasts with a high power factor.

Ballasts available today have a power factor as high as 0.99. In layman's terms, this means that the ballast is consuming literally no power of its own.

Older style installation might have used a F40T12 lamp with a magnetic core and coil ballast. The lamp consumed 40 watts; the ballast consumed another 10 watts. What was thought of as a 40-watt lamp actually used 50 watts or more because of the low power factor of the ballast.

A similar installation today would use a high efficiency F32 lamp that consumes 26 watts. The ballast would consume very little. Checking one ballast manufacturer's literature, the total input wattage was found to be less than 32 watts. Some high-efficiency lamps are marked F32 but actually consume 24 watts, 28 watts, or 30 watts.

High-efficiency electronic ballasts and energy-saving lamps have been designed into the Commercial Building.

Compact Fluorescent Lamps (CFL)

Compact fluorescent lamps were introduced in the 1980s. They have become very popular for a couple of reasons. A CFL has a rated life 10 times that of a typical incandescent lamp and provides about three to four times the light per watt. The long life also makes CFLs attractive for hard-to-reach luminaires.

Some states have enacted energy-saving laws requiring CFLs as opposed to standard incandescent lamps. This is discussed a little later in this chapter.

CFLs might have a twin-tube arrangement, as shown in Figure 14-21. Some are available that have two sets of tubes; these are called double twin-tube or quad-tube lamps, Figure 14-22. Some of these low-wattage CFLs plug into a socket. A typical socket, along with a low-power-factor ballast, is shown in Figure 14-23.

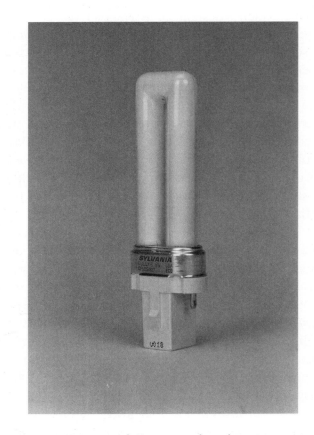

FIGURE 14-21 A 5-watt, twin-tube, compact fluorescent lamp. (*Photo Courtesy Sylvania*)

FIGURE 14-22   A 10-watt, quad-tube, compact fluorescent lamp. (*Delmar/Cengage Learning*)

The most common CFLs are of the twisted (spiral) tube configuration, as shown in Figure 14-24. These CFLs have a standard medium base and are self-ballasted because they have a built-in electronic ballast. This type of lamp can directly replace an incandescent lamp.

HIGH-INTENSITY DISCHARGE (HID) LAMPS

Two of the lamps in this category, mercury and metal halide, are similar in that they use mercury as an element in the light-producing process.

The other HID lamp, high-pressure sodium, uses sodium in the light-producing process. In all three lamps, the light is produced in an arc tube that is enclosed in an outer glass bulb. This bulb serves to protect the arc tube from the elements and to protect the elements from the arc tube. An HID lamp will continue to give light after the lamp is broken, but it should be promptly removed from service. When the outer lamp is broken, people can be exposed to harmful ultraviolet radiation. Mercury and metal halide lamps produce a light with a strong blue content. The light from sodium lamps is orange in color.

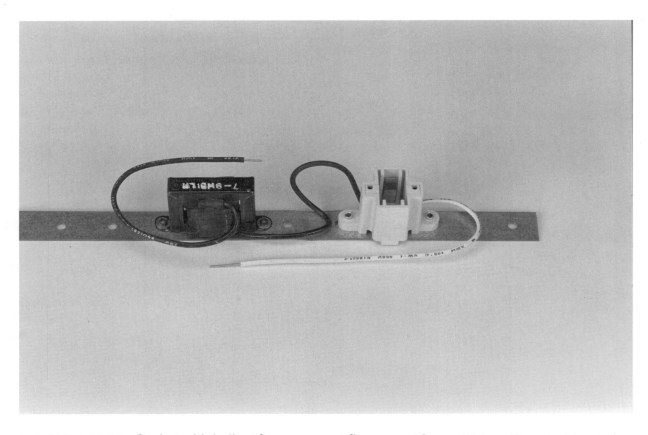

FIGURE 14-23 Socket with ballast for a compact fluorescent lamp. (*Delmar/Cengage Learning*)

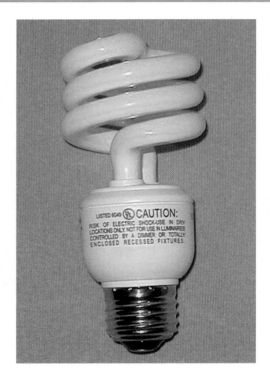

FIGURE 14-24 Typical compact fluorescent lamp. (*Delmar/Cengage Learning*)

Mercury Vapor Lamps

Many people consider the mercury vapor lamp to be obsolete. Because they contain mercury, they have been banned since 2008 in the United States for new installations. Replacement mercury vapor lamps are still available as replacements for existing installations.

It is easy to recognize a mercury vapor lamp because when first turned on, they are dark blue. When fully lit, they still give off a bluish light.

They have the lowest efficacy of the HID family, which ranges from 30 lumens per watt for the smaller-wattage lamps to 65 lumens per watt for the larger-wattage lamps. They have a long average life of 24,000 hours.

Mercury vapor lamps can be dangerous if the outer glass is broken. Serious skin burns and eye inflammation from ultraviolet radiation can result. Some of the newer mercury vapor lamps are designed so that the lamp will automatically extinguish itself if the outer glass is broken.

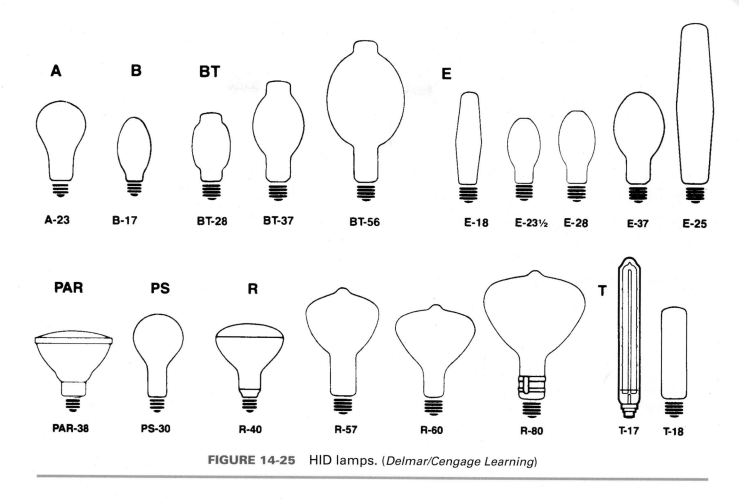

A   A-23

B   B-17

BT   BT-28   BT-37   BT-56

E   E-18   E-23½   E-28   E-37   E-25

PAR   PAR-38

PS   PS-30

R   R-40   R-57   R-60   R-80

T   T-17   T-18

FIGURE 14-25   HID lamps. (*Delmar/Cengage Learning*)

Catalog Designations

Catalog designations vary considerably for HID lamps, depending on the manufacturer. In general the designation for a mercury lamp will begin with an *H*; for metal halide lamps it will begin with an *M*; and for high-pressure sodium lamps, with either an *L* or *C*. See Figure 14-25.

Metal Halide Lamps

A metal halide lamp is closely related to a mercury vapor lamp. The startup time can be up to 5 minutes to reach full output. If they are shut off, they must cool off for 5 to 10 minutes before they will restart.

The metal halide lamp has the disadvantage of a relatively short life and a rapid drop-off in light output as the lamp ages. Rated lamp life varies from 5000 hours to 24,000 hours. During this period the light output can be expected to drop by 30% or more. These lamps are considered to have good color-rendering characteristics and are often used in retail clothing and furniture stores. The lamp has a high efficacy rating, which ranges from 65 to 120 lumens per watt. One lamp manufacturer lists metal halide lamps from 22 watts to 1500 watts, depending on the type. Operating position (base up, base down, horizontal) is critical with many of these lamps and should be checked before a lamp is installed. The design of the luminaire will probably dictate the position of the lamp.

Common applications are for lighting football and baseball fields, warehouses, and industrial plants.

High-Pressure Sodium (HPS) Lamps

This type of lamp is ideal for applications in warehouses, parking lots, and similar places where color recognition is necessary but high-quality color rendition is not required. The light output is orange in color. The lamp has a life rating equal to or better than that of any other HID lamp and has very stable light output over the life of the lamp. The efficacy is very good, ranging as high as 140 lumens per watt. A HPS lamp starts with a dark pink glow, then changes to a pinkish-orange light when fully on.

Low-Pressure Sodium (LPS) Lamps

This lamp has a high efficacy rating, ranging from 130 to greater than 180 lumens per watt. The light is monochromatic, containing energy in only a very narrow band of yellow. This lamp is usually used only in street lighting, security lighting, parking, and storage areas where no color recognition is required. An LPS lamp starts with a dim reddish-pink glow. When fully on in a few minutes, it turns bright yellow. The lamp has a good life rating of 18,000 hours. It maintains a very constant light output throughout its life.

Ballasts for HID Lighting

The Energy Independence & Security Act of 2007 mandates that energy-efficient ballasts be used on all 150W–500W metal halide luminaires. Efficiency requirements vary from 88% to 94%, depending on the type of ballast. Metal halide luminaires that meet the EISA mandate are marked with a circled "E" on the product packaging and on the ballast label.

Pulse-start ballasts meet the EISA mandates. They have a high-voltage igniter that starts the lamp by using a series of high-voltage pulses (typically 3 to 5 kilovolts).

⏤⌁⊢ ENERGY SAVINGS

The name of the game today is to **GO GREEN**. In the electrical field, we are concerned with the equipment we install. We have discussed quite a bit of this earlier in this chapter. Here is more information to consider.

There is much concern about energy savings today. There are many federal, state, and local organizations relating to energy conservation. To name a few:

- ANSI/ASHRAE/IESNA Standard 90.1 (www .ansi.org) (www.ashrae.org) (www.iesna.org)
- California's Title 24 (www.energy.ca.gov/ title24)
- Energy Star (www.energystar.gov)
- Environmental Protection Association (EPA) (www.epa.gov)
- EPAct (www.energy.gov)
- International Code Council (ICC) (www .iccsafe.org)
- International Energy Conservation Code (www .internationalcodes.net)
- NEMA (www.nema.org) Look for the NEMA Premium Electronic Ballast Program.
- U.S. Department of Energy (DOE) mandates (www.DOE.gov) (www.energycodes.gov)
- U.S. Green Building Council (www.usgbc.org)

The Energy Policy Act of 1992 (EPAct) was drastically amended in 2005, setting up state building energy standards for lighting. Now, instead of only thinking about watts, volts, and amperes, we must now concern ourselves with *lighting power allowance* (LPA). The requirements for LPA cover different building types and different space types.

The lighting allowance for offices in *NEC Table 220.12*, $3\frac{1}{2}$ VA per square foot, is considerably different from the lighting power allowance stipulated in the Energy Policy Act and the State of California's Title 24 Act.

The *NEC* presently is not concerned with energy savings. Its main objective is safety. This is apparent when reading *NEC 90.1(A)*, which states:

> *Practical Safeguarding. The purpose of this Code is the practical safeguarding of persons and property from hazards arising from the use of electricity.*

For example, in a 10,000-square-foot (ft²) office, the lighting allowance is 1.3 watts per square foot (W/ft²). Thus, $10,000 \times 1.3 = 13,000$ watts (13 kW). For a manufacturing facility, the allowable energy is 2.2 W/ft². For a warehouse facility, the allowable energy is 1.2 W/ft².

It becomes a matter of matching the desired lighting level (footcandles) for a given occupancy with luminaires that will produce the desired lighting level and, at the same time, meet the energy standards. After all of this is accomplished, the proper branch circuits must be provided in conformance to the *NEC*.

As you can readily see, lighting energy requirements are quite complicated. It is not as simple as it used to be. If you are involved in a major lighting project, it would be a good idea for you to check with a consulting engineer familiar with current EPAct requirements in your state. This person has the computer programs needed to calculate the desired lighting needed and the energy requirements for the specific type of installation.

The following briefly discusses some of the more common energy-savings products associated with lighting.

Energy-Saving Ballasts

The market for magnetic (core-and-coil) ballasts is shrinking. Their obsolescence is a result of federal, state, and local energy mandates, as previously discussed.

The National Appliance Energy Conservation Amendment of 1988, Public Law 100-357 prohibited manufacturers from producing ballasts having a power factor of less than 90%. Ballasts that meet or exceed the federal standards for energy savings are marked with the letter *E* in a circle. Dimming ballasts and ballasts designed specifically for residential use were exempted.

As we have pointed out earlier in this chapter, electronic ballasts are much more energy efficient than the older magnetic ballasts (core-and-coil). Energy-saving ballasts might cost more initially, but the payback is in the energy consumption saving over time.

The older, antiquated fluorescent magnetic core-and-coil ballasts become very warm and might consume 14 to 16 watts, whereas an electronic ballast might consume 2 or 3 watts. Combined with energy-saving fluorescent lamps that use 25, 28, 30, or 32 watts instead of 40 watts, there is a considerable energy savings. You are buying light—not heat.

When installing fluorescent luminaires, check the label on the ballast that shows the actual volt-amperes or total input wattage that the ballast and lamp will draw in combination. DO NOT attempt to use only lamp wattage when making load calculations, as this could lead to an overloaded branch circuit. For example, a high-efficiency ballast might have a total input wattage of 32, whereas an older magnetic ballast might draw 102 volt-amperes.

The higher the power factor rating of a ballast, the more energy efficient is the ballast. Look for a power factor rating in the mid- to high 90s.

Energy-Saving Lamps

U.S. EPAct of 1992 enacted restrictions on lamps. In October 1995, the common 4-ft, 40-watt T12 linear medium bipin fluorescent lamp was eliminated. These discontinued lamps were directly replaced by energy-efficient 34-watt T12 lamps and 32-watt T8 lamps.

Some lamps may be designated F40T12/ES but draw 34 watts instead of 40 watts. The *ES* stands for "energy-saving." *ES* is a generic designation. Manufacturers may use other designations such as *SS* for SuperSaver, *EW* for Econ-o-Watt, *WM* for Watt-Miser, and others.

The older, high-wattage incandescent R30, R40, and PAR38 lamps were also discontinued and replaced with lower-wattage lamps with the same shape.

T12 lamps are still found in 4-ft shop lights and square luminaires that use U-tube lamps. Most newer square luminaires have U-tube T8 lamps. In new commercial installations, the T8 lamp has taken over from the T12 lamp. You will see in Chapter 15 that the commercial building lighting consists mainly of energy efficient T8, CFLs, halogen, and LED lamps.

Energy-saving fluorescent lamps use up to 80% less energy than an incandescent lamp of similar brightness. Fluorescent lamps can last 13 times longer than incandescent lamps.

It has been estimated that the total electric bill savings across the country will exceed $250 billion over the next 15 years.

Fluorescent lamps and ballasts are a moving target. In recent years there have been dramatic improvements in both lamps and electronic ballast efficiency. First, the somewhat antiquated T12 fluorescent lamps (40 watts) were replaced by energy-saving T8 fluorescent lamps (34 watts). Later, these original T8 lamps reached the point when they needed to be replaced. The replacement lamps are the latest T8 high-efficiency, energy-saving lamps (25, 28, 30, or 32 watts vs. 34 watts) that have an expected life that is 50% longer than the original T8 lamps.

The newer T8 lamps use approximately 40% less energy than the older T12 lamps. At $0.06 per kWh, one manufacturer claims a savings of $27.00 per lamp over the life (30,000 hours) of the lamp. At $0.10 per kWh, the savings is said to be $45.00 per lamp over the life of the lamp. Using the newer energy-saving T8 lamps on new installations and as replacements for existing installations makes the payback time pretty attractive. One electronic ballast can operate up to four lamps, whereas the older style magnetic ballast could operate only two lamps. For a 3- or 4-lamp luminaire, one ballast instead of two ballasts results in quite a savings. Some electronic ballasts can operate six lamps.

A 3-lamp high-efficiency luminaire with T8 lamps gives out the same or more light than older style 4-lamp luminaires with T12 lamps.

You will also see increased use of T5 lamps. Depending on luminaire design, a single T5 lamp might provide more light than a 2-lamp luminaire with T8 lamps. A T5 lamp has a smaller diameter than a T8 or T12 lamp. A T5 has a diameter of 5/8 in., a T8 has a diameter of 1 in., and a T12 has a diameter of 1½ in.

The T5 retains its light output for many more hours of rated lamp life than a T8 or T12. T5 lamps require a different ballast than ballasts for T8 and T12 lamps. T5 ballasts are available for instant start, rapid start, and programmed start.

A T5 lamp cannot replace a T8 or T12 lamp. A T5 lamp is slightly shorter than a T8 or T12. For example, a T8 or T12 lamp might be 48 inches long, whereas the similar T5 lamp is 45.23 inches long. The luminaire must be designed for T5 lamps for proper ballast as well as proper spacing of the lampholders.

High-efficiency electronic ballasts have efficiencies of 98% to 99%.

Hard to believe! You can now have increased light output, significantly reduced power consump-

tion, and longer life using energy-efficient lamps and energy-efficient ballasts.

The only way you can stay on top of these rapid improvements is to check out the Web sites of the various lamp and ballast manufacturers.

Today's magnetic and electronic ballasts handle most of the fluorescent lamp types sold, including standard and energy-saving preheat, rapid start, slimline, high output, and very high output. Again, check the label on the ballast to make sure the lamp and ballast are compatible.

Light-Emitting Diodes (LEDs)

LED is pronounced "ell-ee-dee."

Reduce the electric bill! Save energy! Reduce energy consumption! Reduce the air-conditioning load!

It has been said that the incandescent lamp is from the dinosaur age, having been around since Thomas Edison applied for a patent on May 4, 1880.

Coming on strong is a new concept for lighting with low power consumption—LED lighting. Electricians had better get ready for this new type of light source in luminaires. LEDs for lighting are rapidly coming on the scene. LED lighting is not here yet for general lighting of a room, but that will come as the light output of LEDs increases. You could call it "solid-state lighting."

As this text is being written, LED technology is rapidly changing: better color rendition, lower energy costs, and more lumens per watt. More luminaires with LED lamps are coming on the market faster than electrical distributors can print their catalogs.

LEDs are solid-state devices and have been around since 1962. When connected to a dc source, the electrons in the LED smash together, creating light. Think of an LED as a tiny light bulb. Figure 14-26A is a typical LED. Figure 14-26B is a cluster of LEDs in one small enclosure; the size and number of LEDs determine the amount of lumen output of the device.

An LED lamp needs a device to turn it on and to control its output. This device is called a "driver." LED drivers are tested and listed in conformance to UL Standard 8750. LED luminaires are tested and listed in conformance to UL Standard 1598. Today, LEDs are all around us.

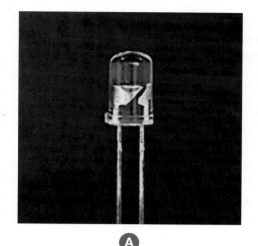

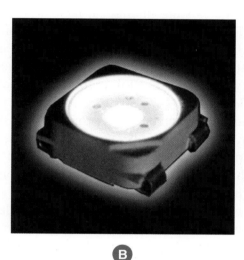

FIGURE 14-26 (A) A typical single light-emitting diode (LED); (B) A cluster of LEDs in one device. (*Delmar/Cengage Learning*)

They are commonly recognized by the tiny white, red, yellow, green, purple, orange, and blue lights found in the digital displays in TVs, radios, DVD and CD players, remotes, computers, printers, fax machines, telephones, answering machines, Christmas light strings, night lights, "locator" switches, traffic lights, digital clocks, meters, testers, tail lights on automobiles, strobe lights, occupation sensors, and other electronic devices, equipment, and appliances.

Using LEDs for lighting is a rather recent concept because the lumens per watt in these devices is on the increase. Individual LEDs are rather small. Putting a cluster (called an array) of LEDs together (i.e., 5, 20, 30, 60, 120) produces a lot of light. The result can be an LED lamp, usable in a luminaire in the same way as a typical medium Edison-base incandescent lamp. Figure 14-27 shows three different styles of LED lamps.

The 2008 *NEC* recognized light-emitting diode (LED) lighting for the first time in *Article 410*. The 2011 *NEC* contains many more references to LEDs and LED lighting. LED lighting is beginning to appear in exit, recessed, surface, and under-cabinet luminaires, desk lamps, wall luminaires, down lights, and stoplights in automobiles, and floor egress path lighting on airplanes, to name a few. LEDs in flashlights have been around for a while.

A few lamp manufacturers have come out with many different types of lamps for accent, task, conventional shape, flood, spots, and so on. They are available with the standard Edison and candelabra bases to replace existing incandescent lamps.

FIGURE 14-27 Three light bulbs (lamps) powered by a number of individual LEDs. These lamps screw into a conventional medium Edison-base lampholder. (*Delmar/Cengage Learning*)

The light is white, but with some LEDs, other colors are available.

Today's LEDs last 60,000 to 100,000 hours. They produce virtually no heat, have no filament to burn out, can withstand vibration and rough usage, contain no mercury, operate better when cool, lose life and lumen output at extremely high temperatures, and can operate at $-40°F$ ($-40°C$).

Their lumen output slowly declines over time. The industry seems to be settling on a lumen output rating of 70% after 50,000 hours of use.

LEDs use a tiny amount of electricity. Virtually all of the power consumption converts to light, whereas for an incandescent lamp, 5% of the power consumption converts to light and the rest converts to heat. When the purpose is to have illumination, heat is a waste.

Some LEDs may be dimmed, but others may not. Check with the manufacturer of the luminaire to verify the dimming capability. LEDs start instantly with no flickering.

LEDs put out directional light, as compared to the conventional incandescent lamp that shoots light out in almost all directions.

A study was recently made to compare an LED's predicted life of 60,000 hours (that's almost 7 years of continuous burning, or 21 years at 8 hours per day of usage) to a standard 60-watt incandescent lamp that has a rated life of 1000 hours. Over the 60,000-hour life:

- 60 standard incandescent lamps would be used compared to one LED lamp.

- the standard incandescent lamps would use 3600 kWhr, resulting in an electricity cost of $360. The LED lamp would use 120 kWhr, resulting in an electricity cost of $12.

- the total cost (lamps and cost of electricity) for the incandescent lamps was $400 and for the LED lamp it was $47.

Today, the lumen output of LED lamps is similar to that of compact fluorescent lamps (CFL), producing approximately 50 lumens per watt. One manufacturer recently announced an LED producing in excess of 1000 lumens—over 100 lumens per watt. This is expected to increase to 150 lumens per watt by 2010. Compare the LED lumen output to a typical incandescent lamp that has a lumen output of 14–18 lumens per watt. That's a significant increase in lumens per watt!

Repeating what we have stated many times over: Always install electrical equipment that has been "listed" by a NRTL (National Recognized Testing Laboratory). They have the know-how to develop standards for testing. Don't "wing it" on your own in deciding what is safe or not safe to install. Follow the requirements in the *National Electrical Code*, and always use and/or install "listed" products; "things" are changing fast.

Recently announced is a line of LED lamps that are direct replacements for the conventional 40-watt fluorescent lamp. Nothing has to be done other than replace the existing fluorescent lamp with an LED lamp. These lamps both work on magnetic and electronic ballasts. When compared to a conventional 40-watt fluorescent lamp, the LED replacement lamp typically has a 10-year life as opposed to a 2- or 3-year life. Their power consumption is 20% less and their lumen output is comparable or slightly greater, and they can operate at $32°F$ ($0°C$). These are currently available in warm white, cool white, daylight, neutral, and bright white. These tubular lamps have an integral "driver."

The use of LED lighting will be accelerating at a rapid pace. The *NEC* Code Making Panels are seeing more and more proposals for changes in the *NEC* relating to LED-type luminaires. Keep an eye out for these changes. As with all electrical equipment, carefully read the label on the luminaire to be sure your installation meets *Code*.

For more information about LEDs and LED lighting, check out the Web site of the LED industry: www.ledsmagazine.com and www.everled.com.

HAZARDOUS WASTE MATERIAL

Fluorescent Lamps

Mercury (sometimes referred to as "quicksilver") is hazardous to human health and the environment.

The federal government's Environmental Protection Agency (EPA) and all 50 states have enacted rules and regulations to manage the disposal/recycling of lamps that contain mercury. These are fluorescent, high-intensity discharge, mercury vapor, high-pressure sodium, metal halide, and some neon lamps. These lamps are considered to be hazardous and must be handled as such. Mercury is also found in mercury switches, mercury relays, and some thermostats, thermometers, and button cell batteries.

When these lamps are broken, the mercury slowly evaporates into the air and eventually finds its way back into the earth. Proper disposal of these lamps is critical.

The mercury content in lamps depends on the type of lamp as well as the year it was manufactured. Conventional T12 and T8 fluorescent lamps manufactured pre-1992 contained considerably more mercury than those manufactured in the period from 1997 to 2007.

Manufacturers of lamps meet the requirements of the EPA regarding the lamp's mercury content.

Originally, the EPA's Green Lights Program encouraged group re-lamping of older installations with higher-efficiency lamps. This brought about having to dispose of the older lamps that contained high amounts of mercury into a waste management system.

Today, the EPA's Energy Star program goes far beyond replacing T12 lamps with T8 lamps and replacing incandescent lamps with halogen lamps. The Energy Star program includes such things as replacing old style magnetic ballasts with high-efficiency electronic ballasts, controlling lighting with timers and occupation sensors, and replacing low-efficiency appliances with high-efficiency appliances. The EPA even goes as far as specifying building materials, lighting, and overall building design.

Energy-efficient luminaires, appliances, and electronic equipment such as computers that have a power-down "sleep mode" feature that meets the requirements of the EPA and Department of Energy (DOE) will have the label shown in Figure 14-28. To meet these requirements, energy-efficient recessed luminaires will be airtight and sometimes double-walled; holes will be plugged and housing will be gasketed.

Some mercury-containing lamps are recycled. In the recycling (reclamation) process, the lamps are crushed, then separated into glass, end caps, and phosphor powder. The phosphor powder contains most of the mercury. Mercury is recovered from the phosphor powder.

Incandescent lamps are generally not considered to contain hazardous materials.

FIGURE 14-28 Energy Star is a government-backed program dedicated to helping protect the environment through energy efficiency. (*Delmar/Cengage Learning*)

Waste management companies provide 4-ft (1.22-m) and 8-ft (2.44-m) boxes and fiber lamp drums for suitable pickup of discarded lamps.

Some states have banned the disposal of lamps that contain mercury.

If you are involved with the disposal of mercury-containing lamps, be sure to contact a waste management company and/or the environmental protection agency in your area that is familiar with the local, state, and federal laws regarding the disposal of these lamps.

The following are typical instructions issued by most states' environmental protection agencies:

What to Do If You Have a Mercury Spill

All mercury spills, regardless of quantity, should be treated seriously.

DO

- leave the area if you are not involved in the cleanup.

- open windows and doors to ventilate the area.

- collect very small amounts of mercury with adhesive tape or an eyedropper. Store it in a sealed plastic container for transport to a household hazardous waste collection.

DO NOT

- use a vacuum cleaner to clean up mercury. A vacuum cleaner will spread mercury vapors and tiny droplets and increase the area of contamination.

Health Problems Associated with Exposure to Mercury

This depends on how much mercury has entered your body, how it entered your body, and how long you have been exposed. Exposure to even small amounts of mercury over a long period may cause negative health effects including damage to the brain, kidney, lungs, and a developing fetus. Brief contact with high levels of mercury can cause immediate health effects including loss of appetite, fatigue, insomnia, and changes in behavior or personality.

Old Ballasts

Pre-1979 ballasts contained hazardous polychlorinated biphenyls (PCBs) in their capacitors and in the potting compound. PCBs are considered to be a probable carcinogen—a cancer-causing substance. Later on, through 1991, ballast manufacturers used diethylhexyl phthalate (DEHP) instead of PCBs—also a hazardous chemical.

More Information

For much more information, check out the Web sites of the EPA (www.epa.gov), (www.lamprecycle .org), (www.energystar.gov), and your state's environmental protection agency relating to the safe disposal of fluorescent and other lamps that contain mercury, as well as ballasts. You can also check lamp and ballast manufacturers' Web sites, where you will find a list of all states, and how to be redirected to their recycling requirements.

⎍ SUMMARY

A well-designed lighting system is a combination of installing energy-efficient lamps and energy-efficient ballasts and using luminaires that direct the light in the desired direction. And, as always, the wiring must conform to the requirements found in the *National Electrical Code*.

REVIEW

Refer to the *National Electrical Code* or the working drawings when necessary.

1. What article in the *NEC* covers luminaires, lampholders, and lamps? *Article*

   _____

2. a. What does the term *LED* stand for? _____

   b. What does the term *CFL* stand for? _____

3. Complete the following sentences: _____ .

   a. A fluorescent lamp requires a _____ .

   b. An LED lamp requires a _____ .

   c. To assure that luminaire is safe, always check its label to make sure it is _____ .

   d. An LED is a solid-state device. When connected to a dc source, the _____ in the LED _____ together, creating _____ .

4. The *NEC* defines a fluorescent lamp as a (electric discharge lamp) (filament lamp) (resistive lamp). (Circle the correct answer.)

5. Light output of a lamp is measured in (lumens) (amperes) (volts). (Circle the correct answer.)

6. The efficacy of a lamp is measured in (watts) (lumens per watt) (volt-amperes). (Circle the correct answer.)

7. Circle the lamp that has the highest lumens per watt.
   (incandescent)   (LED)   (fluorescent)

8. Circle the lamp that has the lowest lumens per watt.
   (incandescent)   (LED)   (fluorescent)

9. All lamps have a Kelvin rating. The lower the K rating, the (warmer) (cooler) its color. The higher the K rating, the (warmer) (cooler) its color. (Circle the correct answers.)

10. A halogen lamp provides a (whiter) (redder) color than a conventional tungsten filament lamp. (Circle the correct answer.)

11. Low voltage lighting is trendy, but because the lamps operate at low voltage (typically 12 volts), the lamps draw significantly more current than a 120-volt lamp of the same wattage.
    Calculate the amperes for a 300-watt 12-volt load and for a 300-watt 120-volt load.

12. What is the diameter of each of the following lamps?
    a. A19 _____
    b. T8 _____
    c. T12 _____
    d. R40 _____

13. Is it permissible to discard old ballasts and lamps in the normal trash?

14. What advantages are there to installing ballasts that have a rating of 120/277 volts?
    _____
    _____
    _____

15. The power factor of high efficiency electronic ballasts is generally in the range of (50%) (80%) (98–99%). (Circle the correct answer.)

16. Nonlinear loads have harmonic components that (increase) (decrease) current of the neutral conductor. (Circle the correct answer.)

17. Most of the ballasts in the commercial building are (rapid start) (trigger start) (instant start) (series rated). (Circle the correct answers.)

18. What is a Class P ballast? _____

_____

_____

19. A ballast is marked Sound Rating "D." Another ballast is marked Sound Rating "A." Which ballast is the quietist? _____

20. What do the letters *HID* stand for? _____

21. Does the *NEC* have requirements that mandate energy savings? _____

22. Which ballast has the highest power factor? (magnetic) (electronic) (Circle the correct answer.)

23. Give a brief explanation of the following terms:

    a. Efficacy _____

    _____

    _____

    b. Footcandle _____

    _____

    _____

    c. Lumen _____

    _____

    _____

24. What are the approximate lumens per watt for an

    a. incandescent lamp? _____ lumens per watt.

    b. fluorescent lamp? _____ lumens per watt.

    c. LED lamp? _____ lumens per watt.

25. To combat our country's growing energy problem, the federal government established a building energy standard that stipulates, among other things, the amount of electricity allowed for various types of buildings. The types of lamps permitted and the lumens per watt are two of the items addressed in the standard. What is the name of the regulation?

_____

_____

_____

26. The Environmental Protection Agency (EPA) has a program that sets standards for the energy efficiency of appliances, lamps, and similar electrical equipment. This program is referred to as

_____ .

27. Let's review our knowledge of where to find things in the National Electrical Code. What Article in the *National Electrical Code* covers

    a. luminaires?   *Article* _____

    b. services?   *Article* _____

    c. grounding?   *Article* _____

    d. overcurrent protection?   *Article* _____

    e. branch circuits?   *Article* _____

    f. motors?   *Article* _____

helpful in providing the lighting layout for the Commercial Building in this text.

ENERGY SAVINGS BY CONTROL

We have all heard the words, "Shut off the lights when you leave the room." Sometimes this works . . . sometimes not.

Electrical Wiring—Commercial is used in all 50 states in this country and in Canada. It is not the intent of this text to go into the detail of specific state and/or local energy requirements. Just be aware that requirements are being implemented relative to energy conservation, and you need to be aware of them. As an example, California recently put into place Title 24, Part 6. The intent is to reduce power consumption in new homes. The states of Washington and Wisconsin also have stringent energy conservation laws. As time passes, other states will follow. Do you know what the energy requirements are in your city or state?

As discussed in Chapter 14, you need to be familiar with the terms *efficacy* and *lumens per watt*. Simply stated, efficacy is measured in lumens per watt.

The new laws in these states simply require that lamps in permanently installed luminaires

- have high efficacy rating (high lumens per watt), or
- be controlled by occupancy sensors, or
- have a combination of lamp efficacy, occupancy sensor control, and/or dimmers.

Cord-connected floor or table lamps are not governed by current energy conservation laws.

Most fluorescent lamps with energy savings ballasts, compact fluorescent lamps (CFLs), LEDs, and halogen lamps meet this high-efficacy requirement. Conventional incandescent lamps do not. As you study the lighting layout for the Commercial Building, you will note that incandescent lamps are not specified.

The following chart shows what is required to meet the lamp requirement in the state of California.

Lamp Wattage	Lamp Efficacy
Less than 15 watts	40 lumens per watt
15–40 watts	50 lumens per watt
Over 40 watts	60 lumens per watt

The Bottom Line

In a nutshell, today's lighting incorporates a combination of well-designed, high-efficiency (airtight) luminaires, high-efficiency lamps, high-efficiency ballasts, and various levels of control such as occupancy sensors, timers, and dimmers.

Occupancy sensors, timers, and dimmer switches are available that install the same way as conventional toggle switches. They provide manual on–off as well as automatic shut-off control when no motion is detected after a pre-set time. As soon as motion is detected, they turn the lamps on. See Figure 15-1.

To learn more about the California energy conservation law, check out http://www.energy.ca.gov/title24 and http://www.haloltg.com. Browse for their summary of the California Title 24 law. You might also want to check out the Cooper Lighting Web site www.cooperlighting.com to learn more about lighting issues. Other lighting manufacturers' Web sites should also be checked out.

The installation of luminaires is a frequent part of the work required for new building construction and for remodeling projects in which customers are upgrading the illumination of their facilities. To execute work of this sort, the electrician must know how to install luminaires and, in some cases, select the luminaires.

The luminaires required for the Commercial Building are described in the specifications and indicated on the plans. The installation of luminaires, lighting outlets, and supports is covered in

FIGURE 15-1 An occupancy (motion) sensor. (*Courtesy of Legrand/Pass & Seymour*)

Article 410 of the *NEC. Article 410* sets forth the basic requirements for the installation, commonly referred to as the "rough-in." *Article 410* also covers grounding, wiring, construction, lampholder installation and construction, lamps and ballasts, and special rules for recessed luminaires.

The rough-in must be completed before the ceiling material can be installed. The exact location of the luminaires is rarely dimensioned on the electrical plans and, in some remodeling situations, there are no plans. In either case, the electrician should be able to rough in the outlet boxes and supports so that the luminaires will be correctly spaced in the area. If a single luminaire is to be installed in an area, the center of the area can be found by drawing diagonals from each corner, Figure 15-2. When more than one luminaire is required in an area, the procedure shown in Figure 15-3 should be followed. Uniform light distribution is achieved by spacing the luminaires so that the distances between the luminaires and between the luminaires and the walls follow these recommended

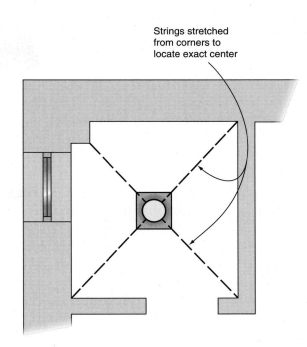

FIGURE 15-2   Luminaire location.
(*Delmar/Cengage Learning*)

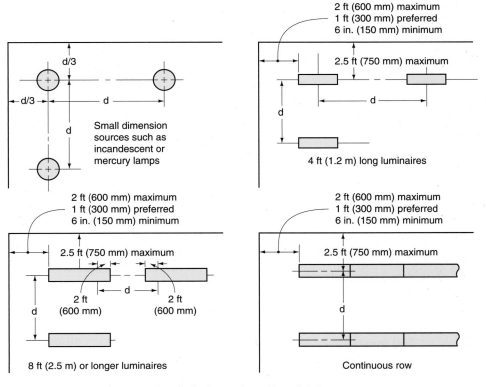

Luminaire layout for uniform lighting
Note: d should not exceed spacing ratio times mounting height. Mounting height is from luminaire (fixture) to work plane for direct, semidirect, and general diffuse luminaires and from ceiling to work plane for semi-indirect and indirect luminaires.

FIGURE 15-3   Luminaire spacing. (*Delmar/Cengage Learning*)

spacing guides. The spacing ratios for specific luminaires are given in the data sheets published by each manufacturer. This number, usually between 0.5 and 1.5, when multiplied by the mounting height, gives the maximum distance that the luminaires may be separated and provide uniform illuminance on the work surface. Again, the lighting recommendations of the IESNA and the requirements of federal, state, and local laws must be followed.

Although the above installation steps might seem simple, remember that there will always be obstacles such as ceiling joists, beams, ductwork, piping, and so on. Be ready to be flexible.

While doing luminaire layouts, it may be necessary to apply good judgment in making the final decision. Consider the following situation where uniform illuminance is desired:

- The space is 20 ft (6 m) square.
- One-ft (300-mm) square ceiling tile will be installed.
- The floor-to-ceiling height is 9 ft (2.7 m).
- The work plane height is 2.5 ft (750 mm).
- The luminaire is 1 ft (300 mm) square.
- The luminaire spacing ratio is 1.1.

The work plane to luminaire (fixture) distance is

$$9 \text{ ft} - 2.5 \text{ ft} = 6.5 \text{ ft}$$
$$(2.7 \text{ m} - 0.7 \text{ m} = 2 \text{ m})$$

The maximum center-to-center spacing distance is

$$6.5 \text{ ft} \times 1.1 = 7.15 \text{ ft}$$
$$(2 \text{ m} \times 1.1 = 2.2 \text{ m})$$

If the ceiling is a 1-ft (300-mm) square grid, the maximum spacing becomes 7 ft (2.1 m).

The maximum distance from the center of the luminaire to the wall should not be greater than

$$\frac{7 \text{ ft}}{3} = 2 \text{ ft } 4 \text{ in.}$$

$$\frac{2.1 \text{ m}}{3} = 700 \text{ mm}$$

Working with the installer of the ceiling system, it is a given that the ceiling will be installed in a uniform layout. There will be either a full or a half grid at the wall. This limits the luminaire placement at 1.5 ft (450 mm), 2 ft (600 mm), or 2.5 ft (750 mm) from center to wall.

Usually it will be decided, when there is an odd number of rows, that the center row of luminaires will be installed in the center of the space. This would place a 6-in. (150-mm) grid along each wall. For a uniform luminaire layout using nine luminaires, they would be centered at 2 ft (600 mm), 10 ft (3 m), and 18 ft (5.5 m).

This layout exceeds the recommended spacing distance. An alternative is to place the luminaires centered on 3 ft, 10 ft, and 17 ft (900 mm, 3 m, and 5.2 m), which would violate the recommended wall spacing distance. Another alternative is to exchange the luminaires for a style with a higher spacing ratio or to install additional luminaires. Using 16 luminaires, they could be placed at 2 ft (600 mm), 7 ft (2.7 m), 13 ft (4 m), and 18 ft (5.5 m). This would not violate any of the recommendations. This type of compromise is a common occurrence in luminaire layout where there is a grid ceiling system.

Supports

The suspended ceiling grid components are required to be securely fastened to one another. These ceilings are most often installed by contractors that specialize in such installations. Lay-in luminaires are normally installed by the electrician.

Both the lighting outlet and the luminaire must be supported from a structural member of the building. To provide this support, a large variety of clamps and clips are available, Figures 15-4 and 15-5. The selection of the type of support depends on the way in which the building is constructed.

The Underwriters *Guide Information for Electrical Equipment (White Book)* in the category for Fluorescent Recessed Luminaires (IEVV) contains the following information on Suspended-Ceiling Luminaires: "All recessed luminaires, except those marked for use in concrete only, are suitable for use in suspended ceilings and may be marked 'SUITABLE FOR SUSPENDED CEILING.'

"Recessed luminaires intended for use in suspended ceilings and provided with integral clips are marked for use with particular grid systems. When installed in accordance with this marking, they comply with 410.16(C) of the NEC. Instructions for using clips to secure the luminaire to the grid

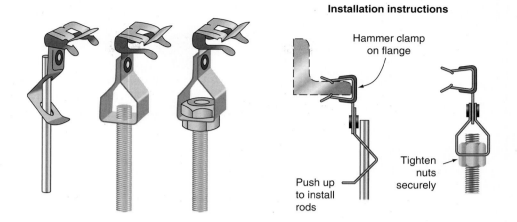

Installation instructions

Hammer clamp on flange

Push up to install rods

Tighten nuts securely

FIGURE 15-4 Rod hangers for a connection to a flange. (*Delmar/Cengage Learning*)

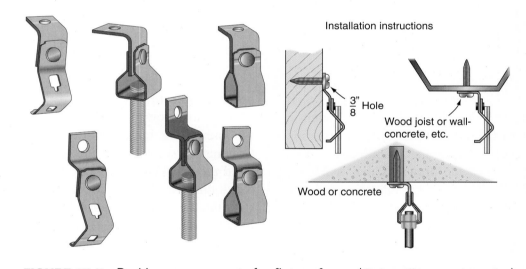

Installation instructions

$\frac{3}{8}$" Hole

Wood joist or wall-concrete, etc.

Wood or concrete

FIGURE 15-5 Rod hanger supports for flat surfaces. (*Delmar/Cengage Learning*)

are provided with the luminaire. The ability of these clips to withstand seismic disturbances has not been investigated."

The luminaires have to be securely fastened to the ceiling framing members using bolts, screws, or rivets; see *NEC 410.36(B)*.

Some states have requirements that the lay-in luminaires be additionally supported from the structural ceiling by slack wires or chains from at least two opposite diagonal corners of the luminaire. This provides additional protection against falling luminaires.

Why is there so much concern about securely fastening the suspended ceiling grid components together and securely fastening the lay-in luminaires to the grid? In the event of a fire, earthquake, or similar catastrophe, we don't want luminaires falling down on people's heads, possibly causing more injury than the fire or earthquake. If you merely laid the luminaires into the openings of a ceiling grid, should the ceiling grid fail, not only would the ceiling grid collapse, the luminaires would fall.

Surface-Mounted Luminaires

For surface-mounted and pendant-hung luminaires, the outlets and supports must be roughed in so the luminaire can be installed after the ceiling is finished. Support rods should be placed so that they extend about 1 in. (25 mm) below the finished ceiling. The support rod may be either a

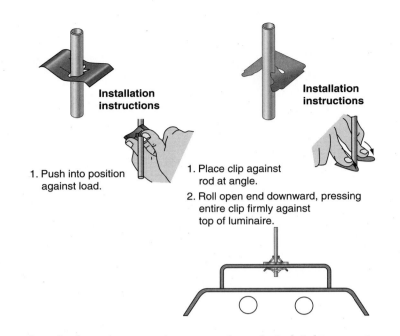

1. Push into position against load.

Installation instructions

1. Place clip against rod at angle.
2. Roll open end downward, pressing entire clip firmly against top of luminaire.

FIGURE 15-6 Luminaire support using an unthreaded rod. (*Delmar/Cengage Learning*)

threaded rod or an unthreaded rod, Figure 15-6. If the luminaires are not available when the rough-in is necessary, luminaire construction information should be requested from the manufacturer. The manufacturer can provide drawings that will indicate the exact dimensions of the mounting holes in the back of the luminaire, Figure 15-7.

Recessed Luminaires

For recessed luminaires, the outlet or junction box will be either mounted directly on the luminaire or located above the ceiling. This junction box must be accessible to the luminaire opening by means of a metal raceway or cable that is at least 18 in. (450 mm) long, but not more than 6 ft (1.8 m) long. Conductors suitable for the temperatures encountered are necessary. This information is marked on the luminaire. Branch-circuit conductors may be run directly into the junction box on listed prewired luminaires, Figure 15-8.

Recessed luminaires are usually supported by adjustable rails (hangers) that are attached to the luminaires' rough-in box. These adjustable rails (hangers) are clearly shown in Figures 15-9(A) and 15-9(B).

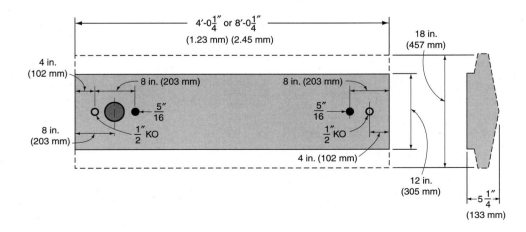

FIGURE 15-7 Luminaire shop drawing indicating mounting holes. (*Delmar/Cengage Learning*)

All recessed incandescent luminaires are required to have thermal protection and be so marked. See *NEC 410.115(C)*.

At least 18 in. (450 mm) but not more than 6 ft (1.8 m) of suitable type raceway or Type AC or MC cable with conductors having insulation suitable for temperature encountered. See *NEC 410.117(C)*.

Branch-circuit conductors with insulation suitable for the temperature requirements marked on the "listed" prewired recessed luminaire may be run directly to junction box, per *NEC 410.117*.

Junction box at least 1 ft (300 mm) from luminaire, per *NEC 410.117(C)*.

"Fixture whip" not required to be supported if less than 6 ft (1.8 m). See *Section 30* in *NEC Articles 320, 330, 348, 350, 356, 362*.

May be size ³/₈ in. (10 mm), per *NEC 348.20(A)(2)(c)*.

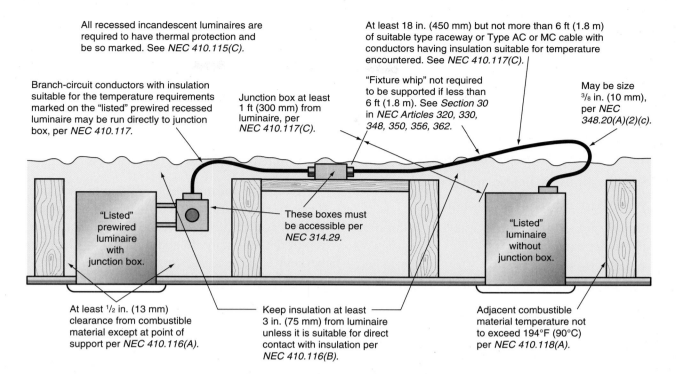

"Listed" prewired luminaire with junction box.

These boxes must be accessible per *NEC 314.29*.

"Listed" luminaire without junction box.

At least ¹/₂ in. (13 mm) clearance from combustible material except at point of support per *NEC 410.116(A)*.

Keep insulation at least 3 in. (75 mm) from luminaire unless it is suitable for direct contact with insulation per *NEC 410.116(B)*.

Adjacent combustible material temperature not to exceed 194°F (90°C) per *NEC 410.118(A)*.

FIGURE 15-8 Requirements for installing recessed luminaires. Recessed luminaires are either Type IC or Type Non-IC. (*Delmar/Cengage Learning*)

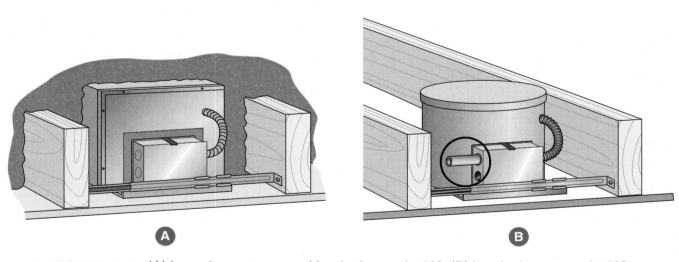

Ⓐ   Ⓑ

FIGURE 15-9 (A) Incandescent recessed luminaire marked IC. (B) Luminaire not marked IC. (*Delmar/Cengage Learning*)

LABELING

Always carefully read the label(s) on the luminaires. The labels will provide, as appropriate, information similar to the following:

- Wall mount only
- Ceiling mount only

- Maximum lamp wattage
- Type of lamp
- Access above ceiling required
- Suitable for air-handling use
- For chain or hook suspension only

- Suitable for operation in ambient not exceeding ___°F (___°C) temperature
- Suitable for installation in poured concrete
- For installation in poured concrete only
- For line volt-amperes, multiply lamp wattage by ___ volts.
- Suitable for use in suspended ceilings
- Suitable for use in uninsulated ceilings
- Suitable for use in insulated ceilings
- Suitable for damp locations (such as bathrooms and under eaves)
- Suitable for wet locations
- Suitable for use as a raceway
- Suitable for mounting on low-density cellulose fiberboard

Underwriters Laboratories produces and distributes a *Luminaire Marking Guide* containing valuable information on how luminaires are marked and what the markings mean.

If followed, the information on these labels, together with conformance with *NEC Article 410*, should result in a safe installation.

The Underwriters Laboratories *Electrical Construction Materials Directory* (known as the *Green Book*), the *Guide Information for Electrical Equipment Directory* (known as the *White Book*), and the luminaire manufacturers' catalogs and literature are also excellent sources of information on how to install luminaires properly. Underwriters Laboratories, along with other qualified electrical product testing laboratories, tests luminaires for safety, and allows manufacturers to place a label (listing mark) on luminaires. Directories of listed products are either published annually or available on the Underwriters Laboratories' Internet site.

Type IC Incandescent Recessed Luminaires

Type IC luminaires are permitted to be installed in direct contact with thermal insulation. They are recommended for homes. These luminaires operate under 194°F (90°C) when covered with insulation. Integral thermal protection is required, and they must

be marked that thermal protection is provided. Watch the markings showing maximum lamp wattage and temperature rating for supply conductors. These luminaires may also be installed in non-insulated locations. See Figure 15-9(A).

Incandescent Recessed Luminaires Not Marked IC

Recessed incandescent luminaires not marked Type IC and those marked for installing directly in poured concrete must not have insulation over the top of the luminaire. In addition, the insulation must be kept back at least 3 in. (75 mm) from the sides of the luminaire. Other combustible material must be kept at least ½ in. (13 mm) from the luminaire except at the point of support. Watch for markings on the luminaire for maximum clearances, maximum lamp wattage, and temperature ratings for the supply conductors. See Figure 15-9(B).

Installing Recessed Luminaires

The electrician must follow very carefully the requirements given in *NEC Article 410 Parts XI and XII* for the installation and construction of recessed luminaires. Of particular importance are the restrictions on conductor temperature ratings, luminaire clearances from combustible materials, and maximum lamp wattage.

Recessed luminaires generate a considerable amount of heat within the enclosure and are a definite fire hazard if not wired and installed properly, Figure 15-8, Figure 15-9, and Figure 15-10. In addition, the excess heat will have an adverse effect on lamp and ballast life and performance.

If other than a Type IC luminaire is being installed, the electrician should work closely with the installer of the insulation to be sure that the clearances, as required by the *NEC*, are followed.

To ensure that clearances are maintained, at least one manufacturer catalogs a box that fits around a recessed luminaire, Figure 15-11. The box prevents thermal insulation from coming into contact with the luminaire. Spacing insulation away from the recessed luminaire obviously creates an uninsulated space around the luminaire. Considerable heat loss can be expected, and in all

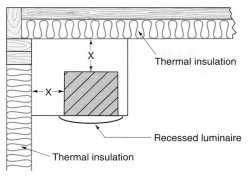

A luminaire shall be installed to permit air circulation unless the luminaire is identified for installation within thermal insulation. At least 3 in. (75 mm) of distance (see "**X**" in illustration) shall separate the luminaire and the insulation.

FIGURE 15-10 Clearances for a recessed luminaire. (*Delmar/Cengage Learning*)

FIGURE 15-11 A box built of fireproof material (not asbestos) can be built around the luminaire to prevent thermal insulation from coming into contact with the luminaire. (*Delmar/Cengage Learning*)

likelihood an installation of this type would not be in conformance with EPA requirements. The availability of Type IC recessed luminaires makes this type of box unnecessary.

According to *NEC 410.117(C)*, a tap conductor is permitted to be run from the luminaire terminal connection to an outlet box. These taps are often installed to provide a conductor with an

insulation rated for the high temperature required for connection to the luminaire housing. The term "tap" as used here, does not imply that the conductor connecting to the luminaire is smaller in size than the branch circuit conductor. For this installation, the following conditions must be met:

- The conductor insulation must be suitable for the temperatures encountered.
- The outlet box must be at least 1 ft (300 mm) from the luminaire.
- The tap conductor must be installed in a suitable raceway or cable such as Type AC or MC.
- The raceway shall be at least 18 in. (450 mm) but not more than 6 ft (1.8 m) long.

The branch-circuit conductors are run to the junction box. Here, they are connected to conductors from the luminaire. These wires have an insulation suitable for the temperature encountered at the lampholder. Locating the junction box at least 1 ft (300 mm) from the luminaire ensures that the heat radiated from the luminaire cannot overheat the wires in the junction box. The conductors must run through at least 18 in. (450 mm) of the metal raceway (but not to exceed 6 ft [1.8 m]) between the luminaire and the junction box. Thus, any heat that is being conducted in the metal raceway will be dissipated considerably before reaching the junction box. Many recessed luminaires are factory equipped with a flexible metal raceway containing high-temperature wires that meet the requirements of *NEC 410.117*.

Some recessed luminaires have a box mounted on the side of the luminaire so the branch-circuit conductors can be run directly into the box and then connected to the conductors entering the luminaire.

Additional wiring is unnecessary with these pre-wired luminaires, Figure 15-12. It is important to note that *NEC 410.21* states that branch-circuit wiring shall not be passed through an outlet box that is an integral part of an incandescent luminaire unless the luminaire is identified for through wiring.

A luminaire may not be used as a raceway unless it is identified for that use. The issue here is conductors being routed through a luminaire to supply other luminaires. See *NEC 410.64* and *410.65*.

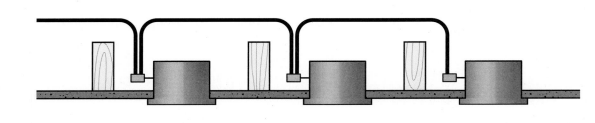

FIGURE 15-12 Installation permissible with prewired luminaire. (*Delmar/Cengage Learning*)

If a recessed luminaire is not prewired, the electrician must check the luminaire for a label indicating what conductor insulation temperature rating is required.

Recessed luminaires are inserted into the rough-in opening and fastened in place by various devices. One type of support and fastening method for recessed luminaires is shown in Figure 15-13. The flag hanger remains against the luminaire until the screw is turned. The flag then swings into position and hooks over the support rail as the screw is tightened.

Sloped Ceilings

Conventional recessed luminaires are suitable for installation in flat ceilings, not sloped ceilings. When selecting luminaires for installation in a sloped ceiling, be certain they are suitable for the particular slope such as a $^2/_{12}$, $^6/_{12}$, or $^{12}/_{12}$ pitch. This information is available in the manufacturers' catalogs. Look for the marking "Sloped Ceilings."

Thermal Protection

All recessed incandescent luminaires must be equipped with factory-installed thermal protection, *NEC 410.115(C)*, Figure 15-14. Marking on these luminaires must indicate this thermal protection. The ONLY exceptions to the *Code* rule are

1. if the luminaire is identified for use and installed in poured concrete.

2. if the construction and design of the luminaire are such that temperatures would be no greater than if the luminaires had thermal protection. See *NEC 410.115(C)* for the exact phrasing of the exceptions.

Thermal protection, Figure 15-14, will cycle on–off–on–off repeatedly until the heat problem is removed. These devices are factory-installed by the manufacturer of the luminaire.

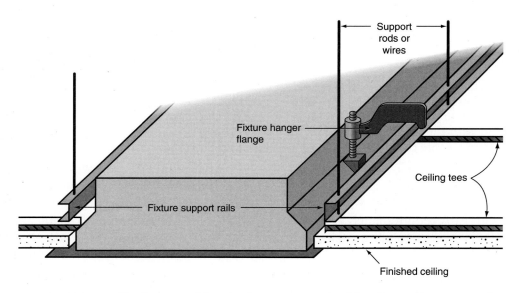

FIGURE 15-13 Recessed luminaire supported with a flange hanger and support rails. (*Delmar/Cengage Learning*)

FIGURE 15-14 A thermal protector.
(*Delmar/Cengage Learning*)

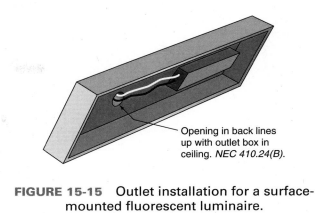

Opening in back lines up with outlet box in ceiling. *NEC 410.24(B).*

FIGURE 15-15 Outlet installation for a surface-mounted fluorescent luminaire.
(*Delmar/Cengage Learning*)

Both incandescent and fluorescent recessed luminaires are marked with the temperature ratings required for the supply conductors if more than 60°C. In the case of fluorescent luminaires, branch-circuit conductors within 3 in. (76.2 mm) of a ballast must have a temperature rating of at least 90°C. All fluorescent ballasts, including replacement ballasts, installed indoors must have integral thermal protection to be in compliance with *NEC 410.130(E)(1).* These thermally protected ballasts are designated as "Class P ballasts." This device will provide protection during normal operation, but it should not be expected to provide protection from the excessive heat that will be created by covering the luminaire with insulation. For this reason, fluorescent luminaires, just as incandescent luminaires, must have ½ in. (13 mm) clearance from combustible materials and 3 in. (75 mm) clearance from thermal insulation.

Because of the inherent risk of fire due to the heat problems associated with recessed luminaires, always read the markings on the luminaire and any instructions furnished with the luminaire, and consult *NEC Article 410.*

Wiring

As previously stated, it is very important to provide an exact rough-in for surface-mounted luminaires. *NEC 410.24(B)* emphasizes this fact by requiring that a concealed outlet box be accessible without having to remove the surface-mounted luminaire.

A typical large surface-mounted luminaire is never supported solely by the outlet box. Instead,

screws, toggle bolts, bolts and anchors, and similar fastening devices are used to support the luminaire independently from the outlet box. Figure 15-15 illustrates how the large opening in the back of the luminaire is located directly over the outlet box, meeting the intent of *NEC 410.24(B).* To assure proper grounding of the luminaire, a separate equipment grounding conductor must be provided between the concealed metallic outlet box and the luminaire.

NEC Article 410, Part IV, contains all of the requirements for supporting luminaires.

Figure 15-16 shows a 3-phase, 4-wire multiwire branch circuit supplying a row of end-to-end mounted luminaires. The four conductors of the multiwire branch circuit are run in a single raceway to the first luminaire, and then continue on through the remaining luminaires to accomplish the desired switching arrangement. When luminaires are designed for end-to-end mounting, or if the luminaires are connected together by a recognized wiring method, they do not have to be marked as a raceway. If not designed for end-to-end mounting, the luminaires must be marked "SUITABLE FOR USE AS A RACEWAY." Refer to *NEC 410.64* and *410.65.* As always, read the label and installation instructions.

Conductors Supplying Luminaires

NEC 410.117(B) requires that branch-circuit conductors that terminate in luminaires have an insulation suitable for the temperature encountered. Type THHN conductors have a temperature rating of 90°C and generally meet the temperature requirement for connecting luminaires.

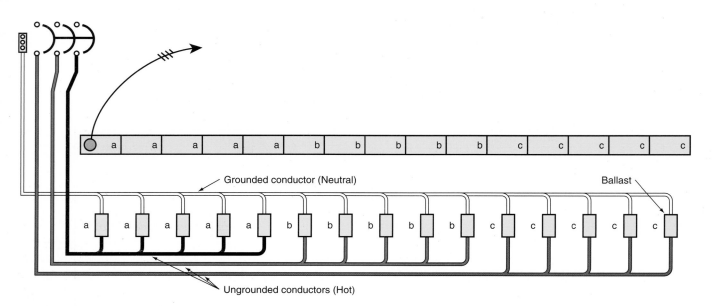

FIGURE 15-16 Multiwire circuits supplying fluorescent luminaires and receptacles. A means must be provided to simultaneously disconnect all ungrounded conductors of the circuit, *NEC 210.4(B).* This is accomplished by installing 2-pole or 3-pole circuit breakers in the panel. (*Delmar/Cengage Learning*)

Disconnecting Means

NEC 410.130(G) requires a means shall be provided for disconnecting fluorescent luminaires that have double-ended lamps and ballast(s). This requirement is for the safety of those having to work on the luminaire, such as changing out a ballast or socket. The disconnecting means is permitted to be inside or outside of the luminaire. If outside of the luminaire, it must be within sight of the luminaire.

Two-wire branch circuit: The disconnecting means shall disconnect the ungrounded conductor supplying the ballast.

Multiwire branch circuit: The disconnecting means shall simultaneously disconnect all conductors supplying the ballast, including the grounded conductor.

Figure 15-17 shows a simple disconnecting means that meets the intent of *NEC 410.130(G).*

Multiwire Branch Circuits

NEC 210.4 contains the requirements for multiwire branch circuits. See Figure 15-16 for an illustration of the definition of *Multiwire Branch Circuit.* Multiwire branch circuits are defined in *NEC Article 100* as *A branch circuit that consists of two or more ungrounded conductors that have a voltage between*

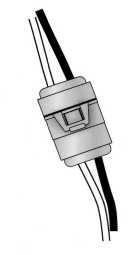

FIGURE 15-17 A simple disconnecting means that allows anyone working on a luminaire to snap the device apart, killing power to the luminaire. Note that the disconnect breaks both the ungrounded and grounded conductors. (*Delmar/Cengage Learning*)

them, and a grounded conductor that has equal voltage between it and each ungrounded conductor of the circuit and that is connected to the neutral or grounded conductor of the system.*

---

*Reprinted with permission from NFPA 70-2011.

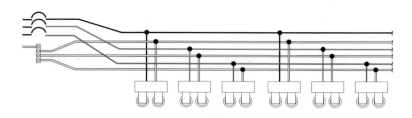

FIGURE 15-18   Installing an individual neutral for each ungrounded conductor avoids simultaneous disconnecting means as the circuits are no longer *multiwire*. (*Delmar/Cengage Learning*)

1. All conductors of the multiwire branch circuit must originate in the same panelboard.

2. The disconnecting means must simultaneously disconnect all ungrounded conductors at the panelboard. In most cases, this is a circuit breaker(s).

3. The multiwire branch circuit must supply only line-to-neutral loads, as illustrated in Figure 15-16, if the luminaires are connected line-to-neutral.

4. In the panelboard, the conductors that constitute a multiwire branch circuit shall be grouped by tie wraps, tape, or similar means. This grouping of conductors does not have to be done when it is obvious where the conductors belong, such as a raceway that contains only one multiwire circuit with one grounded neutral conductor, or Type AC or Type MC cable.

When multiwire branch circuits are used, a means must be provided to disconnect simultaneously (at the same time) all of the "hot" ungrounded conductors at the panelboard where the branch circuit originates. Never use single-pole circuit breakers on multiwire branch circuits. To do so could lead to personal injury via electrical shock as well as the possibility of falling off a ladder from a shock. Multiwire branch circuits are also used to connect appliances such as electric clothes dryers, electric ranges, and electric furnaces and boilers. See *NEC 210.4(A)*, *210.4(B)*, and *210.7(B)*.

If simultaneous disconnecting means is undesirable for all three ungrounded conductors of a multiwire branch circuit, this can be avoided by installing an individual neutral conductor for each ungrounded or hot conductor. As a result, the branch circuits no longer meet the definition of a multiwire branch circuit and simultaneous

disconnecting means is not required. This is illustrated in Figure 15-18. One disadvantage is each ungrounded and neutral conductor count as a current-carrying conductor. Derating is required in accordance with *NEC 310.15(B)(3)(a)*. In addition, space must be provided in raceways and junction boxes. Manufacturers of Type MC cable produce a cable with a dedicated neutral for each ungrounded or hot conductor.

LOADING ALLOWANCE CALCULATIONS

The branch circuits are usually determined when the luminaire layout is completed. For incandescent luminaires, the volt-ampere allowance for each luminaire is based on the wattage rating of the luminaire. *NEC 220.14(D)* stipulates that recessed incandescent luminaires be included at maximum rating. If an incandescent luminaire is rated at 300 watts, it must be included at 300 watts even though a smaller lamp is to be installed. For fluorescent and HID luminaires, the volt-ampere allowance is based on the rating of the ballast. In the past it was a general practice to estimate this value, usually on the high side. With recent advances in ballast manufacture and increased interest in reducing energy usage, the practice is to select a specific ballast type and to base the allowance on the operation of that ballast. The contractor is required to install an "as good as," or "better than," ballast. Several types of ballasts have been discussed in Chapter 14.

The design volt-ampere and the watts for the luminaires selected for the Commercial Building are shown in Tables 15-1(A), 15-1(B), and 15-1(C), one table for each floor.

COMMERCIAL BUILDING LUMINAIRES

TABLE 15-1(A)

Luminaire/lamp/ballast/schedule for the basement.

BASEMENT

Style	Catalog Number	Lamp Type & Color	No. of Lamps	Ballast Type	Ballast Voltage	Mounting	Description	Qty.	Input Watts per Luminaire/ Total Input Watts
A	WSC-232-UNV-ER81 (See Figure 15-19)	T8 Bipin 32 watt 2800 lumens per lamp Rapid start 48" 3500°K	2	Note 1	Note 2	Surface	Decorative wrap. Snap-fit nonglare acrylic lens. Ballast cover removable without tools. Suitable for damp locations.	2	64/128
A-EM	WSC-232-UNV-EL4-ER81 (See Figure 15-19)	T8 Bipin 32 watt 2800 lumens per lamp Rapid start 48" 3500°K	2	Note 1	Note 2	Surface	Decorative wrap. Snap-fit nonglare acrylic lens. Ballast cover removable without tools. Suitable for damp locations. See Note 3.	2	70/140
B	IAF-232-UNV-ER81 (See Figure 15-20)	T8 Bipin 32 watt 2800 lumens per lamp Rapid start 48" 3500°K	2	Note 1	Note 2	Surface or chain	Industrial. White enamel finish. 8% uplight. Ballast cover removable without tools. Suitable for damp locations.	14	64/896
B-EM	IAF-232-UNV-EL4-ER81 (See Figure 15-20)	T8 Bipin 32 watt 2800 lumens per lamp Rapid start 48" 3500°K	2	Note 1	Note 2	Surface or chain	Industrial. White enamel finish. 8% uplight. Ballast cover removable without tools. Suitable for damp locations. See Note 3.	12	70/840
E-EM	ML709835 ICAT120D-EM-494P01 (See Figure 15-22)	LED module 3500°K	LED module	Universal LED driver 120–277 Volts	Not applicable	Recessed	Recessed downlight with emergency battery backup. Dimmable. May be installed in insulated (IC) or noninsulated (non-IC) ceilings. See Note 3.	1	14/14
EXIT	EEX7 (See Figure 15-31)	LED	LED module	LED driver	Note 2	Surface	Suitable for ceiling, wall, or corner end mounting	1	4.6/4.6

Area = 2378 Sq. Ft. (See Note 4)	Total Watts = 2023	Lighting Power Density = 0.851 Watts/Sq. Ft.

Note 1: High-efficiency Class P electronic. Power Factor >98%. Ballast Factor 0.88.

Note 2: Universal input voltage: 120–277 V ±10%.

Note 3: Luminaire has an emergency nickel cadmium battery pack. In the event of a power outage, electronic circuitry instantly senses the loss of power and keeps the luminaire partially illuminated for a minimum of 90 minutes. When normal power is restored, battery pack automatically goes into charge mode. Battery charges while luminaire is energized. Full recharge time: 24 hours.

Note 4: Square footage based on outside building dimensions minus one foot for each of the outside walls.

TABLE 15-1(B)

Luminaire/lamp/ballast/schedule for the first floor.

FIRST FLOOR

Style	Catalog Number	Lamp Type & Color	No. of Lamps	Ballast Type	Ballast Voltage	Mounting	Description	Qty.	Input Watts per Luminaire/Total Input Watts
C	VT3-232DR-UNV-EL4-EB81-WL-8L-M4-GT4 (See Figure 15-21)	T8 Bipin 32 watt 2800 Lumens per lamp. Instant start 48" 3500°K	2	Note 1	Note 2	Surface or chain	Vapor tight. Fiberglass housing. Acrylic lens. Excellent choice for indoor or outdoor applications where weathering, high temps, humidity, dust, or corrosive fumes are present.	15	61/915
C-EM	VT3-232DR-UNV-EL4-EB81-WL-8L-M4-GT4 (See Figure 15-21)	T8 Bipin 32 watt 2800 Lumens per lamp. Instant start 48" 3500°K	2	Note 1	Note 2	Surface or chain	Vapor tight. Fiberglass housing. Acrylic lens. Excellent choice for indoor or outdoor applications where weathering, high temps, humidity, dust, or corrosive fumes are present. See Note 3.	3	67/201
E	ML709835I CAT120D-494P06 (See Figure 15-22)	LED module 3500°K	LED module	Universal LED driver 120–277 Volts	Not applicable	Recessed	Recessed downlight. Dimmable May be installed in insulated (IC) or non-insulated (non-IC) ceilings. Lumens: 813	10	14/140
E-EM	ML709835 CAT120D-EM-494P06 (See Figure 15-22)	LED module 3500°K	LED module	Universal LED driver 120–277 Volts	Not applicable	Recessed	Recessed downlight with emergency battery backup. Dimmable May be installed in insulated (IC) or non-insulated (non-IC) ceilings Lumens: 813 See Note 3.	4	20/80
E2	H750 TD010-ML712835-TUNVD010-494P06 (See Figure 15-22)	LED module 3500°K	LED module	Universal driver 120–277 Volts	Not applicable	Recessed	Recessed downlight. Dimmable For installation in Non-insulated (non-IC) ceilings Lumens: 1356	3	26/78

(continues)

TABLE 15-1(B)

(*Continued*)

FIRST FLOOR

Style	Catalog Number	Lamp Type & Color	No. of Lamps	Ballast Type	Ballast Voltage	Mounting	Description	Qty.	Input Watts per Luminaire/ Total Input Watts
G	2GR8-332A-UNV-EB82 (See Figure 15-24)	T8 Bipin 32 watt 2800 Lumens per lamp. Instant start 48" 3500°K	3	Note 1	Note 2	Recessed Troffer for dropped ceiling grid-lay-in. 3-3/4" depth	Acrylic prismatic lens.	7	80/560
G-EM	2GR8-332A-UNV-EL4-EB82 (See Figure 15-24)	T8 Bipin 32 watt 2800 Lumens per lamp. Instant start 48" 3500°K	3	Note 1	Note 2	Recessed Troffer for dropped ceiling grid-lay-in/ 3-3/4" depth	Acrylic prismatic lens. See Note 3.	2	86/172
I	RCG-SS-232-UNV-EB82 (See Figure 15-25)	T8 Bipin 32 watt 2800 Lumens per lamp. Instant start 48" 3500°K	2	Note 1	Note 2	Recessed Strip light for dropped ceiling grid-lay-in 5 1/8" deep	Recessed strip light. Recommended for mass merchandise and retail sales areas. Reflector reduces glare. Suitable for damp locations. See Note 3.	27	61/1647
K	LDWP-FC-2A-ED (See Figure 15-26)	Light-emitting diodes (LED)	1	Electronic LED Driver	Not applicable	Surface	Lumens: 2400 Exterior LED wall light. Cast aluminum. Wall Pack. Suitable for wet locations. Full light cut-off door.	8	29/232
N	BC-217A-UNV-ER81 (See Figure 15-27)	T8 Bipin 17 watt 1400 Lumens per lamp. Rapid start 24" 3500°K	2	Note 1	Note 2	Surface reflector.	Wall light. Opal acrylic. All purpose wall bracket.	3	34/102
P	2RDI-217FLS-UNV-EB81 (See Figure 15-28)	T8 Bipin 17 watt 1400 Lumens per lamp. Rapid start 24" 3500°K	2	Note 1	Note 2	Recessed.	Direct/indirect luminaire. Suitable for damp locations. Indirect frosted white acrylic lamp shields. Ballast removable from above or below.	4	34/136

(*continues*)

TABLE 15-1(B)

(Continued)

FIRST FLOOR

Style	Catalog Number	Lamp Type & Color	No. of Lamps	Ballast Type	Ballast Voltage	Mounting	Description	Qty.	Input Watts per Luminaire/ Total Input Watts
Q	CEL7035SD (See Figure 15-29)	75 watt MR16 Halogen	2	No ballast.	Dual-voltage 120/277 VAC	Recessed.	Lumens: 1137 Emergency lighting. Solid state charger for nickel cadmium battery. Gives appearance of a ceiling. When power is lost, doors open automatically. When power is restored, doors close automatically. Maintains emergency lighting for 90 minutes. Self-tests every 28 days.	3	24/72
T	L805SML FL830P (See Figure 15-30)	LED module 3000°K	Three 3-watt White LEDs	LED driver	120	Flexible track mounting.	Lumens: 312 Track light. Adjustable heads. White housing. Flood, narrow flood, spot lighting available.	7	9/63
EXIT	EEX7 (See Figure 15-31)	LED	LED module	LED Driver	Note 2	Surface	Suitable for ceiling, wall, or corner end mounting	6	4.6/27.6

Area = 2378 Sq. Ft. (See Note 4)	Total Watts = 4426	Lighting Power Density = 1.861 Watts/Sq. Ft.

Note 1: High efficiency Class P electronic. Power Factor >98%. Ballast Factor 0.88.

Note 2: Universal input voltage: 120–277 V ±10%.

Note 3: Luminaire has an emergency nickel cadmium battery pack. In the event of a power outage, electronic circuitry instantly senses the loss of power and keeps the luminaire partially illuminated for a minimum of 90 minutes. When normal power is restored, battery pack automatically goes into charge mode. Battery charges while luminaire is energized. Full recharge time: 24 hours.

Note 4: Square footage based on outside building dimensions minus one foot for each of the outside walls.

TABLE 15-1(C)

Luminaire/lamp/ballast schedule for the second floor.

SECOND FLOOR

Style	Catalog Number	Lamp Type & Color	No. of Lamps	Ballast Type	Ballast Voltage	Mounting	Description	Qty.	Input Watts per Luminaire/ Total Input Watts
E	ML709835I CAT120D-494P06 (See Figure 15-22)	LED module 3500°K	LED module	Universal driver 120–277 Volts	120	Recessed	Lumens: 813 Dimmable May be installed in insulated (IC) or non-insulated (non-IC) ceilings.	24	14/336

(continues)

TABLE 15-1(C)

(Continued)

SECOND FLOOR

Style	Catalog Number	Lamp Type & Color	No. of Lamps	Ballast Type	Ballast Voltage	Mounting	Description	Qty.	Input Watts per Luminaire/ Total Input Watts
E-EM	ML709835I CAT120D-EM-494P06 (See Figure 15-22)	LED module 3500°K	LED module	Universal driver 120–277 Volts	120	Recessed	Lumens: 813 Recessed downlight with emergency battery backup. Dimmable May be installed in insulated (IC) or non-insulated (non-IC) ceilings. See Note 3	4	14/56
E2	H750 TD010-ML712835-TUNVD010-494P06 (See Figure 15-22)	LED module 3500°K	LED module	Universal driver 120–277 Volts	120	Recessed	Lumens: 813 Recessed downlight. Dimmable For non-insulated (non-IC) ceilings.	8	26/208
E2-EM	H750 TD010-ML 12835-TUND010-EM-494P06 (See Figure 15-22)	LED module 3500°K	LED module	Universal driver 120-277 Volts	120	Recessed	Lumens: 813 Recessed downlight. with emergency battery backup. Dimmable For non-insulated (non-IC) ceilings. See Note 3.	3	32/96
F	GR8-232A-UNV-EB81 (See Figure 15-23)	T8 Bipin 32 watt 2800 Lumens per lamp. Instant start 48" 3500°K	2	Note 1	Note 2	Recessed Troffer for dropped ceiling grid-lay-in 3-3/4" depth	Acrylic lens.	6	61/366
F-EM	GR8-232A-UNV-EL4-EB81 (See Figure 15-23)	T8 Bipin 32 watt 2800 Lumens per lamp. Instant start 48" 3500°K	2	Note 1	Note 2	Recessed Troffer for dropped ceiling grid-lay-in 3-3/4" depth	Acrylic lens. See Note 3.	3	58/174
G	2GR8-332A-UNV-EB82 (See Figure 15-24)	T8 Bipin 32 watt 2800 Lumens per lamp. Instant start 48" 3500°K	3	Note 1	Note 2	Troffer for dropped ceiling grid-lay-in 3-3/4"depth	Lumens: 2800 Acrylic lens. See Note 3.	14	80/1120

(continues)

TABLE 15-1(C)

(*Continued*)

SECOND FLOOR

Style	Catalog Number	Lamp Type & Color	No. of Lamps	Ballast Type	Ballast Voltage	Mounting	Description	Qty.	Input Watts per Luminaire/Total Input Watts
G-EM	2GR8-332A-UNV-EL4-EB82 (See Figure 15-24)	T8 Bipin 32 watt 2800 Lumens per lamp. Instant start 48" 3500°K	3	Note 1	Note 2	Troffer for dropped ceiling grid-lay-in 3-3/4" depth	Lumens: 2800 Acrylic lens. See Note 3.	1	86/86
N	BC-217A-UNV-ER81 (See Figure 15-27)	T8 Bipin 17 watt 1400 Lumens per lamp. Rapid start 24" 3500°K	2	Note 1	Note 2	Surface	Wall light. Opal acrylic reflector. All purpose wall bracket.	3	34/102
Q	CEL 7035SD (See Figure 15-29)	75 watt MR16 Halogen	2	No ballast.	Dual-voltage 120/277 VAC	Recessed.	Lumens: 1137 Emergency lighting. Solid state charger for nickel cadmium battery. Gives appearance of a ceiling. When power is lost, doors open automatically. When power is restored, doors close automatically. Maintains emergency lighting for 90 minutes. Self-tests every 28 days.	2	75/150
EXIT	EEX7 (See Figure 15-31)	LED	LED module	LED Driver	Note 2	Surface	Suitable for ceiling, wall, or corner end mounting	2	4.6/9.2

Area = 2378 Sq. Ft. (See Note 4)	Total Watts = 2702	Lighting Power Density = 1.136 Watts/Sq. Ft.

Note 1: High efficiency Class P electronic. Power Factor >98%. Ballast Factor 0.88.

Note 2: Universal input voltage: 120–277 V ±10%.

Note 3: Luminaire has an emergency nickel cadmium battery pack. In the event of a power outage, electronic circuitry instantly senses the loss of power and keeps the luminaire partially illuminated for a minimum of 90 minutes. When normal power is restored, battery pack automatically goes into charge mode. Battery charges while luminaire is energized. Full recharge time: 24 hours.

Note 4: Square footage based on outside building dimensions minus one foot for each of the outside walls.

FIGURE 15-19 Style A and A-EM fluorescent luminaire. (*Courtesy Cooper Lighting*)

FIGURE 15-23 Style F and F-EM fluorescent luminaire. (*Courtesy Cooper Lighting*)

FIGURE 15-20 Style B and B-EM fluorescent luminaire. (*Courtesy Cooper Lighting*)

FIGURE 15-24 Style G and G-EM fluorescent luminaire. (*Courtesy Cooper Lighting*)

FIGURE 15-21 Style C and C-EM fluorescent luminaire. (*Courtesy Cooper Lighting*)

FIGURE 15-25 Style I fluorescent luminaire. (*Courtesy Cooper Lighting*)

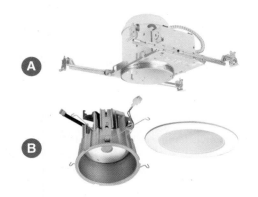

FIGURE 15-22 (A) Style E2 and E2-EM, (B) Style E and E-EM downlight. (*Courtesy Cooper Lighting*)

FIGURE 15-26 Style K HID luminaire. (*Courtesy Cooper Lighting*)

FIGURE 15-27 Style N linear fluorescent luminaire.
(*Courtesy Cooper Lighting*)

FIGURE 15-28 Style P fluorescent luminaire.
(*Courtesy Cooper Lighting*)

FIGURE 15-29 Style Q luminaire.
(*Courtesy Cooper Lighting*)

FIGURE 15-30 Style T luminaire.
(*Courtesy Cooper Lighting*)

FIGURE 15-31 Style X (Exit) luminaire.
(*Courtesy Cooper Lighting*)

LUMINAIRES IN CLOTHES CLOSETS

The *NEC* defines a clothes closet as *A non-habitable room or space intended primarily for storage of garments and apparel.**

Clothing, boxes, and other material normally stored in clothes closets are a potential fire hazard. These items may ignite on contact with the hot surface of an exposed light bulb. The bulb, in turn, may shatter and spray hot sparks and hot glass onto combustible materials. *NEC 410.16* addresses the special requirements for installing in clothes closets. See *NEC 410.2* for the definition of *storage space* as it relates to clothes closets.

It is significant to note that these rules cover ALL clothes closets—residential, commercial, and industrial. In Figures 15-32 and 15-33 "A" represents the width of the storage space above the rod or 12 in. (300 mm), whichever is greater; "B" represents the depth of the storage space below the rod, which is 24 in. (600 mm) in depth and 6 ft (1.8 m), or to the highest rod, in height. Figure 15-34 shows a recessed incandescent luminaire that is acceptable for use in clothes closets if provided with a completely enclosed lamp. Figure 15-35 shows luminaires that are and are not permitted in clothes closets. This figure illustrates the requirements in *NEC 410.16(A)* and *(B)*. Figure 15-36 illustrates required clearances for luminaires in clothes closets.

*Reprinted with permission from NFPA 70-2011.

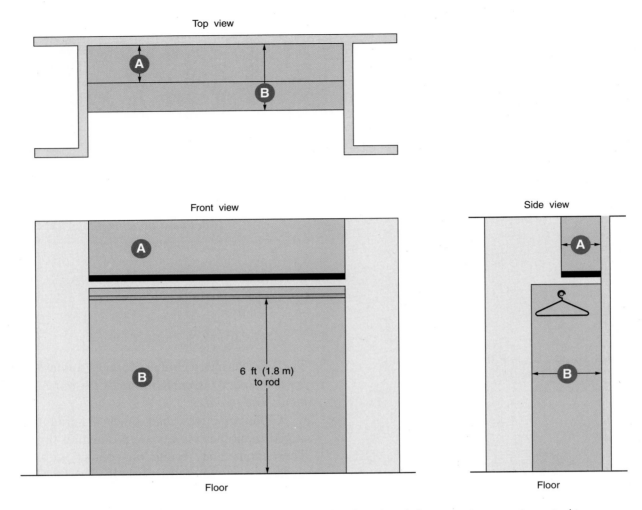

Top view

Front view

Side view

6 ft (1.8 m)
to rod

Floor

Floor

FIGURE 15-32 Clothes closet with one shelf and rod. (*Delmar/Cengage Learning*)

WATTS PER UNIT AREA CALCULATIONS

As stated earlier in this chapter, it is necessary that the illumination of a building comply with energy codes. These codes usually evaluate the illumination on the basis of the average watts per square foot (meter) of illumination load. Incandescent lamps are rated in watts, so it is just a matter of counting each type of luminaire, then multiplying each by the appropriate lamp watts, and adding the products. Fluorescent and HID luminaires are rated by the volt-amperes required for circuit design and the watts required to operate the luminaire. The watt figure is never higher than the volt-ampere and is used to determine the watts per unit area for the ing. The difference between watts and volt-eres is determined by the quality of the ballast,

the efficiency of the luminaire, and the efficacy of the lamp. Where the quality, the efficiency, and the efficacy are high, more light will be produced per watt, and thus fewer luminaires will be required to achieve the desired illumination at a given electrical load. After determining the total watts of load, this value is divided by the area to find the watts per unit area.

For simplicity, most luminaire manufacturers provide the total input wattage for their incandescent, fluorescent, LED, and HID luminaires. Generally, this is how the IESNA provides its lighting recommendations and does its calculations. This results in accurate loading because today's electronic ballast have a power factor of almost 100%. In the Commercial Building, the wattage values supplied by the lighting manufacturer were used for the lighting load calculations.

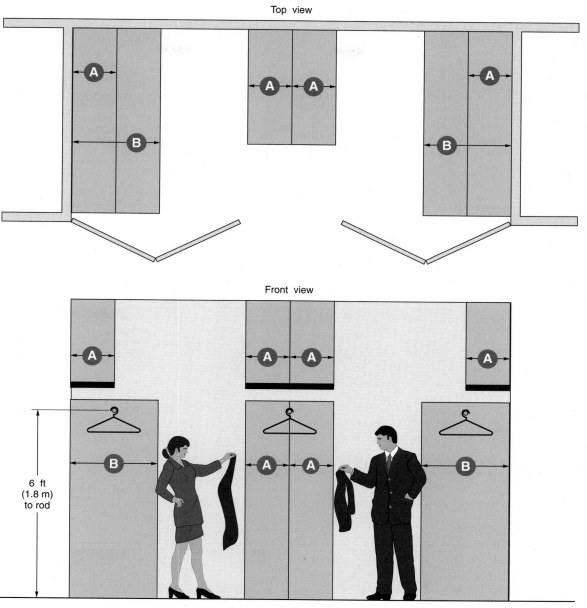

Top view

Front view

6 ft
(1.8 m)
to rod

Floor

FIGURE 15-33 Walk-in clothes closet. (*Delmar/Cengage Learning*)

FIGURE 15-34 Recessed luminaire. (*Delmar/Cengage Learning*)

Luminaires in clothes closets

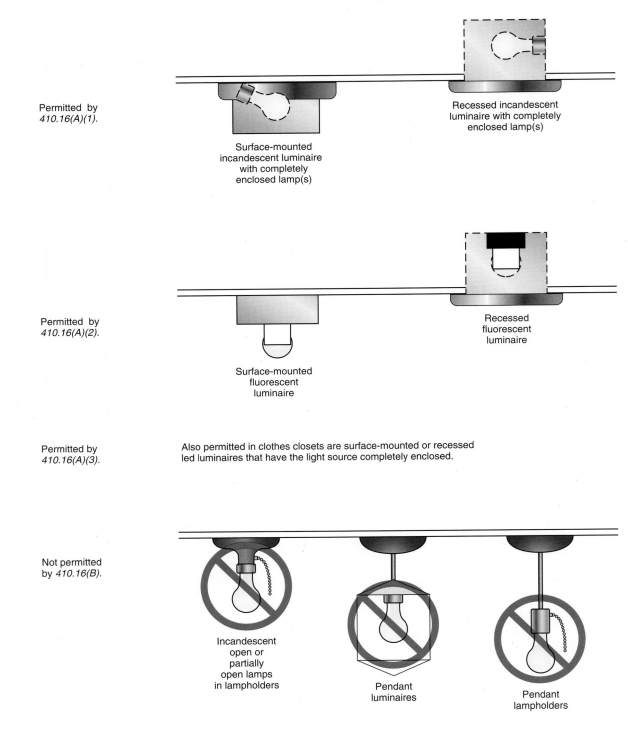

Permitted by
410.16(A)(1).

Surface-mounted
incandescent luminaire
with completely
enclosed lamp(s)

Recessed incandescent
luminaire with completely
enclosed lamp(s)

Permitted by
410.16(A)(2).

Surface-mounted
fluorescent
luminaire

Recessed
fluorescent
luminaire

Permitted by
410.16(A)(3).

Also permitted in clothes closets are surface-mounted or recessed
led luminaires that have the light source completely enclosed.

Not permitted
by *410.16(B).*

Incandescent
open or
partially
open lamps
in lampholders

Pendant
luminaires

Pendant
lampholders

FIGURE 15-35 Luminaires permitted and not permitted in clothes closets. (*Delmar/Cengage Learning*)

Location of luminaires in clothes closets

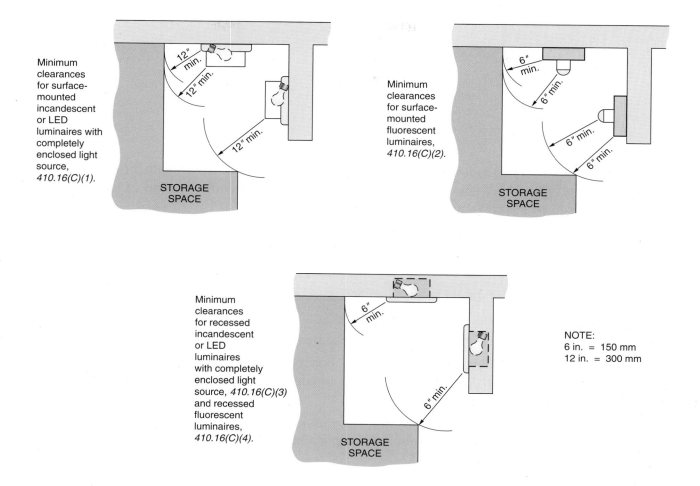

Minimum clearances for surface-mounted incandescent or LED luminaires with completely enclosed light source, *410.16(C)(1)*.

STORAGE SPACE

Minimum clearances for surface-mounted fluorescent luminaires, *410.16(C)(2)*.

STORAGE SPACE

Minimum clearances for recessed incandescent or LED luminaires with completely enclosed light source, *410.16(C)(3)* and recessed fluorescent luminaires, *410.16(C)(4)*.

STORAGE SPACE

NOTE:
6 in. = 150 mm
12 in. = 300 mm

FIGURE 15-36 Required clearances for luminaires in clothes closets. (*Delmar/Cengage Learning*)

REVIEW

Refer to the *National Electrical Code* or the working drawings when necessary. Where applicable, responses should be written in complete sentences.

For Questions 1 and 2, six luminaires, similar to Style E used in the Commercial Building, are to be installed in a room that is 12 ft × 18 ft (~3.7 m × ~5.5 m) with a 9-ft (2.8-m) floor-to-ceiling height. The spacing ratio for the luminaire is 1:0.

1. The maximum distance that the luminaires can be separated and achieve uniform illuminance is _____ ft (_____ m).

2. Draw a sketch indicating center-to-center and center-to-wall distances of how the six luminaires are to be installed.

For Questions 3–6, two luminaires, 8 ft (2.5 m) and 4 ft (1.2 m) in length with dimensions as shown in Figure 15-7, are to be installed in tandem (end to end). The end of the long luminaire is to be 2 ft (600 mm) from the wall.

3. The center of the outlet box should be roughed in at _____ ft (_____ m) from the wall.

4. The first support should be installed at _____ ft (_____ m) from the wall.

5. The second support should be installed at _____ ft (_____ m) from the wall.

6. The final support should be installed at _____ ft (_____ m) from the wall.

Questions 7 and 8 concern the illumination in the Beauty Salon.

7. The maximum loading allowance for the Beauty Salon illumination in accordance with *NEC Table 220.12* is _____ volt-amperes per square foot (_____ volt-amperes per square meter).

8. Complete the information in the following table for the luminaires installed in the Beauty Salon. Refer to the Luminaire/Lamp/Ballast Schedules in this chapter and the Second Floor Electrical Plan E3.

Luminaire Style	Count	Watts/ Luminaire	Total Wattage
Total			

9. A Style Q luminaire is installed in the ceiling of the Beauty Salon. Give a brief description of this luminaire.

_____

_____

_____

The remaining questions pertain to the installation of luminaires in a clothes closet. Circle the correct answer (compliance or violation) and cite the *NEC* reference. In each case, the luminaire is located on the ceiling with 10 in. (250 mm) clearance from the storage area.

10. A porcelain socket with a PL fluorescent lamp.   Compliance   Violation

11. A totally enclosed incandescent luminaire.   Compliance   Violation

12. A fluorescent strip luminaire (bare lamp).   Compliance   Violation

Emergency, Legally Required Standby, and Optional Standby Power Systems

OBJECTIVES

After studying this chapter, you should be able to

- achieve a basic understanding of emergency, legally required standby, and optional standby power systems.

- select and size a generator for an emergency or standby power system.

- understand how to connect typical emergency and standby power equipment.

- understand requirements for transfer switches and equipment.

Chapter 16 is an introduction to the basics of emergency, legally required standby, and optional standby power systems that electricians need to be aware of. In just about all applications, the type and sizing of the generator equipment is left to the electrical consulting engineer, who in turn often relies on the engineering know-how of the technical support staff and technical sales force of generator and transfer switch manufacturers.

Many state and local codes require that equipment be installed in public buildings to ensure that electric power is provided automatically if the normal power source fails. The electrician should be aware of the special installation requirements of these systems.

Several systems addressed by the *NEC* ensure power is available in critical conditions. Some of these systems are mandated in the *NEC*, whereas other systems are optional as needed or required by the owner. These systems include the following:

1. *Article 700: Emergency Systems.* An emergency system consists of the circuits and equipment that supply, distribute, and control electricity for illumination, power, or both to required facilities when the normal electrical supply is interrupted. Emergency systems are installed where life safety is involved.

   Emergency systems will be legally required by a municipality, state, federal, other governmental agency, or other codes for specific types of buildings, such as health care facilities, hotels, theaters, and similar occupancies. See the definition of emergency systems in *NEC 700.2*.

   The main purpose of an emergency system is to automatically come on to *supply, distribute, and control power and illumination essential for safety to human life* in the event that the normal supply of power is interrupted. This would include such things as elevators, fire pumps, fire detection and alarm systems, communication systems, ventilation, lighting for the safe exiting of the building, and similar critical loads.

   In order to minimize the possibility of power outages, overcurrent devices for the emergency system(s) shall be selectively coordinated with all supply-side overcurrent protective devices; see *NEC 700.27*. Achieving selectivity of overcurrent devices is covered in Chapter 18.

2. *Article 701: Legally Required Standby Systems.* A legally required standby system automatically supplies power to selected loads such as illumination, power, or both. Loss of power to such loads is not life threatening. These systems will be required by a municipality, state, federal, or other governmental agency, or other codes.

   In order to minimize the possibility of unwanted power outages, overcurrent devices for the legally required standby system(s) shall be selectively coordinated with all supply-side overcurrent protective devices; see *NEC 701.27*. Achieving selectivity of overcurrent devices is covered in Chapter 18.

   The *Informational Note* following *NEC 701.2* gives an insight into the types of loads typically served by these systems. It states, *Legally required standby systems are typically installed to serve loads, such as heating and refrigeration systems, communications systems, ventilation and smoke removal systems, sewage disposal, lighting systems, and industrial processes, that, when stopped during any interruption of the normal electrical supply, could create hazards or hamper rescue or fire-fighting operations.**

3. *Article 702: Optional Standby Systems.* An optional standby system is just that—optional. Life safety is not an issue. No laws require it. Should loss of the normal supply of power occur, an optional standby system will reestablish power to selected loads until the normal power source comes back on. An optional standby system might be connected to the alternate power source manually or automatically.

4. *Article 517: Health Care Facilities.* Extensive rules are found in *NEC Article 517* for emergency systems in a variety of health-care facilities. These facilities include *Buildings or portions of buildings in which medical, dental,*

---

*Reprinted with permission from NFPA 70–2011.

psychiatric, nursing, obstetrical, or surgical care are provided. Health care facilities include, but are not limited to, hospitals, nursing homes, limited care facilities, clinics, medical and dental offices, and ambulatory care centers, whether permanent or movable.*

As can be seen, many types of healthcare facilities are covered in *NEC Article 517*. Necessarily, the requirements for the backup power requirements vary significantly based on the type of procedure being performed as well as on the capability for the person to evacuate the building without the assistance of others. Hospitals generally have the most stringent requirements, and clinics and medical and dental offices have the least stringent requirements.

5. *Article 708, Critical Operations Power Systems (COPS):* This was a new Article added to the 2008 *NEC*. The term *Critical Operations Power Systems* is defined in *NEC 708.2* as *Power systems for facilities or parts of facilities that require continuous operation for the reasons of public safety, emergency management, national security, or business continuity.**

An explanation for how and when these systems may be required is contained in *NEC 708.1*. It reads, *Critical operations power systems are those systems so classed by municipal, state, federal, or other codes by any governmental agency having jurisdiction or by facility engineering documentation establishing the necessity for such a system. These systems include but are not limited to power systems, HVAC, fire alarm, security, communications, and signaling for designated critical operations areas.**

Informational Note No. 1 goes on to explain the operational purposes of these systems and reads, *Critical operations power systems are generally installed in vital infrastructure facilities that, if destroyed or incapacitated, would disrupt national security, the economy, public health or safety; and where enhanced electrical infrastructure for continuity of operation has been deemed necessary by governmental authority.**

The preceding *NEC* articles contain the how-to for on-the-job wiring requirements. The who, what, when, where, and why these systems are required are found in

NFPA 99, *Standard for Health Care Facilities*
NFPA 101, *Life Safety Code*
NFPA 110, *Standard for Emergency and Standby Power Systems*
ANSI/IEEE Standard 446, *Recommended Practice for Emergency and Standby Power Systems for Industrial and Commercial Applications*

SOURCES OF POWER

When the need for emergency power is confined to a definite area, such as a stairway or exit hallway, and the power demand in this area is low, then self-contained battery-powered units are a convenient and efficient means of providing power. See *NEC 700.12(F)*. These are referred to as "unit equipment." In general, these units are wall mounted and are connected to the normal source by permanent wiring methods or at times by flexible cord if so supplied by the manufacturer. Under normal conditions, this regular source powers a regulated battery charger to keep the battery at full power. When the normal power fails, the circuit is completed automatically to one or more lamps that provide enough light to the area to permit its use, such as lighting a stairway sufficiently to allow people to leave the building. Battery-powered units are commonly used in stairwells, hallways, shopping centers, supermarkets, and other public structures. This equipment is also used for exit signs. See Figure 16-1 for an example of a pathway lighting luminaire and an exit sign that have internal battery backup.

As you study the plans for the Commercial Building, you will see that a number of the luminaires are identified as "EM." These are the luminaires that have an integral emergency battery pack that continually "floats" on the circuit, maintaining a full charge and ready to go into action should a power outage occur. These emergency battery packs look very similar to a fluorescent ballast. They consume 1.5 to 6 watts, depending on the type and quantity of lamps served.

---

*Reprinted with permission from NFPA 70–2011.

FIGURE 16-1 Unit equipment for pathway lighting and exit sign. (*Delmar/Cengage Learning*)

CLASSIFICATION OF SYSTEMS

First things first! Before planning the installation, and assuming the installation is for an *Emergency System*, be certain that an Authority Having Jurisdiction (AHJ) has actually classified the building or structure as requiring an *Emergency System*. The building may be required to have a *Legally Required Standby* or the owner may choose to install an *Optional Standby* system.

The electrical wiring and equipment performance requirements differ significantly depending on the type of system that is either legally required or is optional. Table 16-1 summarizes these requirements.

SPECIAL WIRING ARRANGEMENTS

NEC 700.10(B) states that the wiring from an emergency source wiring shall be kept entirely independent of all other wiring and equipment and shall not enter the same raceway, cable, box, or cabinet with other wiring. Installing a separate service for the emergency system is permitted in *NEC 700.12(D)* only if doing so is approved by the AHJ. Many AHJs will not permit a separate service for an emergency system from the same transformer as the normal service as a single transformer or primary feeder would cause both services to be without power. Likewise, other AHJs will not permit the emergency service to be supplied from the same utility substation as the normal service. A separate service is illustrated in Figure 16-2. Wiring for the feeders and branch circuits for emergency systems are similarly required to be maintained entirely separate from circuits for normal, legally required standby systems and optional standby systems.

NEC 701.10 allows the wiring of a legally required standby system to occupy the same raceways, cables, boxes, and cabinets with other general wiring. Furthermore, *NEC 701.12(E)* permits the connection for this system to be made ahead of, but not within, the main service disconnect. Figure 16-3 illustrates this arrangement. The connections could be made in a junction box or to the service conductors outside of the building. Because the size of the tap conductors would be much smaller than the service conductors, cable limiters are installed at the point of the tap. These will limit the available fault current to the withstand rating of the tap conductors, and the interrupting rating of the disconnect device.

TABLE 16-1

Rules for emergency, legally required standby, and optional standby systems.

Feature or Requirement	*Article 700*	*Article 701*	*Article 702*
Rules apply only if legally required	Yes	Yes	No
Equipment approval required	Yes	No, but subject to AHJ rules	No, but subject to AHJ rules
Restoration of power	10 seconds	60 seconds	No rule
Transfer equipment	Automatic, dedicated and listed	Automatic, listed for emergency use	Suitable for intended use
Priority for load shedding or peak shaving	Highest	Second	Lowest
Permitted to supply other systems	Emergency, LR standby, optional standby	LR standby, optional standby	No
Circuit conductors isolated	Yes	No	No
Connection ahead of normal service disconnecting means	No	Yes, if acceptable to AHJ	No rule
Separate service permitted	Yes, if acceptable to AHJ	Yes, if approved	Yes
Selective coordination required	Yes	Yes	No
Installation and periodic testing required	Yes	Yes	No
Adequate capacity for all loads to be carried at same time	Yes	Yes	Yes, but owner can select loads to be supplied
Signals and signs required	Yes	Yes	No
Ground-fault protection of equipment required	No	No	Yes
Ground-fault indication required (if protection is not provided)	Yes	No	Not permitted

⎍⎍ GENERATOR SOURCE

Generator sources may be used to supply power to emergency, legally required standby and optional systems [see *NEC 700.12(B) 701.12(B)*] and by default to optional standby systems. The selection of such a source for a specific installation involves a consideration of the following factors:

- Engine type
- Generator capacity
- Load transfer controls

A typical totally enclosed generator for emergency or standby power is shown in Figure 16-4.

Engine Types and Fuels

The type of fuel to be used in the driving engine of a generator is an important consideration in the installation of the system. Fuels that may be used are LPG (liquid propane) gas, natural gas, gasoline, or diesel fuel. Factors affecting the selection of the fuel to be used include the availability of the fuel, the size and type of the prime mover and local regulations governing the storage of the fuel.

Generator prime movers that power generators and use gasoline, natural gas, or LPG usually have lower installation and operating costs than a diesel-powered source. However, the problems of fuel storage can be a deciding factor in the selection of the type of fuel. Gasoline is not only dangerous, but becomes stale after a relatively short period of time and thus cannot be stored for long periods. Gasoline is rarely used as a fuel for the prime mover for larger generators used in standby systems today. If natural gas is used, the British thermal unit (Btu) content must be greater than 1100 Btu per cubic foot. Diesel-powered generators require less maintenance and have a longer

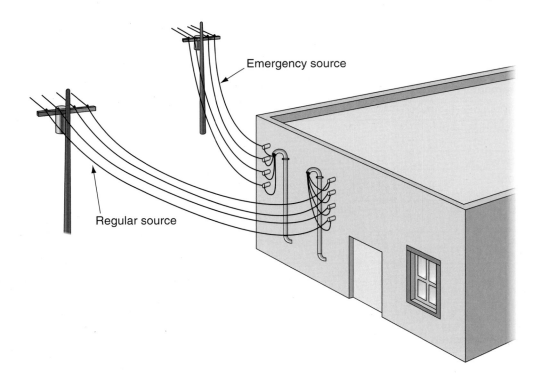

FIGURE 16-2 Separate services permitted for emergency and legally required standby systems. See *NEC 700.12(D)* and *701.12(D)*. (*Delmar/Cengage Learning*)

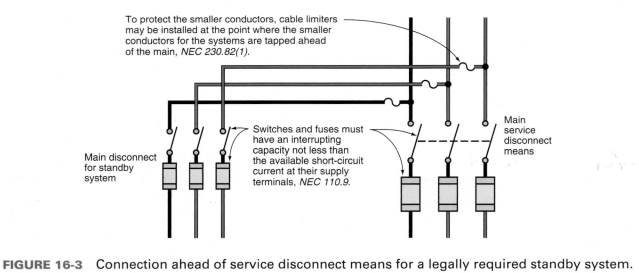

FIGURE 16-3 Connection ahead of service disconnect means for a legally required standby system. (*Delmar/Cengage Learning*)

life. This type of diesel system is usually selected for installations having large power requirements because the initial costs closely approach the costs of systems using other fuel types. Generally, a time-delay is required after the utility system is reconnected, to prevent damage to the prime mover, often referred to as "cold stacking," caused by the engine not getting up to normal operating temperature.

Fuel consumption is based on the loading of the generator. Manufacturer's technical data generally provides fuel consumption based on loading the generator at 100%, 75%, 50%, and 25%.

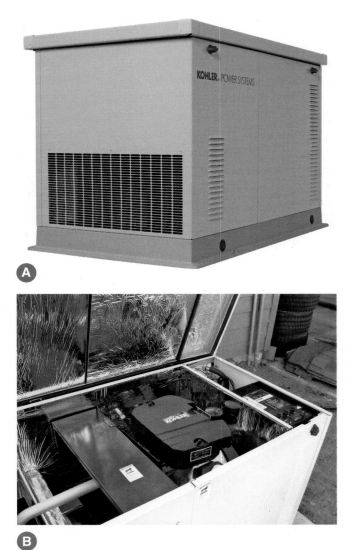

A

B

FIGURE 16-4   Generator suitable for emergency, legally required or standby power of the type generally specified for typical small commercial applications. [*Photo 16-4(A) courtesy of Kohler Co., Photo 16-4(B) Delmar/Cengage Learning*]

Cooling

Smaller generator units are available with either air or liquid cooling systems. Units having a capacity greater than 15 kilowatts generally use liquid cooling. Generators for most commercial applications are installed outdoors, as is the generator for the Commercial Building. For air-cooled units installed indoors, it is recommended that the heated air be exhausted to the outside. In addition, a provision should be made to bring in fresh air so that the room where the generator is installed can be kept from becoming excessively hot. A generator suitable for typical commercial buildings to serve emergency, legally required or optional standby systems is shown in Figure 16-4.

Generator Voltage Characteristics

Generators having any required voltage characteristic are available. A critical factor in the selection of a generator for a particular application is that it must have the same voltage output as the normal building supply system. For the Commercial Building covered in this text, the generator selected provides 208Y/120-volt, 3-phase, 60-Hz power.

Capacity

The size (capacity) of a generator is simply that it must be able to adequately and safely handle the loads that are to be operated simultaneously by the generator. The requirement for emergency systems is found in *NEC 700.4(A)*. Similar capacity requirements for legally required standby systems can be found in *NEC 701.4* and in *702.4* for optional standby systems.

How is the size (kW and kVA rating) determined? Someone has to decide what loads are to be picked up in the event of a power outage. It very well could be a combination of local, state, and/or Federal laws, the owner's wishes, the electrical contractor, the architect, the consulting engineer, or the tenant. All will have input to the final selection. For emergency and legally required systems, it becomes a matter of what is required to be served by emergency power or by the legally required system. If additional capacity is available and proper transfer switches are provided, other loads that are optional are permitted to be supplied from the emergency or legally required systems.

Additional requirements are imposed by *NEC 702.4(B)* for optional standby systems. If a manual transfer switch is installed, the generator is required to have capacity for all equipment intended to be operated at the same time. If automatic transfer equipment is installed, the generator is required to have capacity for all loads to be transferred to the generator. If load management equipment is installed, the generator can be sized for the load to be connected at the same time by the load management equipment.

For the Commercial Building, the loads shown in the Owner's panelboard schedule with a superscript "G" (^{G}) will be picked up by the generator should a power outage occur. This includes the following loads:

Receptacles Basement^G	720
Sump Pump^G	1127
Receptacles 1st and 2nd Fl^G	900
Elevator^G	9000
Elevator Circuits Feeder^G	9600
Total Running Load	21,347 VA

The lighting loads are not included, as every area has one or more battery packs to provide lighting should the electric utility power fail.

Manufacturers of generators and their authorized distributors must always be involved in the selection of their generating equipment. They know the legal requirements. They know the requirements of the UL, NEMA-MG-1, IEEE, ANSI, NFPA, and EPA standards. Without question, they have the know-how to select the proper size and type generator for a given installation. They can perform the necessary calculations to assure proper sizing. They provide detailed wiring diagrams and installation requirements relating to the connections for the generator, transfer switch, sensing devices, and derangement devices.

Is it important to maintain a certain amount of lighting should there be a loss of power during bad weather conditions or utility company's equipment failure? How about keeping your computers running? How about maintaining power for communication and information systems? How about keeping refrigeration running to prevent food from thawing and spoiling? How about keeping the sump pump(s) running to avoid flooding? These are some of the questions that must be answered. It is an involved but extremely important task to determine the correct size of the engine-driven power system so it has the minimum capacity necessary to supply the selected equipment. If the system is oversized, additional costs are involved in the installation, operation, and maintenance of the system. However, an undersized system may fail at the critical period when it is being relied on to provide emergency

legally required or optional standby power. To size the generator system, it is necessary to determine initially all of the equipment to be supplied. If motors are to be supplied by the standby system, it must be determined whether all of the motors or other loads can be started at the same time. This information is essential to ensure that the standby power system has the capacity to provide the total starting inrush kilovolt-amperes required.

One manufacturer suggests that after the loads to be picked up by the generator have been determined, an additional capacity of at least 20% be added to provide an extra "cushion" for unforeseen situations.

The starting kilovolt-amperes for a motor is equivalent to the locked-rotor kilovolt-amperes of the motor. This value is determined by selecting the appropriate value from Table 16-2 in this text (*Table 430.7(B)* in the *NEC*] and then multiplying this value by the horsepower. The locked-rotor kilovolt-amperes value is independent of the voltage characteristics of the motor. A 5-horsepower, Code Letter E, 3-phase motor requires a generator capacity of

$$5 \text{ HP} \times 4.99 \text{ kVA/HP} - 24.95 \text{ kVA}$$

TABLE 16-2

Table 430.7(B) Locked-Rotor Indicating Code Letters

Code Letter	Kilovolt-Amperes per Horsepower with Locked Rotor
A	0–3.14
B	3.15–3.54
C	3.55–3.99
D	4.0–4.49
E	4.5–4.99
F	5.0–5.59
G	5.6–6.29
H	6.3–7.09
J	7.1–7.99
K	8.0–8.99
L	9.0–9.99
M	10.0–11.19
N	11.2–12.49
P	12.5–13.99
R	14.0–15.99
S	16.0–17.99
T	18.0–19.99
U	20.0–22.39
V	22.4 and up

Reprinted with permission from NFPA 70-2011.

TABLE 16-3

Approximate motor kVA values.

HP	Type	Locked-Rotor kVA
1/6	C	1.85
1/6	S	2.15
1/4	C	2.50
1/4	S	2.55
1/3	C	3.00
1/3	S	3.25

C = Capacitor-start motor

S = Split-phase motor

If two motors are to be started at the same time, the standby power system must have the capacity to provide the sum of the starting kilovolt-ampere values for the two motors.

For a single-phase motor rated at less than $1/2$ horsepower, Table 16-3 lists the approximate kilovolt-ampere values that may be used if exact information is not available.

A check of one manufacturer's data shows that a small 12-kilovolt-ampere unit is available having a 20-kilovolt-ampere maximum rating for motor starting purposes and a 12-kilovolt-ampere continuous rating. For our installation, the largest motor load is the elevator motor. Although it draws approximately 25-A, 208-V, 3-phase to run, it draws only 40 amperes to start because it is provided with a solid-state soft-start feature. This should allow the 12-kVA generator to provide adequate energy for the elevator motor to start without a problem. Without the soft start, a typical squirrel cage, 3-phase motor would require about 6 times the 25-A running current, or 150 amperes, to start. If a number of electric motors are to be served by the generator, it is possible to install time delays on the motor contactors, to prevent the motors from starting at the same time. See Table 16-2 for approximate locked-rotor kilovolt-amperes of small fractional horsepower, single-phase, capacitor start, and split-phase motors.

You can readily see from this that sizing generators and generator-associated equipment requires knowing maximum and continuous kVA values.

When sizing conductors for branch-circuits, feeders, and service-entrances, we have to convert kVA to amperes, using the following formulae:

For 3-phase:

$$I = \frac{kVA \times 1000}{E \times 1.732}$$

For single-phase:

$$I = \frac{kVA \times 1000}{E}$$

You might want to review Chapter 7, where motors are discussed in greater detail. Be sure to study the section on "Motor Starting Currents/*Code* Letters."

Derangement Signals

According to *NEC 700.6(A)*, a derangement signal is required for emergency systems to consist of a signal device having both audible and visual alarms. The signal annunciator device is required to be installed outside the generator room in a location where it can be readily and regularly observed. The purposes of a device such as the one shown in Figure 16-5 are to continually monitor the equipment and to indicate any malfunction in the generator unit, any load on the system, or the correct operation of a battery charger. Similar requirements are found in *NEC 701.6(A)* for legally required standby systems and *702.6(1)* for optional standby systems.

FIGURE 16-5 Audible and visual signal device as required by *NEC 445.12 Exception, 700.6, 701.6,* and *702.6. (Photo courtesy of Kohler Co.)*

TRANSFER SWITCHES AND EQUIPMENT

One of the major functions of transfer switches and equipment is to prevent the dangerous interconnection of two electrical systems that are not intended to be operated in parallel. Ask a power company's lineman about backfeed. He knows all too well of its hazards. There have been many injuries and deaths as a result of backfeed; it has been called "the Hidden Killer."

Think about it. The normal source of power might be coming through a step-down transformer; for example, a 2400/4160-volt primary stepped down to a 120/208-volt secondary. Now imagine that the normal source of power is lost. The generator kicks in. And guess what? A transformer works both ways: it can step down the voltage, or it can step up the voltage. Unless there is a properly connected transfer switch that disconnects the normal service conductors, the power from the generator will feed back into the service conductors, to the transformer, through the transformer, and onto the primary conductors. We now have a transformer that steps up the voltage from 120/208 volts to 2400/4160 volts. A lineman working on what he thinks is de-energized equipment or conductors is now unknowingly working on a "hot" system. Other services connected to the secondary of that transformer will also be live. Always install listed transfer switches that perform their job of transferring power safely!

Transfer switches are generally available from 30–4000 amperes. Typically they are available in 2-, 3-, and 4-pole configurations for ranges of 208–600 volts ac. Transfer equipment is available with ratings over 600 volts.

Transfer switches are marked with a maximum available RMS symmetrical short circuit that they may be connected to. Simply stated, this is the transfer switches short-circuit rating.

Position of Transfer Equipment

Generally, transfer switches and equipment are required to be located on the load side of overcurrent protection both from the normal utility source and alternate power. Figures 16-6(A), 16-6(B),

and 16-6(C) illustrate correct positions for transfer switches.

NEC 230.82 gives a list of equipment permitted to be located on the supply side of the service equipment. Transfer switches and equipment are absent from the list. Transfer switches located on the supply side of the service equipment become the service disconnecting means and are required to be marked "Suitable For Use As Service Equipment." This equipment must have three positions: (1) connected to normal power, (2) off, and (3) connected to emergency or standby power. As mentioned before, the transfer switch or equipment must be suitable for the short-circuit current available on the supply terminals of the equipment.

Types of Transfer Equipment

The two broad types of transfer equipment are automatic transfer switches and manual transfer equipment. The *NEC Articles* that provide requirements for the emergency, legally required, or optional system contains rules on the type of equipment that is required or permitted. These provisions are included in Table 16-4.

Automatic Transfer Switches

If the main power source fails, equipment must be provided to start the engine of the emergency generator and transfer the supply connection from the regular source to the emergency source; see *NEC 700.5*. Similar requirements are found in *NEC 701.5* for legally required standby systems and *NEC 702.5* for optional standby power systems.

These operations can be accomplished by installing individual components, or by an automatic transfer switch such as the one shown in Figure 16-7. This panel contains all of the devices needed to monitor and sense a failure of the main power source, activate the transfer switch, and instruct the generator to start. A voltage sensitive relay is connected to the main power source. This relay activates the control cycle when the utility source voltage fails. The generator motor is started when the control cycle is activated.

As soon as the generator reaches the correct speed, the transfer switch disconnects the load

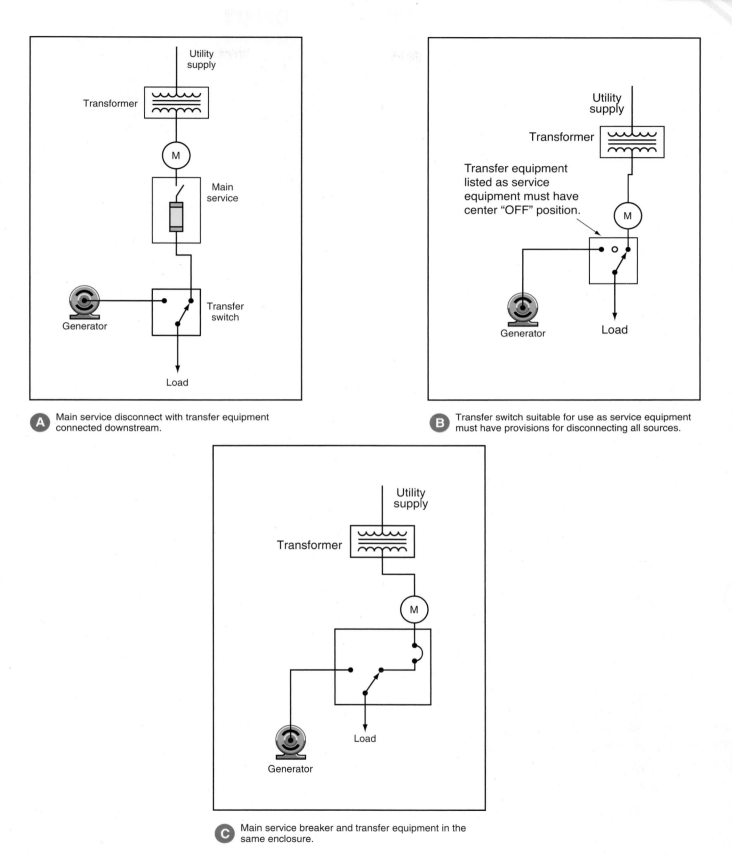

A Main service disconnect with transfer equipment connected downstream.

B Transfer switch suitable for use as service equipment must have provisions for disconnecting all sources.

C Main service breaker and transfer equipment in the same enclosure.

FIGURE 16-6   Variations of transfer switch connections. The equipment shall be suitable for the maximum available fault current at its terminals, *NEC 700.4(A)*, *701.4*, and *702.4(A)*. (*Delmar/Cengage Learning*)

TABLE 16-4

Requirements for transfer switches and equipment.

NEC Article	Switch or Equipment	Listing Requirements
517 (Emergency System—Hospitals)	Automatic	Emergency Systems [UL *White Book* (WPWR)]
700 Emergency Systems	Automatic	Emergency Systems [UL *White Book* (WPWR)]
701 Legally Required Standby Systems	Automatic	Legally Required Standby Systems [See UL *White Book* (WPWR)]
702 Optional Standby Systems	Automatic or manual	Suitable for Intended Use [See UL *White Book* (WPYV)]
708 Critical Operations Power Systems	Automatic	Emergency Systems [UL *White Book* (WPWR)]

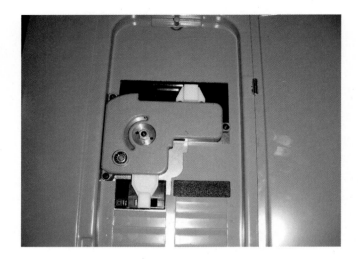

FIGURE 16-8 Panelboard that contains automatic transfer equipment. (*Delmar/Cengage Learning*)

from its normal source and connects the load to the generator output.

A transfer switch is basically a "double-throw" switch that is "break before make."

Features available for monitoring a system are to sense load values, over and under voltage sensing, voltage unbalance, frequency, and anti-single phasing protection.

Specifying and installing "packaged" equipment instead of individual components results in lower installation costs and is neater in appearance.

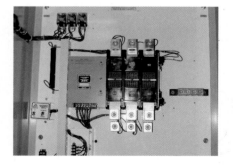

FIGURE 16-7 Automatic transfer switch suitable for optional standby systems. (*Delmar/Cengage Learning*)

Manufacturers of generators, transfer switches, and signaling devices also provide panelboards that have the transfer switch built right into the panelboard. This is a nice feature for those installations where one panelboard is to serve the emergency or standby power requirements. There is no need for a separate transfer switch. This makes the installation neater in appearance. Another optional feature available is a load management system, which allows a reduced capacity generator to supply a larger capacity panelboard by selecting specific loads to be transferred to the generator. An example of the panelboard/transfer equipment is shown in Figure 16-8.

Although it might seem like an automatic transfer switch is "just another switch," high-quality automatic transfer switches conform to UL 508 (*Standard for Industrial Control*), UL 1008 (*Standard for Automatic Transfer Switches for Use in Emergency Standby Systems*), Canadian Standards Association *C22.2*, NFPA *NEC 70*, NFPA 99 (*Essential Electrical Systems for Health Care Facilities*), NFPA 110 (*Emergency and Standby Power Systems*), IEEE 446 (*Recommended Practice for Emergency and Standby Power Systems for Commercial and Industrial Applications*), and NEMA ICS 10 (*Automatic Transfer Switches, Parts 1 and 2*). There are also International Standards for ATS devices, but this is not our concern for the Commercial Building discussed in this text.

Wiring

The branch-circuit wiring of emergency systems must be separated from the standard system except for the special conditions noted. Figure 16-9 shows a typical branch-circuit installation for an emergency system. Key-operated switches, Figure 16-10, are installed to prevent unauthorized personnel from operating the luminaires.

Under certain conditions, emergency circuits are permitted to enter the same junction box as

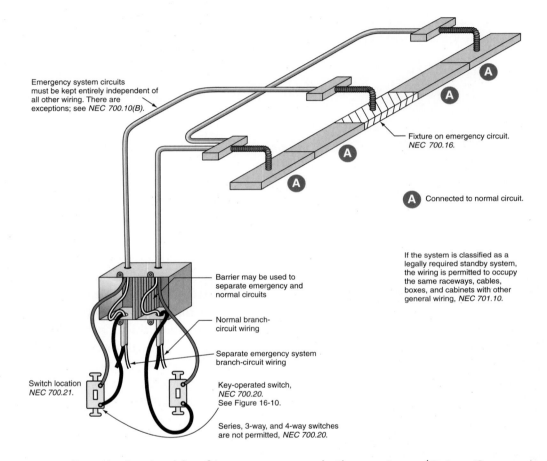

Emergency system circuits must be kept entirely independent of all other wiring. There are exceptions; see *NEC 700.10(B)*.

Fixture on emergency circuit. *NEC 700.16.*

(A) Connected to normal circuit.

If the system is classified as a legally required standby system, the wiring is permitted to occupy the same raceways, cables, boxes, and cabinets with other general wiring, *NEC 701.10.*

Barrier may be used to separate emergency and normal circuits

Normal branch-circuit wiring

Separate emergency system branch-circuit wiring

Switch location *NEC 700.21.*

Key-operated switch, *NEC 700.20.* See Figure 16-10.

Series, 3-way, and 4-way switches are not permitted, *NEC 700.20.*

FIGURE 16-9 Branch-circuit wiring for emergency and other systems. (*Delmar/Cengage Learning*)

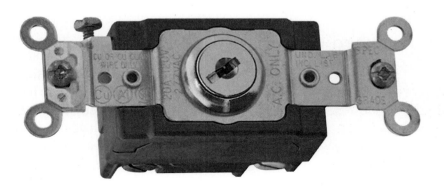

FIGURE 16-10 Key-operated switch for emergency lighting control. (*Delmar/Cengage Learning*)

FIGURE 16-11 Wiring from normal and emergency source permitted for some applications. *NEC 700.10(B)*. (*Delmar/Cengage Learning*)

normal circuits, *NEC 700.10(B)*. Figure 16-11 shows an exit luminaire that is supplied by two branch circuits: one from the normal circuit and the other connected to the emergency circuit. Lighting manufacturers also provide luminaires that have an integral battery that automatically provides the power to keep that particular luminaire operating in the event of a power outage. Many times in smaller buildings, this type of luminaire is the choice over a generator. It all depends on how complicated the installation is.

Branch circuits for legally required standby systems are not permitted in the same raceway or enclosure with emergency systems, but are permitted to be installed with normal branch circuits.

Sizing Generators When UPS Systems Are Involved

Uninterruptible power supply (UPS) systems are often installed to provide "clean power" to critical loads such as computers. If the normal main power supply is lost, the UPS system kicks in with no interruption of power to the critical loads.

However, loss of utility power also means that the standby generator will start. If the UPS system "sees" unstable power from a generator supply, it will reject that source of power, and the batteries will furnish the critical power until they are run down. The system will then shut down.

UPS systems loads are nonlinear loads. Nonlinear loads are heavy in harmonic currents that cause overheating of the neutral conductor in a multiwire circuit where the neutral conductor is common to more than one ungrounded phase conductor. Likewise, there will be overheating of the generator windings.

When nonlinear loads such as UPS systems are involved, leave the sizing of the generator to the experts! Generators must be oversized when supplying nonlinear loads to reduce heating and voltage waveform distortion. Contact the manufacturer of the generator for sizing recommendations. Different manufacturers have recommendations that vary from over-sizing the generator two times, three times, and even as much as five times the UPS load. Call the manufacturer. Better to be safe than sorry!

Refer to the *National Electrical Code* or the working drawings when necessary. Where applicable, responses should be written in complete sentences.

There are three options for providing standby electrical power to building systems. The options are listed here. Discuss the advantages and disadvantages of each.

1. A local source of power such as a battery or a motor-generator

   _____

   _____

   _____

   _____

   _____

2. A separate service

   _____

   _____

   _____

   _____

   _____

3. A connection ahead of the service main

   _____

   _____

   _____

   _____

   _____

For Questions 4–7, perform the following calculations and cite *NEC* references.

4. The minimum starting kilovolt-amperes for a $7\frac{1}{2}$-horsepower, 3-phase, 230-volt, *Code* letter F motor is _____ .

5. The maximum starting kilovolt-amperes for a $7\frac{1}{2}$-horsepower, 3-phase, 230-volt, *Code* letter F motor is _____ .

6. The minimum kilovolt-amperes of a generator that will be required to start, simultaneously, two $7\frac{1}{2}$-horsepower, 3-phase, 230-volt, *Code* letter F motors is

_____ .

7. The minimum kilovolt-amperes of a generator that will be required to start two selectively controlled, $7\frac{1}{2}$-horsepower, 3-phase, 230-volt, *Code* letter F motors is

_____ .

8. Branch circuits from a normal system are permitted to be installed in the same conduit with circuits from an emergency system. True _____ False _____

9. All generators connected to a building wiring system are considered "emergency systems." True _____ False _____

10. By placing a 1, 2, or 3 in the blank, identify the highest to the lowest level of standby system:

Emergency system _____

Legally required standby _____

Optional standby _____

11. The generator is required to be sized for all the loads that can be connected to it. True _____ False _____

12. A tap ahead of the main is permitted as the source of power for emergency systems. True _____ False _____

13. When an automatic transfer switch is installed for an optional standby system, the owner is permitted to manually select the loads to be supplied. True _____ False _____

Overcurrent Protection: Fuses and Circuit Breakers

OBJECTIVES

After studying this chapter, you should be able to

- list and identify the types, classes, and ratings of fuses and circuit breakers.

- describe the operation of fuses and circuit breakers.

- develop an understanding of switch sizes, ratings, and requirements.

- define *interrupting rating, short-circuit currents, I^2t, I_p, RMS,* and *current limitation.*

- apply the *National Electrical Code* to the selection and installation of overcurrent protective devices.

- use the time-current characteristics curves and peak let-through charts.

Overcurrent protection is one of the most important components of an electrical system. The overcurrent device opens an electrical circuit whenever an overload or short circuit occurs. Overcurrent devices in an electrical circuit may be compared to pressure relief valves on a boiler. If dangerously high pressures develop within a boiler, the pressure relief valve opens to relieve the high pressure. In a similar manner, the overcurrent device in an electrical system also acts as a "safety valve."

NEC Article 240 sets forth the requirements for overcurrent protection of conductors and for overcurrent devices. *NEC 240.1 (Informational Note)* states that overcurrent protection for conductors and equipment is provided to open the circuit if the current reaches a value that will cause an excessive or dangerous temperature in conductors or conductor insulation. *NEC 110.9* and *110.10* set forth requirements for interrupting rating and protection against fault current.

As shown in *NEC 240.3* and *Table 240.3,* several other *NEC* articles contain requirements for protection of equipment. For example, the rules on overcurrent protection of air-conditioning equipment are found in *NEC Article 440*. Rules on overcurrent protection of appliances are found in *NEC Article 422* and in *Article 430* for motors and conductors that supply motors.

Two types of overcurrent protective devices are commonly used: fuses and circuit breakers. The Underwriters Laboratories, Inc. (UL), and the National Electrical Manufacturers Association (NEMA) establish standards for the ratings, types, classifications, and testing procedures for fuses and circuit breakers.

As indicated in *NEC 240.6(A)*, the standard ampere ratings for fuses and nonadjustable circuit breakers are 15, 20, 25, 30, 35, 40, 45, 50, 60, 70, 80, 90, 100, 110, 125, 150, 175, 200, 225, 250, 300, 350, 400, 450, 500, 600, 700, 800, 1000, 1200, 1600, 2000, 2500, 3000, 4000, 5000, and 6000. Additional standard ratings for fuses are 1, 3, 6, 10, and 601 amperes.

Why does the *NEC* list "standard ampere ratings"? There are many sections in the *Code* where permission is given to use *the next standard higher ampere rating overcurrent device*, or a requirement to use *a lower than standard ampere rating overcurrent device*. For example, in *NEC 240.4(B)*, we find that if the ampacity of a conductor does not match a standard ampere rating fuse or breaker, it is allowable to use the *next higher standard rating* for ratings of 800 amperes or less provided the three conditions in the section are met. Therefore, for a conductor that has an ampacity of 115 amperes (greater than the standard 110-ampere overcurrent device and lower than a 125-ampere overcurrent device), the *Code* permits the overcurrent protection to be sized at 125 amperes. A very important concept must be mentioned here; the fact that the rule in *NEC 240.4(B)* allows us to increase the rating of the overcurrent device above the allowable ampacity of the conductor, we are never permitted to overload the conductor. As a result, the conductor in our example can carry no more than 115 amperes while being protected from overcurrent at 125 amperes.

When protecting conductors where the overcurrent protection is greater than 800 amperes, the overcurrent protection must not be greater than the conductor's allowable ampacity. An example of this would be if three 500-kcmil copper conductors with 75°C insulation were installed in parallel. Each has an ampacity of 380 amperes in the 75°C column of *Table 310.15(B)(16)*. Therefore,

$$380 \times 3 = 1140 \text{ amperes}$$

It would be a *Code* violation to protect these conductors with a 1200-ampere overcurrent device for other than motor circuits. The correct thing to do is to round down to an ampere rating less than 1140 amperes. Fuse manufacturers provide 1100-ampere-rated fuses that are permitted to be used, as are adjustable-trip circuit breakers as provided in *NEC 240.6(C)*.

DISCONNECT SWITCHES

Fused switches are available in ratings of 30, 60, 100, 200, 400, 600, 800, 1200, 1600, 2000, 2500, 3000, 5000, and 6000 amperes in both 250 and 600 volts. They are for use with copper conductors unless marked to indicate that the terminals are suitable for use with aluminum conductors. The switch's rating is based on 60°C insulated wire allowable ampacities in sizes 14 AWG through 1 AWG and 75°C insulation ampacities for wires sized 1/0 AWG and larger, unless the switch is marked otherwise; see *NEC 110.14(C)(1)*. Switches also may be equipped with ground-fault sensing devices. These will be clearly marked as such.

Switches may have labels that indicate their intended application, such as "Continuous Load Current Not To Exceed 80% Of The Rating Of The Fuses Employed In Other Than Motor Circuits."

- Switches intended for isolating use only are marked "For Isolation Use Only—Do Not Open Under Load."

- Switches that are suitable for use as service disconnecting means are marked "Suitable For Use As Service Equipment."

- When a switch is marked "Motor-Circuit Switch," it is for use only in motor circuits.

- Switches of higher quality may have markings such as "Suitable For Use On A Circuit Capable Of Delivering Not More Than 100,000 [root mean squared] RMS Symmetrical Amperes, 600 Volts Maximum: Use Class T Fuses Having An Interrupting Rating Of No Less Than The Maximum Available Short Circuit Current Of The Circuit."

- Enclosed switches with horsepower ratings in addition to ampere ratings are suitable for use in motor circuits as well as for general use.

- Some switches have a dual-horsepower rating. The larger horsepower rating is applicable when using dual-element, time-delay fuses. Dual-element, time-delay fuses are discussed later in this chapter (see "Types of Fuses").

- Fusible bolted pressure contact switches are tested for use at 100% of their current rating, at 600 volts ac, and are marked for use on systems having available fault currents of 100,000, 150,000, and 200,000 RMS symmetrical amperes.

For more data regarding fused disconnect switches and fused power circuit devices (bolted pressure contact switches), see UL publications referred to in Chapter 1 and the *National Electrical Code* under "*Disconnecting Means, Switchboards, and Switches.*"

In the case of externally adjustable trip circuit breakers, the rating is considered to be the breaker's

maximum trip setting, *NEC 240.6(B)*. The exceptions to this are found in *NEC 240.6(C)*:

1. If the breaker has a removable and sealable cover over the adjusting screws

2. If it is located behind locked doors accessible only to qualified personnel

3. If it is located behind bolted equipment enclosure doors

In these cases, the adjustable setting is considered to be the breaker's ampere rating (long-time pickup setting). This is an important consideration when selecting proper size ungrounded (hot) and neutral conductors, equipment grounding conductors, overload relays in motor controllers, and other situations where sizing is based on the rating or setting of the overcurrent protective device.

Accessibility of Overcurrent Devices

The occupants of the various businesses in the commercial building must have access to the overcurrent devices for their branch circuits, *240.24(B)*.

Note that the panelboards in the commercial building are located to give proper access to the occupants. The occupants also have ready access to the service and feeder disconnects that are located outside the commercial building. The service disconnects are located outside to save valuable space inside the structure as well as to give the electric utility ready access to the metering equipment. Should vandalism or tampering be a concern, all equipment suitable for installation outdoors are equipped by the manufacturer with a means for locking.

Additional major concerns relating to the location of overcurrent devices include that they shall not

- be exposed to physical damage.

- be installed in the vicinity of easily ignitable material.

- be installed in clothes closets.

- be installed in bathrooms. This requirement is for dwelling units and guest room of hotels and motels, not for commercial and industrial buildings.

- be installed over the steps of a stairway.

How High Should Disconnect Switches Be Mounted?

NEC 240.24(A) requires that overcurrent devices be readily accessible. What does "readily accessible" mean? The term is defined in *NEC Article 100* and reads *Accessible, Readily (Readily Accessible). Capable of being reached quickly for operation, renewal, or inspections without requiring those to whom ready access is requisite to climb over or remove obstacles or to resort to portable ladders, and so forth.**

NEC 240.24(A) also requires that the center of the grip of a disconnect switch handle in its highest position be not more than 6 ft 7 in. (2.0 m) above the floor or working platform. The same rule applies to circuit breakers.

⎓⎓⎓ FUSES AND CIRCUIT BREAKERS

For general applications, the voltage rating, the continuous current rating, the interrupting rating, and the speed of response are factors that must be considered when selecting the proper fuses and circuit breakers.

Voltage Rating

According to *NEC 110.4* and *110.9*, the voltage rating of a fuse or circuit breaker shall be equal to or greater than the voltage of the circuit in which the fuse or circuit breaker is to be used. In the commercial building, the system voltage is 208Y/120-volt, wye connected. Therefore, fuses rated 250 volts or greater may be used. For a 480Y/277-volt, wye-connected system, fuses rated 600 volts would be used. Fuses and circuit breakers will generally work satisfactorily when used at any voltage less than the fuse's or circuit breaker's rating. For example, there would be no problem using a 600-volt fuse on a 208-volt system, although this practice may not be economically sound.

CAUTION: Fuses or circuit breakers that are marked ac should not be installed on dc circuits. Fuses and breakers that are suitable for use on dc circuits will be so marked.

*Reprinted with permission from NFPA 70-2011.

Continuous Current Rating

The continuous current rating of a protective device is the amperes that the device can continuously carry without interrupting the circuit. The *continuous current rating* of the overcurrent device must be distinguished from its *capability to carry current continuously*. Unless the overcurrent device is rated for operation at 100% of its rating, continuous loads (operating for three hours or more) are limited to not more than 80% of the rating. See *NEC 210.20(A)* for branch circuits and *215.3* for feeders. The standard ratings of overcurrent devices are listed in *NEC 240.6(A)*. When applied to a circuit, the selection of the rating is usually based on the ampacity of the circuit conductors, although there are notable exceptions to this rule.

For example, referencing *NEC Table 310.15(B) (16)*, we find that a size 8 AWG, Type THHN conductor has an allowable ampacity of 55 amperes. The proper overcurrent protection for this conductor would be 40 amperes if the terminals of the connected equipment are suitable for 60°C conductors or 50 amperes if the terminals of the connected equipment are suitable for 75°C conductors.

As mentioned previously, *NEC 240.4* lists several situations that permit the overcurrent protective device rating to exceed the ampacity of the conductor.

For example, the rating of a branch-circuit short-circuit and ground-fault protection fuse or circuit breaker on a motor circuit may be greater than the current rating of the conductors that supply the electric motor (175% for a dual element fuse and 250% for an inverse-time circuit breaker). This is permitted because properly sized motor overload protection (i.e., 125% of the motors' full-load current rating) will match the conductors' ampacity, which are also sized at 125% of the motors' full-load current rating. This overcurrent protection "scheme" can be thought of as a "systems approach" as the fuse or circuit breaker is sized to allow the motor to start and the running overload protection protects the branch circuit conductors and motor windings from dangerous overheating effects that would be caused by overloading.

Protection of Conductors

- When the overcurrent device is rated 800 amperes or less:

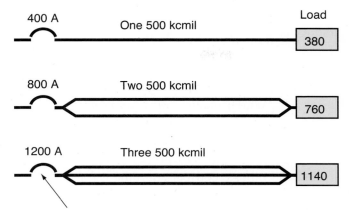

VIOLATION as round-up not permitted above 800 amperes.

FIGURE 17-1 *NEC 240.4(B)* allows round-up provisions through an 800-ampere overcurrent device. (*Delmar/Cengage Learning*)

FIGURE 17-2 Circuit breaker is marked for 22 kAIC. Verify rating of panelboard assembly. (*Delmar/Cengage Learning*)

When the ampacity of a cond uctor does not match the ampere rating of a standard fuse or does not match the ampere rating of a standard circuit breaker that does not have an overload trip adjustment above its rating, the *Code* permits the use of the next higher standard ampere-rated fuse or circuit breaker. The three conditions in *NEC 240.4(B)* must be complied with. Adjustable trip circuit breakers were discussed previously. See *NEC 240.4(B)* and Figure 17-1.

- When the overcurrent device is rated above 800 amperes:

When the ampere rating of a fuse or of a circuit breaker that does not have an overload trip adjustment exceeds 800 amperes, the conductor ampacity must be equal to or greater than the rating of the fuse or circuit breaker. The *Code* allows the use of fuses and circuit breakers that have ratings less than the standard sizes as listed in *NEC 240.6(A)*. Adjustable trip circuit breakers were discussed previously. See *NEC 240.4* and Figure 17-1.

Interrupting Rating

Interrupting rating is defined in the *Article 100* of the *NEC* as *the highest current at rated voltage that a device is intended to interrupt under standard test conditions.**

NEC 110.9 states that *Equipment intended to interrupt current at fault levels shall have an interrupting rating sufficient for the nominal circuit voltage and the current that is available at the line terminals of the equipment.**

An *interrupting rating* is a maximum rating, which is a measure of the fuse's or circuit breaker's ability to safely open an electrical circuit under fault conditions, such as an overload, short circuit, or ground fault. See Figure 17-2.

Think about *interrupting rating* this way. You would not use a 20-ampere switch to handle a 100-ampere load. The contacts would not have an adequate interrupting rating. You would not install a circuit breaker with an interrupting rating of 10,000 amperes to interrupt a fault of 50,000 amperes. In either case, the devices do not have proper interrupting rating for the amount of current they might be called on to interrupt.

Interrupting rating requirements are found in *NEC 240.60(C)*, *240.83(C)*, and in *240.86* for series combination ratings.

Short-Circuit Current Rating

Electrical equipment may also be marked with a *short-circuit current rating*. A short-circuit current rating is determined by the manufacturer of the equipment in conformance to specific UL Standards. Electrical equipment shall not be connected to a system capable of delivering more fault current than the equipment's short-circuit current rating. Short-circuit current ratings, impedance, and other characteristics are referenced in *NEC 110.10*.

*Reprinted with permission from NFPA 70–2011.

Short-circuit rating is the ability of the equipment (the entire assembly) to withstand a fault current equal to or less than the short-circuit rating for the length of time it takes the specified overcurrent device to open the circuit to the equipment.

Think about short-circuit rating this way. You would not connect a panelboard having a short-circuit rating of 22,000 amperes to a system capable of delivering 50,000 amperes. Another example: A series-rated panelboard having a short-circuit rating of 22,000 amperes may be connected to a system capable of delivering 22,000 amperes where the main breaker has an interrupting rating of 22,000 amperes and the branch-circuit breakers have an interrupting rating of 10,000 amperes.

The short circuit current rating of all electrical equipment is based on *how much current will flow* and *how long the current will flow*. Chapters 17 and 18 of this text discuss the important issues of current limitation and peak let-through current.

Overload currents have the following characteristics:

- They are greater than the normal current.
- They are contained within the normal conducting current path.
- If allowed to continue, they will cause overheating of the equipment, conductors, and the insulation of the conductors.

Short-circuit and ground-fault currents have the following characteristics:

- They flow outside the normal current path.
- They may be greater than the normal current.
- They may be less than the normal current.

Short-circuit and ground-fault currents, which flow outside the normal current paths, can cause conductors to overheat. In addition, mechanical damage to equipment can occur as a result of the magnetic forces of the large current and arcing. Some short-circuit and ground-fault currents may be no larger than the normal load current, or they may be thousands of times larger than the normal load current.

The terms *interrupting rating* and *interrupting capacity* are used interchangeably in the electrical

industry, but there is a subtle difference. Interrupting rating is the value of the test circuit capability. Interrupting capacity is the actual current that the contacts of a circuit breaker see when opening under fault conditions. You will also hear the term *AIC*, which means *amps interrupting capacity*. In Chapter 18, you will learn how to calculate available short-circuit currents (fault currents) so that you will be able to properly apply breakers, fuses, and other electrical equipment for the available short-circuit current to which they might be subjected.

For example, the test circuit in Figure 17-3 is calibrated to deliver 14,000 amperes of fault current. The standard test circuit allows 4 ft (1.2 m) of conductor to connect between the test bus and the breaker's terminals. The standard test further allows 10 in. (25 mm) of conductor per pole to be used as the shorting wire. Therefore, when the impedance of the connecting conductors is taken into consideration, the actual fault current that the contacts of the breaker see as they open is approximately 9900 amperes. The label on this circuit breaker is marked "14,000 amperes interrupting rating," yet its interrupting capacity as tested is 9900 amperes.

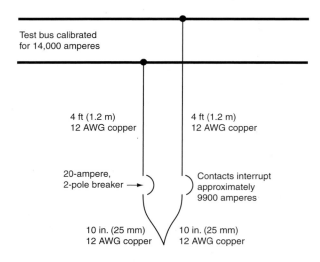

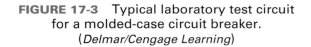

The label on this breaker is marked 14,000 amperes interrupting rating. Its actual interrupting capacity is approximately 9900 amperes. The connecting conductors have an ampacity that is the same as the breaker's ampere rating.

FIGURE 17-3 Typical laboratory test circuit for a molded-case circuit breaker.
(*Delmar/Cengage Learning*)

That is why it is important to adhere to the requirement of *NEC 110.3(B)*, which states:

> *Installation and Use. Listed or labeled equipment shall be installed and used in accordance with any instructions included in the listing or labeling.**

Speed of Response

The time required for the fusible element of a fuse to open varies inversely with the magnitude of the current that flows through the fuse. In other words, as the current increases, the time required for the fuse to open decreases. The time-current characteristic of a fuse depends on its rating and type. A circuit breaker also has a time-current characteristic. For the circuit breaker, however, there is a point at which the opening time cannot be reduced further due to the inertia of the moving parts within the breaker. The time-current characteristic of a fuse or circuit breaker should be selected to match the connected load of the circuit to be protected. Time-current characteristic curves are available from the manufacturers of fuses and circuit breakers.

TYPES OF FUSES

Dual-Element, Time-Delay Fuse

The dual-element, time-delay fuse provides a time delay in the low-overload range to eliminate unnecessary opening of the circuit because of harmless overloads. This type of fuse is extremely responsive in opening on short circuits. A dual-element, time-delay fuse has one or more fusible elements (links) connected in parallel. Each of these links is then connected in series with an overload trigger assembly. The number of links depends on the ampere rating of the fuse. The overload element(s) open when the current reaches a value of approximately 110% of the fuse's rating for a sustained period. The short-circuit element(s) opens when a short circuit or ground fault occurs. Dual-element, time-delay fuses carry 500% of their rating for at least 10 seconds. For 250-volt fuses rated 30 amperes or less, the minimum time delay is 8 seconds. Figure 17-4 illustrates the fuse element's configuration and operation.

FIGURE 17-4 Dual-element, time-delay, current-limiting Class RK1 fuses. They provide time delay in the overload range and are fast acting in the short-circuit range. (*Photo courtesy of Cooper Bussmann, Division of Cooper Industries*)

This fuse is a UL Class RK1 fuse. Classes of fuses are discussed later in this chapter.

Dual-element fuses are ideal for use on motor circuits and other circuits having high-inrush characteristics. This type of fuse can be used as well for mains, feeders, subfeeders, and branch circuits. Dual-element fuses may be used to provide backup protection for circuit breakers, bus duct, and other circuit components that lack an adequate interrupting rating, bracing, or withstand rating (see types of fuses covered later in this chapter).

Using Fuses for Motor Overload Protection

Dual-element fuses can provide both motor overload protection and branch-circuit protection. Table 17-1 shows overload protection percentages per the *NEC*.

Sizing dual-element fuses slightly larger than the overload relay provides backup protection. This tight sizing of fuses means that if the overload relays fail to operate, the dual-element fuses will provide backup overload protection.

- Motor overload protection is based on the motor's nameplate rating in compliance with *NEC 430.6(A)(2)*.

---

*Reprinted with permission from NFPA 70–2011.

	TABLE 17-1

Motor overload protection.

	Maximum Size in Percentage of the Motor's Nameplate Current Rating, NEC 430.32	If the Percentage in the Column to the Left Does Not Result in a Standard Size, the Absolute Maximum Is
Service factor not less than 1.15	125%	140%
Marked not more than 40°C temperature rise	125%	140%
All other motors	115%	130%

- Conductor size, disconnect switch size, controller size, branch-circuit, short-circuit, and ground-fault protection are based on the motor's full-load current values in *NEC Tables 430.247, 430.248, 430.249,* and *430.250.* This is required by *NEC 430.6(A)(1).*

NEC 430.32 sets forth motor overload protection requirements. There are many questions to ask when searching for the proper level of motor overload protection. What are the motor's starting characteristics? Is the motor manually or automatically started? Is the motor continuous duty or intermittent duty? Does the motor have integral built-in protection? Is the motor impedance protected? Is the motor larger than 1 horsepower or is the motor 1 horsepower or less? Is the motor fed by a general-purpose branch circuit? Is the motor cord and plug connected? Will the motor automatically restart if the overload trips? When these questions are answered, then look at *NEC 430.32* for the level of overload protection needed.

EXAMPLE

What is the ampere rating of a dual-element, time-delay fuse that is to be installed to provide branch-circuit protection as well as overload protection for a motor with a 1.15 service factor and a nameplate current rating of 16 amperes?

$$16 \times 1.25 = 20 \text{ amperes}$$

Applying Fuses and Breakers on Motor Circuits

In recent years, energy-efficient motors have entered the scene. Energy-efficient motors have higher starting current characteristics than older style motors. These motors are referred to as Design B motors. Their high starting currents can cause nuisance opening of fuses and nuisance tripping of circuit breakers. Different manufacturers' same-size energy-efficient motors will have different starting current characteristics. Therefore, be careful when applying nontime-delay, fast-acting fuses on motor circuits. Because these fuses do not have the ability to handle the high inrush currents associated with energy-efficient motors, they cannot be sized at 125% of a motor's full-load ampere rating. They may have to be sized at 150%, 175%, 225%, or as high as 300% to 400%. Instant-trip circuit breakers will also experience nuisance tripping. Always check the time-current curves of fuses and breakers to make sure that they will handle the momentary motor starting inrush currents without nuisance opening or tripping. Additional information is found in Chapter 7 of this text.

Using Fuses for Motor Branch-Circuit, Short-Circuit, and Ground-Fault Protection

NEC Table 430.52 shows that for a typical motor branch-circuit, short-circuit, and ground-fault protection, the maximum size permitted for dual-element fuses is based on a maximum of 175% of the full-load current of the motor. An exception to the 175% value is given in *NEC 430.52,* where permission is given to go as high as 225% of the motor full-load current if the lower value is not sufficient to allow the motor to start.

EXAMPLE

$$16 \times 1.75 = 28 \text{ amperes (maximum)}$$

The next higher standard rating is 30 amperes. If for some reason the 30-ampere fuse cannot handle the starting current, *NEC 430.52(C)(1)*

Exception 2 allows the selection of a dual-element, time-delay fuse not to exceed

$$16 \times 2.25 = 36 \text{ amperes (maximum)}$$

The next lower standard rating is 35 amperes.

Dual-Element, Time-Delay, Current-Limiting Fuses

Figure 17-4 shows dual-element, time-delay, current-limiting Class RK1 fuses that come under the UL Class RK1 category. They operate in the same manner as the Class RK5 dual-element, time-delay fuses previously discussed. That is, they can handle five times their ampere rating for at least 10 seconds (8 seconds for 250-volt, 30-ampere or less ratings), which is desirable for motor and other high inrush loads. They open extremely quickly under short-circuit or ground-fault situations.

This type of fuse is available with a small indicator that clearly shows whether the fuse is good or whether it has opened and needs to be replaced.

Fast-Acting, Current-Limiting Fuses (Nontime-Delay)

The straight current-limiting fuse, shown in Figure 17-5, has an extremely fast response in both the low-overload and short-circuit ranges. When compared to other types of fuses, this type of fuse has the lowest energy let-through values. Current-limiting fuses are used to provide better protection to mains, feeders and subfeeders, circuit breakers, bus duct, switchboards, and other circuit components that lack an adequate interrupting rating, bracing, or withstand rating.

Current-limiting fuse elements can be made of silver or copper surrounded by a quartz sand arc-quenching filler.

FIGURE 17-5   Current-limiting, fast-acting, single-element fuse. (*Photo courtesy of Cooper Bussmann, Division of Cooper Industries*)

A fast-acting, current-limiting fuse does not have the spring-loaded or loaded-link overload assembly found in dual-element fuses. To apply straight current-limiting fuses for motor circuits, refer to *NEC Table 430.52* under the column *Nontime Delay Fuse*.

To be classified as *current limiting*, *NEC 240.2* states that when a fuse or circuit breaker is subjected to heavy (high-magnitude) fault currents, the fuse or breaker must reduce the fault current flowing into the circuit to a value less than the fault current that could have flowed into the circuit had there been no fuse or breaker in the circuit.

When used on motor circuits or other circuits having high current-inrush characteristics, the current-limiting nontime-delay fuses must be sized at a much higher rating than the actual load. That is, for a motor with a full-load current rating of 10 amperes, a 30- or 40-ampere fast-acting, current-limiting fuse may be required to start the motor. In this case, the fuse is considered to be the motor branch-circuit short-circuit protection as required in *NEC Table 430.52* (Table 17-2).

Types of Cartridge Fuses

The requirements governing cartridge fuses are contained in *NEC Article 240, Part VI*. According to the *Code*, all cartridge fuses must be marked to show

- ampere rating.
- voltage rating.
- interrupting rating when greater than 10,000 amperes.
- current-limiting type, if applicable.
- trade name or name of manufacturer.

The UL and Canadian Standards Association (CSA) standards list Classes G, H, J, K1, K5, K9, L, RK1, RK5, T, and CC fuses.

All fuses carrying the UL/CSA Class listings (Class G, H, J, K1, K5, K9, L, RK1, RK5, T, and CC) and plug fuses are tested on alternating-current circuits and are marked for ac use. When fuses are to be used on direct-current systems, the electrician should consult the fuse manufacturer because it may be necessary to reduce the fuse voltage rating and the interrupting rating to ensure safe operation.

TABLE 17-2

Table 430.52 Maximum Rating or Setting of Motor Branch-Circuit Short-Circuit and Ground-Fault Protective Devices

Type of Motor	Percentage of Full-Load Current			
	Nontime Delay Fuse[1]	Dual Element (Time-Delay) Fuse[1]	Instantaneous Trip Breaker	Inverse Time Breaker[2]
Single-phase motors	300	175	800	250
AC polyphase motors other than wound-rotor	300	175	800	250
Squirrel cage — other than Design B energy-efficient	300	175	800	250
Design B energy-efficient	300	175	1100	250
Synchronous[3]	300	175	800	250
Wound rotor	150	150	800	150
Direct current (constant voltage)	150	150	250	150

Note: For certain exceptions to the values specified, see 430.54.
[1]The values in the Nontime Delay Fuse column apply to Time-Delay Class CC fuses.
[2]The values given in the last column also cover the ratings of nonadjustable inverse time types of circuit breakers that may be modified as in 430.52(C)(1), Exception No. 1 and No. 2.
[3]Synchronous motors of the low-torque, low-speed type (usually 450 rpm or lower), such as are used to drive reciprocating compressors, pumps, and so forth, that start unloaded, do not require a fuse rating or circuit-breaker setting in excess of 200 percent of full-load current.

Reprinted with permission from NFPA 70-2011.

The variables in the physical appearance of fuses include length, ferrule diameter, and blade length/width/thickness, as well as other distinctive features. For these reasons, it is difficult to insert a fuse of a given ampere rating into a fuseholder rated for less amperage than the fuse. The differences in fuse construction also make it difficult to insert a fuse of a given voltage into a fuseholder with a higher voltage rating, *NEC 240.60(B)*. Figure 17-6 indicates several examples of the method of ensuring that fuses and fuseholders are not mismatched.

NEC 240.60(B) also specifies that fuseholders for current-limiting fuses shall be designed so that they cannot accept fuses that are noncurrent limiting, Figure 17-7.

In general, low-voltage fuses may be used at system voltages that are less than the voltage rating

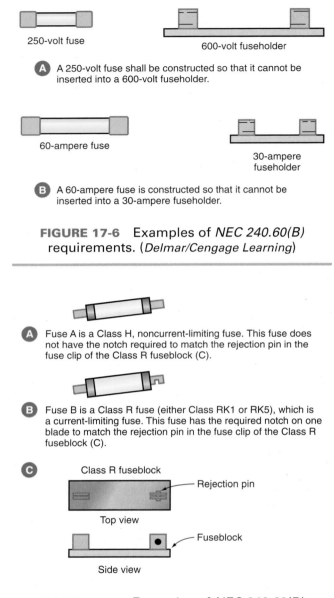

A A 250-volt fuse shall be constructed so that it cannot be inserted into a 600-volt fuseholder.

B A 60-ampere fuse is constructed so that it cannot be inserted into a 30-ampere fuseholder.

FIGURE 17-6 Examples of *NEC 240.60(B)* requirements. (*Delmar/Cengage Learning*)

A Fuse A is a Class H, noncurrent-limiting fuse. This fuse does not have the notch required to match the rejection pin in the fuse clip of the Class R fuseblock (C).

B Fuse B is a Class R fuse (either Class RK1 or RK5), which is a current-limiting fuse. This fuse has the required notch on one blade to match the rejection pin in the fuse clip of the Class R fuseblock (C).

C Class R fuseblock
Rejection pin
Top view
Fuseblock
Side view

FIGURE 17-7 Examples of *NEC 240.60(B)* requirements. (*Delmar/Cengage Learning*)

of the fuse. For example, a 600-volt fuse combined with a 600-volt switch may be used on 575-volt, 480Y/277-volt, 208Y/120-volt, 240-volt, 50-volt, and 32-volt systems.

Class H. Class H fuses formerly were called *NEC* or *Code* fuses. Most low-cost, common, standard nonrenewable one-time fuses are Class H fuses. Renewable-type fuses also come under the Class H classification. *NEC 240.60(D)* requires Class H cartridge fuses of the renewable type to be used only for replacement in existing installations where there is no evidence of overfusing or tampering. Neither the interrupting rating nor the notation *Class H* appears on the label of a Class H fuse. This type of fuse is tested by the Underwriters Laboratories on circuits that deliver 10,000 amperes ac. Class H fuses are available with ratings ranging from 1 ampere to 600 amperes in both 250-volt ac and 600-volt ac types. Class H fuses are not current limiting.

A higher-quality, nonrenewable, one-time fuse is also available, called a Class K5 fuse, which has a 50,000-ampere interrupting rating.

Class K.   Class K fuses are grouped into three categories: K1, K5, and K9, Figures 17-8A through 17-8D. These fuses may be UL listed or CSA certified with interrupting ratings in RMS symmetrical amperes in values of 50,000, 100,000, or 200,000 amperes. For each K rating, UL has assigned a maximum level of peak let-through current (I_p) and energy as given by I^2t. Class K fuses have varying degrees of current-limiting ability, depending on the K rating. Class RK1 fuses have the greatest current-limiting ability and Class K9 fuses have the least current-limiting ability. A check of various fuse manufacturers' literature reveals that Class K9 fuses are no longer being manufactured.

Class K fuses may be listed as time-delay fuses as well. In this case, UL requires that the fuses have a minimum time delay of 10 seconds at 500% of the rated current (8 seconds for 250 volts, 30 amperes). Class K fuses are available in ratings ranging from $\frac{1}{10}$ ampere to 600 amperes at 250 or 600 volts ac. Class K fuses have the same dimensions as Class H fuses.

Class J.   Class J fuses are current limiting and are so marked, Figures 17-9(A) and 17-9(B). They are listed by UL/CSA with a minimum interrupting rating of 200,000 RMS symmetrical amperes. Some have a special listing identified by the letters *SP*, and have an interrupting rating of 300,000 RMS symmetrical amperes. Certain Class J fuses are also considered to be dual-element, time-delay fuses, and are marked "time-delay." Class J fuses are physically smaller than Class H fuses. Therefore, when a fuseholder is installed to accept a Class J fuse, it will be impossible to install a Class H fuse in the fuseholder, *NEC 240.60(B)*.

The Underwriters Laboratories has assigned maximum values of I^2t and I_p that are slightly less than those for Class RK1 fuses. Both fast-acting, current-limiting Class J fuses and time-delay, current-limiting

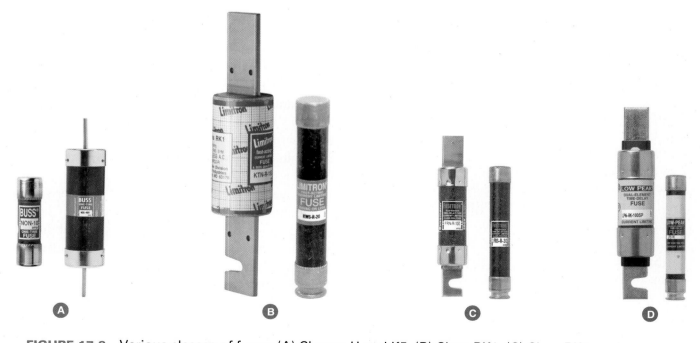

FIGURE 17-8   Various classes of fuses: (A) Classes H and K5, (B) Class RK1, (C) Class RK5, and (D) Class RK1 fuses. (*Photo courtesy of Cooper Bussmann, Division of Cooper Industries*)

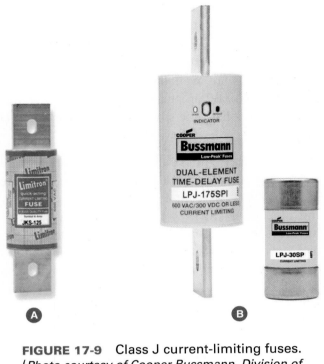

FIGURE 17-9 Class J current-limiting fuses. (*Photo courtesy of Cooper Bussmann, Division of Cooper Industries*)

FIGURE 17-10 Class L fuses. All Class L fuses are rated 600 volts. Listed in 601- to 6000-ampere rating. (*Photo courtesy of Cooper Bussmann, Division of Cooper Industries*)

Class J fuses are available in ratings ranging from 1 ampere to 600 amperes at 600 volts ac.

Some Class J fuses are available with a small indicator that clearly shows whether the fuse is good or whether the fuse has opened and needs to be replaced.

Class L. Class L fuses, Figures 17-10(A) and 17-10(B), are listed by UL/CSA in sizes ranging from 601 amperes to 6000 amperes at 600 volts ac. These fuses have specified maximum values of I^2t and I_p. They are current-limiting fuses and have a minimum interrupting rating of 200,000 RMS symmetrical amperes. These bolt-type fuses are used in bolted pressure contact switches. Class L fuses are available in both a fast-acting, current-limiting type and a time-delay, current-limiting type. Both types of Class L fuses meet UL requirements.

Some Class L fuses have a special interrupting rating of 300,000 RMS symmetrical amperes. The fuse's label will indicate the part number followed by the letters "SP."

Class T. Class T fuses, Figure 17-11, are current-limiting fuses and are so marked. These fuses are UL listed with an interrupting capacity of 200,000 RMS symmetrical amperes. Class T

fuses are physically smaller than Class H or Class J fuses. The configuration of this type of fuse limits its use to fuseholders and switches that will reject all other types of fuses.

Class T fuses rated 600 volts have electrical characteristics similar to those of fast-acting Class J fuses and are tested in a similar manner by Underwriters Laboratories. Class T fuses rated at 300 volts have lower peak let-through currents and I^2t values than comparable Class J fuses. Many series-rated panelboards are listed by Underwriters Laboratories with Class T mains. Because Class T fuses do not have a lot of time delay (the ability to hold momentary inrush currents such as occurs when a motor is started), they are sized according to the nontime-delay fuse column in *NEC Table 430.52*.

UL/CSA presently lists Class T fuses in sizes from 1 ampere to 1200 amperes. Common applications for Class T fuses are for mains, feeders, and branch circuits.

Class T 300-volt fuses may be used on 120/240-volt, single-phase; 208Y/120-volt, 3-phase, 4-wire

FIGURE 17-11   Class T current-limiting, fast-acting fuse; 200,000-ampere interrupting rating.
(*Photo courtesy of Cooper Bussmann, Division of Cooper Industries*)

wye; and 240-volt, 3-phase, 3-wire, delta systems (ungrounded or corner grounded). The *NEC* permits 300-volt Class T fuses to be installed in single-phase, line-to-neutral circuits supplied from 3-phase, 4-wire solidly grounded neutral systems where the line-to-neutral voltage does not exceed 300 volts. The *NEC* does not permit the use of 300-volt Class T fuses for line-to-line or line-to-line-to-line applications on 480Y/277-volt, 3-phase, 4-wire wye systems. Class T 600-volt fuses may be used on 480Y/277-volt, 3-phase, 4-wire wye; 480-volt, 3-phase, 3-wire; and any of the systems where Class T 300-volt fuses are permitted; see *NEC* 240.60(A).

Class G.   Class G fuses, Figure 17-12, are cartridge fuses with small physical dimensions. Ratings ½ through 20 amperes are rated 600 volts ac. Ratings 25 through 60 are rated 480 volts ac. They are UL and CSA listed at an interrupting rating of 100,000 amperes. To prevent overfusing, Class G fuses are size limiting within the four categories assigned to their ampere ratings. For example, a 15-ampere

FIGURE 17-12   Class G fuses.
(*Photo courtesy of Cooper Bussmann, Division of Cooper Industries*)

fuseholder will accept ½-through 15-ampere Class G fuses. A 20-ampere fuseholder will accept 20-ampere Class G fuses; a 30-ampere fuseholder will accept 25- and 30-ampere Class G fuses; and a 60-ampere fuseholder will accept 35-, 40-, 50-, and 60-ampere Class G fuses.

Class G fuses are current limiting. They may be used for the protection of ballasts, electric heat, and similar loads. They are UL listed for branch-circuit protection.

Class R. The Class R fuse is a nonrenewable cartridge type and has a minimum interrupting rating of 200,000 RMS symmetrical amperes. The peak let-through current (I_p) and the total clearing energy (I^2t) values are specified for the individual case sizes. The values of I^2t and I_p are specified by UL based on short-circuit tests at 50,000, 100,000, and 200,000 amperes.

Class R fuses are divided into two subclasses: Class RK1 and Class RK5. The Class RK1 fuse has characteristics similar to those of the Class K1 fuse. The Class RK5 fuse has characteristics similar to those of the Class K5 fuse. These fuses must be marked either Class RK1 or RK5. In addition, they are marked to be current limiting.

Some Class RK1 fuses have a special interrupting rating of 300,000 RMS symmetrical amperes. The fuse's label will indicate the part number followed by the letters *SP*.

The ferrule-type Class R fuse has a rating range of ¹⁄₁₀ ampere to 60 amperes and can be distinguished by the annular ring on one end of the case, Figure 17-13(A). The knife blade-type Class

R fuse has a rating range of 61 amperes to 600 amperes and has a slot in the blade on one end, Figure 17-13(B). When a fuseholder is designed to accept a Class R fuse, it will be impossible to install a standard Class H or Class K fuse (these fuses do not have the annular ring or slot of the Class R fuse). The requirements for noninterchangeable cartridge fuses and fuseholders are covered in *NEC 240.60(B)*. However, the Class R fuse can be installed in older style fuse clips on existing installations. As a result, the Class R fuse may be called a one-way rejection fuse.

Electrical equipment manufacturers will provide the necessary rejection-type fuseholders in their equipment, which is then tested with a Class R fuse at short-circuit current values such as 50,000, 100,000, or 200,000 amperes. Each piece of equipment will be marked accordingly.

Cube Fuse. A new type of fuse, shown in Figure 17-14, is finding its way into original equipment manufacturers' (OEM) products. It is physically smaller than ordinary fuses and is dual element, time delay, having a 10-second opening time at 500% of its ampere rating. The cube fuse also has a 300,000-ampere interrupting capacity, has the characteristics of UL/CSA listed Class J fuses (fast acting under fault conditions), is rated 600 volts ac or less, has no exposed live parts, has open-fuse indication, mounts in a special fuseholder and rails designed for the purpose, provides Type 2

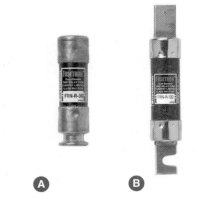

FIGURE 17-13 Class R cartridge fuses. (*Photo courtesy of Cooper Bussmann, Division of Cooper Industries*)

FIGURE 17-14 A cube fuse. Note the blades for easy insertion into a fuseholder. (*Photo courtesy of Cooper Bussmann, Division of Cooper Industries*)

FIGURE 17-15 Class CC fuse with a rejection feature. (*Photo courtesy of Cooper Bussmann, Division of Cooper Industries*)

protection (no damage) for motor controllers, and is available in many ampere ratings from 1 ampere through 100 amperes.

Class CC. Class CC fuses are primarily used for control-circuit protection, for the protection of motor control circuits, ballasts, small transformers, and so on. They are UL listed as branch-circuit fuses. Class CC fuses are rated at 600 volts or less and have a 200,000-ampere interrupting rating in sizes from $\frac{1}{10}$ ampere through 30 amperes. These fuses measure 1½ in. × $\frac{13}{32}$ in. (38.1 mm × 10.3 mm) and can be recognized by a button on one end of the fuse Figure 17-15. This button is unique to Class CC fuses. When a fuseblock or fuseholder that has the matching Class CC rejection feature is installed, it is impossible to insert any other 1½ in. × $\frac{13}{32}$ in. (38.1 mm × 10.3 mm) fuse. Only a Class CC fuse will fit into these special fuseblocks and fuseholders. A Class CC fuse can be installed in a standard fuseholder.

Class CC fuses are available in time-delay (to hold on momentary inrush current), time-delay, current-limiting (to hold on momentary inrush current but open very fast on a short circuit), and fast acting types (not much time-delay on momentary inrush current, but they open very fast on a short-circuit).

Plug Fuses

NEC Article 240, Part V, spells out the requirements for plug fuses.

Conventional plug fuses have a base referred to as an Edison base. This base is of the same shape as the base on a light bulb. Type S fuses have a special base.

The opening characteristics of plug fuses are available in three types, each serving a purpose:

1. The standard fuse link type does not have much time delay. It opens fast. It is used on lighting circuits and other nonmotor loads.

2. The loaded link type has a metal bead (heat sink) on the fuse element that gives it time delay to hold motor inrush starting currents.

3. The dual-element, time-delay has a spring-loaded short-circuit element plus an overload element connected in series with the short-circuit element. It is excellent for motor circuits.

Electricians will be most concerned with the following requirements for plug fuses, fuseholders, and adapters:

- They shall not be used in circuits exceeding 125 volts between conductors, except on systems having a grounded neutral conductor, with no conductor having more than 150 volts to ground. This situation is found in the 208Y/120-volt system in the commercial building covered in this text, or in the case of a 120/240-volt, single-phase system.

- They shall have ampere ratings of 0 to 30 amperes.

- They shall have a hexagonal configuration of the fuse window for ratings of 15 amperes and below.

- The screw shell must be connected to the load side of the circuit.

- Edison-base plug fuses may be used only as replacements in existing installations where there is no evidence of overfusing or tampering.

- All new installations shall use fuseholders requiring Type S plug fuses or fuseholders with a Type S adapter inserted to accept Type S fuses only.

- Type S plug fuses are classified 0 to 15 amperes, 16 to 20 amperes, and 21 to 30 amperes.

Prior to the installation of fusible equipment, the electrician must determine the ampere rating of the various circuits. An adapter of the proper size is then inserted into the Edison-base fuseholder; finally, the proper size of Type S fuse can be inserted into the fuseholder. The adapter makes the fuseholder non-tamperable and noninterchangeable. For example,

FIGURE 17-16 Plug fuse Types T and W. (*Photo courtesy of Cooper Bussmann, Division of Cooper Industries*)

after a 15-ampere adapter is inserted into a fuseholder for a 15-ampere circuit, it is impossible to insert a Type S or an Edison-base fuse having a larger ampere rating into this fuseholder without removing the 15-ampere adapter. Adapters are designed so they are extremely difficult to remove.

Type S fuses and suitable adapters, Figure 17-16, are available in a large selection of ampere ratings ranging from ¼ ampere to 30 amperes. When a Type S fuse has dual elements and a time-delay feature, it does not open unnecessarily under momentary overloads, such as the current surge caused by the startup of an electric motor. On heavy overloads or short circuits, this type of fuse opens very rapidly.

Summary of Fuse Applications

Table 17-3 provides rating and application information for the fuses supplied by one manufacturer. The table indicates the class, voltage rating, current range, interrupting rating, and typical applications for the various fuses. A table of this type may be used to select the proper fuse to meet the needs of a given situation.

TESTING FUSES

As mentioned at the beginning of this text, the Occupational Safety and Health Act (OSHA) clearly states that electrical equipment must not be worked on when it is energized. There have been too many injuries to those intentionally working on equipment hot, or thinking the power is off, only to find that it is still energized. If equipment is to be worked on hot, then proper training, fire-resistant clothing, and protective gear (rubber blankets, insulated tools,

goggles, rubber gloves, etc.) needs to be used. A second person should be present when working on electrical equipment hot. OSHA has specific lockout and tag-out rules for working on energized electrical equipment.

When Power Is Turned On. On live circuits, extreme caution must be exercised when checking fuses. There are many different voltage readings that can be taken, such as line to line, line to ground, line to neutral, and so on.

Using a voltmeter, the first step is to be sure to set the scale to its highest voltage setting and then change to a lower scale after you are sure you are within the range of the voltmeter. For example, when testing what you believe to be a 120-volt circuit, it is wise to first use the 600-volt scale, then try the 300-volt scale, and then use the 150-volt scale—JUST TO BE SURE!

Taking a voltage reading across the bottom (load side) of fuses—either fuse to fuse, fuse to neutral, or fuse to ground—can show voltage readings because even though a fuse might have opened, there can be feedback through the load. You could come to a wrong conclusion. Taking a voltage reading from the line side of a fuse to the load side of a fuse will show open-circuit voltage if the fuse has blown and the load is still connected. This also can result in a wrong conclusion.

Reading from the line-to-load side of a good fuse should show zero voltage or else an extremely small voltage across the fuse.

Always read carefully the instructions furnished with electrical test equipment such as voltmeters, ohmmeters, and so on.

When Power Is Turned Off. This is the safest way to test fuses. Remove the fuse from the switch, and then take a resistance reading across the fuse using an ohmmeter. A good fuse will show zero or a very minimal resistance. An open (blown) fuse will generally show a high resistance reading or an open circuit indication.

Cable Limiters

Cable limiters are used quite often in commercial and industrial installations where parallel cables are used on service entrances and feeders.

TABLE 17-3

Bussmann fuses: Their ratings, class, and application.**

Fuse and Ampere Rating Range	Voltage Rating	Symbol	Industry Class	Interrupting Rating in Sym. RMS Amperes	Application All Types Recommended for Protection of General Purpose Circuits and Components
LOW-PEAK time-delay FUSE 601–6000 A	600 V	KRP-C (SP)	L	300,000 A	• Will hold 500% of its rating for at least 4 seconds. High capacity main, feeder, branch circuits, and large motors circuits. Has more time-delay than KTU. For motors, can be sized 150–225% of motor FLA.
LIMITRON fast-acting FUSE 601–6000 A	600 V	KTU	L	200,000 A	• High-capacity main, feeder, branch circuits, and circuit breaker protection. For motors, size 300% of motor FLA. Holds 500% of its rating for 5 seconds.
LIMITRON time-delay FUSE 601–4000 A	600 V	KLU	L	200,000 A	• High-capacity mains, feeders, and circuits for large motors. Has good time-delay characteristics. Will hold 500% of its rating for 10 seconds. Copper links, thus not as current limiting as KTU or KRPC type.
FUSETRON dual-element FUSE 1/10–600 A	250 V 600 V	FRN-R FRS-S	RK5 RK5	200,000 A	• Main, feeder, branch circuits, and circuit-breaker protection. Especially recommended for motors, welders, and transformers and any loads having high inrush characteristics. Will hold 500% for 10 seconds. Size 115%–175% of motor FLA.
LOW-PEAK dual-element FUSE 1/10–600 A	250 V 600 V	LPN-RK (SP) LPS-RK (SP)	RK1 RK1	300,000 A	• Main, feeder, branch circuits, and circuit-breaker protection. Especially recommended for motors, welders, and transformers. (More current limiting than Fusetron dual-element fuse.) Will hold 500% for 10 seconds. Size at 115%–175% of motor FLA.
LIMITRON fast-acting FUSE 1–600 A	250 V 600 V	KTN-R KTS-R	RK1 RK1	200,000 A	• Main, feeder, branch circuits, and circuit-breaker protection. Especially recommended for circuit-breaker protection. (High degree of current limitation.) Because these fuses are fast acting, size at 300% or more for motors.
TRON fast-acting FUSE 1–1200 A (300 V) 1–800 A (600 V)	300 V 600 V	JJN JJS	T	200,000 A	• Main, feeder, branch circuits, and circuit-breaker protection. Circuit-breaker protection, small physical dimensions. Smaller than Class J. (High degree of current limitation.) Because these fuses are fast acting, size at 300% or more for motors.

(continues)

TABLE 17-3

Bussmann fuses: Their ratings, class, and application.** *(continued)*

LOW-PEAK time-delay FUSE 1–600 A	600 V	LPJ (SP)	J	300,000 A	Main, feeder, branch circuits, and circuit-breaker protection. Motor and transformer circuits. Size at 150%–225% for motors.
LIMITRON fast-acting FUSE 1–600 A	600 V	JKS	J	200,000 A	Main, feeder, and branch circuits. Circuit-breaker protection. Because these fuses are fast acting, size at 300% or more for motors.
ONE-TIME FUSE ⅛–600 A	250 V 600 V	NON NON	1–60 A, K5 65–600 A, H 1–60 A, K5 65–600 A, H	50,000 A 10,000 A 50,000 A 10,000 A	General purpose—where fault currents are low. For motors, size at 300% or more.
SUPERLAG renewable (replaceable) link fuse 1–600 A	250 V 600 V	REN RES	H	10,000 A	• Renewable fuses are only permitted to be used as replacements on existing installations where there is no evidence of overfusing or tampering, *NEC 240.60(D)*.
SC FUSE 1–60 A	600 V, 1–20 A 480 V, 25–60 A	SC	G	100,000 A	• General-purpose branch circuits. For motors, size at 300% or more.
Dual-element CUBE fuse 1–100 A	600 V	TCF	J	300,000 A	Control panels on equipment. 10-second time-delay at 500% load.
FUSETRON dual-element PLUG FUSE ³⁄₁₀–30 A, Edison Base	125 V	T	***	10,000 A	• General-purpose sizes ³⁄₁₀ A through 14 A, excellent for motors. Size at 115%–125% for motors. Branch-circuit sizes 15, 20, 25, 30 A.
FUSTAT dual-element type S PLUG FUSE ³⁄₁₀–30 A	125 V	S	S****	10,000 A	General purpose, sizes ³⁄₁₀ A through 14 A excellent for motors. Size at 115%–125% for Branch-circuit sizes 15, 20, 25, 30 A.
ORDINARY PLUG FUSE Edison Base, ½–2 A	125 V	W	***	10,000 A	General purpose. Fast acting. Less time delay than types T and S. Not recommended for motor circuits unless sized at 300% or more of motor FLA.

Note: For motor circuits, refer to *NEC Article 430*, especially *Table 430.52*.

** This table gives general application recommendations from one manufacturer's technical data. For critical applications, consult electrical equipment and specific fuse manufacturer's technical data.

*** Listed as Edison-Base Plug Fuse.

**** Difficult to change ampere ratings because of different thread on screw base for different ampere ratings.

Cable limiters differ in purpose from fuses, which are used for overload protection. Cable limiters are short-circuit devices that can *isolate* a faulted cable rather than having the fault open the entire phase. They are selected on the basis of conductor size; for example, a 500-kcmil cable limiter would be used on a 500-kcmil conductor.

Cable limiters are available for cable-to-cable or cable-to-bus installation for either aluminum or copper conductors. The type of cable limiter illustrated

FIGURE 17-17 Cable limiter for 500-kcmil copper conductor. (*Photo courtesy of Cooper Bussmann, Division of Cooper Industries*)

in Figure 17-17 is for use with a 500-kcmil copper conductor and is rated at 600 volts with an interrupting rating of 200,000 amperes.

Cable limiters are also used where taps are made ahead of the mains, such as in the Commercial Building where the emergency system is tapped ahead of the main fuses (as shown in Figure 16-2).

Figure 17-18 is an example of how cable limiters may be installed on the service of a commercial building. A cable limiter is installed at each end of each 500-kcmil conductor. Three conductors per phase terminate in the main switchboard where the service is then split into two 800-ampere bolted pressure switches.

Figure 17-19 shows how cable limiters are used where more than one customer is connected to one transformer.

Cable limiters are permitted to be installed on the supply side of the service, *NEC 230.82(1)* and for transformer tie circuits in *450.6(A)(3)*. Cable limiters are installed in the hot phase conductors only. They are not installed in a grounded conductor.

DELTA, 3-PHASE, CORNER-GROUNDED "B" PHASE SYSTEM

See Chapter 13 for a discussion of delta-connected, 3-phase systems.

An overcurrent device shall be installed in series with ungrounded conductors for overcurrent

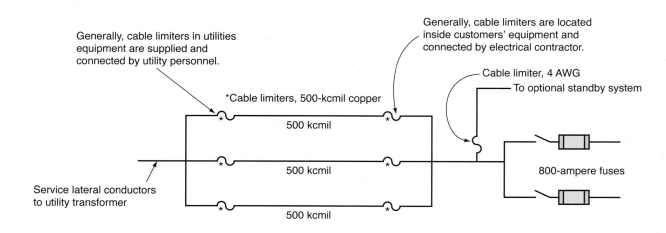

FIGURE 17-18 Use of cable limiters in supply to a service. (*Delmar/Cengage Learning*)

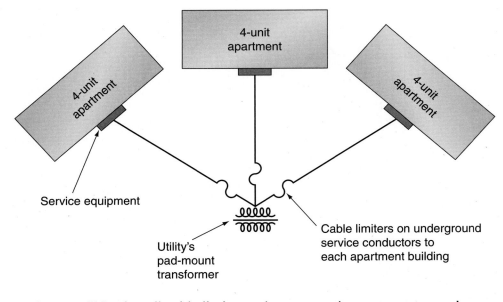

FIGURE 17-19   Some utilities install cable limiters where more than one customer is connected to one transformer. Thus, if one customer has problems in the main service equipment, the cable limiter on that service will open, isolating the problem from other customers on the same transformer. This is common for multiunit apartments, condos, and shopping centers. (*Delmar/Cengage Learning*)

protection, *NEC 240.15(A)*. This connection is sometimes called a corner ground. The *Code* generally prohibits overcurrent devices from being connected in series with any conductor that is intentionally grounded, *NEC 240.22*.

However, there are certain instances where a 3-pole, 3-phase switch may be installed, where it is permitted to install fuses in two poles only. This would be in the case of a delta, 3-phase, corner-grounded "B" phase system, Figure 17-20. A device called a *solid neutral* can be installed in the "B" phase, as shown in the figure. Note that the switch has two fuses and one solid neutral installed.

Figure 17-20 shows a 3-pole service-entrance switch. *NEC 230.75* permits a busbar to be used as a disconnecting means for the grounded conductor. Thus, a 2-pole service-entrance switch is permitted if the grounded conductor can be disconnected from the service-entrance conductors, Figure 17-21.

Here is a section in the *Code* that requires a fuse to be installed in a grounded conductor. It is *NEC 430.36*. It reads, *if fuses are used for a motor overload protection, a fuse shall be inserted in each ungrounded conductor and also in the grounded* *conductor if the supply system is 3-wire, 3-phase ac with one conductor grounded.*

Solid Neutrals

A solid neutral, Figure 17-22, is made of copper bar that has exactly the same dimensions as a fuse for a given ampere rating and voltage rating. For example, a solid neutral rated 100 amperes, 600 volts would be installed in a switch rated at 100 amperes, 600 volts.

Substitutes for solid neutrals such as pipe, tubing or wire should never be used.

Solid neutrals are installed in situations, for example, where a 3-pole, 3-fuse disconnect switch is used on a grounded "B" phase system, Figure 17-22. *NEC 240.15(A)* requires that *a fuse shall be connected in series with each ungrounded conductor.* The solid neutral would be inserted into the grounded "B" phase leg instead of a fuse. Other examples of where a solid neutral might be used is where a 3-pole, 3-fuse disconnect switch is used on a 3-wire, 120/240-volt, single-phase system, or where a fusible disconnect is used only as a disconnecting means, and where the user does not want any fuses to be installed in that particular switch.

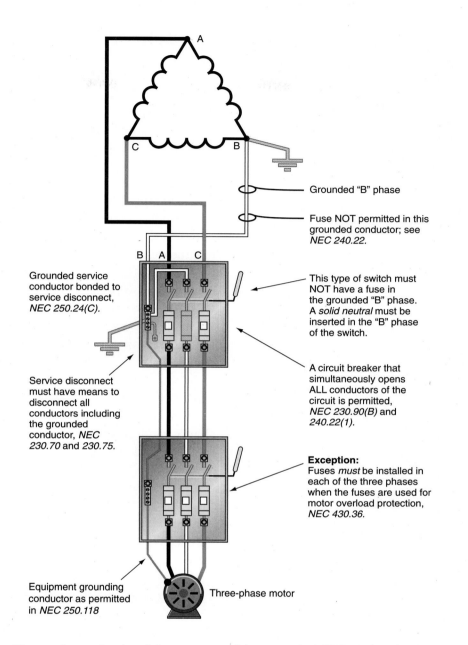

Grounded "B" phase

Fuse NOT permitted in this grounded conductor; see *NEC 240.22.*

Grounded service conductor bonded to service disconnect, *NEC 250.24(C).*

This type of switch must NOT have a fuse in the grounded "B" phase. A *solid neutral* must be inserted in the "B" phase of the switch.

Service disconnect must have means to disconnect all conductors including the grounded conductor, *NEC 230.70* and *230.75.*

A circuit breaker that simultaneously opens ALL conductors of the circuit is permitted, *NEC 230.90(B)* and *240.22(1).*

Exception: Fuses *must* be installed in each of the three phases when the fuses are used for motor overload protection, *NEC 430.36.*

Equipment grounding conductor as permitted in *NEC 250.118*

Three-phase motor

FIGURE 17-20   Three-phase, 3-wire delta system with grounded "B" phase. (*Delmar/Cengage Learning*)

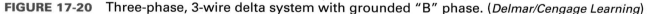

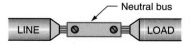

Neutral bus

LINE   LOAD

FIGURE 17-21   *NEC 230.75* allows this for disconnecting the grounded conductor at the service disconnect. (*Delmar/Cengage Learning*)

TIME-CURRENT CHARACTERISTIC CURVES AND PEAK LET-THROUGH CHARTS

The electrician must have a basic amount of information concerning fuses and their application to be able to make the correct choices and decisions for everyday situations that arise on an installation.

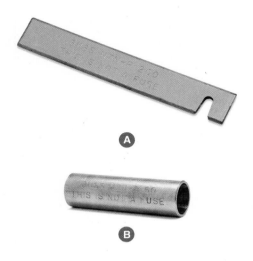

FIGURE 17-22   Examples of a solid neutral.
(*Delmar/Cengage Learning*)

The electrician must be able to use the following types of fuse data:

- Time-current characteristic curves, including total clearing, minimum melting, and average melting curves
- Peak let-through charts

Fuse manufacturers furnish this information for each of the fuse types they produce.

The Use of Time-Current Characteristic Curves

The use of the time-current characteristic curves shown in Figures 17-23 and 17-24 can be demonstrated by considering a typical problem. Assume that an electrician must select a fuse to protect a motor that is governed by *NEC 430.32(A)*. This section indicates that the running overcurrent protection shall be based on not more than 125% of the full-load running current of the motor. The motor in this example is assumed to have a full-load current of 24 amperes. Therefore, the size of the required protective fuse is determined as follows:

$$24 \times 1.25 = 30 \text{ amperes}$$

Now the question must be asked: Is this 30-ampere fuse capable of holding the inrush current of the motor, which is approximately four to five times the running current, for a sufficient length of time for the motor to reach its normal speed? For this problem, the inrush current of the motor is assumed to be 100 amperes.

Refer to Figure 17-23. From the 100-ampere bottom horizontal axis (base line of the chart), draw a vertical line upward until it intersects the 30-ampere fuse curve. Then draw a horizontal line to the left side (axis) of the chart. At this point, it can be seen that the 30-ampere fuse will hold 100 amperes for slightly less than 30 seconds. In this amount of time, the motor can be started. In addition, the fuse will provide running overload protection as required by *NEC 430.32(A)*.

If the time-current curve for thermal overload relays in a motor controller is checked, it will show that the overload element will open for the same current in much less than 40 seconds. Therefore, in the event of an overload, the thermal overload elements will operate before the fuse opens. If the overload elements do not open for any reason, or if the contacts of the controller weld together, the properly sized dual-element fuse will open to reduce the possibility of motor burnout. The preceding method is a simple way to obtain backup or double-motor protection.

Referring again to Figure 17-23, assume that a short circuit of approximately 250 amperes occurs. Find the 250-ampere line on the bottom horizontal axis of the chart and then, as before, draw a vertical line upward until it intersects the 30-ampere fuse curve. From this intersection point, draw a horizontal line to the left side of the chart. This value indicates that the fuse will clear the fault in approximately $\frac{2}{100}$ (0.02) second. On a 60-hertz system (60 cycles per second), one cycle equals 0.016 second. Therefore, this particular 30-ampere fuse will clear a 250-ampere fault in just over one cycle.

The Use of Peak Let-Through Charts

It is important to understand the use of peak let-through charts if one is to properly match the short-circuit current rating of electrical equipment with the let-through current values of the overcurrent protective devices (fuses or circuit breakers). Short-circuit current rating can be defined as "at a specified nominal voltage, the highest current at the equipment or component line terminals that the equipment or component can safely carry until an overcurrent device opens the circuit."

The short-circuit current ratings of electrical equipment (such as panelboards, bus duct, switchboard

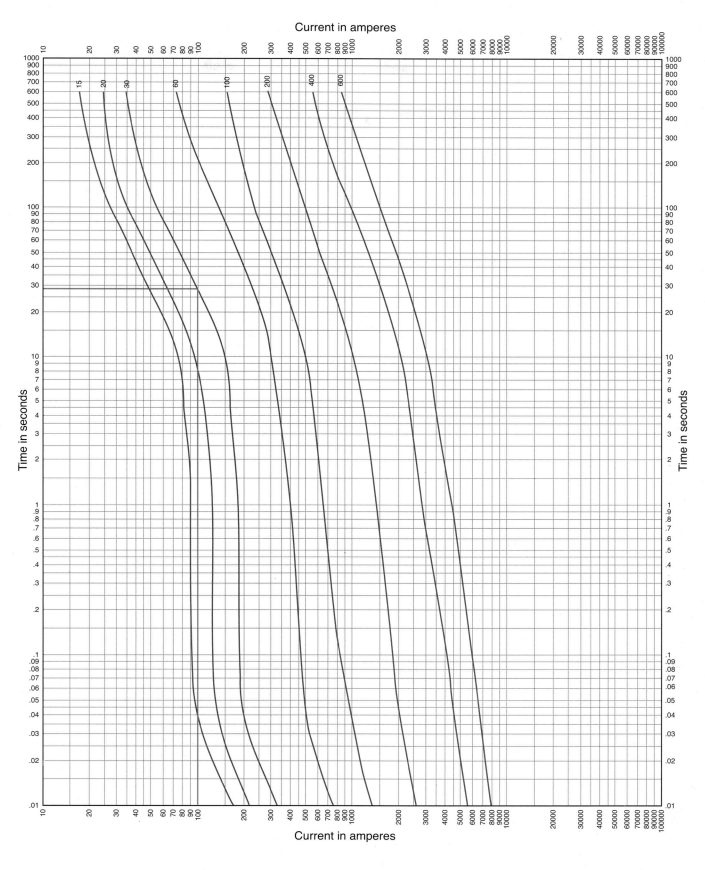

Current in amperes

Time in seconds

FIGURE 17-23   Average melting time-current curves for one family of current-limiting, dual-element, time-delay fuses rated 15 through 600 amperes. (*Courtesy of Cooper Bussmann, Division of Cooper Industries*)

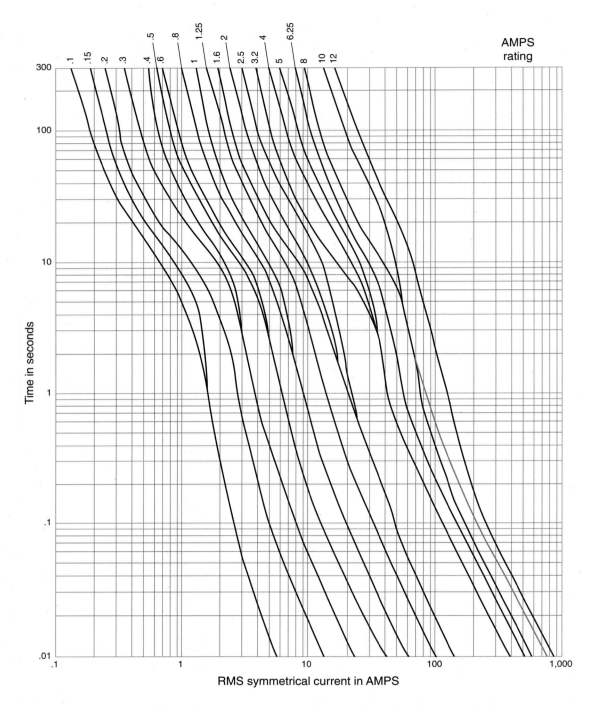

FIGURE 17-24 Average melting time-current curves for one family of current-limiting, dual-element, time-delay fuses rated 0.1 through 12 amperes. (*Courtesy of Cooper Bussmann, Division of Cooper Industries*)

bracing, controllers, and conductors) and the interrupting ratings of circuit breakers are given in the published standards of the Underwriters Laboratories, NEMA, and the Insulated Cable Engineering Association (ICEA). The withstand ratings may be based either on the peak current (I_p) or on the RMS current.

Peak let-through charts give a good indication of the current-limiting effects of a current-limiting fuse or circuit breaker under "bolted fault" conditions. Current limitation is one of the principles used to obtain a series-rating listing for electrical equipment. Another principle used to obtain a series

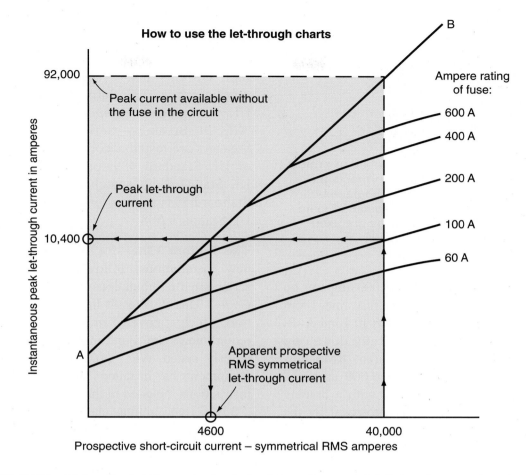

How to use the let-through charts

Peak current available without the fuse in the circuit

Ampere rating of fuse:

600 A

400 A

200 A

100 A

60 A

Peak let-through current

92,000

10,400

Apparent prospective RMS symmetrical let-through current

A

B

Instantaneous peak let-through current in amperes

4600

40,000

Prospective short-circuit current – symmetrical RMS amperes

FIGURE 17-25 Using the let-through charts to determine peak let-through current and apparent prospective RMS symmetrical let-through current. (*Courtesy of Cooper Bussmann, Division of Cooper Industries*)

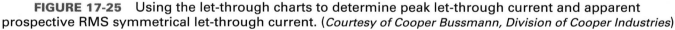

rating is referred to as "dynamic impedance," where the available fault current is further reduced because of the added impedance of the arc between the contacts of a circuit breaker as it is opening. Series rating of panels is discussed later in this chapter.

To use charts such as Figure 17-25, find the 40,000-ampere point on the base line of the chart, draw a vertical line upward to Line A–B, and then move horizontally to the left margin where the value of 92,000 amperes is found. This is the peak current, I_p, at the first half cycle.

To find the peak let-through current of the 100-ampere fuse, start again at the 40,000-ampere point on the base line, go vertically upward to Line A–B, and then horizontally to the left margin, where the value of 10,400 is found. This is the peak let-through current, I_p, of this particular 100-ampere fuse.

To find the apparent RMS let-through current of this fuse, start again at the 40,000-ampere point

on the base line, go vertically upward to Line A–B, and then horizontally to the left until Line A–B is reached, and then vertically downward to the base line. The value at this point on the base line is 4600 amperes. This is the apparent RMS amperes let-through current of this particular 100-ampere fuse.

Figure 17-26 shows how a current-limiting main overcurrent device can be applied to "listed" series-rated panels, discussed in more detail a little later in this chapter.

Figure 17-27 shows a peak let-through chart for a family of dual-element fuses. Fuse and circuit-breaker manufacturers provide charts of this type for various sizes and types of their fuses and circuit breakers.

What are these charts used for? Let's take an example of a 400-ampere busway that is marked as having 22,000-ampere bracing. The available fault current of the system has been determined to be approximately 50,000 amperes. In Figure 17-27,

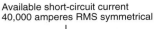

FIGURE 17-26 Example of a fuse used to protect circuit breakers. This installation meets the requirements of *NEC 110.9* when the panelboard is listed as series-rated and the specified main fuses and branch-circuit breakers are used. (*Delmar/Cengage Learning*)

draw a vertical line upward from 50,000 amperes on the base line up to the 400-ampere fuse line. From this point, go horizontally to the left to Line A–B. From this point, draw a line vertically downward to the base line and read an apparent RMS current value of slightly more than 10,000 amperes. Therefore, the 400-ampere busway with a bracing of 22,000 amperes is adequately protected by the 400-ampere, fast-acting, current-limiting fuse.

Table 17-4 shows how a manufacturer might indicate the maximum size and type of fuse that will protect given circuit breakers against fault currents that exceed the breakers' interrupting rating. This information is found in the UL *Recognized Components Directory* under "Circuit Breakers—Series Connected." Circuit breakers are tested according to the requirements of UL Standard 489; panelboards are tested to UL Standard 67; switchboards are tested to UL Standard 891; and motor controllers are tested to UL Standard 508.

The Bottom Line—for New Installations. For new installations, following the manufacturer's series-rated combinations is the only way to "meet *Code*." This will result in a tested series combination, which is covered in *NEC 240.86(B)*. Marking requirements are found in *NEC 110.22(C)*. As previously stated, *NEC 110.3(B)* requires that listed equipment be installed in accordance with any instructions included in the listing.

A current-limiting class J, L, R, RK, or T on the supply side of a panelboard will often provide

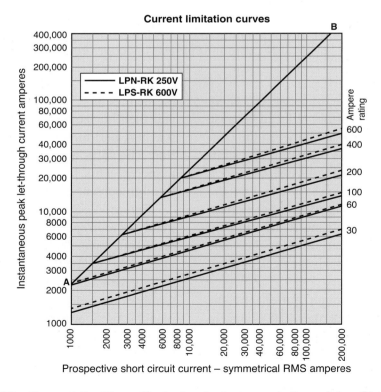

FIGURE 17-27 Current-limiting effect of a dual-element, time-delay Class RK1 fuse. (*Courtesy of Cooper Bussmann, Division of Cooper Industries*)

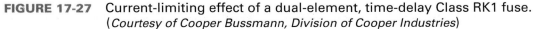

TABLE 17-4

Typical table showing sizes and types of fuses to protect circuit breakers.

| | BRANCH BREAKER | | MAIN FUSE | | INTERRUPTING RATING | | |
Frame	Poles	Max. Amp.	Type	Max. Amp.	VAC	KA	Phase
ABC	2	15–100	J, T	200	120/240	200	1
DEF	2, 3	15–100	R	100	120/240	200	1
GHI	2, 3	70–225	J, T	400	600	100	1,3
JKL	1	15–100	J, T	200	277	200	1
MNO	2, 3	15–150	R	200	600	200	1,3
PQR	2, 3	70–400	L	1200	480	100	1,3
STU	2, 3	400	J, T, R	600	480	100	1,3
VWX	2, 3	400	J, T, R	400	480	200	1,3
SYX	1, 2	30	RK5	200	240	100	3

Note: Typical table published by manufacturers showing the maximum sizes and types of fuses that may be used to protect circuit breakers against fault currents that exceed the breakers' maximum interrupting rating. This table was modeled from various manufacturers' "series-connected" technical information. Always refer to the specific manufacturer's data.

adequate short-circuit protection for the panelboard and the circuit breakers in the panelboard. See Table 17-4. A new panelboard listed for use with a tested series-rated combination must be used.

The Bottom Line for Existing Installations. Another way to verify that a current-limiting fuse ahead of the panelboard will limit the fault current to equal to or less than the short-circuit rating of the panelboard is by referring to a chart similar to Figure 17-27. We have already discussed how to use this type of chart. This relates to "*under engineering supervision*" as covered in *NEC 110.22(B)* and *240.86(A)*.

It is significant to note that the up-over-down method applies to the current-limiting effect of current-limiting overcurrent devices when protecting equipment that does not attempt to open the circuit during the time it takes for the upstream current limiting device to open. An example of this would be using the up-over-down method for protecting bus duct. It does not apply to the protection of circuit breakers that open within one-half cycle or less, which would include most thermal-magnetic molded case circuit breakers made in recent decades.

For protection of an existing old panelboard, if it can be determined that the older circuit breakers will not open within a one-half cycle, the up-over-and-down method can be used. The only valid way to determine whether older downstream circuit breakers will open

within a one-half cycle is to contact the manufacturer or have a test run at one of the high-current test facilities that can be rented for such testing.

CIRCUIT BREAKERS

In *NEC Article 100,* a circuit breaker is defined as *a device that is designed to open and close a circuit by nonautomatic means and to open the circuit automatically on a predetermined overcurrent without being damaged itself when properly applied within its rating.*

Molded-case circuit breakers are the most common type in use today, Figure 17-28. The tripping mechanism of this type of breaker is enclosed in a molded plastic case. The thermal-magnetic type of circuit breaker is covered in this chapter.

Another type is the power circuit breaker, often referred to as an air-frame circuit breaker. They are larger and of heavier construction than molded-case and insulated-case breakers. They are available with many options such as draw-out mounting, electronic adjustable tripping and rating settings, and integral current limiters or fuses. Power circuit breakers are used in large industrial applications. Power circuit breakers can be maintained in the field such as replacing contacts. Molded-case circuit breakers are not intended to be maintained in the field, but to be replaced if there is a problem.

FIGURE 17-28 Molded-case circuit breakers. (*Delmar/Cengage Learning*)

Still another type of circuit breaker is the insulated-case circuit breaker. It is a hybrid between a molded-case and a power (air-frame) circuit breaker, and is less costly than a power (air-frame) circuit breaker. Most of the features available on power breakers are also available on insulated-case circuit breakers.

Low-voltage power circuit breakers are listed under UL Standard 1066. Molded-case and insulated-case circuit breakers are listed under UL Standard 489. *NEC 240.80* through *240.86* state the basic requirements for circuit breakers and are summarized as follows:

- Breakers shall be trip free so that if the handle is held in the On position, the internal mechanism trips the breaker to the "Off" position.

- The breaker shall clearly indicate whether it is in the "On" or "Off" position.

- The breaker shall be nontamperable; that is, it cannot be readjusted (to change its trip point or time required for operation) without dismantling the breaker or breaking the seal.

- The rating shall be durably marked on the breaker. For the smaller breakers with ratings of 100 amperes or less and 600 volts or less, this marking must be molded, stamped, or etched on the handle or other portion of the breaker that will be visible after the cover of the panel is installed.

- Every breaker having an interrupting rating other than 5000 amperes shall have its interrupting rating shown on the breaker.

- *NEC 240.83(D)* and UL Standard 489 require that when a circuit breaker is to be used as a switch, the breaker must be so evaluated, listed, and marked. A typical use of a circuit breaker as a switch is in a panel to turn 120- or 277-volt lighting circuits on and off instead of installing separate switches. This is very common in commercial buildings. To switch fluorescent lighting circuits, breakers shall be marked SWD or HID. To switch high-intensity discharge (HID) lighting circuits, breakers shall be marked HID. Breakers not marked in this manner shall not be used as switches.

- A circuit breaker shall not be loaded to more than 80% of its current rating for loads that are likely to be on for 3 hours or more, unless the breaker is marked as being suitable for operation at 100% of its rating.

- If the voltage rating on a circuit breaker is marked with a single voltage (example: 240 volts), the breaker may be used in grounded or ungrounded systems where the voltage between any two conductors does not exceed the breaker's marked voltage rating.

CAUTION: Do not use a 2-pole, single-voltage-marked breaker to protect a 3-phase, 3-wire, corner-grounded circuit unless the breaker is marked with both 1-phase and 3-phase markings.

- If the voltage rating on a circuit breaker is marked with a slash voltage (Example: 480/277 volts or 120/240 volts), the breaker may be used on a solidly grounded system where the voltage to ground does not exceed the breaker's lower voltage marking and where the voltage between any two conductors does not exceed the breaker's higher voltage marking.

- A cautionary *Informational Note* has been added, stating that *Proper application of molded case circuit breakers on 3-phase systems, other than solidly grounded wye, particularly on corner-grounded delta systems, considers the circuit breaker's individual interrupting capacity.**

*Reprinted with permission from NFPA 70–2011.

Thermal-Magnetic Circuit Breakers

A thermal-magnetic circuit breaker contains a bimetallic element. On a continuous overload, the bimetallic element heats and bends until it unlatches the inner tripping mechanism of the breaker. Harmless momentary overloads do not cause the tripping of the bimetallic element. If the overload is heavy, or if a short circuit occurs, then the magnetic mechanism within the circuit breaker causes the breaker to interrupt the circuit instantly. The time required for the breaker to open the circuit completely depends on the magnitude of the fault current and the mechanical condition of the circuit breaker. This time may range from approximately one-half cycle to several cycles.

Circuit-breaker manufacturers calibrate and set the tripping characteristic for most molded-case breakers. Breakers are designed so that it is difficult to alter the set tripping point, *NEC 240.82*. For certain types of breakers, however, the trip coil can be changed physically to a different rating. Adjustment provisions are made on some breakers to permit the magnetic trip range to be changed. For example, a breaker rated at 100 amperes may have an external adjustment screw with positions marked HI-MED-LO. The manufacturer's application data for this breaker indicates that the magnetic tripping occurs at 1500 amperes, 1000 amperes, or 500 amperes, respectively, for the indicated positions. These settings usually have a tolerance of ±10%.

Circuit breakers are available with electronic or solid-state trip units, adjustments and accessories. This technology allows many sophisticated features to be performed. See Figure 17-29 for an example. Some features require a consulting engineer or the factory to adjust the settings or to give specific instructions to be followed. Here's a list of several features some electronically controlled circuit breakers may have:

- Ammeter/trip indicator (monitors current in phases A, B, and C and ground-fault current)
- Memory feature (used to record intermittent overload or ground-fault conditions)
- Adjustments for time-current curve shaping include these:
  - Long-time pickup switch
  - Long-time-delay switch
  - Short-time pickup switch
  - Short-time delay switch
  - Instantaneous trip switch
  - Ground-fault pickup switch
  - Ground-fault delay switch
  - Ground-fault alarm switch
- Zone-selective interlocking (allows circuit breakers to communicate with each other so the downstream circuit breaker is allowed to clear a fault before the upstream overcurrent device will open. Carefully consider all ramifications of deliberately introducing a time delay on opening an overcurrent device. Doing so may increase the level of personal protective equipment required should electricians be required to work on energized electrical equipment.)
- Accessories that may be added to electronic or solid-state circuit breakers include these:
  - Shunt trip
  - Undervoltage trip
  - Auxiliary switch
  - Alarm switch
  - Trip indicator
  - Ammeter/trip indicator

FIGURE 17-29   Circuit breaker with solid-state controls to shape tripping curve. (*Delmar/Cengage Learning*)

Figure 17-30 shows a drawing of a ground-fault system that is built into an electronic controlled circuit breaker. As can be seen, the circuit breaker monitors the current on each of the three ungrounded phase conductors and provides connections for an

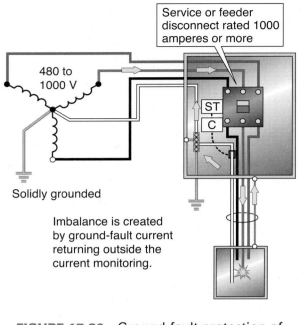

480 to 1000 V

Service or feeder disconnect rated 1000 amperes or more

ST

C

Solidly grounded

Imbalance is created by ground-fault current returning outside the current monitoring.

FIGURE 17-30 Ground-fault protection of equipment with solid-state circuit breaker. (*Delmar/Cengage Learning*)

external current transformer that the neutral conductor passes through. This, then, represents a zero-sequence ground-fault protection system, Additional information on how these systems function is found in Chapter 13 of this text.

Ambient-Compensated Circuit Breakers

Thermal-magnetic circuit breakers are temperature sensitive. Some circuit breakers are ambient (surrounding temperature) compensated. This means that the tripping characteristic of the breaker is partially or completely neutralized by the effect of the surrounding temperature. Heat is generated by circuit breakers and conductors in a panelboard. UL Standard 489 specifies calibration testing at various loads and ambient temperatures. The continuous test is run at 104°F (40°C). One manufacturer states that no rerating is necessary when the circuit breaker is installed in an ambient in the range of 14°F to 140°F (−10°C to 60°C). If you are installing thermal circuit breakers in extremely hot or extremely cold temperatures, consult the manufacturers' literature. Look for thermal characteristic time-current curves.

It is a good practice to turn the breaker on and off periodically to "exercise" its moving parts.

Factors that can affect the proper operation of a circuit breaker include moisture, dust, vibration, corrosive fumes and vapors, and excessive tripping and switching. As a result, care must be taken when locating and installing circuit breakers and all other electrical equipment.

The interrupting rating of a circuit breaker is marked on the breaker label. The electrician should check the breaker carefully for the interrupting rating because the breaker may have several voltage ratings with a different interrupting rating for each. For example, assume that it is necessary to select a breaker having an interrupting rating at 240 volts of at least 50,000 amperes. A close inspection of the breaker may reveal the following data:

Voltage	Interrupting Rating
240 volts	65,000 amperes
480 volts	25,000 amperes
600 volts	18,000 amperes

Recall that for a fuse, the interrupting rating is marked on the fuse label. This rating is the same for any voltage up to and including the maximum voltage rating of the fuse.

The standard full-load ampere ratings of nonadjustable circuit breakers are the same as those for fuses according to *NEC 240.6*. As previously mentioned, additional standard ratings of fuses are 1, 3, 6, 10, and 601 amperes.

The time-current characteristics curves for circuit breakers are similar to those for fuses. A typical circuit breaker time-current curve is shown in Figure 17-31. Note that the current is indicated in percentage values of the breaker trip unit rating. Therefore, according to this graph, the 100-ampere breaker being considered will

1. carry its 100-ampere (100%) rating indefinitely.

2. carry 300 amperes (300%) for a minimum of 25 seconds and a maximum of 70 seconds.

3. unlatch its tripping mechanism in 0.0032 seconds (approximately one-quarter cycle) with a current of 5000 amperes.

4. interrupt the circuit in a maximum time of 0.016 seconds (one cycle) with a current of 5000 amperes (5000%).

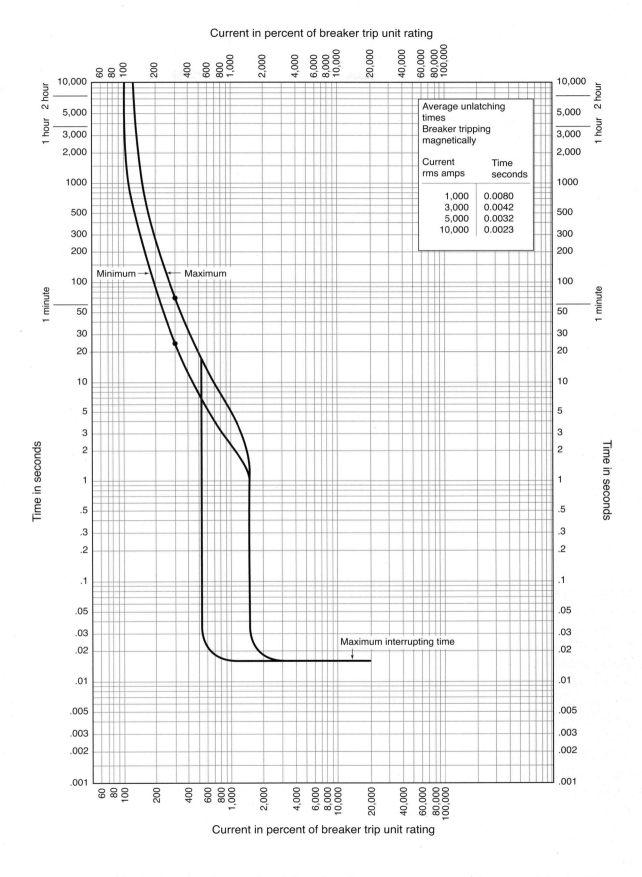

FIGURE 17-31   Typical molded-case circuit-breaker time-current curve. (*Courtesy of Cooper Bussmann, Division of Cooper Industries*)

The same time-current curve can be used to determine that a 200-ampere circuit breaker will

1. carry its 200-ampere (100%) rating indefinitely.

2. carry 600 amperes (300%) for a minimum of 25 seconds and a maximum of 70 seconds.

3. unlatch its tripping mechanism in 0.0032 seconds (approximately one-quarter cycle) with a current of 5000 amperes.

4. interrupt the circuit in a maximum time of 0.016 seconds (one cycle) with a current of 5000 amperes (2500%).

This example shows that if a short circuit occurs in the magnitude of 5000 amperes, then both the 100-ampere breaker and the 200-ampere breaker installed in the same circuit will open together because they have the same unlatching times. In many instances, this action is the reason for otherwise unexplainable power outages (Chapter 18). This is rather common when heavy (high value) fault currents occur on circuits protected with molded-case circuit breakers.

Common Misapplication

A common violation of *NEC 110.9* and *110.10* is the installation of a main circuit breaker (such as

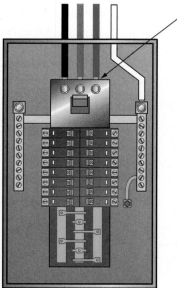

Short circuit current 40k at main breaker

- Panelboard is NOT series rated.
- 100-ampere main breaker with 50 kAIC rating
- Branch circuit breakers with 10 kAIC rating

FIGURE 17-32 Violation of *NEC 110.9*.
(*Delmar/Cengage Learning*)

a 100-ampere breaker) that has a high interrupting rating (such as 50,000 amperes) while making the assumption that the branch-circuit breakers (with interrupting ratings of 10,000 amperes) are protected adequately against the 40,000-ampere short circuit, Figure 17-32.

Standard molded case circuit breakers with high interrupting ratings cannot protect standard molded case circuit breakers or other end-use equipment having a lower interrupting rating. You will find markings such as the following on electrical equipment based on UL labeling requirements:

- *The short-circuit rating of a panelboard is equal to the lowest interrupting rating of any installed circuit breaker or fused switch.*

- *If the marked interrupting rating of the breaker exceeds the marked short-circuit rating of the end-use equipment, such as a panelboard, in which the breaker is installed, the interrupting rating of the overall combination is still considered to be the lesser rating marked on the end-use equipment.*

SERIES-RATED APPLICATIONS

When you hear and see the term "series-rated" equipment, the term means exactly what is says. Simply stated, it generally means two circuit breakers that are connected in series. They can consist of a main and feeder breakers or the feeder and the branch-circuit circuit breakers. Generally, the upstream breaker is fully rated for the short-circuit current that is available at its line terminals and the downstream circuit breaker is not. Tested systems are available as indicated in *NEC 240.86(B)* and field-engineered systems as covered in *NEC 240.86(B)*.

Series-rated systems are less costly than fully rated systems because downstream circuit breakers have a lower interrupting rating than the upstream overcurrent device. It is safe to say that high interrupting rated circuit breakers cost more than low interrupting-rated circuit breakers.

The upstream high interrupting-rated overcurrent device is permitted to be in the same enclosure as the lower interrupting-rated devices, or is permitted to be remote, such as at a main switchboard, distribution switchboard, or panelboard.

Series-rated equipment is tested and listed by a Nationally Recognized Testing Laboratory (NRTL) as a total assembly. To establish a series rating, the testing includes the panelboard and the fuses and/or the breakers in place. Many series-rating combinations are available. Read the label on the panelboard or other equipment to find the series combination short-circuit rating.

A series-rated system basically is where the available fault current does not exceed the interrupting rating of the line-side overcurrent device but does exceed the interrupting rating of the load-side overcurrent device. The series-rating short-circuit current values will be found on the marking of the equipment.

To properly select a series-rated system, you must use equipment that is tested and listed by a NRTL and is marked by the manufacturer. For existing installations only, an engineered series-combination system is permitted to be selected under engineering supervision. This means selected by a licensed professional engineer engaged primarily in the design or maintenance of electrical installations; see *NEC 240.86(A)*.

The selection of the series-rated components must be documented and stamped by a professional engineer.

As discussed a little later, you must also consider motor contribution to the available fault current, because a motor acts much like a generator when a system becomes short-circuited. Instead of drawing current from the system, the motor acting as a generator "pumps" fault current back into the system. Roughly, the motor contribution is approximately

the amount of current the motor would draw on start-up. Adding motor contribution to fault-current calculations is not unique to series-rated systems—it must be considered with all systems.

Where high available fault currents indicate the need for high interrupting breakers or fuses, a fully rated system is generally used. In a fully rated system, all breakers and fuses are rated for the fault current available at the point of application.

Another less costly way to safely match a main circuit breaker or main fuses ahead of branch-circuit breakers is to use listed *series-rated* equipment. Series-rated equipment is accepted by most electrical inspectors across the country. Series-rated equipment is also referred to as *series-connected*.

Underwriters Laboratories lists series-rated panelboards with breaker main/breaker branches and fused main/breaker branches. Figure 17-33 shows a series-rated panel that may be used where the available fault current does not exceed 22,000 amperes. The panel has a 22,000 AIC main breaker and branch-circuit breakers rated 10,000 AIC. This series arrangement of breakers results in a cushioning effect when the upstream breaker contacts and the downstream breaker contacts simultaneously open under fault conditions that exceed the interrupting rating of the lower rated breaker. The impedance of the two arcs in series reduces the fault current that the downstream breakers actually see. The impedance of the arc is called "dynamic impedance."

A disadvantage of series-rated systems is that should a heavy fault current occur on any of the downstream breaker circuits, the upstream breaker will also trip off, causing a complete loss of power

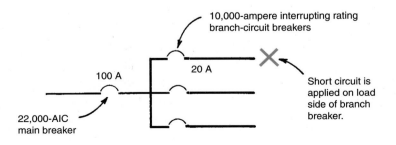

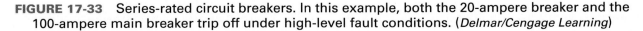

FIGURE 17-33   Series-rated circuit breakers. In this example, both the 20-ampere breaker and the 100-ampere main breaker trip off under high-level fault conditions. (*Delmar/Cengage Learning*)

to the panel. Likewise, a series-rated main fuse/breaker branches panel would also experience a power outage, but the possibility exists that only one fuse would open, leaving one-half (for a single-phase panel) or two-thirds (for a 3-phase, 4-wire panel) of the power on.

For normal overload conditions on a branch-circuit, the branch breaker only will trip off, and should not cause the main breaker to trip or main fuses to open. Refer to Chapter 18 for more data on the subject of selectivity and nonselectivity.

CAUTION: Be extremely careful when installing series-rated breakers. Do not mix different manufacturers' breakers. Do not use any breakers that have not been UL recognized as being suitable for use in combination with one another.

The UL *Recognized Component Directory* (*Yellow Book*) has this to say about series-connected breakers:

> *The devices covered under this category are incomplete in certain constructional features or restricted in performance capabilities and are intended for use as components of complete equipment submitted for investigation rather than for direct separate installation in the field. The final acceptance of the component is dependent upon its installation and use in complete equipment submitted to Underwriters Laboratories Inc.*

NEC 240.86(B) requires that the manufacturer of a series-rated panel must legibly mark the equipment. The label will show the many ampere ratings and catalog numbers of those breakers that are suitable for use with that particular piece of equipment.

***NEC 110.22(B)* and *(C)* extends *240.86* by requiring that there be "field marking" of series-rated equipment.** See Figure 17-34 for the required marking.

Electrical inspectors require the electrical contractor to attach this labeling to the equipment. For example, the main distribution equipment of the commercial building is located remote from the individual tenants' panelboards. The electrical contractor will affix legible and durable labels to each disconnect switch in the main distribution equipment, identifying the size and type of overcurrent protection that has been installed for the

CAUTION
ENGINEERED SERIES COMBINATION SYSTEM
RATED __________ AMPERES. IDENTIFIED
REPLACEMENT COMPONENTS REQUIRED.

► Field labeling as required by *NEC 110.22(B)* for engineered series combination systems.◄

CAUTION
SERIES COMBINATION SYSTEM
RATED __________ AMPERES. IDENTIFIED
REPLACEMENT COMPONENTS REQUIRED

► Field labeling as required by *NEC 110.22(C)* for tested series combination systems.◄

CAUTION
SERIES COMBINATION SYSTEM
RATED __________ AMPERES.REPLACE WITH:
__________ TYPE BREAKER
__________ TYPE FUSE

► Field labeling for existing installations selected under engineering supervision as required by *NEC 240.86(A)*.◄

FIGURE 17-34 Typical labels for required series combination systems. (*Delmar/Cengage Learning*)

proper protection of the downstream panelboards, Figure 17-34.

This CAUTION—SERIES RATED SYSTEM labeling will alert future users of the equipment that they should not indiscriminately replace existing overcurrent devices or install additional overcurrent devices with sizes and types of fuses or circuit breakers not compatible with the series-rated system as originally designed and installed. Otherwise, the integrity of the series-rated system might be sacrificed, creating a hazard to life and property.

SERIES-RATED SYSTEMS WHERE ELECTRIC MOTORS ARE CONNECTED

Series-rated systems have an available fault current higher than the interrupting rating of the load-side breakers and have an acceptable overcurrent device of higher interrupting rating ahead of the load-side breakers. The series combinations' interrupting rating is marked on the end-use equipment, *NEC 240.86(B)*.

Because electric motors contribute short-circuit current under fault conditions on an electrical system, *NEC 240.86(C)* sets forth two requirements when series-rated systems are applied where electric motors are involved.

1. Do not connect electric motors between the load side of the higher rated overcurrent device and the line side of the lower rated overcurrent device if,

2. The sum of the connected motor full-load currents exceeds 1% of the interrupting rating of the lower rated circuit breaker. For example, if the interrupting rating of the lower rated breakers in a series-rated panel is 10,000 amperes, then the 1% motor load limitation is 100 amperes. Therefore, the full-load currents of all of the motors connected between the series-rated equipment must not exceed 100 amperes.

CURRENT-LIMITING CIRCUIT BREAKERS

A current-limiting circuit breaker will limit the let-through energy (I^2t) to something less than the I^2t of a one-half-cycle symmetrical wave.

The label on this type of breaker will show the words "CURRENT-LIMITING." The label will also indicate the breaker's let-through characteristics or will indicate where to obtain the let-through characteristic data. This let-through data is necessary to ensure adequate protection of downstream components (wire, breakers, controllers, busbar bracing, and so on). Therefore, it is important when installing circuit breakers to ensure that not only the circuit breakers have the proper interrupting rating (capacity), but also that all of the components connected to the circuit downstream from the breakers are capable of withstanding the let-through current of the breaker. To use the previous problem as an example, this means that the branch-circuit conductors must be capable of withstanding 40,000 amperes

for approximately one-half cycle (the opening time of the breaker).

COST CONSIDERATIONS

As previously discussed, there are a number of different types of circuit breakers to choose from. The selection of the type to use depends on a number of factors, including the interrupting rating, selectivity, space, and cost. When an installation requires something other than the standard molded-case-type circuit breaker, the use of high interrupting types or current-limiting types may be necessary. The electrician and/or design engineer must complete a "short-circuit study" to ensure that the overcurrent protective devices provide proper protection against short-circuit conditions per *National Electrical Code* requirements.

The list price cost comparisons in Table 17-5 are for various 100-ampere, 240-volt, panelboard-type circuit breakers.

Molded-case circuit breakers with an integral current-limiting fuse are available. The thermal element in this type of breaker is used for low overloads; the magnetic element is used for low-level short circuits; and the integral fuse is used for short circuits of high magnitude. The interrupting capacity of this breaker/fuse combination (sometimes called a limiter) is higher than that of the same breaker without the fuse.

The selection of the ampacity of circuit breakers for branch circuits is governed by the requirements of *NEC Article 240.*

MOTOR CIRCUITS

NEC Table 430.52 shows the general requirement that for motor branch circuits, the maximum setting of a conventional inverse-time circuit breaker must not exceed 250% of the full-load current of the motor.

For an instantaneous-trip circuit breaker, permitted only in listed combination controllers, the maximum setting is 800% of the motors' full-load current, except for Design B motors. For Design B motors, the maximum setting is 1100%. The design data is found on the nameplate of the motor.

TABLE 17-5

List price cost comparisons for various types of 100-ampere, 240 volt, 2-pole molded-case panelboard-type circuit breakers.

Type	Interrupting Rating	List Price
Standard thermal magnetic plug-on	10,000	$189
Standard thermal magnetic bolt-on	22,000	$315
Standard thermal magnetic bolt-on	42,000	$651
Standard thermal magnetic bolt-on	65,000	$1521
Standard thermal magnetic bolt-on	100,000	$1890
Current-limiting thermal magnetic bolt-on	65,000	$1497
Current-limiting thermal magnetic bolt-on	100,000	$1721
Electronic w/adjustable trips (600 A frame, 600 VAC, 3-pole)	65,000	$5081
Electronic w/adjustable trips (600 A frame, 600 VAC, 3-pole)	100,000	$8478

Note: Prices 3/2010.

The exceptions in *NEC 430.52(C)* recognize that the 800% and 1100% settings might not allow the motor to start, particularly in the case of energy-efficient Design B motors. Design B motors have a high starting inrush current that can cause an instantaneous-trip circuit breaker to nuisance trip or a fuse to open. If an "engineering evaluation" can demonstrate the need to exceed the percentages shown in *NEC Table 430.52*, then

- the 800% setting may be increased to a maximum of 1300%.
- the 1100% setting may be increased to a maximum of 1700%.

Additional discussion regarding motor circuit design is found in Chapter 7.

*Reprinted with permission from NFPA 70-2011.

HEATING, AIR-CONDITIONING, AND REFRIGERATION OVERCURRENT PROTECTION

Check the nameplate carefully—and do what it says! The nameplate on HACR (heating air-conditioning and refrigeration) equipment might indicate "maximum size fuse," "maximum size fuse or circuit breaker," or "maximum size fuse or HACR circuit breaker."

In the past, circuit breakers were subjected to specific tests unique to HVAC equipment and were marked with the letters *HACR*. Today, no special additional tests are made. All currently listed circuit breakers are now suitable for HVAC application. These circuit breakers and HVAC equipment may or may not show the letters *HACR*. Because of existing inventories, it will take many years for the marking HACR to disappear from the scene. In the meantime, read and follow the information found on the nameplate of the equipment. See Figures 17-35, 17-36, and 17-37.

NEC 440.14 has a requirement that seems unnecessary, yet apparently there were too many violations relative to the location of the disconnecting means for air-conditioning equipment. The requirement is that *The disconnecting means shall not be located on panels that are designed to allow access to the air-conditioning or refrigeration equipment or to obscure the equipment nameplate(s).* *

It is worth repeating that in accordance with *NEC 310.10(C)*, conductors installed in wet locations shall have their type designation include the letter "W". Examples of this are Type THWN and Type THHN/THWN.

The *NEC* in *Definitions* clearly indicates that a location subject to saturation with water is considered to be a wet location.

The *NEC* in *300.9* tells us that the interior of a raceway installed in a wet location above grade is considered to be a wet location.

A disconnect mounted outdoors, such as shown in Figures 17-35, 17-36, and 17-37, without question is a wet location, requiring that the conductors running in and out of the switch be rated as suitable for wet location. The typical liquidtight flexible conduit connection from the switch to the air-conditioner likewise is considered to be in a wet location.

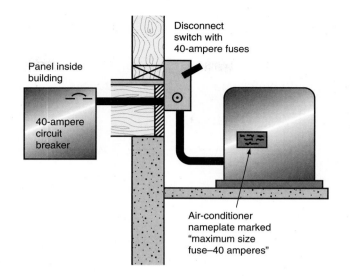

FIGURE 17-35 This installation complies with *NEC 440.14.* The disconnect switch is within sight of the unit and contains the 40-ampere fuses called for on the air-conditioner nameplate as the branch-circuit protection.
(*Delmar/Cengage Learning*)

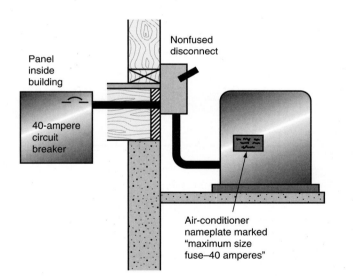

FIGURE 17-36 This installation violates *NEC 110.3(B).* Although the disconnect switch is within sight of the air conditioner, it does not contain fuses. Note that the branch-circuit protection is provided by the 40-ampere circuit breaker inside the building. Note also that the nameplate requires that the branch-circuit protection be via 40-ampere fuses maximum. If fused branch-circuit protection were provided at the panel inside the building, the installation would meet the *Code* requirements.
(*Delmar/Cengage Learning*)

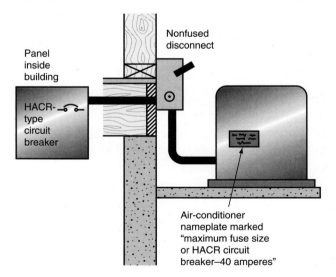

FIGURE 17-37 This installation complies with *NEC 110.3(B)* if the circuit breaker installed in the indoor panelboard is rated 40-amperes. It also satisfies the requirement that the disconnecting means is within sight of the air-conditioner equipment.
(*Delmar/Cengage Learning*)

REVIEW

Refer to the *National Electrical Code* or the working drawings when necessary. Where applicable, responses should be written in complete sentences.

1. Why must overcurrent protection be provided?

_____

_____

_____

_____

Several factors should be understood when selecting overcurrent protective devices. In your own words, define each of the following as they apply to overcurrent devices.

2. Voltage rating:

_____

_____

_____

_____

3. Continuous current rating:

_____

_____

_____

_____

4. Interrupting rating:

_____

_____

_____

_____

5. Speed of response:

_____

_____

_____

_____

6. Current limitation:

_____

_____

_____

_____

7. Peak let-through:

_____

_____

_____

_____

NEC Table 430.52 gives four options for selecting the *maximum* size motor branch-circuit, short-circuit, and ground-fault protective devices. These options are listed here. For each option, determine the correct rating or setting for a 25-horsepower, 3-phase, 230-volt motor. The motor is marked "Design B."

8. Nontime-delay fuses:

_____

_____

9. Dual-element, time-delay fuses:

_____

_____

10. Instantaneous-trip breakers:

_____

_____

11. Inverse time breakers:

_____

_____

12. Define the following terms:
    a. I^2t _____
    b. I_p _____
    c. RMS _____

13. Class J fuses (will fit) (will not fit) into standard fuse clips. (Circle the correct answer.)

Using the time-current curves from Figure 17–25, find the opening times for the following conditions:

14. A 60-ampere fuse subjected to a 300-ampere overload will cause the fuse to open in approximately _____ seconds.

15. A 60-ampere fuse subjected to a 600-ampere load will cause the fuse to open in approximately _____ seconds.

16. A 200-ampere fuse subjected to a 1000-ampere overload will cause the fuse to open in approximately _____ seconds.

17. A 600-ampere fuse subjected to an 8000-ampere short circuit will cause the fuse to open in approximately _____ seconds.

For Questions 18 and 19, use the chart from Figure 17-27 to determine the approximate current values for a 60-ampere, 250-volt fuse when the prospective short-circuit current is 40,000 amperes.

18. The instantaneous peak let-through current is approximately _____ amperes.

19. The apparent RMS current is approximately _____ amperes.

20. A section of plug-in busway nameplate indicates that the busway is braced for 14,000 amperes. The available fault current is 30,000 amperes. Using Figure 17-29, determine whether a 200-ampere, 250-volt, dual-element fuse will limit the current sufficiently to protect the busway against a possible fault of 30,000 amperes. Explain.

_____

_____

_____

21. In a thermal-magnetic circuit breaker, overloads are sensed by the _____ element, and short circuits are sensed by the _____ element.

22. Circle the correct answer. A cable limiter is

    a. a short-circuit device only.   True   False

    b. not to be used for overload protection of conductors.   True   False

    c. generally connected to both ends of large paralleled conductors so that if a fault occurs on one of the conductors, that faulted cable is isolated from the system.   True   False

23. Listed series-rated equipment allows the branch-circuit breakers to have a lower interrupting rating than the main. What two sections of the *National Electrical Code* cover the marking of series-rated systems with the words CAUTION—SERIES RATED SYSTEM? _____

_____

Using the chart in Figure 17-31, provide the unlatching time and interrupting time (also referred to as *opening time*) for a 50-ampere circuit breaker under the following stated conditions. Unlatching time and interrupting time are critical when designing a selectively coordinated electrical system.

24. For a load of 300 amperes, the minimum interrupting time is _____ seconds.

25. For a load of 300 amperes, the maximum interrupting time is _____ seconds.

26. For a 1000-ampere fault, the average unlatching time is _____ seconds.

27. For a 5000-ampere fault, the average unlatching time is _____ seconds.

28. The ampere range of Class L fuses is from _____ to _____ amperes.

29. The voltage rating of a plug fuse is _____ volts.

30. The interrupting rating of a Class J fuse is _____ amperes.

31. When the nameplate of HVAC equipment is marked "Maximum Size Fuse 50-Amperes," is it permitted to connect the equipment to a 50-ampere circuit breaker? (Explain.) _____

_____

CHAPTER 100

Short-Circuit Calculations and Coordination of Overcurrent Protective Devices

OBJECTIVES

After studying this chapter, you should be able to

- perform short-circuit calculations using the point-to-point method.
- calculate short-circuit currents using the appropriate tables and charts.
- define the terms *coordination*, *selective systems*, and *non-selective systems*.
- use time-current curves.

The student must understand the intent of *NEC 110.9* and *110.10*. That is, ensure that the fuses and/or circuit breakers selected for an installation are capable of interrupting the current at the rated voltage under any condition (overload, short circuit, or ground fault) with complete safety to personnel and without extensive damage to the panel, load center, switch, or electrical equipment in which the protective devices are installed.

An overloaded condition resulting from a miscalculation of load currents will cause a fuse to open or a circuit breaker to trip in a normal manner. However, a miscalculation, a guess, or ignorance of the magnitude of the available short-circuit currents may result in the installation of breakers or fuses having inadequate interrupting ratings. Such a situation can occur even though the load currents in the circuit are checked carefully. Breakers or fuses having inadequate interrupting ratings need only be subjected to a short circuit to cause them to explode, resulting in injury to personnel and serious damage to the electrical equipment. The interrupting rating of an overcurrent device is its maximum rating and must not be exceeded.

In any electrical installation, individual branch circuits are calculated as has been discussed previously in this text. After the quantity, size, and type of branch circuits are determined, these branch-circuit loads are then combined to determine the size of the feeder conductors to the respective panelboards. Most consulting engineers will specify that a certain number of spare branch-circuit breakers be installed in the panelboard, plus a quantity of spaces that can be used in the future.

For example, a certain calculation may require a minimum of sixteen 20-ampere branch circuits. The specification might call for a 24-circuit panelboard with 16 active circuits, 4 spares, and 4 spaces.

The next step is to determine the interrupting rating requirements of the fuses or circuit breakers to be installed in the panel. *NEC 110.9* is an all-encompassing section that covers the interrupting rating requirements for services, mains, feeders, subfeeders, and branch-circuit overcurrent devices. For various types of equipment, normal currents can be determined by checking the equipment nameplate current, voltage, and wattage ratings. In addition, an ammeter can be used to check for normal and overloaded circuit conditions.

A standard ammeter must not be used to read short-circuit current, as this practice will result in damage to the ammeter and possible injury to personnel.

SHORT-CIRCUIT CALCULATIONS

The following sections will cover several of the basic methods of determining available short-circuit currents. As the short-circuit values given in the various tables are compared with the actual calculations, it will be noted that there are slight variances in the results. These differences are due largely to (1) rounding off the numbers in the calculations and (2) variations in the resistance and reactance data used to prepare the tables and charts. For example, the value of the square root of three (1.732) is used frequently in 3-phase calculations. Depending on the accuracy required, values of 1.7, 1.73, or 1.732 can be used.

In actual practice, the available short-circuit current at the load side of a transformer is less than the values shown in Problem 1. However, this simplified method of finding the available short-circuit currents will result in values that are conservative.

The actual impedance value on a UL-listed 25-kilovolt-ampere or larger transformer can vary plus or minus 10% from the transformer's marked impedance. The actual impedance for a transformer with a marked impedance of 2% could be as low as 1.8% and as high as 2.2%. This will affect the available fault-current calculations.

For example, in Problem 1, the marked impedance is reduced by 10% to reflect the transformer's possible actual impedance. The calculations show this worst-case scenario. All short-circuit examples in this text have the marked transformer impedance values reduced by 10%.

Another factor that affects fault-current calculations is voltage. Utility companies are allowed to vary voltage to their customers within a certain range. This might be plus or minus 10% for power services and 5% for lighting services. The higher voltage will result in a larger magnitude of fault current.

Determining the Short-Circuit Current at the Terminals of a Transformer Using the Impedance Formula

PROBLEM 1:

The three-phase transformer installed by the utility company to serve the Commercial Building has a rating of 225 kVA at 208Y/120 volts. The marked impedance on the nameplate is 2%. The available short-circuit current at the secondary terminals of the transformer must be determined. To simplify the calculations, assume that the utility can deliver unlimited short-circuit current to the primary of the transformer. This is referred to as an *infinite bus* or an *infinite primary*.

The first step is to determine the normal full-load current rating of the transformer:

$$I = \frac{kVA \times 1000}{E \times 1.732} = \frac{225 \times 1000}{208 \times 1.732} = 625 \text{ amperes}$$

Using the impedance value given on the nameplate of the transformer, the next step is to find a multiplier that can be used to determine the short-circuit current available at the secondary terminals of the transformer.

The factor of 0.9 shown in the following calculations reflects the fact that the transformer's actual impedance might be 10% less than marked on the nameplate and would be a worst-case condition. In electrical circuits, the lower the impedance, the higher the short-circuit current.

For simplicity, carry the results to two places beyond the decimal point, and round up or down as necessary. It is not necessary to be a rocket scientist. The goal is to simply derive the *approximate* available short-circuit current at the secondary terminals of the transformers.

- If the transformer is marked 2% impedance:

$$\text{the multiplier is } \frac{100}{2 \times 0.9} = 55.66$$

then, $625 \times 55.56 = 34,725$ amperes of short-circuit current.

- If the transformer had been marked 1% impedance:

$$\text{the multiplier would have been } \frac{100}{1 \times 0.9} = 111.11$$

then, $625 \times 111.11 = 69,444$ amperes of short-circuit current.

- If the transformer had been marked 4% impedance:

$$\text{the multiplier would have been } \frac{100}{4 \times 0.9} = 27.78$$

then, $625 \times 27.78 = 17,363$ amperes of short-circuit current.

Another source of short-circuit current comes from electric motors that are running at the time the fault occurs. This is covered later in this chapter.

Thus, it can be seen that no matter how much data we plug into our fault-current calculations, there are many variables that are out of our control. What we hope for is to arrive at a result that is reasonably accurate so that our electrical equipment is reasonably safe insofar as interrupting ratings and withstand ratings are concerned.

In addition to the methods of determining available short-circuit currents that are provided in the following discussion, there are computer software programs that do the calculations. These programs are fast, particularly when there are many points in a system to be calculated.

Determining the Short-Circuit Current at the Terminals of a Transformer Using Tables

Table 18-1 is a table of the short-circuit currents for a typical transformer. NEMA and transformer manufacturers publish short-circuit tables for many sizes of transformers having various impedance values. Table 18-1 provides data for a 300-kilovolt-ampere, 3-phase transformer with an impedance of 2%. According to the table, the symmetrical short-circuit current is 42,090 amperes at the secondary terminals of a 208Y/120-volt transformer (refer to the zero-foot row of the table). This value is on the low side because the manufacturer that developed the table did not allow for the ±10% impedance variation allowed by the UL Standard 1561. Table 18-2 indicates that the available fault current at the secondary of a 120/208-volt, 3-phase, 300-kilovolt-ampere transformer with a 1.11% impedance is 83,383 amperes.

Determining the Short-Circuit Current at Various Distances from a Transformer Using the Table

The amount of available short-circuit current decreases as the distance from the transformer increases, as indicated in Table 18-1. See Problem 2.

Determining Short-Circuit Currents at Various Distances from Transformers, Switchboards, Panelboards, and Load Centers Using the Point-to-Point Method

A simple method of determining the available short-circuit currents (also referred to as fault current) at various distances from a given location is the point-to-point method. Reasonable accuracy is obtained when this method is used with 3-phase and single-phase systems.

The following procedure demonstrates the use of the point-to-point method:

Step 1.　Determine the full-load rating of the transformer in amperes from the transformer nameplate, tables, or the following formulas:

  a.　For 3-phase transformers:

$$I_{FLA} = \frac{kVA \times 1000}{E_{L-L} \times 1.732}$$

    where E_{L-L} = Line-to-line voltage

  b.　For single-phase transformers:

$$I_{FLA} = \frac{kVA \times 1000}{E_{L-L}}$$

Step 2.　Find the percent impedance (Z) on the name plate of the transformer.

Step 3.　Find the transformer multiplier M_1:

$$M_1 = \frac{1000}{transformer\ \%\ impedance\ (Z) \times 0.9}$$

Note: Because the marked transformer impedance can vary ±10% per the UL Standard 1561, the 0.9 factor takes this into consideration to show worst-case conditions.

Step 4.　Determine the transformer let-through short-circuit current at the secondary terminals of the transformer. Use tables or the following formula:

  a.　For 3-phase transformers (L–L):

$$I_{SCA} = transformer_{FLA} \times multiplier\ M_1$$

  b.　For single-phase transformers (L–L):

$$I_{SCA} = transformer_{FLA} \times multiplier\ M_1$$

  c.　For single-phase transformers (L–N):

$$I_{SCA} = transformer_{FLA} \times multiplier\ M_1 \times 1.5$$

Note: The 1.5 factor is explained in the explanatory note below Step 5 in the paragraph marked "Explanatory Note."

PROBLEM 2:

For a 300-kilovolt-ampere transformer with a secondary voltage of 208 volts, find the available short-circuit current at a main switch that is located 25 ft (7.5 m) from the transformer. The main switch is supplied by four 750-kcmil copper conductors per phase in steel conduit.

Refer to Table 18-1 and read the value of 39,090 amperes in the column on the right-hand side of the table for a distance of 25 ft (7.5 m).

TABLE 18-1

Symmetrical short-circuit currents at various distances from a 300-kVA, 3-phase, 2% impedance transformer.

WIRE-SIZE (COPPER)

VOLTS	Dist (FT)	(MM)	#14	#12	#10	#8	#6	#4	#1	0	00	000	2-000	0000	250 kcmil	2-250 kcmil	3-300 kcmil	350 kcmil	2-350 kcmil	3-350 kcmil	3-400 kcmil	500 kcmil	2-500 kcmil	750 kcmil	4-750 kcmil
208	0	0	42090	42090	42090	42090	42090	42090	42090	42090	42090	42090	42090	42090	42090	42090	42090	42090	42090	42090	42090	42090	42090	42090	42090
	5	1.5	6910	10290	14730	19970	25240	29840	34690	35770	36640	37340	39610	37930	38270	40100	40870	38840	40410	40960	41030	39300	40650	39650	41460
	10	3	3640	5610	8460	12350	17090	22230	29030	30760	33240	33410	37340	34420	35030	38270	39710	36040	38840	39870	40010	36850	39300	37480	40840
	25	7.5	1500	2360	3670	5650	8430	12150	18930	21170	23240	25090	31170	26750	27780	33590	36560	29550	34780	37020	37230	31020	35730	32190	39090
	50	15	760	1200	1890	2950	4530	6810	11740	13670	15610	17510	25090	19320	20520	27780	32250	22660	29550	32850	33340	24520	31020	26050	36480
	100	30	380	600	960	1510	2350	3610	6610	7920	9320	10810	17510	12320	13380	20520	26010	15400	22660	26850	27530	17250	24520	18860	32190
	200	60	190	300	480	760	1190	1860	3510	4280	5140	6090	10810	7110	7860	13380	18660	9360	15400	19590	20370	10820	17250	12150	26050
	500	150	80	120	190	310	480	760	1460	1800	2180	2630	4990	3130	3500	6510	10030	4290	7820	10770	11400	5100	9120	5870	16570
	1000	300	40	60	100	150	240	380	740	910	1110	1350	2630	1620	1820	3500	5650	2250	4290	6140	6560	2710	5100	3160	10310
	5000	1500	10	10	20	30	50	80	150	180	230	280	550	330	380	740	1260	470	930	1380	1490	570	1130	670	2560
240	0	0	37820	37820	37820	37820	37820	37820	37820	37820	37820	37820	37820	37820	37820	37820	37820	37820	37820	37820	37820	37820	37820	37820	37820
	5	1.5	7750	11330	15810	20720	25260	28940	32560	33340	33960	34460	36080	34870	35120	36420	36960	35520	36640	37020	37070	35840	36800	36090	37370
	10	3	4140	6320	9400	13430	18040	22670	28230	29560	30660	31550	34460	32290	32730	35120	36140	33470	35520	36260	36350	34060	35840	34510	36930
	25	7.5	1720	2700	4180	6360	9360	13190	19640	21620	23380	24920	30240	26260	27090	31650	33860	28480	32530	34130	34340	29610	33230	30510	35680
	50	15	870	1380	2160	3360	5130	7620	12730	14630	16480	18230	24920	19850	20900	27090	30610	22740	28480	31060	31430	24300	29610	25570	33780
	100	30	440	700	1100	1730	2680	4100	7380	8770	10220	11720	18230	13220	14240	20900	25600	16150	22740	26280	26630	17860	24300	19310	30510
	200	60	220	350	550	880	1370	2130	3990	4830	5770	6790	11720	7880	8650	14240	19200	10190	16150	19200	20710	11660	17860	12960	25570
	500	150	90	140	220	350	560	870	1670	2050	2490	2990	5610	3540	3960	7230	10890	4820	8590	11620	12250	5700	9920	6520	17200
	1000	300	40	70	110	180	280	440	850	1050	1280	1550	2990	1850	2080	3960	6300	2560	4820	6820	7270	3070	5700	3570	11130
	5000	1500	10	10	20	40	60	90	170	210	260	320	630	380	430	860	1440	540	1070	1580	1710	660	1290	770	2910
480	0	0	18910	18910	18910	18910	18910	18910	18910	18910	18910	18910	18910	18910	18910	18910	18910	18910	18910	18910	18910	18910	18910	18910	18910
	5	1.5	10450	12820	14750	16150	17080	17690	18200	18310	18400	18470	18690	18520	18550	18730	18800	18610	18760	18810	18810	18650	18780	18690	18850
	10	3	6750	9170	11630	13780	15400	16530	17540	17740	17910	18040	18470	18150	18210	18550	18690	18320	18610	18710	18720	18400	18650	18470	18800
	25	7.5	3180	4740	6770	9150	11520	13570	15690	16160	16540	16840	17830	17100	17250	18040	18380	17490	18180	18410	18440	17690	18280	17840	18630
	50	15	1680	2590	3900	5680	7840	10170	13190	13960	14600	15120	16840	15560	15820	17250	17870	16260	17490	17940	18000	16610	17690	16890	18360
	100	30	860	1350	2090	3180	4680	6600	9820	10810	11690	12460	15120	13130	13540	15820	16930	14240	16260	17060	17170	14810	16610	15260	17840
	200	60	440	690	1080	1680	2560	3810	6370	7320	8240	9110	12460	9930	10450	13540	15300	11370	14240	15530	15710	12150	14810	12780	16890
	500	150	180	280	440	700	1080	1670	3040	3640	4290	4960	8010	5560	6140	9360	11820	7050	10320	12190	12500	7880	11150	8600	14550
	1000	300	90	140	220	350	550	860	1620	1970	2370	2800	4960	3270	3610	6140	8520	4300	7050	8940	9290	4960	7880	5560	11830
	5000	1500	20	30	40	70	110	180	340	420	510	620	1210	750	840	1610	2600	1040	1980	2820	3020	1250	2350	1450	4730
600	0	0	15130	15130	15130	15130	15130	15130	15130	15130	15130	15130	15130	15130	15130	15130	15130	15130	15130	15130	15130	15130	15130	15130	15130
	5	1.5	10210	11790	12920	13690	14180	14500	14770	14820	14870	14900	15010	14930	14940	15040	15070	14970	15050	15080	15080	15000	15060	15010	15100
	10	3	7270	9270	11010	12350	13280	13890	14410	14520	14610	14680	14900	14730	14770	14940	15020	14820	14970	15020	15030	14870	15000	14900	15070
	25	7.5	3740	5370	7280	9230	10920	12200	13410	13670	13870	14040	14570	14170	14250	14680	14850	14380	14750	14870	14890	14490	14800	14570	14980
	50	15	2040	3080	4500	6270	8170	9950	11940	12400	12770	13060	14040	13310	13460	14250	14590	13700	14380	14620	14650	13900	14490	14050	14840
	100	30	1060	1650	2510	3730	5290	7080	9650	10350	10930	11420	13060	11840	12090	13460	14080	12510	13700	14150	14210	12850	13900	13120	14570
	200	60	540	850	1330	2040	3040	4390	6840	7640	8390	9050	11420	9640	10010	12090	13160	10640	12510	13290	13390	11160	12850	11580	14050
	500	150	220	350	550	860	1330	2010	3550	4180	4830	5480	8180	6110	6530	9210	10960	7310	9900	11210	11400	7990	10470	8560	12700
	1000	300	110	170	280	440	680	1050	1950	2360	2800	3270	5480	3760	4110	6530	8540	4780	7310	8860	9120	5410	7990	5970	10940
	5000	1500	20	40	60	90	140	220	420	520	640	770	1470	910	1030	1930	3030	1260	2340	3270	3480	1510	2750	1740	5180

TABLE 18-2

Short-circuit currents available from various size transformers.

Voltage and Phases	kVA	Full Load Amps	% Impedance[††] (Nameplate)	Short-Circuit Amps[†]
120/240 single-phase[*]	25	104	1.58	11,574
	37½	156	1.56	17,351
	50	209	1.54	23,122
	75	313	1.60	32,637
	100	417	1.60	42,478
	167	695	1.80	60,255
120/208 3-phase[**]	25	69	1.60	4,791
	50	139	1.60	9,652
	75	208	1.11	20,821
	100	278	1.11	27,828
	150	416	1.07	43,198
	225	625	1.12	62,004
	300	833	1.11	83,383
	500	1388	1.24	124,373
	750	2082	3.50	66,095
	1000	2776	3.50	88,167
	1500	4164	3.50	132,190
	2000	5552	5.00	123,377
	2500	6950	5.00	154,444
277/480 3-phase	112½	135	1.00	15,000
	150	181	1.20	16,759
	225	271	1.20	25,082
	300	361	1.20	33,426
	500	601	1.30	51,362
	750	902	3.50	28,410
	1000	1203	3.50	38,180
	1500	1804	3.50	57,261
	2000	2406	5.00	53,461
	2500	3007	5.00	66,822

*Single-phase values are L–N values at transformer terminals. These figures are based on change in turns ratio between primary and secondary, 100,000 kVA primary, zero feet from terminals of transformer, 1.2 (%X) and 1.5 (%R) multipliers for L–N versus L–L reactance and resistance values and transformer X/R ratio = 3.

**Three-phase, short-circuit currents based on "infinite" primary.

††UL listed transformers 25 kVA or greater have a ±10% impedance tolerance. Short-circuit amps reflect a worst-case condition.

†Fluctuations in system voltage will affect the available short-circuit current. For example, a 10% increase in system voltage will result in a 10% increase in the available short-circuit currents shown in the table.

Step 5. Determine the f factor:

a. For 3-phase faults:

$$f = \frac{1.732 \times L \times I_{SCA}}{N \times C \times E_{L-L}}$$

b. For single-phase, line-to-line (L–L) faults on single-phase, center-tapped transformers:

$$f = \frac{2 \times L \times I_{SCA}}{N \times C \times E_{L-L}}$$

c. For single-phase, line-to-neutral (L–N) faults on single-phase, center-tapped transformers:

$$f = \frac{1.732 \times L \times I_{SCA}}{N \times C \times E_{L-N}}$$

where

L = the length of the circuit to the fault, in feet.

I_{SCA} = the available fault current in amperes at the beginning of the circuit.

C = the constant derived from Table 18-3 for the specific type of conductors and wiring method.

E = the voltage, line-to-line or line-to-neutral. See Step 4a, b, and c to decide which voltage to use.

N = the number of conductors in parallel.

Explanatory note: At the secondary terminals of a single-phase, center-tapped transformer, the L–N fault current is higher than the L–L fault current. At some distance from the terminals, depending on the wire size and type, the L–N fault current is lower than the L–L fault current. This can vary from 1.33 to 1.67 times. These figures are based on the change in the turns ratio between the primary and secondary, infinite source impedance, a distance of zero feet from the terminals of the transformer, and 1.2% reactance (X) and 1.5% resistance (R) for the L–N versus L–L resistance and reactance values. For simplicity, in Step 4c we used an approximate multiplier of 1.5. First, do the L–N calculation at the transformer secondary terminals, Step 4c, and then proceed with the point-to-point method. See Figure 18-1 for an example.

Step 6. After finding the f factor, refer to Table 18-4 and locate in Chart M the appropriate value of the multiplier M_2 for the specific f value. Or, calculate as follows:

$$M_2 = \frac{1}{1 + f}$$

Step 7. Multiply the available fault current at the beginning of the circuit by the multiplier M_2 to determine the available symmetrical fault current at the fault.

$$I_{SCA} \text{ at fault} = I_{SCA} \text{ at beginning of circuit} \times M_2$$

Motor Contribution. All motors running at the instant a short circuit occurs contribute to the short-circuit current. The amount of current from the motors is equal approximately to the starting (locked-rotor) current for each motor. This current value depends on the type of motor, its characteristics, and its code letter. Refer to *NEC Table 430.7(B)*. It is common practice to multiply the full-load ampere rating of the motor by 4 or 5 to obtain a close approximation of the locked-rotor current and provide a margin of safety. For energy-efficient motors, multiply the motor's full-load current rating by 6 to 8 times for a reasonable approximation of fault-current contribution. The current contributed by running motors at the instant a short circuit occurs is added to the value of the short-circuit current at the main switchboard prior to the start of the point-to-point calculations for the rest of the system. To simplify the following problems, motor contributions have not been added to the short-circuit currents.

SHORT-CIRCUIT CURRENT VARIABLES

Phase-to-Phase-to-Phase Fault (L–L–L)

The 3-phase fault-current value determined in Step 7 is the *approximate* current that will flow if the three hot phase conductors of a 3-phase system are shorted together in what is commonly referred to as a bolted fault. This is the worst-case condition.

TABLE 18-3

C Values.

| AWG or kcmil | Copper Conductors Three Single Conductors | | | | | | Copper Conductors Three Conductor Cable | | | | | |
| | Steel Conduit | | | Nonmagnetic Conduit | | | Steel Conduit | | | Nonmagnetic Conduit | | |
	600V	5KV	15KV	600V	5KV	15KV	600V	5KV	15KV	600V	5KV	15KV
14	389	—	—	389	—	—	389	—	—	389	—	—
12	617	—	—	617	—	—	617	—	—	617	—	—
10	981	—	—	981	—	—	981	—	—	981	—	—
8	1557	1551	1557	1556	1555	1558	1559	1557	1559	1559	1558	1559
6	2425	2406	2389	2430	2417	2406	2431	2424	2414	2433	2428	2420
4	**4779**	3750	3695	3825	3789	3752	3830	3811	3778	3837	3823	3798
3	4760	4760	4760	4802	4802	4802	4760	4790	4760	4802	4802	4802
2	5906	5736	5574	6044	5926	5809	5989	5929	5827	6087	6022	5957
1	7292	7029	6758	7493	7306	7108	7454	7364	7188	7579	7507	7364
1/0	8924	8543	7973	9317	9033	8590	9209	9086	8707	9472	9372	9052
2/0	10755	10061	9389	11423	10877	10318	11244	11045	10500	11703	11528	11052
3/0	12843	11804	11021	13923	13048	12360	13656	13333	12613	14410	14118	13461
4/0	15082	13605	12542	16673	15351	14347	16391	15890	14813	17482	17019	16012
250	16483	14924	13643	18593	17120	15865	18310	17850	16465	19779	19352	18001
300	18176	16292	14768	20867	18975	17408	20617	20051	18318	22524	21938	20163
350	**19529**	17385	15678	22736	20526	18672	**22646**	21914	19821	**24904**	24126	21982
400	20565	18235	16365	24296	21786	19731	24253	23371	21042	26915	26044	23517
500	22185	19172	17492	26706	23277	21329	26980	25449	23125	30028	28712	25916
600	22965	20567	17962	28033	25203	22097	28752	27974	24896	32236	31258	27766
750	24136	21386	18888	28303	25430	22690	31050	30024	26932	32404	31338	28303
1000	25278	22539	19923	31490	28083	24887	33864	32688	29320	37197	35748	31959

| AWG or kcmil | Aluminum Conductors Three Single Conductors | | | | | | Aluminum Conductors Three Conductor Cable | | | | | |
| | Steel Conduit | | | Nonmagnetic Conduit | | | Steel Conduit | | | Nonmagnetic Conduit | | |
	600V	5KV	15KV	600V	5KV	15KV	600V	5KV	15KV	600V	5KV	15KV
14	236	—	—	236	—	—	236	—	—	236	—	—
12	375	—	—	375	—	—	375	—	—	375	—	—
10	598	—	—	598	—	—	598	—	—	598	—	—
8	951	950	951	951	950	951	951	951	951	951	951	951
6	1480	1476	1472	1481	1478	1476	1481	1480	1478	1482	1481	1479
4	2345	2332	2319	2350	2341	2333	2351	2347	2339	2353	2349	2344
3	2948	2948	2948	2958	2958	2958	2948	2956	2948	2958	2958	2958
2	3713	3669	3626	3729	3701	3672	3733	3719	3693	3739	3724	3709
1	4645	4574	4497	4678	4631	4580	4686	4663	4617	4699	4681	4646
1/0	5777	5669	5493	5838	5766	5645	5852	5820	5717	5875	5851	5771
2/0	7186	6968	6733	7301	7152	6986	7327	7271	7109	7372	7328	7201
3/0	8826	8466	8163	9110	8851	8627	9077	8980	8750	9242	9164	8977
4/0	10740	10167	9700	11174	10749	10386	11184	11021	10642	11408	11277	10968
250	12122	11460	10848	12862	12343	11847	12796	12636	12115	13236	13105	12661
300	13909	13009	12192	14922	14182	13491	14916	14698	13973	15494	15299	14658
350	15484	14280	13288	16812	15857	14954	15413	15490	15540	16812	17351	16500
400	16670	15355	14188	18505	17321	16233	18461	18063	16921	19587	19243	18154
500	18755	16827	15657	21390	19503	18314	21394	20606	19314	22987	22381	20978
600	20093	18427	16484	23451	21718	19635	23633	23195	21348	25750	25243	23294
750	21766	19685	17686	23491	21769	19976	26431	25789	23750	25682	25141	23491
1000	23477	21235	19005	28778	26109	23482	29864	29049	26608	32938	31919	29135

| Ampacity | Plug-In Busway | | Feeder Busway | | High Imped. Busway |
	Copper	Aluminum	Copper	Aluminum	Copper
225	28700	23000	18700	12000	—
400	38900	34700	23900	21300	—
600	41000	38300	36500	31300	—
800	46100	57500	49300	44100	—
1000	69400	89300	62900	56200	15600
1200	94300	97100	76900	69900	16100
1350	119000	104200	90100	84000	17500
1600	129900	120500	101000	90900	19200
2000	142900	135100	134200	125000	20400
2500	143800	156300	180500	166700	21700
3000	144900	175400	204100	188700	23800
4000	—	—	277800	256400	—

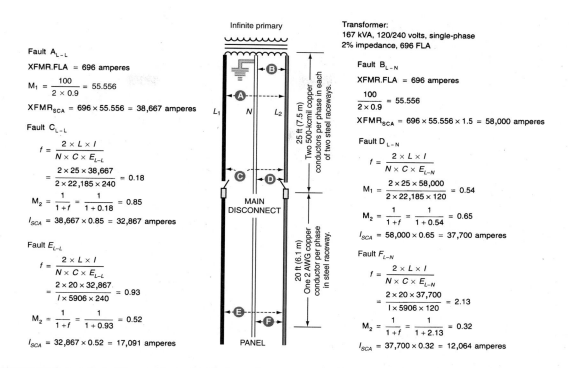

FIGURE 18-1   Point-to-point calculations for a single-phase, center-tapped transformer. Calculations show L–L and L–N values. (*Delmar/Cengage Learning*)

Phase-to-Phase Fault (L–L)

To obtain the *approximate* short-circuit current values when two hot conductors of a 3-phase system are shorted together, use 87% of the 3-phase current value. In other words, if the 3-phase current value is 20,000 amperes when the three hot lines are shorted together (L–L–L value), then the short-circuit current to two hot lines shorted together (L–L) is approximately

$$20,000 \times 0.87 = 17,400 \text{ amperes}$$

Phase-to-Neutral (Ground) (L–N) or (L–G)

For solidly grounded 3-phase systems, such as the 208Y/120-volt system that supplies the commercial building, the phase-to-neutral (ground) bolted short-circuit value can vary from 25% to 125% of the L–L–L bolted short-circuit current value.

It is rare that the L–N (or L–G) fault current would exceed the L–L–L fault current. Therefore, it is common practice to consider the L–N (or L–G) short-circuit current value to be the same as the L–L–L short-circuit current value.

In summary:

L–L–L	bolted short-circuit current = 100%
L–L	bolted short-circuit current = 87%
L–N	bolted short-circuit current = 100%

EXAMPLE

If the 3-phase (L–L–L) fault current has been calculated to be 20,000 amperes, then the L–N fault current is approximately

$$20,000 \times 1.00 = 20,000 \text{ amperes}$$

The main concern is to provide the proper interrupting rating for the overcurrent protective devices and adequate withstand rating for the equipment. Therefore, for most 3-phase electrical systems, the line-line-line bolted fault-current value will provide the desired level of safety.

Arcing Fault Multipliers (Approximate)

There are times when one wishes to know the values of arcing faults. These values vary

TABLE 18-4

Chart to convert f values to M_2 values when using the point-to-point method.

f	M_2	f	M_2
0.01	0.99	1.20	0.45
0.02	0.98	1.50	0.40
0.03	0.97	2.00	0.33
0.04	0.96	3.00	0.25
0.05	0.95	4.00	0.20
0.06	0.94	5.00	0.17
0.07	0.93	6.00	0.14
0.08	0.93	7.00	0.13
0.09	0.92	8.00	0.11
0.10	0.91	9.00	0.10
0.15	0.87	10.00	0.09
0.20	0.83	15.00	0.06
0.30	0.77	20.00	0.05
0.40	0.71	30.00	0.03
0.50	0.67	40.00	0.02
0.60	0.63	50.00	0.02
0.70	0.59	60.00	0.02
0.80	0.55	70.00	0.01
0.90	0.53	80.00	0.01
1.00	0.50	90.00	0.01
		100.00	0.01

$$M_2 = \frac{1}{1 + f}$$

considerably. The following are acceptable approximations.

Type of Fault	480 Volts	208 Volts
L–L–L	0.89*	0.12
L–L	0.74*	0.02
L–G (L–N)	0.38*	0.0

*Some reference books indicate this value to be 0.19.

EXAMPLE

What is the approximate line-to-ground arcing fault value on a 480/277-volt system where the line–line–line fault current has been calculated to be 30,000 amperes?

$$30,000 \times 0.38 = 11,400 \text{ amperes}$$

Fault Current at Main Service

As shown in Problem 3, the fuses or circuit breakers located at the six service disconnecting means for the commercial building must have an interrupting capacity of at least 31,391 RMS symmetrical amperes. This is based on a 225-kVA transformer and 25 feet of three parallel 350-kcmil underground service conductors. It is good practice to install protective devices having an interrupting rating at least 25% greater than the actual calculated available fault current. This practice generally provides a margin of safety to permit the rounding off of numbers, as well as compensating for a reasonable amount of short-circuit contribution from any electrical motors that may be running at the instant the fault occurs.

Another quick way to do a short-circuit current calculation is to refer to Table 18-4. Approximate the multiplier M_2 by noting that an f value of 0.106 is a small amount over 0.10. For reasonable results, it is acceptable to use the M_2 value of 0.10, which is 0.91. Remember that as stated numerous times in the text, there are many approximations involved when doing short-circuit current calculations.

Easy to use on-line programs to make short-circuit current calculations are available from various electrical equipment manufacturers.

The fuses specified for the Commercial Building have an interrupting rating of 200,000 amperes (see the Specifications). These are installed in fusible switches.

After calculation with an estimated 30 ft of feeder conductors to the Bakery panelboard, it is found 22,446 RMS symmetrical amperes are available at the panelboard. The circuit breakers will

PROBLEM 3:

Determine the approximate available short-circuit current at the termination box located on the back of the Commercial Building.

It is essential to know the available short-circuit current value at the termination box in order to select the service equipment and associated overcurrent devices that have adequate short-circuit and interrupting ratings. See *NEC 110.3(B), 110.9,* and *110.10.*

The size, type, and length of the conductors, and the type of raceway (magnetic or non-magnetic) must be considered.

Figure 18-2 shows the details of the Commercial Building's service and metering equipment as well as the feeders to the various tenants.

The service conductors to the Commercial Building consist of three 350-kcmil Type XHHW-2 copper conductors per phase connected in parallel. The service raceways are PVC (nonmagnetic). Assume the length of the service conductors to be 25 feet (7.6 m).

Use the available short-circuit current value of 34,725 amperes at the secondary terminals of the transformer as calculated in Problem 1.

Refer to Table 18-3 and Table 18-4 to find the values of C, f, and M. Use the point-to-point method to make the short-circuit current calculation.

Step 1. $f = \dfrac{1.732 \times L \times I_{SCA}}{3 \times C \times E_{L-L}} = \dfrac{1.732 \times 25 \times 34{,}725}{3 \times 22{,}736 \times 208} = 0.106$

Step 2. $M_2 = \dfrac{1}{1+f} = \dfrac{1}{1+0.106} = 0.904$

Step 3.   The approximate available short-circuit current at the 1000-ampere terminal box:

$I = ISCA \times M_2 = 34{,}725 \times 0.904 = 31{,}391$ RMS symmetrical amperes

need to have a rating of approximately 24,000 AIC or the panelboard must be protected by the use of the series-combination concept. Series-rated equipment is required to comply with *NEC 240.86.*

In the Commercial Building, series-rated equipment may be installed. Current-limiting fuses not to exceed the maximum size and type specified by the manufacturer of the equipment are installed in the main switchboard for the protection of the feeder conductors and panelboards. Circuit breakers having an interrupting rating of 10,000 amperes are installed in the panelboards. The requirements for this arrangement are set in the Specifications for the Commercial Building.

If series-rated equipment is not installed, the panelboards and circuit breakers must be fully rated for the available fault current at each location. Series-rated equipment may be selected under engineering supervision or by using tested combinations.

If noncurrent-limiting overcurrent devices (standard molded-case circuit breakers) are to be installed in the main switchboard, breakers having adequate interrupting ratings must be installed in the panelboards. A short-circuit current study must be made for each panelboard location to determine the value of the available short-circuit current.

The cost of circuit breakers increases as the interrupting rating of the breaker increases. The most economical protection system generally results when current-limiting fuses are installed in the main switchboard to protect the breakers in the panelboards.

See Problem 4 as an example of short-circuit calculations for a single-phase transformer.

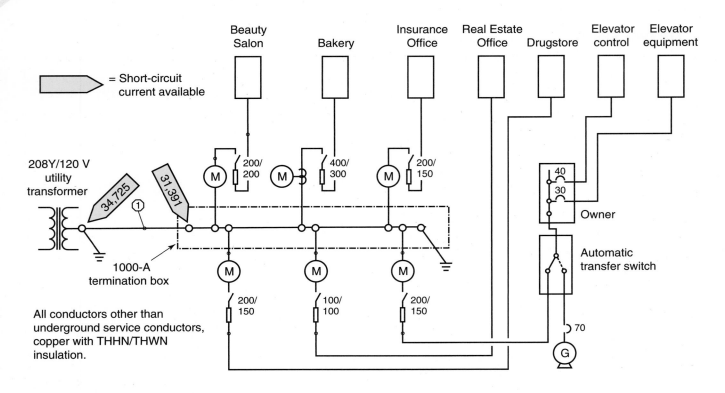

1 - Service: (3) Trade size 2½ PVC (Metric designator 63) with three 350 kcmil and one 1/0 AWG

FIGURE 18-2 Electrical system for the Commercial Building. See Figure 13-17, and blueprint sheet E–4 for a complete one-line diagram. (*Delmar/Cengage Learning*)

Review of Short-Circuit Requirements

1. To meet the requirements of *NEC 110.9* and *110.10,* it is absolutely necessary to determine the available fault currents at various points on the electrical system. If a short-circuit study is not done, the selection of overcurrent devices may be in error, resulting in a hazard to life and property.

2. To install a fully rated system, use fuses and circuit breakers that have an interrupting rating not less than the available fault current at the point of application.

3. To install a series-rated system, use listed series-rated electrical equipment where the available fault current exceeds the short-circuit rating of the downstream lower-rated circuit breakers to be installed in the system. Series-rated electrical equipment is available with fuse/breaker and breaker/breaker combinations. The lower interrupting rated circuit

breakers are protected by a higher interrupting rated main device. Tested in combination, this arrangement is called series-rated and is addressed in *NEC 110.22(B), 110.22(C),* and *240.86.* Series-rated electrical equipment is marked with its maximum short-circuit rating and the type of overcurrent devices permitted to be used in the specific series-rated combination. Series-rated systems may also be designed and selected under engineering supervision.

To properly select a series-rated system, you must use equipment that is tested and listed by a NRTL, or for existing installations only, select the components under engineering supervision. This means they are selected by a licensed professional engineer engaged primarily in the design or maintenance of electrical installations. Refer to *NEC 240.86.*

For an easy-to-use, computer version of doing short-circuit calculations, visit the Bussmann Web site at http://www.bussmann.com. Point to "Application

PROBLEM 4: Single-Phase Transformer

This problem is illustrated in Figure 18-1. The point-to-point is used to determine the currents for both line-to-line and line-to-neutral faults for a 167-kilovolt-ampere, 2% impedance transformer on a 120/240-volt, single-phase system. The marked transformer impedance can vary by ±10% per UL Standard 1561. To show the worst-case condition, this is accounted for at the beginning of the calculations by reducing the impedance 10% using this formula:

$$\frac{100}{2 \times 0.9} = M_2$$

Info" in the left column, and in the drop-down window click on "Software." In the following drop-down window click on "Short Circuit Calculations."

Problem 4 presents an excellent exercise in doing a short-circuit calculation, whether manually or by using a computer program.

COORDINATION OF OVERCURRENT PROTECTIVE DEVICES

Although this text cannot cover the topic of electrical system coordination (selectivity) in detail, the following material will provide the student with a working knowledge of this important topic. See Problem 4.

What Is Coordination?

Electrical system overcurrent protective devices can be coordinated "selective" or "nonselective."

The *NEC* defines selective coordination as *Localization of an overcurrent condition to restrict outages to the circuit or equipment affected, accomplished by the choice of overcurrent protective devices and their ratings or settings.**

You will need to become familiar with the terms *selective*, *selectivity*, and *coordination*. These words appear in many places in the *NEC*.

A situation known as nonselective coordination occurs when a fault on a branch circuit opens not only the branch-circuit overcurrent device but also the feeder overcurrent device, Figure 18-3. Nonselective systems are installed unknowingly and cause needless power outages in portions of an electrical system that should not be affected by a fault.

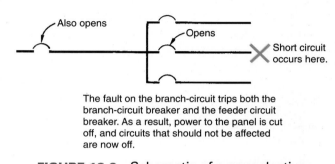

The fault on the branch-circuit trips both the branch-circuit breaker and the feeder circuit breaker. As a result, power to the panel is cut off, and circuits that should not be affected are now off.

FIGURE 18-3 Schematic of a nonselective system. (*Delmar/Cengage Learning*)

A selectively coordinated system, Figure 18-4, is one in which only the overcurrent device immediately upstream from the fault opens. Obviously, the installation of a selective system is much more desirable than a nonselective system.

The importance of selectivity in an electrical system is covered extensively throughout *NEC Article 517*. This article pertains to health care facilities, where maintaining electrical power is extremely important. The unexpected loss of power in certain areas of hospitals, nursing care centers, and similar health care facilities can be catastrophic.

The importance of selectivity is also emphasized in *NEC 240.12* (Electrical Systems Coordination); *NEC 230.95* (Ground-Fault Protection of Equipment);

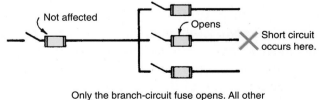

Only the branch-circuit fuse opens. All other circuits and the main feeder remain on.

FIGURE 18-4 Schematic of a selective system. (*Delmar/Cengage Learning*)

---

*Reprinted with permission from NFPA 70–2011.

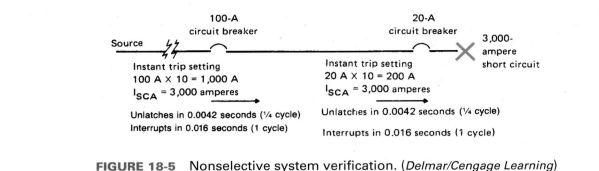

FIGURE 18-5 Nonselective system verification. (*Delmar/Cengage Learning*)

Informational Note No. 2, NEC 620.62 (Selective coordination for elevators, escalators, and moving walks); *NEC 700.4(B)* (Emergency Systems); *NEC 700.27* (coordination for emergency circuits); and *NEC 701.4(2)* and *701.27* (Legally Required Standby Systems; *NEC 708.22(B), 708.52(D),* and *708.54* (Critical Operations Power Systems . . . COPS). These sections refer to installations where additional hazards would be introduced should a nonorderly shutdown occur. The *Code* defines coordination, in part, as the proper localizing of a fault condition to restrict outages to the equipment affected.

Some local electrical codes require that all circuits, feeders, and mains in buildings such as schools, shopping centers, assembly halls, nursing homes, retirement homes, churches, restaurants, and any other places of public occupancy be selectively coordinated so as to minimize the dangers associated with total power outages.

Nonselectivity of an electrical system is not considered good design practice and is generally accepted only as a trade-off for a low-cost installation.

It is advisable to check with the authority enforcing the *Code* before proceeding too far in the selection of overcurrent protective devices for a specific installation.

By knowing how to determine the available short-circuit current and ground-fault current, the electrician can make effective use of the time-current curves and peak let-through charts (Chapter 17) to find the length of time required for a fuse to open or a circuit breaker to trip.

What Causes Nonselectivity?

In Figure 18-5, a short circuit in the range of 3000 amperes occurs on the load side of a 20-ampere breaker. The magnetic trip of the breaker is adjusted permanently by the manufacturer to unlatch at a current value equal to 10 times its rating or 200 amperes. The feeder breaker is rated at 100 amperes; the magnetic trip of this breaker is set by the manufacturer to unlatch at a current equal to 10 times its rating or 1000 amperes. This type of breaker generally cannot be adjusted in the field. Therefore, a current of 200 amperes or more will cause the 20-ampere breaker to trip instantly. In addition, any current of 1000 amperes or more will cause the 100-ampere breaker to trip instantly.

For the breakers shown in Figure 18-5, a momentary fault of 3000 amperes will trip (unlatch) both breakers. Because the flow of current in a series circuit is the same in all parts of the circuit, the 3000-ampere fault will trigger both magnetic trip mechanisms. The time-current curve shown in Figure 17-31 indicates that for a 3000-ampere fault, the unlatching time for both breakers is 0.0042 second and the interrupting time for both breakers is 0.016 second.

The term *interrupting time* refers to the time it takes for the circuit breakers' contacts to open, thereby stopping the flow of current in the circuit. Refer to Figure 18-5 and Figure 18-6.

This example of a nonselective system should make apparent to the student the need for a thorough study and complete understanding of time-current curves, fuse selectivity ratios, and unlatching time data for circuit breakers. Otherwise, a blackout may occur, such as the loss of exit and emergency lighting. The student must be able to determine available short-circuit currents (1) to ensure the proper selection of protective devices with adequate interrupting ratings and (2) to provide the proper coordination as well.

Figure 18-6 shows an example of a selective circuit. In this circuit, a fault current of 500 amperes

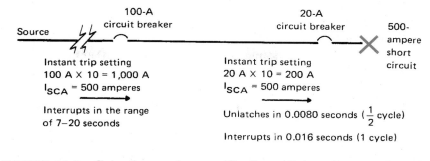

FIGURE 18-6   Selective system verification. (*Delmar/Cengage Learning*)

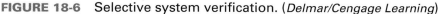

trips the 20-ampere breaker instantly (the unlatching time of the breaker is approximately 0.0080 second, and its interrupting time is 0.016 second). The graph in Figure 17-31 indicates that the 100-ampere breaker interrupts the 500-ampere current in a range from 7 to 20 seconds. This relatively lengthy trip time range is due to the fact that the 500-ampere fault acts on the current thermal trip element only and does not affect the magnetic trip element, which operates on a current of 1000 amperes or more.

The commercial building is a typical building. But you should at least be aware that there are electrical systems in certain types of buildings that are classified as *Critical Operations Power Systems*. New requirements for these types of electrical systems first appeared in the 2008 *NEC* as *Article 708*.

Critical operations power systems are power systems that *are generally installed in vital infrastructure facilities that, if destroyed or incapacitated, would disrupt national security, the economy, public health or safety, and where enhanced electrical infrastructure for continuity of operation has been deemed necessary by governmental authority.* In these types of installations, selective coordination of overcurrent devices is critical. These are highly technical installations and involve the knowledge and skills of a consulting electrical engineer(s) trained in these kinds of electrical system

requirements. The following National Fire Protection Association codes cover these types of installations: NFPA 99 (*Standard for Health Care Facilities*), NFPA 101 (*Life Safety Code*), NFPA 110 (*Standard for Emergency and Standby Power Systems*), and NFPA 1600 (*Standard on Disaster/Emergency Management and Business Continuity Programs*).

Selective System Using Fuses

The proper choice of the various classes and types of fuses is necessary if selectivity is to be achieved, Figure 18-7. Indiscriminate mixing of fuses of different classes, time-current characteristics, and even manufacturers may cause a system to become nonselective.

To ensure selective operation under low-overload conditions, it is necessary only to check and compare the time-current characteristic curves of fuses. Selectivity occurs in the overload range when the curves do not cross one another. See Problem 5.

Fuse manufacturers publish selectivity guides, similar to the one shown in Table 18-5, to be used for short-circuit conditions. When using these guides, selectivity is achieved by maintaining a specific amperage ratio between the various classes and types

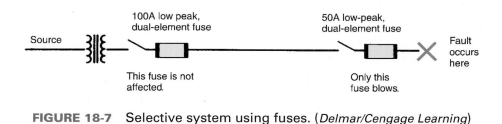

FIGURE 18-7   Selective system using fuses. (*Delmar/Cengage Learning*)

of fuses. A selectivity chart is based on any fault current up to the maximum interrupting ratings of the fuses listed in the chart.

Selective System Using Circuit Breakers

Circuit-breaker manufacturers publish time-current characteristic curves and unlatching information, Figure 17-31.

For normal overload situations, a circuit breaker having an ampere rating lower than the ampere rating of an upstream circuit breaker will trip. The upstream breaker will not trip. The system is "selective."

For low-level faults less than the instantaneous-trip setting of an upstream circuit breaker, a circuit breaker having an ampere rating lower than the ampere rating of the upstream breaker will trip, and the upstream breaker will not trip. The system is selective, Figure 18-6.

For fault-current levels above the instantaneous-trip setting of the upstream circuit breaker, both the branch-circuit breaker and the upstream circuit breaker will trip off. Series-rated systems, by their very nature, are nonselective, Figure 18-5.

⌁⌁ SINGLE PHASING

The *National Electrical Code* requires that all 3-phase motors be provided with running overcurrent protection in each phase in the motor circuit, as stated in *NEC Article 430, Part III*. A line-to-ground fault generally will open one fuse. There will be a resulting increase of 173%–200% in the line current in the remaining two connected phases. This increased current will be sensed by the motor fuses and overload relays when such fuses and relays are sized at 125% or less of the full-load current rating of the motor. Thus, the fuses and/or the overload relays will open before the motor windings are damaged. When properly matched, the overload relays will open before the fuses. The fuses offer backup protection if the overload relays fail to open for any reason.

A line-to-line fault in a 3-phase motor will open two fuses. In general, the operating coil of the motor controller will drop out, thus providing protection to the motor winding.

PROBLEM 5:

It is desired to install 100-ampere, dual-element fuses in a main switch and 50-ampere, dual-element fuses in the feeder switch. Is this combination of fuses "selective"?

Refer to the selectivity guide in Table 18-5 using the *line* and *column* marked Ⓐ. Because 100:50 is a 2:1 ratio, the installation is "selective." In addition, any fuse of the same type having a rating of less than 50 amperes will also be "selective" with the 100-ampere main fuses. Thus, for any short circuit or ground fault on the load side of the 50-ampere fuses, only the 50-ampere fuses will open.

To reduce single-phasing problems, each 3-phase motor must be provided with individual overload protection through the proper sizing of the overload relays and fuses. Phase-failure relays are also available. However, can other equipment be affected by a single-phasing condition?

In general, loads that are connected line-to-neutral or line-to-line, such as lighting, receptacles, and electric heating units will not burn out under a single-phasing condition. In other words, if one main fuse opens, then two-thirds of the lighting, receptacles, and electric heat will remain on. If two main fuses open, then one-third of the lighting remains on, and a portion of the electric heat connected line-to-neutral will stay on.

Nothing can prevent the occurrence of single phasing. What must be detected is the increase in current that occurs under single-phase conditions.

This is the purpose of motor overload protection as required in *NEC Article 430, Part III*. Motor overload, branch-circuit short-circuit, ground-fault protection as well as single-phasing issues were discussed in detail in Chapter 7.

It is essential to maintain some degree of lighting in occupancies such as stores, schools, offices, and health care facilities (such as nursing homes). A total blackout in these public structures has the potential for causing extensive personal injury due to panic. A loss of one or two phases of the system supplying a building should not cause a complete power outage in the building.

TABLE 18-5

Ratios for selectivity.

	RATIOS FOR SELECTIVITY									
LINE-SIDE FUSE	**LOAD-SIDE FUSE**								Ⓐ	
	KRP-CSP LOW-PEAK time-delay Fuse 601–6000A Class L	KTU LIMITRON fast-acting Fuse 601–6000A Class L	KLU LIMITRON time-delay Fuse 601–4000A Class L	KTN-R, KTS-R LIMITRON fast-acting Fuse 0–600A Class RK1	JJS, JJN TRON fast-acting Fuse 0–1200A Class T	JKS LIMITRON quick-acting Fuse 0–600A Class J	FRN-R, FRS-R FUSETRON dual-element Fuse 0–600A Class RK5	LPN-RK-SP, LPS-RK-SP LOW-PEAK dual-element Fuse 0–600A Class RK1	LPJ-SP LOW-PEAK time-delay Fuse 0–600A Class J	SC Type Fuse 0–60A Class G
KRP-CSP LOW-PEAK time-delay Fuse 601–6000A Class L	2:1	2:1	2.5:1	2:1	2:1	2:1	4:1	2:1	2:1	N/A
KTU LIMITRON fast-acting Fuse 601–6000A Class L	2:1	2:1	2.5:1	2:1	2:1	2:1	6:1	2:1	2:1	N/A
KLU LIMITRON time-delay Fuse 601–4000A Class L	2:1	2:1	2:1	2:1	2:1	2:1	4:1	2:1	2:1	N/A
KTN-R, KTS-R LIMITRON fast-acting Fuse 0–600A Class RK1	N/A	N/A	N/A	3:1	3:1	3:1	8:1	3:1	3:1	4:1
JJN, JJS TRON fast-acting Fuse 0–1200A Class T	N/A	N/A	N/A	3:1	3:1	3:1	8:1	3:1	3:1	4:1
JKS LIMITRON quick-acting Fuse 0–600A Class J	N/A	N/A	N/A	3:1	3:1	3:1	8:1	2:1	2:1	4:1
FRN-R, FRS-R FUSETRON dual-element Fuse 0–600A Class RK5	N/A	N/A	N/A	1.5:1	1.5:1	1.5:1	2:1	1.5:1	1.5:1	1.5:1
LPN-RK-SP, LPS-RK-SP LOW-PEAK dual-element Fuse 0–600A Class RK1 Ⓐ	N/A	N/A	N/A	3:1	3:1	3:1	8:1	2:1	2:1	4:1
LPJ-SP LOW-PEAK time-delay Fuse 0–600A Class J	N/A	N/A	N/A	3:1	3:1	3:1	8:1	2:1	2:1	4:1
SC Type Fuse 0–60A Class G	N/A	N/A	N/A	2:1	2:1	2:1	4:1	3:1	3:1	2:1

N/A = NOT APPLICABLE

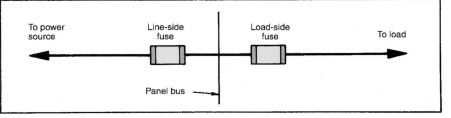

Selectivity Guide. This chart is one manufacturer's selectivity guide. All manufacturers of fuses have similar charts, but the charts may differ in that the ratios for given combinations of fuses may not be the same as in this chart. It is important to use the technical data for the particular type and manufacturer of the fuses being used. Intermixing fuses from various manufacturers may result in a nonselective system.

Refer to the *National Electrical Code* or the working drawings when necessary. Where applicable, responses should be written in complete sentences.

Refer to Table 18-1, which shows the symmetrical short-circuit currents for a 300-kilovolt-ampere transformer with 2% impedance.

1. The short-circuit current on a 208-volt system at a distance of 50 ft using 1 AWG conductors will be _____ amperes.

2. The short-circuit current on a 480-volt system at a distance of 100 ft using 500-kcmil conductors will be _____ amperes.

3. In your own words, define and give a hypothetical example of a selectively coordinated system.

   _____

   _____

   _____

   _____

   _____

   _____

4. Identify some common causes of nonselectivity.

   _____

   _____

   _____

   _____

Indicate whether the following systems would be selective or nonselective. Support your answer.

5. A KRP-C 2000-ampere fuse is installed on a feeder that serves an LPS-R 600-ampere fuse.

   _____

   _____

   _____

   _____

6. A KTS-R 400-ampere fast-acting fuse is installed on a feeder that serves an FRS-R 200-ampere dual-element fuse.

   _____

   _____

   _____

   _____

7. A panelboard is protected by a main 225-ampere circuit breaker and contains several 20-ampere breakers for branch-circuit protection. The breakers are all factory set to unlatch at 10 times their rating. Circle your response.

   a. For a 500-ampere short circuit or ground fault on the load side of a 20-ampere breaker, only the 20-ampere breaker will trip off.   (True) (False)

   b. For a 3000-ampere short circuit or ground fault on the load side of the 20-ampere breaker, the 20-ampere breaker and the 225-ampere breaker will trip off.   (True) (False)

Refer to the following drawing to make the requested calculations. Use the point-to-point method and show all calculations.

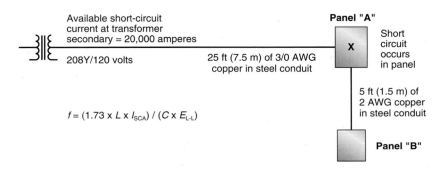

Available short-circuit
current at transformer
secondary = 20,000 amperes

208Y/120 volts

25 ft (7.5 m) of 3/0 AWG
copper in steel conduit

Panel "A"

X

Short
circuit
occurs
in panel

5 ft (1.5 m) of
2 AWG copper
in steel conduit

Panel "B"

$$f = (1.73 \times L \times I_{SCA}) / (C \times E_{L\text{-}L})$$

8. Calculate the short-circuit current at panelboard A.

9. Calculate the short-circuit current at panelboard B.

Calculate the following fault-current values for a 3-phase L–L–L bolted short circuit that has been calculated to be 40,000 amperes.

10. The approximate value of a line-to-line fault will be _____ amperes.

11. The approximate value of a line-to-ground fault will be _____ amperes.

12. a. Calculate the available fault-current at the Bakery panelboard. Refer to Figure 18-2 and Electrical Plan E4. Note that the feeder conductors are 300 kcmil copper installed in electrical metallic tubing. For simplicity, do not include motor contribution. Show your work.

   b. The Bakery panelboard is marked as having a short-circuit rating of 10,000 amperes RMS symmetrical. The circuit breakers in the panelboard have an interrupting rating of 10,000 amperes.

      Because the available fault current on the line side lugs of the Bakery panelboard is 21,942 amperes, explain in your own words how this installation can be made to be in compliance with the *NEC*.

Equipment and Conductor Short-Circuit Protection

OBJECTIVES

After studying this chapter, you should be able to

- understand that all electrical equipment has a withstand rating.

- discuss the withstand rating of conductors.

- understand important *National Electrical Code* sections that pertain to interrupting rating, available short-circuit current, current-limitation, effective grounding, bonding, and temperature limitation of conductors.

- understand what the term *ampere squared seconds* means.

- perform calculations to determine how much current a copper conductor can safely carry for a specified period of time before being damaged or destroyed.

- refer to charts to determine conductor withstand ratings.

- understand that the two forces present when short circuits or overloads occur are *thermal* and *magnetic*.

- discuss the 10-ft (3-m) tap rule and the 25-ft (7.5 m) conductor maximum overcurrent protection.

All electrical equipment, including switches, motor controllers, conductors, bus duct, panelboards, load centers, switchboards, and so on, has the capability to withstand a certain amount of electrical energy for a given amount of time before damage to the equipment occurs. This gives rise to the term *short-circuit current rating*. This term is used in the *NEC* and in numerous UL Standards. The term is synonymous with the term *withstand rating*.

Underwriters Laboratories standards specify certain test criteria for the preceding equipment. For example, switchboards must be capable of withstanding a given amount of fault current for at least three cycles. Both the amount of current and the length of time for the test are specified.

Simply stated, short-circuit current rating is the ability of the equipment to hold together for the time it takes the overcurrent protective device to respond to the fault condition.

Acceptable damage levels are well defined in the various Underwriters Standards for electrical equipment. Where the equipment is intended to actually break the current, such as with a fuse or circuit breaker, the equipment is marked with its interrupting rating. Electrical equipment manufacturers conduct exhaustive tests to determine the withstand and/or interrupting ratings of their products.

Equipment tested and listed is often marked with the size and type of overcurrent protection required. For example, the label on a motor controller might indicate a maximum size fuse for different sizes of thermal overloads used in that controller. This marking indicates that fuses must be used for the overcurrent protection. A circuit breaker would not be permitted; see *NEC 110.3(B)*.

The capability of a current-limiting overcurrent device to limit the let-through energy to a value less than the amount of energy that the electrical system is capable of delivering means that the equipment can be protected against fault-current values of high magnitude. Equipment manufacturers specify current-limiting fuses to minimize the potential damage that might occur in the event of a high-level short circuit or ground fault.

The electrical engineer and/or electrical contractor must perform short-circuit studies and then determine the proper size and type of current-limiting overcurrent protective device, which can be used ahead of the electrical equipment that does not have an adequate

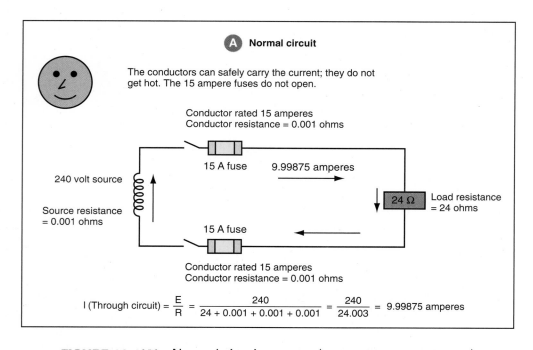

FIGURE 19-1(A) Normal circuit current. (*Delmar/Cengage Learning*)

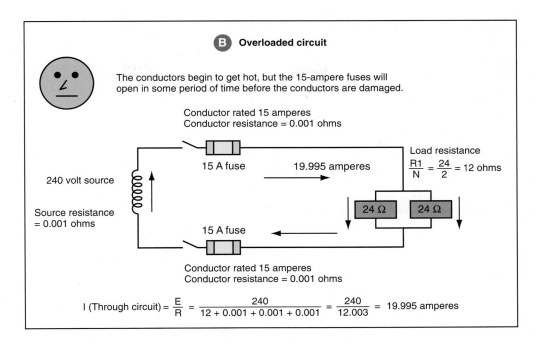

FIGURE 19-1(B)   Overloaded circuit current. (*Delmar/Cengage Learning*)

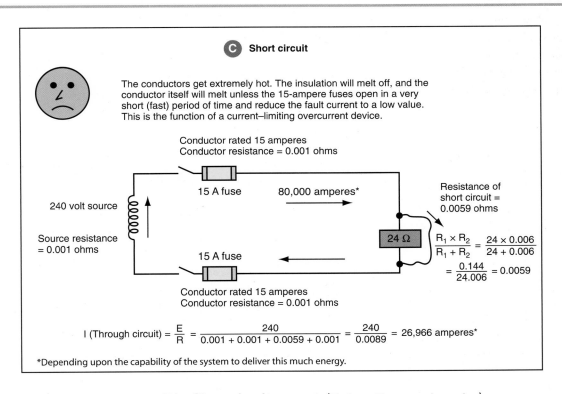

FIGURE 19-1(C)   Short-circuit current. (*Delmar/Cengage Learning*)

withstand or interrupting rating for the available fault current to which it will be subjected. The calculation of short-circuit currents is covered in Chapter 18.

Study the normal circuit, overloaded circuit, short-circuit, ground fault, and open-circuit diagrams [Figures 19-1(A) through 19-1(E)], and observe how Ohm's law is applied to these circuits. The calculations for the ground-fault circuit are not shown, as the impedance of the return ground path can vary considerably.

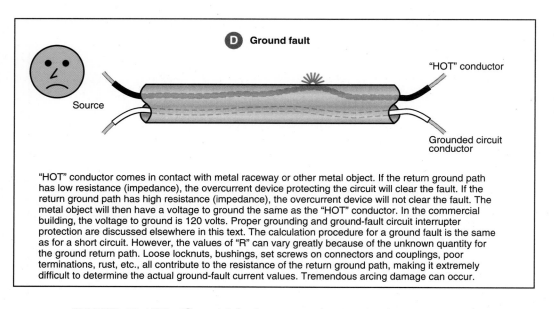

FIGURE 19-1(D) Ground-fault current. (*Delmar/Cengage Learning*)

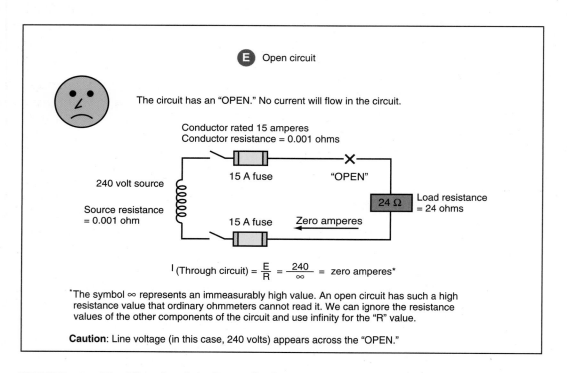

FIGURE 19-1(E) The circuit is "open" where marked "X." (*Delmar/Cengage Learning*)

CONDUCTOR WITHSTAND RATING

Up to this point in the text, we have covered in detail how to calculate conductor ampacities for branch circuits, feeders, and service-entrance conductors, with the basis being the connected load and/or volt-amperes per square foot. Then the demand factors and other diversity factors are applied. If we have followed the *Code* rules for calculating conductor size, the conductors will not be damaged under overloaded conditions because the conductors will have the proper ampacity and will be protected by the proper size and type of overcurrent protective device.

This part of Chapter 19 discusses this topic in a reasonable amount of detail. As difficult as the subject may seem, it is extremely important.

NEC tables such as ▶*Table 310.15(B)(16)*◀ set forth the ampacity ratings for various sizes and insulation types of conductors. These tables consider normal loading conditions.

NEC Article 310, and *Annex B* in the *NEC* relate to a conductor's current-carrying capability. For those of you having the interest to really dig into this subject, you can find additional information in the ANSI/ IEEE Red, Gray, Buff, and Blue books; the Insulated Cable Engineers Association, Inc., documents; the Canadian Electrical Code; and the International Electrotechnical Commission Standards.

We can think of an overload as excessive current that is staying in the intended path. We can think of a short circuit or ground fault as excessive current that is outside of the intended path.

When short circuits and/or ground faults occur, the connected loads are bypassed, thus the term *short circuit*. All that remains in the circuit are the conductors and other electrical devices such as breakers, switches, and motor controllers. The impedance (ac resistance) of these devices is extremely low, so for all practical purposes, the most significant opposition to the flow of current when a fault occurs is the conductor impedance. Fault-current calculations are covered in Chapter 18.

The following discussion covers the actual ability of a conductor to maintain its integrity for the time it is called on to carry fault current, instead of burning off and causing additional electrical problems. It is important that the electrician, electrical contractor, electrical inspector, and consulting engineer have a good understanding of what happens to a conductor under medium- to high-level fault conditions.

The withstand rating of a conductor, such as an equipment grounding conductor, main bonding jumper, or any other current-carrying conductor, reveals that the conductor can withstand a "certain amount of current for a given amount of time." This is the short-time withstand rating of the conductor. The conductor gets hot when excessive current is flowing through it. It is considered a failure in design or application if the temperature

of the conductor increases to the point that the insulation on the conductor is damaged.

Let us review some key *National Electrical Code* sections that focus on the importance of equipment and conductor withstand rating, available fault currents, short-circuit current ratings, and circuit impedance. Recall that we performed short-circuit current studies in Chapter 18.

- The interrupting rating is ▶*the highest current at rated voltage that a device is identified to interrupt under standard test conditions.** ◀ See *NEC Article 100*.

- *Listed or labeled equipment shall be installed and used in accordance with any instructions included in the listing or labeling.** See *NEC 110.3(B)*.

- **Interrupting Rating.** ▶*Equipment intended to interrupt current at fault levels shall have an interrupting rating not less than the nominal circuit voltage and the current that is available at the line terminals of the equipment.*

  Equipment intended to interrupt current at other than fault levels shall have an interrupting rating at nominal circuit voltage not less than the current that must be interrupted. See *NEC 110.9*.* ◀

- ▶**Circuit Impedance, Short-Circuit Current Ratings, and Other Characteristics.** *The overcurrent protective devices, the total impedance, the component short-circuit current ratings, and other characteristics of the circuit to be protected shall be selected and coordinated to permit the circuit protective devices to clear a fault without extensive damage to the electrical components of the circuit. This fault shall be assumed to be either between two or more of the circuit conductors or between any circuit conductor and the equipment grounding conductor or enclosing metal raceway. Listed products applied in accordance with their listing shall be considered to meet the requirements of this section.** ◀ See *NEC 110.10*.

- *110.24* ▶*Available Fault Current (A) Field Marking. Service equipment in other than*

---

*Reprinted with permission from NFPA 70-2011.

dwelling units shall be legibly marked in the field with the maximum available fault current. The field marking(s) shall include the date the fault current calculation was performed and be of sufficient durability to withstand the environment involved.

(B) Modifications. *When modifications to the electrical installation occur that affect the maximum available fault current at the service, the maximum available fault current shall be verified or recalculated as necessary to ensure the service-equipment ratings are sufficient for the maximum available fault current at the line terminals of the equipment. The required field markings in 110.24(A) shall be adjusted to reflect the new level of maximum available fault current.* ◄*

This new rule is fairly self-explanatory. The intention behind the rule is to be certain that service equipment is suitable for the short-circuit current that is available at its line terminals. As can be seen, whenever modifications are made to the electrical installation that might change the maximum available fault current at the service, a new calculation is required, and the new level of short-circuit current must be posted at the service equipment.

Implied in the rule is that the equipment is required to be suitable for the short-circuit current available at its line terminals, *110.9* and *110.10*. So, if the short-circuit current study performed at modifications reveals that the short-circuit current available exceeds the interrupting rating of the equipment, appropriate adjustments or corrections must be performed.

See Chapter 18 for extensive instructions on how to perform short-circuit current studies.

- *Overcurrent protection for conductors and equipment is provided to open the circuit if the current reaches a value that will cause an excessive or dangerous temperature in conductors or conductor insulation.* See NEC 240.1, Informational Note.*

- *A current-limiting overcurrent protective device is a device that, when interrupting currents in its current-limiting range, will reduce the current flowing in the faulted circuit to a magnitude substantially less than that obtainable in the same circuit if the device were replaced*

with a solid conductor having comparable impedance.* See NEC 240.2.

- The required label marking for fuses, one item being the fuse's interrupting rating, is set forth in *NEC 240.60(C)*.

- The required label marking for circuit breakers, one item being the circuit breaker's interrupting rating, is set forth in *NEC 240.83(C)*.

- The requirements for series-rated equipment are contained in *NEC 240.86*.

- Grounding of an electrical system is *to limit the voltage imposed by lightning, line surges, or unintentional contact with higher voltage lines so as to stabilize the voltage to the earth during normal operation.* See NEC 250.4(A)(1).*

- Grounding of conductive electrical conductors and equipment is to limit the voltage to ground on these materials. See *NEC 250.4(A)(2)*.

- Metal piping and other electrically conductive materials that might become energized shall be bonded together so as to establish an effective path for fault currents; see *NEC 250.4(A)(4)*.

- *NEC 250.4(A)(5)* requires that an effective ground-fault current path be established for all electrical equipment, wiring, and other electrically conductive material likely to become energized. This path shall:
  - be of low impedance, and
  - facilitate the operation of the overcurrent device, and
  - be capable of safely carrying the maximum ground fault current likely to be imposed on it.

The earth shall not be considered as an effective ground-fault current path.* See NEC 250.4(A)(5).

- *Bonding shall be provided where necessary to ensure electrical continuity and the capacity to conduct safely any fault current likely to be imposed.* See NEC 250.90.*

- *Metal raceways, cable trays, cable armor, cable sheath, enclosures, frames, fittings, and other metal non-current-carrying parts that are to serve as equipment grounding conductors, with or without the use of supplementary equipment grounding conductors, shall be bonded where necessary to ensure electrical continuity and the capacity to*

---

*Reprinted with permission from NFPA 70-2011.

conduct safely any fault current likely to be imposed on them. Any nonconductive paint, enamel, or similar coating shall be removed at threads, contact points, and contact surfaces or be connected by means of fittings designed so as to make such removal unnecessary. See *NEC 250.96(A)*.

- The size of the grounding electrode conductor for a grounded or ungrounded system shall not be smaller than given in *NEC Table 250.66*. See *NEC 250.66*.

- The size of copper, aluminum, or copper-clad aluminum equipment grounding conductors shall not be smaller than given in *NEC Table 250.122*. See *NEC 250.122*.

- Attention should be given to the requirement that equipment grounding conductors may need to be sized larger than shown in *NEC Table 250.122* because of the need to comply with *NEC 250.4(A)(5)* regarding effective grounding. This would generally be necessary where available fault currents are high, and the possibility of burning off the equipment grounding conductor under fault conditions exists. See *NEC 250.122*.

- *NEC Article 310* is rather lengthy, presenting in detail the requirements for conductors. A very important requirement is found in *NEC 310.15(A)(3)* which states that *No conductor shall be used in such a manner that its operating temperature will exceed that designated for the type of insulated conductor involved.** This section should be read carefully.

CONDUCTOR HEATING

Let's get the picture of why conductors or other electrical equipment get hot.

All electrical equipment is capable of withstanding a certain amount of current for a given period of time. Consider this: Let's agree that a 10 AWG Type THHN copper conductor has an ampacity of 40 amperes. This conductor will get extremely hot if subjected to 80 amperes. At some higher value of current, the insulation on the conductor will melt. At some even higher value of current, the conductor

*Reprinted with permission from NFPA 70-2011.

itself will melt. Simply stated, the amount of heat generated in a conductor or other electrical equipment is a matter of how much current flows, and for how long.

For the most part, we select overcurrent protective devices (fuses and circuit breakers) to protect conductors and other electrical equipment in conformance to the requirements found in *NEC Article 240*. To determine the opening time of overcurrent devices for different values of current, we use standard time-current curves discussed in Chapter 18.

For short periods of time such as the duration of a short circuit or ground fault, where the heat generated in the conductor cannot radiate, conduct, or convect away fast enough, the concept of I^2t is used. I^2t is a measure of heat energy. The I^2t withstand rating of a conductor or other electrical component is compared with the I^2t let-through energy of the overcurrent device.

The value of energy (heat) generated during a fault varies as the square of the root-mean-square (RMS) current multiplied by the time (duration of fault) in seconds. This value is expressed as I^2t.

The term I^2t is called "ampere squared seconds."

Because watts $= I^2R$, we could say that the damage that might be expected under severe fault conditions can be related to

- the amount of current flowing during the fault.

- the time in seconds that the fault current flows.

- the resistance of the fault path.

This relationship is expressed as

$$I^2Rt$$

Because the value of resistance under severe fault conditions is generally extremely low, we can simply think in terms of **how much** current (I) is flowing and for **how long** a time (t) the current will flow.

It is safe to say that whenever an electrical system is subjected to a high-level short circuit or ground fault, less damage will occur if the fault current can be limited to a low value and if the faulted circuit is cleared as fast as possible.

It is important to understand the time-current characteristics of fuses and circuit breakers in order to minimize equipment damage. Time-current characteristic curves were discussed in Chapter 17.

▬〜▬ CALCULATING AN INSULATED 75°C THERMOPLASTIC CONDUCTOR'S SHORT-TIME WITHSTAND RATING

The insulation rated 75°C on copper conductors can withstand

- 1 ampere (rms current)
- for 5 seconds
- for every 42.25 circular mils (cm) of cross-sectional area.

In the preceding statement both current (how much) and time (how long) are included.

If we wish to provide an equipment grounding conductor for a circuit protected by a 60-ampere overcurrent device, we find in *NEC Table 250.122* that a 10 AWG copper conductor is the MINIMUM size permitted.

Referring to *NEC Chapter 9, Table 8,* we find that the cross-sectional area of a 10 AWG conductor is 10,380 circular mils.

This 10 AWG copper conductor has a five-second withstand rating of

$$\frac{10,380 \text{ cm}}{42.25 \text{ cm}} = 246 \text{ amperes}$$

This means that 246 amperes is the maximum amount of current that a 10 AWG copper insulated conductor can carry for five seconds without being damaging the insulation. This is the 10 AWG conductor's short-time withstand rating.

Stating this information using the thermal stress (heat) formula, we have

$$\text{Thermal stress} = I^2 t$$

where

I = RMS current in amperes and
t = time in seconds

Thus, the 10 AWG copper conductor insulation's withstand rating is

$$I^2 t = 246^2 \times 5 = 302,580 \text{ A}^2 \text{ seconds}$$

With this basic information, we can easily determine the short-time withstand rating of this 10 AWG copper insulated conductor for other values of time and/or current. The 10 AWG copper conductor insulation's one-second withstand rating is

$$I^2 t = \text{ampere squared seconds}$$

$$I^2 = \frac{\text{ampere squared seconds}}{t}$$

$$I = \sqrt{\frac{\text{ampere squared seconds}}{t}}$$

$$I = \sqrt{\frac{302,580}{1}} = 550 \text{ amperes}$$

The 10 AWG copper conductor insulation's one-cycle withstand rating (given that the approximate opening time of a typical molded-case circuit breaker is one cycle, or $\frac{1}{60}$ of a second, or 0.0167 second) is

$$I = \sqrt{\frac{302,580}{0.0167}} = 4257 \text{ amperes}$$

The 10 AWG copper conductor insulation's one-quarter-cycle withstand rating (given that the typical clearing time for a current-limiting fuse is approximately one-fourth of a cycle, or 0.004 second) is

$$I = \sqrt{\frac{302,580}{0.004}} = 8697 \text{ amperes}$$

Therefore, a conductor can be subjected to large values of fault current if the clearing time is kept very short.

When applying current-limiting overcurrent devices, it is important to use peak let-through charts to determine the apparent RMS let-through current before applying the thermal stress formula. For example, in the case of a 60-ampere Class RK1 current-limiting fuse, the apparent RMS let-through current with an available fault current of 40,000 amperes is approximately 3000 amperes. This fuse will clear in approximately 0.004 second. See Figures 17-25, 19-2, and 19-4.

$$\begin{aligned} I^2 t \text{ let-through of current-limiting fuse} \\ = 3000 \times 3000 \times 0.004 \\ = 36,000 \text{ ampere squared seconds} \end{aligned}$$

Because a current-limiting overcurrent device is used in this example, the 10 AWG copper equipment

Current Limitation Curves

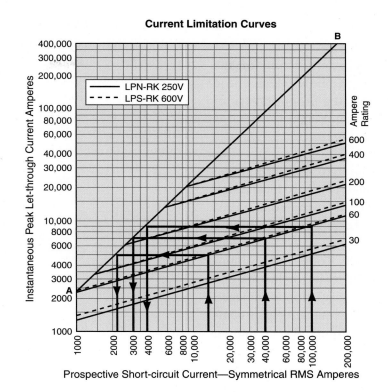

FIGURE 19-2 Current-limiting effect of Class RK1 fuses. The technique required to use these charts is covered in Chapter 17. (*Courtesy Cooper Bussmann*)

grounding conductor could be used where the available fault current is 40,000 amperes.

With an available fault current of 100,000 amperes, the apparent RMS let-through current of the 60-ampere Class RK1 current-limiting fuse is approximately 4000 amperes.

I^2t let-through of current-limiting fuse
$$= 4000 \times 4000 \times 0.004$$
$$= 64,000 \text{ ampere squared seconds}$$

Remember that, as previously discussed, the withstand I^2t rating of a 10 AWG copper conductor is 302,580 ampere squared seconds.

EXAMPLE

A 75°C thermoplastic insulated Type THWN copper conductor can withstand 4200 amperes for one cycle.

a. What is the I^2t withstand rating of the conductor?

b. What is the I^2t let-through value for a noncurrent-limiting circuit breaker that takes one cycle to open? The available fault current is 40,000 amperes.

c. What is the I^2t let-through value for a current-limiting fuse that opens in 0.004 second when subjected to a fault current of 40,000 amperes? The apparent RMS let-through current is approximately 4600 amperes. Refer to Figure 17-27.

d. Which overcurrent device (b or c) will properly protect the conductor under the 40,000-ampere available fault current?

ANSWERS

a. $I^2t = 4200 \times 4200 \times 0.016 = 282,240$ ampere squared seconds.

b. $I^2t = 40,000 \times 40,000 \times 0.016 = 25,600,000$ (2.56×10^7) ampere squared seconds.

c. $I^2t = 4600 \times 4600 \times 0.004 = 84,640$ ampere squared seconds.

d. Comparing the I^2t withstand rating of the conductor to the I^2t let-through values of the breaker (b) and fuse (c), the proper choice of protection for the conductor is (c).

The use of peak let-through, current-limiting charts is discussed in Chapter 17. Peak let-through charts are available from all manufacturers of current-limiting fuses and current-limiting circuit breakers.

Conductors are particularly vulnerable to damage under fault conditions. Circuit conductors can be heated to a point where the insulation is damaged or completely destroyed. The conductor can actually burn off. In the case of equipment grounding conductors, if the conductor burns off under fault conditions, the equipment can become hot, creating an electrical shock hazard.

Even if the equipment grounding conductor does not burn off, it can become so hot that it melts the insulation on the other conductors, shorting out the other circuit conductors in the raceway or cable. This results in further fault-current conditions, damage, and hazards.

Of extreme importance are bonding jumpers, particularly the main bonding jumpers in service equipment. These main bonding jumpers must be capable of safely carrying the fault current that is available at the service.

CALCULATING A BARE COPPER CONDUCTOR AND/OR ITS BOLTED SHORT-CIRCUIT WITHSTAND RATING

A bare conductor can withstand higher levels of current than an insulated conductor of the same cross-sectional area. Do not exceed

- 1 ampere (rms current)
- for 5 seconds
- for every 29.1 circular mils of cross-sectional area of the conductor.

Because bare equipment grounding conductors are often installed in the same raceway or cable as the insulated circuit conductors, the weakest link of the system would be the insulation on the circuit conductors. Therefore, the conservative approach when considering conductor safe withstand ratings is to apply the 1 AMPERE—FOR 5 SECONDS—FOR EVERY 42.25 CIRCULAR MILS formula, or simply refer to a conductor short-circuit withstand rating chart, such as Figure 19-3.

CALCULATING THE MELTING POINT OF A COPPER CONDUCTOR

The melting point of a copper conductor can be calculated by using these values:

- 1 ampere (RMS current)
- for 5 seconds
- for every 16.19 circular mils of cross-sectional area of the conductor.

NEC 250.122(B) states that if the ungrounded conductors are sized larger for any reason, the equipment grounding conductor shall be increased proportionately, based on the cross-sectional area of the ungrounded conductors for that particular branch circuit or feeder. The need to increase the size of the ungrounded conductor might be for voltage drop reasons, or because high values of fault current are possible. In *NEC 250.2* we find a definition of an effective ground-fault current path: *An intentionally constructed, permanent, low-impedance electrically conductive path designed and intended to carry current under ground-fault conditions from the point of a ground fault on a wiring system to the electrical supply source and that facilitates the operation of the overcurrent protective device or ground fault detectors on high-impedance grounded systems.** Under fault-current conditions, the possible burning off of an equipment grounding conductor, a bonding jumper, or any conductor that is dependent on safely carrying fault current until the overcurrent protective device can clear the fault results in a hazard to life, safety, and equipment.

---

*Reprinted with permission from NFPA 70–2011.

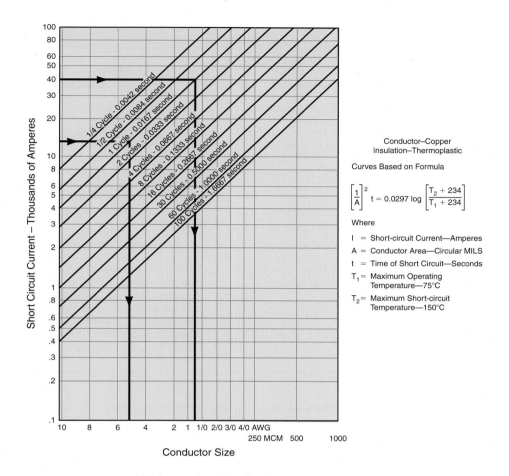

FIGURE 19-3   Withstand rating for insulated copper conductors.
(*Courtesy Insulated Cable Engineers Association*)

USING CHARTS TO DETERMINE A CONDUCTOR'S SHORT-TIME WITHSTAND RATING

The Insulated Cable Engineers Association, Inc., publishes much data on this subject. The graph in Figure 19-3 shows the withstand rating of 75°C thermoplastic insulated copper conductors. Many engineers use these tables for bare or insulated equipment grounding conductors because, in most cases, the equipment grounding conductor is in the same raceway as the ungrounded or grounded conductors. An extremely hot equipment grounding conductor in contact with the phase conductors would damage the insulation on the phase conductors. For example, when nonmetallic conduit is used, the equipment grounding conductor and the phase conductors are in the same raceway.

To use the table, begin on the left side, at the amount of fault current available. Then draw a line to the right, to the time of opening of the overcurrent protective device. Draw a line downward to the bottom of the chart to determine the conductor size.

EXAMPLE

A circuit is protected by a 60-ampere overcurrent device. Determine the minimum size equipment grounding conductor for available fault currents of 14,000 amperes and 40,000 amperes. Refer to *NEC Table 250.122*, the fuse peak let-through chart, and the chart showing allowable short-circuit currents for insulated copper conductors, Figure 19-3.

You can see from Table 19-1 that if the conductor size as determined from the allowable short-circuit current chart is larger than the size given in *NEC Table 250.122*, then you must install the larger size. Installing a conductor too small to handle the available fault current could

TABLE 19-1

Short-circuit currents for insulated copper conductor.

Available Fault Current	Overcurrent Device	Conductor Size
40,000 amperes	Typical one-cycle breaker	1/0 copper conductor
40,000 amperes	Typical RK1 fuse. Clearing time 1/4 cycle or less.	Using a 60-ampere RK1 current-limiting fuse, the apparent RMS let-through current is approximately 3000 amperes. *NEC Table 250.122* shows a minimum 8 AWG equipment grounding conductor. The allowable short-circuit current chart shows a conductor smaller than a 10 AWG. Therefore, 8 AWG is the minimum size EGC permitted.
14,000 amperes	Typical one-cycle breaker	4 AWG copper conductor
14,000 amperes	Typical RK1 fuse. Clearing time 1/4 cycle or less.	Using a 60-ampere RK1 current-limiting fuse, the apparent RMS let-through current is approximately 2200 amperes. *NEC Table 250.122* shows a minimum 8 AWG equipment grounding conductor. The allowable short-circuit current chart shows a conductor smaller than a 10 AWG. Therefore, 8 AWG is the minimum size EGC permitted.

result in insulation damage or, in the worst case, the burning off of the equipment grounding conductor, leaving the protected equipment hot.

If the conductor size as determined from the allowable short-circuit current chart is smaller than the size given in *NEC Table 250.122*, then install an equipment grounding conductor *not smaller* than the minimum size required by *NEC Table 250.122*.

There is another common situation in which it is necessary to install equipment grounding conductors larger than shown in *NEC Table 250.122*. When circuit conductors have been increased in size because of a voltage drop, the equipment grounding conductor size shall be increased proportionately.

MAGNETIC FORCES

Magnetic forces acting on electrical equipment (busbars, contacts, conductors, and so on) are proportional to the square of the PEAK current. This relationship is expressed as I_p^2.

Refer to Figure 19-4 for the case in which there is no fuse in the circuit. The peak current, I_p (available short-circuit current), is indicated as 30,000 amperes. The peak let-through current, I_p, resulting from the current-limiting effect of a fuse is indicated as 10,000 amperes.

EXAMPLE

Because magnetic forces are proportional to the square of the peak current, the magnetic forces (stresses) on the electrical equipment subjected to the full 30,000-ampere peak current are nine times that of the 10,000-ampere peak current let-through by the fuse. Stated another way, a current-limiting fuse or circuit breaker that can reduce the available short-circuit peak current from 30,000 amperes to only 10,000 amperes will subject the electrical equipment to only $1/9$ the magnetic forces.

Visual signs, indicating that too much current was permitted to flow for too long a time, include conductor insulation burning, melting and bending of busbars, arcing damage, exploded overload elements in motor controllers, and welded contacts in controllers.

A current-limiting fuse or current-limiting circuit breaker must be selected carefully. The fuse or breaker must not only have an adequate

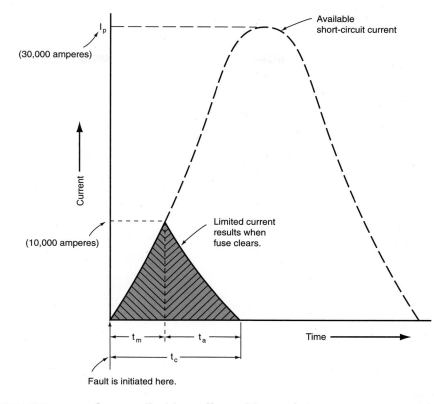

FIGURE 19-4   Current-limiting effect of fuses. (*Delmar/Cengage Learning*)

interrupting rating to clear a fault safely without damage to itself, but it also must be capable of limiting the let-through current and the value of I^2t to the withstand rating of the equipment it is to protect.

The graph shown in Figure 19-4 illustrates the current-limiting effect of fuses. The square of the area under the dashed line is energy (I^2t). I_p is the available peak short-circuit current that flows if there is no fuse in the circuit or if the overcurrent device is not a current-limiting type. For a current-limiting fuse, the square of the shaded area of the graph represents the energy (I^2t), and the peak let-through current, I_p. The melting time of the fuse element is t_m, and the square of the shaded area corresponding to this time is the melting energy. The arcing time is shown as t_a; similarly, the square of the shaded area corresponding to this time is the arcing energy. The total clearing time, t_c, is the sum of the melting time and the arcing time. The square of the shaded area for time, t_c, is the total energy to which the circuit is subjected when the fuse

has cleared. For the graph in Figure 19-4, the area under the dashed line is six times greater than the shaded area. Because energy is equal to the area squared, then $6 \times 6 = 36$; that is, the circuit is subjected to 36 times as much energy when it is protected by noncurrent-limiting overcurrent devices.

Arc-Flash and Arc-Blast Hazards

It is beyond the scope of this text to cover the complicated subject of arc-flash and arc-blast calculations, but a few basics need to be pointed out.

Short circuit calculations are necessary to determine interrupting ratings of fuses and circuit breakers, bracing of switchboards and busways, and short-circuit ratings of panel boards and other electrical equipment. But one must not forget that low levels of fault-current also can be extremely hazardous.

The terms *arc flash* and *arc blast* are used interchangeably, although generally the discussions relating to an arc flash emphasized the heat, whereas discussions about an arc blast focus on the instantaneous expansion of the air.

The unbelievable heat and pressures developed by an arc-flash explosion can destroy electrical equipment but, more importantly, can and will injure anyone in the way of an arc-flash explosion.

The injury might be the loss of one's hearing or sight, or severe burns or death. Damaged electrical equipment can be replaced. A life cannot be replaced.

PPE gear is a requirement by OSHA and is discussed in *NFPA 70E (Electrical Safety in the Workplace)* and IEEE Standard 1584 (Guide for Performing Arc-Flash Hazard Calculations).

A tremendous amount of technical information regarding arc blasts can be found in various manufacturers' literature and on their web sites, such as Bussmann's *Selecting Protective Devices* (SPD) handbook, and Bussmann's Web site: www.cooperbussmann.com.

The important point to be made here is that arcing faults are extremely dangerous! Working electrical equipment "live" requires proper personal protective equipment (PPE).

Don't be complacent, arrogant, cocky, or smug. Be alert! Be careful! Follow all safety requirements! Your life and other lives are at stake.

Summary of Conductor Short-Circuit Protection

When selecting equipment grounding conductors, bonding jumpers, or other current-carrying conductors, the amount of short-circuit or ground-fault current available and the clearing time of the overcurrent protective device must be taken into consideration so as to minimize damage to the conductor and associated equipment. The conductor must not become a fuse. The conductor must remain intact under any values of fault current. Burning off an equipment grounding conductor or a bonding jumper can result in a very serious unsafe, hazardous situation!

It simply boils down to the two important issues that need to be considered when thinking about conductor withstand rating:

1. How much current will flow?
2. How long will the current flow?

The choices are as follows:

- **Limit the current.**

  Install current-limiting overcurrent devices that will limit the let-through fault current and will reduce the time it takes to clear the fault. Then refer to *NEC Table 250.122* to select the minimum size equipment grounding conductor permitted by the *National Electrical Code* and refer to the other tables such as *NEC Table* ▶*310.15(B)(16)*◀ for the selection of circuit conductors.

- **Do not limit the current.**

  Install conductors that are large enough to handle the full amount of available fault current (how much) for the time (how long) it takes a noncurrent-limiting overcurrent device to clear the short circuit or ground fault. Withstand ratings of conductors can be calculated or determined by referring to conductor withstand rating charts.

TAP CONDUCTORS

There are instances where a smaller conductor must be tapped from a larger conductor, in which case the overcurrent protection is greater than the ampacity of the smaller conductor. Because of this, there are special rules relating to a tap.

What Is a Tap?

A tap conductor is defined in *NEC 240.2* as

- a conductor, other than a service conductor,
- having overcurrent protection ahead of its point of supply that exceeds the value permitted for similar conductors that are not taps.

You Are Not Permitted to Tap a Tap

NEC 240.21 prohibits making a tap from one conductor to another smaller conductor unless overcurrent protection for the smaller conductor is provided as spelled out in *240.4*.

Figure 19-5 is a classic example of small conductors tapped to larger conductors. The requirements

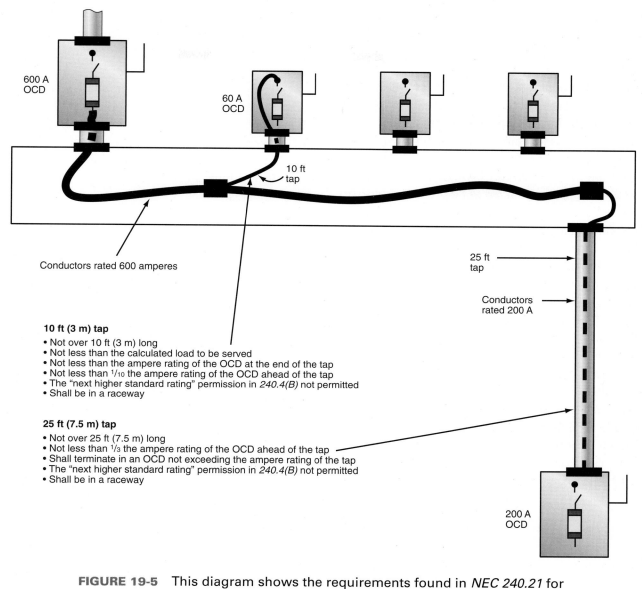

600 A
OCD

60 A
OCD

10 ft
tap

Conductors rated 600 amperes

25 ft
tap

Conductors
rated 200 A

10 ft (3 m) tap
- Not over 10 ft (3 m) long
- Not less than the calculated load to be served
- Not less than the ampere rating of the OCD at the end of the tap
- Not less than 1/10 the ampere rating of the OCD ahead of the tap
- The "next higher standard rating" permission in *240.4(B)* not permitted
- Shall be in a raceway

25 ft (7.5 m) tap
- Not over 25 ft (7.5 m) long
- Not less than 1/3 the ampere rating of the OCD ahead of the tap
- Shall terminate in an OCD not exceeding the ampere rating of the tap
- The "next higher standard rating" permission in *240.4(B)* not permitted
- Shall be in a raceway

200 A
OCD

FIGURE 19-5 This diagram shows the requirements found in *NEC 240.21* for typical 10 ft (3 m) and 25 ft (7.5 m) taps. (*Delmar/Cengage Learning*)

found in *NEC 240.21* for the common 10 ft (3 m) and 25 ft (7.5 m) tap are clearly shown.

Another common example of a tap is for fixture whips that connect a luminaire to a branch circuit. The size of taps to branch circuits is found in *NEC 210.19(A)(4) Exception No. 1* and *Table 210.24*. For instance, for a 20-ampere branch circuit, the minimum size tap is a 14 AWG copper conductor. The tap conductors in a fixture whip shall be suitable for the temperature to be encountered. In most cases, this will call for a 90°C conductor, Type THHN or equivalent.

When Is a Tap Not a Tap?

A common *Code* violation is found where receptacles are connected to a branch circuit.

When connecting a receptacle to a 20-ampere branch circuit, the short conductors from a splice in the box to the receptacle terminal must be rated 20 amperes. These short conductors are not taps; they are part of the branch circuit.

To use 14 AWG conductors with a 15-ampere rating for this connection would be a violation of the *NEC*.

Common Rules of Overcurrent Protection for Conductors

The basic overcurrent protection rule in *NEC 240.4* is that conductors shall be protected at the conductor's ampacity. There are seven subsections to this section, each pertaining to specific applications.

NEC 240.21 is a mirror image of *NEC 240.4* in that it states *Overcurrent protection shall be provided in each ungrounded circuit conductor and shall be located where the conductor to be protected receives its supply.** There are a number of subsections to *NEC 240.21* that are exceptions to the basic overcurrent protection rule. These subsections pertain to taps and transformer primary and secondary conductors.

Common Rules for Permitted Loads on Conductors

Branch-circuit conductors must have an ampacity not less than the maximum load served, *NEC 210.19*. For example, a load on a branch circuit has been determined to be 30 amperes. The conductors therefore must have an ampacity of not less than 30 amperes.

Feeder conductors must have sufficient ampacity to supply the load served, *NEC 220.40*. For

*Reprinted with permission from NFPA 70–2011.

example, the connected load in a panelboard has been calculated to be 100 amperes. The feeder conductors therefore must have an ampacity of not less than 100 amperes.

Service-entrance conductors shall be of sufficient size to carry the loads as calculated, *NEC 230.42*. For example, final load calculations for the service to a commercial building resulted in a calculated load of 850 amperes. Therefore, the service-entrance conductors must have an ampacity of not less than 850 amperes.

⎯⋀⋀⎯ SUMMARY

An electrician must properly size conductors for the load to be served, must select the proper overcurrent protection for the conductors, must take into consideration the possibility of voltage drop, must consider the ambient temperature where the conductors will be installed and select conductors suitable for that temperature, and must check the short-circuit withstand rating of the conductor to be sure that a severe fault will not cause damage to the conductor's insulation or, in the worst case, vaporize the conductor.

Additional *NEC* rules for taps and panelboard protection are found in Chapter 12.

REVIEW

Refer to the *National Electrical Code* or the working drawings when necessary. Where applicable, responses should be written in complete sentences.

1. Define *withstand rating*. _____
   _____
   _____
   _____

2. Define an *effective grounding path.* _____
   _____
   _____
   _____

3. Define I_2t. _____
   _____
   _____
   _____

4. Define a *current-limiting overcurrent device.* _____

_____

_____

_____

5. Define an *equipment grounding conductor.* _____

_____

_____

_____

6. What table in the *National Electrical Code* shows the minimum size equipment grounding conductor? _____

For the following questions, show all required calculations.

7. What is the appropriate action if the label on a motor controller or the nameplate on other electrical equipment is marked "Maximum Fuse Size"? Cite the *Code* section that supports your proposed action.

_____

_____

8. What is the maximum current that a 12 AWG, 75°C thermoplastic insulated conductor can safely carry for 5 seconds?

_____

_____

9. What is the maximum current that an 8 AWG, 75°C thermoplastic insulated conductor can safely carry for 1 second?

_____

_____

10. What is the maximum current that a 12 AWG, 75°C thermoplastic insulated conductor can safely carry for one-quarter cycle (0.004 second)?

_____

_____

11. Referring to Figure 19-3, if the available fault current is 10,000 amperes and the overcurrent device has a total clearing time of one cycle, the minimum permitted copper 75°C thermoplastic insulated conductor would be a _____ AWG.

12. Referring to Figure 19-2, if the available fault current is 40,000 amperes, a 200-ampere, 600-volt fuse of the type represented by the chart will have an instantaneous peak let-through current of approximately _____ amperes.

13. Referring to Figure 19-2, if the available fault current is 40,000 amperes, a 200-ampere, 600-volt fuse of the type represented by the chart will have an apparent RMS let-through current of approximately _____ amperes.

14. A tap conductor that complies with appropriate *Code* requirements has an ampacity of 30 amperes and is 9 ft (2.74 m) long. This is a field installation in which the tap conductor leaves the enclosure where the tap is made (a large junction box). The maximum allowable overcurrent protection for the feeder conductor would be _____ amperes.

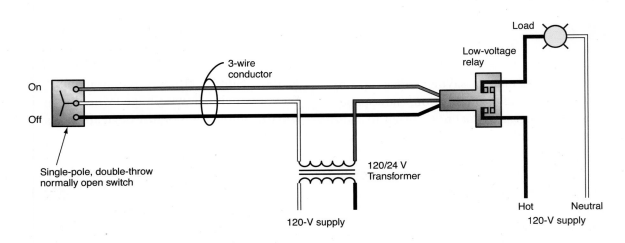

FIGURE 20-1   Connection diagram for low-voltage remote control. (*Delmar/Cengage Learning*)

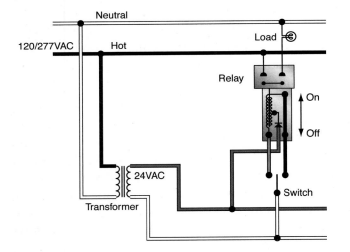

FIGURE 20-2   A cutaway view of a low-voltage relay showing the internal mechanism as well as the external line voltage and low-voltage leads. Some relays have screw terminals or "quick connects" instead of conductor leads. (*Delmar/Cengage Learning*)

FIGURE 20-3   Low-voltage relay. (*Courtesy of General Electric Industrial Systems*)

busbar. As a result, it is not necessary to splice the line voltage leads.

Single Switch

The switch used in the low-voltage remote-control system is a normally open, single-pole, double-throw, momentary contact switch, Figure 20-4. This switch is approximately one-third the size of a standard single-pole switch. In general, this type of switch has slide-on terminals for easy connections, Figure 20-5.

To make connections to the low-voltage switches, the white wire is common and is connected to the 24-volt transformer source. The red wire connects to the "On" circuit and the black wire connects to the "Off" circuit.

Switches are available in single-, 2-, 4-, and 8-switch combinations with features such as pilot light, lighted toggle, and being key operated.

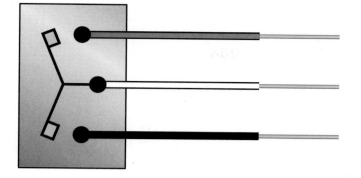

FIGURE 20-4   Single-pole, double-throw, normally open switch. (*Delmar/Cengage Learning*)

FIGURE 20-5   Low-voltage switch. (*Courtesy of General Electric Industrial Systems*)

Master Control

It is often desirable to control several circuits from a single location. Up to eight low-voltage switches, Figure 20-6, can be located in the same area that is required by two conventional switches. This 8-switch combination can be mounted on a 4¹¹⁄₁₆ in. (119 mm) square box using an adapter provided with the switch. Directory strips identifying the switch's functions can be prepared and inserted in the switch cover. An 8-switch combination is used in the Commercial Building Drugstore. The connection diagram is shown in Figure 20-7.

If the control requirements are very complex or extensive, a master sequencer, Figure 20-8, can be installed. There is practically no limit to the number of relays that can be controlled, almost

FIGURE 20-6   Eight-switch remote-control combination. (*Courtesy of General Electric Industrial Systems*)

instantaneously, with this microprocessor-controlled electronic switch.

Master Control with Rectifiers

Several relays can be operated individually or from a single switch through the use of *rectifiers*. The principle of operation of a master control with rectifiers is based on the fact that a rectifier permits current in only one direction, Figure 20-9.

For example, if two rectifiers are connected as shown in Figure 20-10, current cannot exist from A to B or from B to A; however, current can exist from A to C or from B to C. Thus, if a rectifier is placed in one lead of the low-voltage side of the transformer, and additional rectifiers are used to isolate the switches, then a master switching arrangement is achieved, Figure 20-11. This method of master control is used in the Drugstore. Although a switching schedule is included in the Specifications, the electrician may find it necessary to prepare a connection diagram similar to that shown in Figure 20–7. For the Drugstore master control, the transformer, Figure 20-12, and rectifier, Figure 20-13, are located in the low-voltage control panel.

WIRING METHODS

NEC Article 725 governs the installation of remote-control and signal circuits such as are installed in the Drugstore. The provisions of this article apply to remote-control circuits, low-voltage relay switching, low-energy power circuits, and low-voltage circuits.

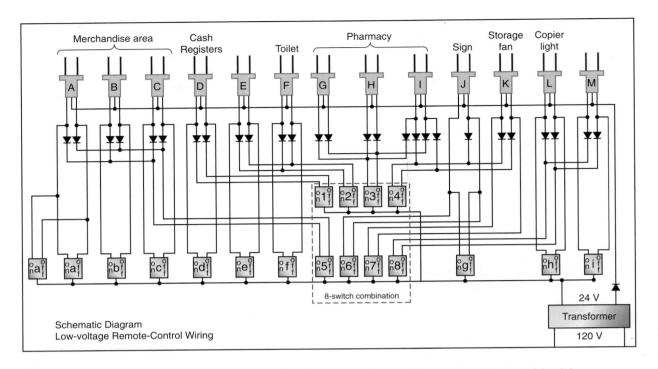

FIGURE 20-7 Connection diagram for Drugstore low-voltage, remote-control wiring. (*Delmar/Cengage Learning*)

FIGURE 20-8 Master sequencer available in three sizes to control 8, 16, or 32 relays. (*Courtesy of General Electric Industrial Systems*)

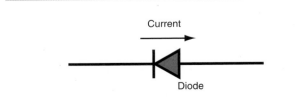

FIGURE 20-9 Diagram of diode showing direction of current. (*Delmar/Cengage Learning*)

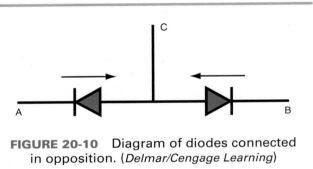

FIGURE 20-10 Diagram of diodes connected in opposition. (*Delmar/Cengage Learning*)

The Drugstore low-voltage wiring is classified as a Class 2 circuit by *NEC 725.2*. Because the power source of the circuit is limited (by the definition of a low-voltage circuit), overcurrent protection is not required.

Power source limitations for alternating current Class 2 circuits are found in *NEC Chapter 9, Tables 11(A)* and *11(B)*.

The circuit transformer, Figure 20-12, is designed so that in the event of an overload, the output voltage decreases and there is less current output. Any overload can be counteracted by these energy-limiting characteristics through the use of a specially designed transformer core. If the transformer is not self-protected, a thermal device may be

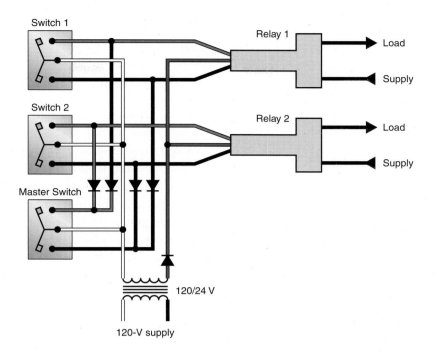

Switch 1

Switch 2

Master Switch

Relay 1

Load

Supply

Relay 2

Load

Supply

120/24 V

120-V supply

FIGURE 20-11 Switch 1 and the master switch control relay 1. Switch 2 and the master switch control relay 2. Rectifiers are connected into the leads of the master switch and the transformer. New systems have rectification as part of the motherboard in the main control panel. (*Delmar/Cengage Learning*)

FIGURE 20-12 Transformer. (*Courtesy of General Electric Industrial Systems*)

FIGURE 20-13 Rectifier. New systems have rectification as part of the motherboard in the main control panel. (*Courtesy of General Electric Industrial Systems*)

used to open the primary side to protect the transformer from overheating. This thermal device resets automatically as soon as the transformer cools. Other transformers are protected with nonresetting fuse links or externally mounted fuses.

Although the *NEC* does not require that the low-voltage wiring be installed in a raceway, the

Specifications for the Commercial Building do contain this requirement. The advantage of using a raceway for the installation is that it provides a means for making future additions at a minimum cost. A disadvantage of this approach is the initial higher construction cost.

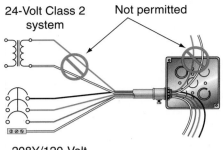

24-Volt Class 2 system

Not permitted

208Y/120-Volt, 3-Phase system

FIGURE 20-14 *NEC 725.136(A)* generally prohibits Class 2 circuits in the same raceway or cable as the wiring for light and power. (*Delmar/Cengage Learning*)

Conductors

NEC 725.136(A) prohibits Class 2 circuits in the same raceway or cable as the wiring for light and power. See Figure 20-14.

Most low-voltage, remote-control systems are wired with 20 AWG conductors, but larger conductors should be used for long runs, to minimize the voltage drop. Cables are available with multiple conductors. To simplify connections, the conductor color-coding combinations are red-black-white, red-black-yellow-white, red-black-yellow plus an 18 AWG common white, red-black-blue-white, and red-black-yellow-blue-white. To install these color-coded cables correctly throughout an entire installation, the wires are connected to the relays, switches, master controls, and panel, like color to like color.

The manufacturers of these types of systems provide detailed wiring diagrams. Follow the installation instructions and wiring diagrams exactly without modifications.

Low-Voltage Panelboard

The low-voltage relays in the Drugstore installation are to be mounted in an enclosure next to the power panelboard, Figure 20-15. A barrier in this low-voltage panelboard separates the 120-volt power lines from the low-voltage control circuits, in compliance with *NEC 725.136(B)*.

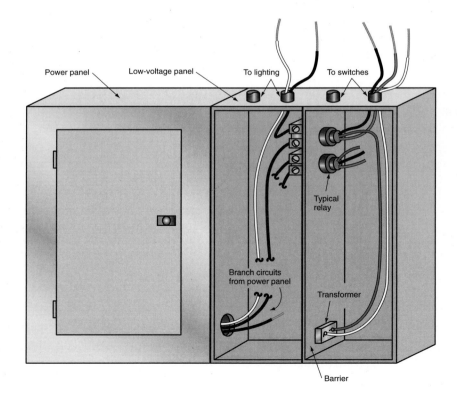

Power panel — Low-voltage panel — To lighting — To switches — Typical relay — Branch circuits from power panel — Transformer — Barrier

FIGURE 20-15 Low-voltage panel installation. Prewired panels are available with 12, 24, or 48 relays. (*Delmar/Cengage Learning*)

REVIEW

Refer to the *National Electrical Code* or the working drawings when necessary. Where applicable, responses should be written in complete sentences.

Refer to Figure 20-1 and respond to the following statements:

1. Describe the action of the low-voltage relay._____

_____

_____

_____

_____

_____

2. Describe the function of the red conductor._____

_____

_____

_____

_____

3. Describe the function of the white conductor._____

_____

_____

_____

_____

_____

4. Describe the function of the black conductor._____

_____

_____

_____

_____

_____

5. Describe the function of the blue conductor._____

_____

_____

_____

_____

_____

Respond to the following statements.

6. The *NEC* article that governs the installation of the low-voltage remote-control system is_____ .

7. Complete the following diagram according to the switching schedule that is given at the bottom of the diagram. Indicate the color of the conductors.

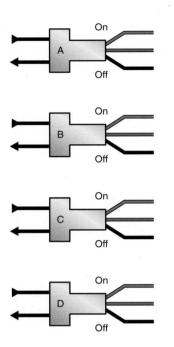

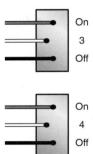

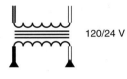

120/24 V

120-V source

Switching schedule

Switch	Control relay
1	A
2	B
3	C & D
4	A & B

The Cooling System

OBJECTIVES

After studying this chapter, you should be able to

- list the parts of a cooling system.
- describe the function of each part of the cooling system.
- make the necessary calculations to obtain the sizes of the electrical components.
- read a typical wiring diagram that shows the operation of a cooling unit.

The electrician working on commercial construction is expected to install the wiring of cooling systems and troubleshoot electrical problems in these systems. Therefore, it is recommended that the electrician know the basic theory of refrigeration and the terms associated with it.

REFRIGERATION

Refrigeration is a method of removing energy in the form of heat from an object. When the heat is removed, the object is colder. An energy balance is maintained, which means that the heat must go somewhere. As long as the locations where the heat is discharged and where it is absorbed are remote from each other, it can be said that the space where the heat was absorbed is cooled. The inside of the household refrigerator or freezer is cold to the touch, but this cold cannot be used to cool the kitchen by having the refrigerator door open. Actually, leaving the door open causes the kitchen to become hotter. This situation demonstrates an important principle

of mechanical refrigeration: to remove heat energy, it is necessary to add energy or power to it.

Mechanical refrigeration relies primarily on the process of evaporation. This process is responsible for the cool sensation that results when rubbing alcohol is applied to the skin or when gasoline is spilled on the skin. Body heat supplies the energy required to vaporize the alcohol or gasoline. It is the removal of this energy from the body that causes the sensation of cold. In refrigeration systems such as those used in the commercial building, the evaporation process is controlled in a closed system. The purpose of this arrangement is to preserve the refrigerant so that it can be reused many times in what is known as the *refrigerant cycle*. As shown in Figure 21-1, the four main components of the refrigerant cycle are as follows:

- *Evaporator*—The refrigerant evaporates here as it absorbs energy from the removal of heat.

- *Compressor*—This device raises the energy level of the refrigerant so that it can be condensed readily to a liquid.

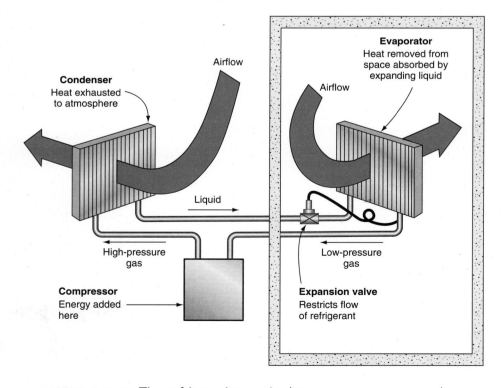

FIGURE 21-1 The refrigeration cycle. (*Delmar/Cengage Learning*)

- *Condenser*—The compressed refrigerant condenses here as the heat is removed.
- *Expansion valve*—This metering device maintains a sufficient imbalance in the system so there is a point of low pressure where the refrigerant can expand and evaporate.

EVAPORATOR

The evaporator in a commercial installation normally consists of a fin tube coil, as in Figure 21-2, through which the building air is circulated by a motor-driven fan. A typical evaporator unit is shown in Figure 21-3. The evaporator may be located inside or outside the building. In either case, the function of the evaporator is to remove heat from the interior of the building or the enclosed space. The air is usually circulated through pipes or ductwork to ensure a more even distribution. The window-type air conditioner, however, discharges the air directly from the evaporator coil. In general, the cooling air from the evaporator is recirculated within the space to be cooled and is passed again across the cooling coil. A certain percentage of outside air is added to the circulating air to replace air lost through exhaust systems and fume hoods, or because of the gradual leakage of air through walls, doors, and windows.

FIGURE 21-3   An evaporator with dual fans. (*Delmar/Cengage Learning*)

COMPRESSOR

The compressor (Figure 21-4) serves as a pump to draw the expanded refrigerant gas from the evaporator. In addition, the compressor boosts the pressure of the gas and sends it to the condenser. The compression of the gas is necessary because this process adds the heat necessary to condense the gas to a liquid. When the temperature of the air or water surrounding the condenser is relatively warm, the gas temperature must be increased to ensure that the temperature around the condenser will liquefy the refrigerant.

Direct-drive compressors are usually used in large installations. For smaller installations, however,

FIGURE 21-2   Fin tube coils. (*Delmar/Cengage Learning*)

FIGURE 21-4   A motor-driven reciprocating compressor. (*Delmar/Cengage Learning*)

the trend is toward the use of hermetically sealed compressors. Due to several built-in electrical characteristics, these hermetic units cannot be used on all installations. (Restrictions on the use of hermetically sealed compressors are covered later in this chapter.)

CONDENSER

Condensers are generally available in three types: as air-cooled units, Figure 21-5, as water-cooled units, Figure 21-6, or as evaporative cooling units, Figure 21-7, in which water from a pump is sprayed on air-cooled coils to increase their capacity.

The function of the condenser in the refrigerant cycle is to remove the heat taken from the evaporator, plus the heat of compression. Thus, it can be seen that keeping the refrigerator door open causes the kitchen to become hotter because the condenser rejects the combined heat load to the condensing medium. In the case of the refrigerator in a residence, the condensing medium is the room air.

Air-cooled condensers use a motor-driven fan to drive air across the condensing coil. Water-cooled condensers require a pump to circulate the water. Once the refrigerant gas is condensed to a liquid state, it is ready to be used again as a coolant.

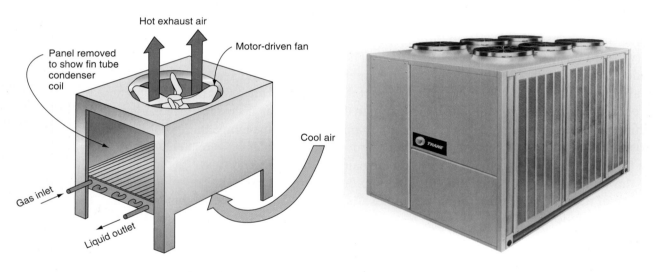

FIGURE 21-5 An air-cooled condenser. (*Delmar/Cengage Learning*)

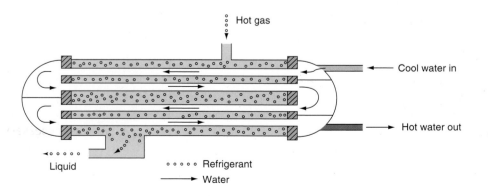

FIGURE 21-6 A water-cooled condenser. (*Delmar/Cengage Learning*)

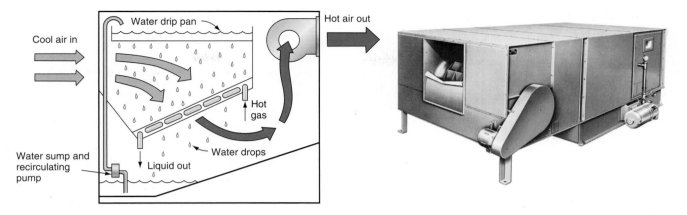

FIGURE 21-7 An evaporative condenser. (*Delmar/Cengage Learning*)

EXPANSION VALVE

It was stated previously that the refrigerant must evaporate or boil if it is to absorb heat. The process of boiling at ordinary temperatures can occur in the evaporator only if the pressure is reduced. The task of reducing the pressure is simplified because the compressor draws gas away from the evaporator and tends to evacuate it. In addition, a restricted flow of liquid refrigerant is allowed to enter the high side of the evaporator. As a result, the pressure remains fairly low in the evaporator coil so that the liquid refrigerant entering through the restriction flashes into a vapor and fills the evaporator as a boiling refrigerant.

The restriction to the evaporator may be a simple orifice. In commercial systems, however, a form of automatic expansion valve is generally used because it is responsive to changes in the heat load. The expansion valve is located in the liquid refrigerant line at the inlet to the evaporator. The valve controls the rate at which the liquid flows into the evaporator. The liquid flow rate is determined by installing a temperature-sensitive bulb at the outlet of the evaporator to sense the heat gained by the refrigerant as it passes through the evaporator. The bulb is filled with a volatile liquid (which may be similar to the refrigerant). This liquid expands and passes through a capillary tube connected to a spring-loaded diaphragm to cause the expansion valve to meter more or less refrigerant to the coil. The delivery of various amounts of refrigerant compensates for changes in the heat load of the evaporator coil.

HERMETIC COMPRESSORS

The tremendous popularity of mechanical refrigeration for household use in the 1930s and 1940s stimulated the development of a new series of nonexplosive and nontoxic refrigerants. These refrigerants are known as chlorinated hydrofluorides of the halogen family and, at the time of their initial production, were relatively expensive. The expense of these refrigerants was great enough so that it was no longer possible to permit the normal leakage that occurred around the shaft seals of reciprocating, belt-driven units. As a result, the hermetic compressor was developed (Figure 21-8). This unit consists of a motor-compressor completely sealed in a gastight, steel casing. The refrigerant gas is circulated through the compressor and over the motor windings, rotor, bearings, and shaft. The circulation of the expanded gas through the motor helps to cool the motor.

The initial demand for hermetic compressors was for use on residential-type refrigerators. Therefore, most of the hermetic compressors were constructed for single-phase service. In other words, it was necessary to provide auxiliary winding and starting devices for the compressor installation. Because the refrigerant gas surrounded and filled the motor cavity, it was necessary to remove the centrifugal switch commonly provided to disconnect the starting winding at approximately 85% of the full speed. The switch was removed because any arcing in the presence of the refrigerant gas caused the formation of an acid from the hydrocarbons in

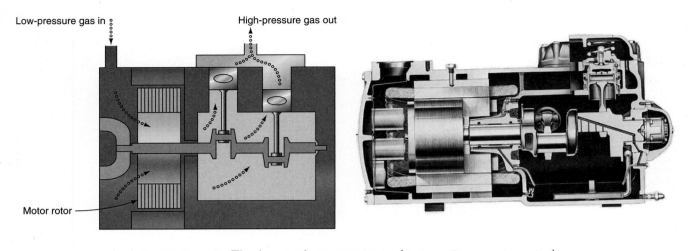

Low-pressure gas in

High-pressure gas out

Motor rotor

FIGURE 21-8 The hermetic compressor. (*Delmar/Cengage Learning*)

the gas. This acid attacked and etched the finished surfaces of the shafts, bearings, and valves. The acid also carbonized the organic material used to insulate the motor winding and caused the eventual breakdown of the insulation. To overcome the problem of the switch, the relatively heavy magnetic winding of a relay was connected in series with the main motor winding. The initial heavy inrush of current caused the relay to lift a magnetic core and energize the starting winding. As the motor speed increased, the main winding current decreased and allowed the relay to remove the starting or auxiliary winding from the circuit.

A later refinement of this arrangement was the use of a voltage-sensitive relay that was wound to permit pickup at voltage values greater than the line voltage. The coil of this voltage-sensitive relay was connected across the starting winding. This connection scheme was based on the principle that the rotor of a single-phase induction motor induces in its own starting winding a voltage that is approximately in quadrature (phase) with the main winding voltage and has a value greater than the main voltage. The voltage-sensitive relay broke the circuit to the starting winding and remained in a sealed position until the main winding was de-energized.

Because the major maintenance problems in hermetic compressor systems were due generally to starting relays and capacitors, it was desirable to eliminate as many of these devices as possible. It was soon realized that small- and medium-sized systems could make use of a different form of refrigerant metering device, thus eliminating the automatic expansion valve. Recall that this valve has a positive shut-off characteristic, which means that the refrigerant is restricted when the operation cycle is finished as indicated by the evaporator reaching the design temperature. As a result, when a new refrigeration cycle begins, the compressor must start against a head of pressure. If a small-bore, open capillary tube is substituted for the expansion valve, the refrigerant is still metered to the evaporator coil, but the gas continues to flow after the compressor stops until the system pressure is equalized. Therefore, the motor is required only to start the compressor against the same pressure on each side of the pistons. This ability to decrease the load led to the development of a new series of motors that contained only a running capacitor (no relay was installed). This type of motor furnished sufficient torque to start the unloaded compressor and, at the same time, greatly improved the overall power factor.

COOLING SYSTEM CONTROL

Figure 21-9 shows a wiring diagram that is representative of standard cooling units.

When the selector switch in the heating-cooling control is set to COOL and the fan switch is placed on AUTO, any rise in temperature above the set point causes the cooling contacts (TC) to close and complete two circuits. One circuit through the fan

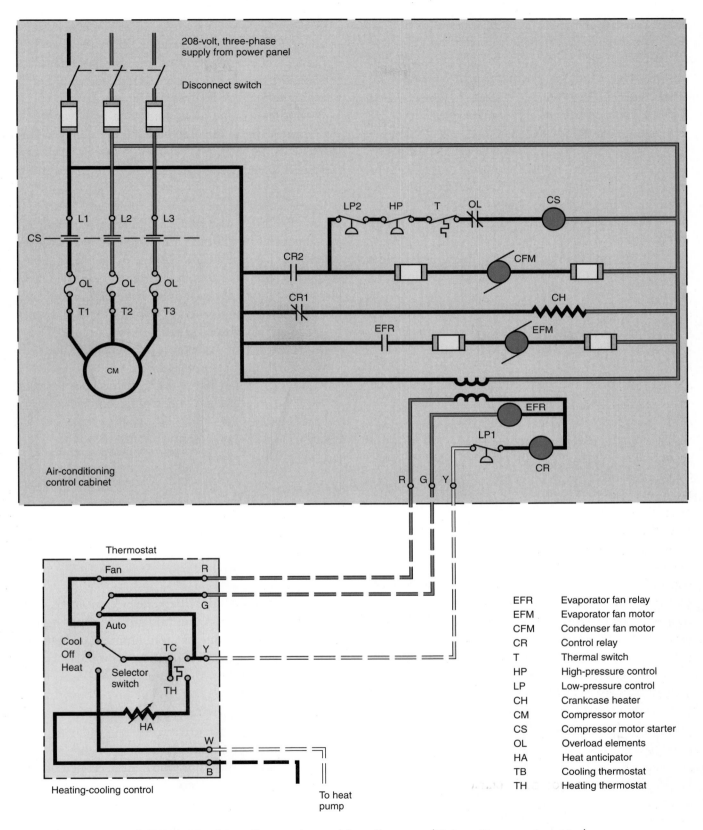

208-volt, three-phase
supply from power panel

Disconnect switch

Air-conditioning
control cabinet

Thermostat

Fan

Auto

Cool
Off
Heat

Selector
switch

Heating-cooling control

To heat
pump

EFR	Evaporator fan relay
EFM	Evaporator fan motor
CFM	Condenser fan motor
CR	Control relay
T	Thermal switch
HP	High-pressure control
LP	Low-pressure control
CH	Crankcase heater
CM	Compressor motor
CS	Compressor motor starter
OL	Overload elements
HA	Heat anticipator
TB	Cooling thermostat
TH	Heating thermostat

FIGURE 21-9 A cooling system wiring diagram. (*Delmar/Cengage Learning*)

switch energizes the evaporator fan relay (EFR), which, in turn, closes EFR contacts to complete the 208-volt circuit to the evaporator fan motor (EFM). The second circuit is to the control relay (CR) and causes the CR1 contacts to open. When the CR1 contacts open, the crankcase heater (CH) is de-energized. (This heater is installed to keep the compressor oil warm and dry when the unit is not running.) In addition, CR2 contacts close to complete a circuit to the condenser fan motor (CFM) and another circuit through low-pressure switch 2 (LP2), the high-pressure switch (HP), the thermal contacts (T), and the overload contacts (OL) to the compressor motor starter coil (CS). When the CR2 contacts close, the three CS contacts in the power circuit to the compressor motor (CM) also close.

Contacts LP1 and LP2 open when the refrigerant pressure drops below a set point. When contacts LP2 open, CS is de-energized. When contacts LP1 open, CR is de-energized, the circuit to CFM opens, and the circuit to CH is completed. The high-pressure switch (HP) contacts open and de-energize the compressor motor starter (CS) when the refrigerant pressure is above the set point. The low-pressure control (LP1) is the normal operating control and the high-pressure control (HP) and LP2 act as safety devices. The T contacts shown in Figure 21-9 are located in the compressor motor and open when the winding temperature of the motor is too high. The OL contacts are controlled by the OL elements installed in the power leads to the motor. The OL elements are sized to correspond to the current draw of the motor. That is, a high current causes the overload elements to overheat and open the OL contacts. As a result, the compressor starter is de-energized and the compressor stops. The evaporator fan motor (EFM) used to circulate air in the store area can be run continuously if the fan switch is turned to FAN. Actually, in many situations, it is recommended that the fan motor run continuously to keep the air in motion.

COOLING SYSTEM INSTALLATION

The owner of the Commercial Building leases the various office and shop areas on the condition that heating and cooling systems will be provided by the owner. However, tenants agree to pay the cost of operating the HVAC systems for their areas. The heating and cooling equipment for the Insurance Office, the Real Estate Office, and the Beauty Salon are single-package heating/cooling units located on the roof. The compressor, condenser, and evaporator for each of these units are constructed within a single enclosure. Electric resistance heating strips are installed in each unit. The system for the Drugstore and Bakery is a split system with the compressor and condenser located on the ground adjacent to the building and the evaporator located in an air-handling unit in the basement, as shown in Figure 21-10. Regardless of the type of heating/cooling system installed, a duct system must be provided to connect the package unit or air handler with the area to be heated and cooled. The environmental duct system shown diagrammatically in Figure 21-10 does not represent the actual duct system, which will be installed by another contractor.

The electrician is expected to provide a power circuit to each heating/air-conditioning unit, as shown on the plans. In addition, it is necessary to provide wiring to the thermostat in each area.

ELECTRICAL REQUIREMENTS FOR AIR-CONDITIONING AND REFRIGERATION EQUIPMENT

NEC Article 440 provides the requirements for installing air-conditioning and refrigeration equipment that involves one or more hermetic refrigerant motor-compressors. Where these compressors are not involved, *NEC 440.3* directs the reader to *NEC Articles 422, 424*, and *430* and gives other references for special provisions.

For the Insurance Office, a packaged unit with both heating and cooling is located on the roof of the Commercial Building. The data furnished on the nameplate with this air-conditioning unit supply the information in Figure 12-11.

Most everything the electrician needs to know about the wiring of the HVAC equipment is provided on the nameplate. Often, the equipment, particularly of size and electrical ratings supplying

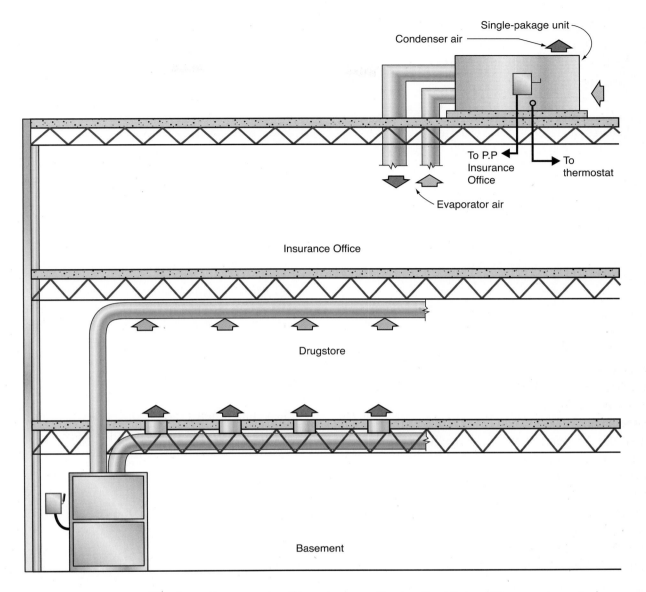

FIGURE 21-10 Single-package and split-system cooling units. (*Delmar/Cengage Learning*)

the Commercial Building, is provided with a single branch circuit. We have reviewed in detail the sizing rules for the branch circuit to this equipment earlier in this text. If the nameplate on equipment specifies fuses and does not give an option for HACR-type circuit breakers, then only fuses are permitted to be used as the branch-circuit overcurrent protection. Circuit breakers would not be permitted; see *NEC 110.3(B)*. This has been covered in Chapter 17.

Air-conditioning and refrigeration equipment is required by *NEC 440.4(A)* and *(B)* to be marked

with a *visible nameplate marked with the maker's name, the rating in volts, frequency and number of phases, the rated-load current in amperes, the locked-rotor current, minimum supply circuit conductor ampacity, the maximum rating of the branch-circuit, short-circuit, and ground-fault protective device, and the short-circuit current rating of the motor controllers or industrial control panel.** The exceptions to the short-circuit current rating requirement are multimotor and

---

*Reprinted with permission from NFPA 70-2011.

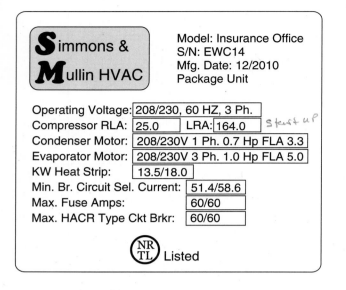

FIGURE 21-11 (*Delmar/Cengage Learning*)

combination-load equipment in one- and two-family dwellings, cord-and plug-connected equipment, or equipment supplied by a branch circuit rated at 60 amperes or less.

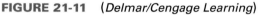

SPECIAL TERMINOLOGY

Understanding the following definitions is important when installing air conditioners, heat pumps, and other equipment using hermetic refrigerant motor-compressors. These terms are found throughout *NEC Article 440* and *UL Standard 1995*.

Locked Rotor Amperes (LRA): The manufacturer is required to furnish the locked rotor current on the compressor nameplate and/or the unit nameplate. One of the reasons the manufacturer is required to supply this information is, the motors in the units are not produced in standard horsepower ratings. This information is needed when sizing the disconnecting means, particularly when a switch is used, as the switches are rated in horsepower. *430.251 A*

Rated Load-Current (RLA or RLC): The RLA is determined by the manufacturer of the hermetic refrigerant motor-compressor through testing at rated refrigerant pressure, temperature conditions, and voltage.

In most instances, the RLA is at least equal to 64.1% of the hermetic refrigerant motor-compressor's maximum continuous current (MCC).

EXAMPLE

The nameplate on an air-conditioning unit is marked:

Compressor RLA: 25.0 amperes

Minimum Branch-Circuit Selection Current (MBCSC): Some hermetic refrigerant motor-compressors are designed to operate continuously at currents greater than 156% of the RLA. In such cases, the unit's nameplate is marked with "Branch-Circuit Selection Current" or "Minimum Supply Circuit Conductor Ampacity." The MBCSC will be no less than at least 64.1% of the MCC rating of the hermetic refrigerant motor-compressor.

Note: 156% and 64.1% have an inverse relationship:

$$\frac{1}{1.56} = 0.641 \text{ and } \frac{1}{0.641} = 1.56$$

EXAMPLE

The MCC of a hermetic refrigerant motor-compressor is 31 amperes. The BCSC will be no less than

$$31 \times 0.641 = 19.9 \text{ amperes}$$

Because the MBCSC value, when marked, is always equal to or greater than the unit's nameplate marked RLA value, the manufacturer of the air-conditioning unit must use the MBCSC value instead of the RLA value, to determine the minimum circuit ampacity (MCA) and the maximum overcurrent protection (MOP). For installation of individual hermetic refrigerant motor-compressors, where the electrician must select the conductors, the controller, the disconnecting means, and the short-circuit and ground-fault protection, the electrician must use the MBCSC, if given, instead of the RLA; see *NEC 440.4(C)*.

This is what the electrician needs to know. The manufacturer determines the MBCSA by multiplying the RLA, or the BCSC, of the hermetic refrigerant motor-compressor by 125%. The current ratings of all other concurrent loads, such as fan motors, transformers, relay coils, and so on, are then added to this value.

 EXAMPLE —————————————

An air-conditioning unit's nameplate is marked:

> Minimum Branch Circuit Selection Ampacity: 26.4 amperes

This value was derived as follows:

> BCSC of 19.9 × 1.25 = 24.9 amperes
> plus ¼-hp fan motor @ <u>1.5</u> amperes
> Therefore, the MBCSA = 26.4 amperes

———————————————————

The electrician must install the conductors, the disconnect switch, and the overcurrent protection based on the MBCSA value. No further calculations are needed; the manufacturer of the equipment has done it all.

Maximum Overcurrent Protective (MOP): The electrician should always check the nameplate of an air-conditioning unit for this information.

The manufacturer is required to mark this value on the nameplate. This value is determined by multiplying the RLA, or the MBCSC, of the hermetic refrigerant motor-compressor by 225%, then adding all concurrent loads such as electric heaters, motors, and so on.

EXAMPLE —————————————

The nameplate of an air-conditioning unit is marked as follows:

> Maximum Time-Delay Fuse: 45 amperes

This value was derived as follows:

> BCSC of 19.9 × 2.25 = 44.8 amperes
> plus ¼-hp fan motor @ <u>1.5</u> amperes
> Therefore, the MOP = 46.3 amperes

———————————————————

Because 46.275 is the maximum, the next lower standard size time-delay fuse (that is, 45 amperes) must be used. See *NEC 240.6* and *Article 440 Part III*.

The electrician installs branch-circuit overcurrent devices with a rating not to exceed the MOP, as marked on the nameplate of the air-conditioning unit. The electrician makes no further calculations; all calculations have been done by the manufacturer of the air-conditioning unit.

Overcurrent Protection Device Selection: How does an electrician know if fuses or circuit breakers are to be used for the branch-circuit overcurrent device? The electrician reads the nameplate and the instructions carefully. *NEC 110.3(B)* requires that *listed or labeled equipment be used or installed in accordance with any instructions included in the listing or labeling.*

EXAMPLE —————————————

If the nameplate of an air-conditioning unit reads "Maximum Size Time-Delay Fuse: 45 amperes," fuses are required for the branch-circuit protection. To install a circuit breaker would be a violation of *NEC 110.3(B)*.

If the nameplate reads "Maximum Fuse or HACR Type Breaker . . . ," then either fuses or an HACR-type circuit breaker is permitted. See Figures 17-33, 17-34, and 17-35.

———————————————————

Disconnecting Means Rating: The *Code* rules for determining the size of the disconnecting means required for HVAC equipment are covered in *NEC Article 440 Part II*.

The horsepower rating of the disconnecting means must be at least equal to the sum of all the individual loads within the equipment, at rated load conditions and at locked-rotor conditions; see *NEC 440.12(B)(1)*.

TABLE 21-1

Single-phase, locked-rotor currents for selection of disconnect means. See *NEC Table 430.251(A)*

MAXIMUM LOCKED-ROTOR CURRENT (AMPERES, SINGLE PHASE)

Rated Horse- power	115 Volts	208 Volts	230 Volts
½	58.8	32.5	29.4
¾	82.8	45.8	41.4
1	96	53	48
1½	120	66	60
2	144	80	72
3	204	113	102
5	336	186	168
7½	480	265	240
10	600	332	300

For use only with *NEC 430.110, 440.12, 440.41 and 455.8(C).*

The ampere rating of the disconnecting means must be at least 115% of the sum of all the individual loads within the equipment. Refer to *NEC 440.12(A)(1)* and *440.12(B)(2)* for details.

Because disconnect switches are horsepower rated, it is sometimes necessary to convert the locked-rotor current information found on the nameplate of air-conditioning equipment to an equivalent horsepower rating; *NEC Article 440 Part II* sets forth the conversion procedure.

Air-Conditioning and Refrigeration Equipment Disconnecting Means

- The disconnecting means for an individual hermetic motor-compressor must not be less than 115% of the rated load current or the branch-circuit selection current, whichever is greater; see *NEC 440.12(A)(1).*

- The disconnecting means for an individual hermetic motor-compressor shall have a horsepower rating equivalent to the horsepower rating shown in *NEC Table 430.251(A)* (Table 21-1) and *Table 430.251(B)* (Table 21-2).

This table is a conversion table to convert locked-rotor current to an equivalent horsepower; see *NEC 440.12(A)(2).*

- For equipment that has at least one hermetic motor-compressor and for other loads, such as fans, heaters, solenoids, and coils, the disconnecting means must be not less than 115% of the sum of the currents of all of the components; see *NEC 440.12(B)(2).*

- For equipment that has at least one hermetic motor-compressor and for other loads, such as fans, heaters, solenoids, and coils, the disconnecting means rating is determined by adding all the individual currents. This total is then considered to be a single motor for the purpose of determining the disconnecting means horsepower rating; see *NEC 440.12(B)(1).*

In our example, for simplicity, we will consider the minimum circuit ampacity to be the full-load current. This is 31.65 amperes. Checking *NEC Table 430.250* (Table 21-3) we find that a full-load current rating of 31.65 amperes falls between a 10-horsepower and a 15-horsepower motor. Therefore, we must consider our example to be equivalent to a 15-horsepower motor.

NEC Table 430.248 (Table 21-4) lists the full-load current ratings for single-phase motors.

- The total locked-rotor current must also be checked to be sure that the horsepower rating of the disconnecting means is capable of safely disconnecting the full amount of locked-rotor current, see *NEC 440.12(B)(1).*

In our example, the locked-rotor current equals 182 amperes. Checking *NEC Table 430.251(B)* (Table 21-2) in the 208-volt column for the conversion of locked-rotor current to horsepower, we find that an LRA of 182 amperes falls between a 10-horsepower and a 15-horsepower motor. Therefore, we again consider our example to be equivalent to a 15-horsepower motor, see *NEC 440.12(A)* and *440.12(B).*

If one of the selection methods (FLA or LRA) indicates a larger size disconnect switch than the other method does, then the larger switch shall be installed. See *NEC 440.12(A)* and *440.12(B).* Also see the last sentence in *NEC 430.110(C)(1).*

TABLE 21-2

Table 430.251(B) Conversion Table of Polyphase Design B, C, and D Maximum Locked-Rotor Currents for Selection of Disconnecting Means and Controllers as Determined from Horsepower and Voltage Rating and Design Letter
For use only with 430.110, 440.12, 440.41 and 455.8(C).

Rated Horsepower	Maximum Motor Locked-Rotor Current in Amperes, Two- and Three-Phase, Design B, C, and D*					
	115 Volts	200 Volts	208 Volts	230 Volts	460 Volts	575 Volts
	B, C, D	B, C, D	B, C, D	B, C, D	B, C, D	B, C, D
½	40	23	22.1	20	10	8
¾	50	28.8	27.6	25	12.5	10
1	60	34.5	33	30	15	12
1½	80	46	44	40	20	16
2	100	57.5	55	50	25	20
3	—	73.6	71	64	32	25.6
5	—	105.8	102	92	46	36.8
7½	—	146	140	127	63.5	50.8
10	—	186.3	179	162	81	64.8
15	—	267	257	232	116	93
20	—	334	321	290	145	116
25	—	420	404	365	183	146
30	—	500	481	435	218	174
40	—	667	641	580	290	232
50	—	834	802	725	363	290
60	—	1001	962	870	435	348
75	—	1248	1200	1085	543	434
100	—	1668	1603	1450	725	580
125	—	2087	2007	1815	908	726
150	—	2496	2400	2170	1085	868
200	—	3335	3207	2900	1450	1160
250	—	—	—	—	1825	1460
300	—	—	—	—	2200	1760
350	—	—	—	—	2550	2040
400	—	—	—	—	2900	2320
450	—	—	—	—	3250	2600
500	—	—	—	—	3625	2900

*Design A motors are not limited to a maximum starting current or locked rotor current.

Reprinted with permission from NFPA 70–2011.

Recap of Air-Conditioning and Refrigerating Equipment Disconnect Requirements. *NEC 440.14* requires that the disconnecting means be

- located within sight from the air-conditioning or refrigerating equipment.
- readily accessible.
- permitted to be installed on or within the air-conditioning or refrigerating equipment. Very large equipment often has the disconnect switch mounted inside the enclosure.
- located to allow access to panels that need to be accessed for servicing the equipment.
- mounted so as not to obscure the equipment nameplate(s).
- provided with a lock-off provision that *remains in place with or without the lock installed.* Snap-on covers for circuit breakers do not meet this requirement.

NEC 440.14 states that if you mount the disconnect on the equipment, do not obscure the equipment's nameplate(s). It's hard to believe that some installers mount the disconnect switch on top of the equipment's nameplate, making it impossible to read the nameplate.

TABLE 21-3

Table 430.250 Full-Load Current, Three-Phase Alternating-Current Motors

The following values of full-load currents are typical for motors running at speeds usual for belted motors and motors with normal torque characteristics.

The voltages listed are rated motor voltages. The currents listed shall be permitted for system voltage ranges of 110 to 120, 220 to 240, 440 to 480, and 550 to 600 volts.

Horsepower	Induction-Type Squirrel Cage and Wound Rotor (Amperes)							Synchronous-Type Unity Power Factor* (Amperes)			
	115 Volts	200 Volts	208 Volts	230 Volts	460 Volts	575 Volts	2300 Volts	230 Volts	460 Volts	575 Volts	2300 Volts
½	4.4	2.5	2.4	2.2	1.1	0.9	—	—	—	—	—
¾	6.4	3.7	3.5	3.2	1.6	1.3	—	—	—	—	—
1	8.4	4.8	4.6	4.2	2.1	1.7	—	—	—	—	—
1½	12.0	6.9	6.6	6.0	3.0	2.4	—	—	—	—	—
2	13.6	7.8	7.5	6.8	3.4	2.7	—	—	—	—	—
3	—	11.0	10.6	9.6	4.8	3.9	—	—	—	—	—
5	—	17.5	16.7	15.2	7.6	6.1	—	—	—	—	—
7½	—	25.3	24.2	22	11	9	—	—	—	—	—
10	—	32.2	30.8	28	14	11	—	—	—	—	—
15	—	48.3	46.2	42	21	17	—	—	—	—	—
20	—	62.1	59.4	54	27	22	—	—	—	—	—
25	—	78.2	74.8	68	34	27	—	53	26	21	—
30	—	92	88	80	40	32	—	63	32	26	—
40	—	120	114	104	52	41	—	83	41	33	—
50	—	150	143	130	65	52	—	104	52	42	—
60	—	177	169	154	77	62	16	123	61	49	12
75	—	221	211	192	96	77	20	155	78	62	15
100	—	285	273	248	124	99	26	202	101	81	20
125	—	359	343	312	156	125	31	253	126	101	25
150	—	414	396	360	180	144	37	302	151	121	30
200		552	528	480	240	192	49	400	201	161	40
250	—	—	—	—	302	242	60	—	—	—	—
300	—	—	—	—	361	289	72	—	—	—	—
350	—	—	—	—	414	336	83	—	—	—	—
400	—	—	—	—	477	382	95	—	—	—	—
450	—	—	—	—	515	412	103	—	—	—	—
500	—	—	—	—	590	472	118	—	—	—	—

*For 90 and 80 percent power factor, the figures shall be multiplied by 1.1 and 1.25, respectively.

Reprinted with permission from NFPA 70–2011.

NEC 110.3(B) requires that *Listed or labeled equipment shall be installed and used in accordance with any instructions included in the listing or labeling.** It is this section that prohibits the use of a circuit breaker to protect this particular air-conditioning unit. If the feeder or branch circuit originates from a circuit-breaker panel, then the requirements of this section can be met by installing a fusible disconnect switch near, on, or within the air-conditioning equipment.

Air-Conditioning and Refrigeration Equipment, Short-Circuit Rating of Controller

NEC 440.4(B) requires that controllers be marked with their short-circuit current rating. Of course, this

requirement is found in many places in the *NEC*. The basis for this is found in *NEC 110.9*, which states that *Equipment intended to interrupt current at fault levels shall have an interrupting rating sufficient for the nominal circuit voltage and the current that is available at the line terminals of the equipment.** You need to know the available fault current at points on an electrical system. This is covered in Chapters 17 and 18.

Air-Conditioning and Refrigeration Equipment Motor-Compressor and Branch-Circuit Overload Protection

This topic is covered in *NEC Article 440, Part VI.*

---

*Reprinted with permission from NFPA 70-2011.

TABLE 21-4

Table 430.248 Full-Load Currents in Amperes, Single-Phase Alternating-Current Motors

The following values of full-load currents are for motors running at usual speeds and motors with normal torque characteristics. The voltages listed are rated motor voltages. The currents listed shall be permitted for system voltage ranges of 110 to 120 and 220 to 240 volts.

Horsepower	115 Volts	200 Volts	208 Volts	230 Volts
⅙	4.4	2.5	2.4	2.2
¼	5.8	3.3	3.2	2.9
⅓	7.2	4.1	4.0	3.6
½	9.8	5.6	5.4	4.9
¾	13.8	7.9	7.6	6.9
1	16	9.2	8.8	8.0
1½	20	11.5	11.0	10
2	24	13.8	13.2	12
3	34	19.6	18.7	17
5	56	32.2	30.8	28
7½	80	46.0	44.0	40
10	100	57.5	55.0	50

Reprinted with permission from NFPA 70–2011.

The manufacturer of this type of equipment provides the overload protection for the various internal components that make up the complete air-conditioning unit. The manufacturer submits the equipment to Underwriters Laboratories for testing and listing of the product. Overload protection of the internal components is covered in the UL Standards. This protection can be in the form of fuses, thermal protectors, certain types of inverse-time circuit breakers, overload relays, and so forth.

NEC 440.52 discusses overload protection for motor-compressor appliances. This section requires that the motor-compressor be protected from overloads and failure to start by

- a separate overload relay that is responsive to current flow and trips at not more than 140% of the rated load current, or

- an integral thermal protector arranged so that its action interrupts the current flow to the motor-compressor, or

- a time-delay fuse or inverse-time circuit breaker rated at not more than 125% of the rated load current.

It is rare for the electrician to have to provide additional overload protection for these components. The electrician's job is to make sure that he or she supplies the equipment with the proper size conductors, the proper size and type of branch-circuit overcurrent protection, and the proper horsepower-rated disconnecting means.

Tons versus Amperes

It is possible to estimate the current required for air-conditioning equipment, heat pumps, or other hermetic motor-compressor-type loads, based on the equipment tons, but the procedure is complex and is not necessary. In conformance to *NEC 440.4(B)*, all units are marked with the *maker's name, the rating in volts, frequency and number of phases, minimum supply circuit conductor ampacity, the maximum rating of the branch-circuit short-circuit and ground-fault protective device, and the short circuit current rating of the motor controllers or industrial control panel.**

---

* Reprinted with permission from NFPA 70-2011

Refer to the *National Electrical Code* or the working drawings when necessary. Where applicable, responses should be written in complete sentences.

Complete the following sentences by writing a component used in a refrigeration system.

1. The _____ raises the energy level of the refrigerant.

2. The heat is removed from the refrigerant in the _____.

3. The refrigerant absorbs heat energy in the _____.

4. The _____ is a metering device.

5. In the typical residential heating/cooling system, the evaporator would be located

   _____.

6. In the typical residential heating/cooling system, the condenser would be located

   _____.

Respond to the following statements.

7. Name three types of condensers. (1) _____ , (2) _____, and
   (3) _____

8. A 230-volt, single-phase, hermetic compressor with an FLA of 21 and an LRA of 250 requires a disconnect switch with a rating of _____ horsepower.

9. A motor-compressor unit with a rating of 32 amperes is protected from overloads by a separate overload relay selected to trip at not more than _____ ampere(s).

10. Which *NEC* section permits the disconnect switch for a large rooftop air conditioner to be mounted inside the unit?

    _____

11. The nameplate of an air-conditioning unit specifies 85 amperes as the minimum ampacity for the supply conductors. Select the conductor size and type.

    _____

12. Where the nameplate of an air-conditioning unit specifies the minimum ampacity required for the supply conductors, should the electrician multiply this ampacity by 1.25 to determine the correct size conductors? Explain.

    _____

    _____

13. Which current rating of an air-conditioning unit is greater? Check the correct answer.

    a. _____ rated load current

    b. _____ branch-circuit selection current

Commercial Utility Interactive Photovoltaic Systems

OBJECTIVES

After studying this chapter, you should be able to

- list the components of a utility interactive solar photovoltaic system.

- describe the function of utility interactive solar photovoltaic system components.

- apply the *National Electrical Code (NEC)* to the design and installation of commercial utility interactive solar photovoltaic system components.

- interpret a typical utility interactive solar photovoltaic system single-line diagram.

Utility interactive solar photovoltaic (PV) systems have been around for many years. The installation of these systems has accelerated in recent years due to energy conservation and environmental concerns. There are some unique challenges involved with the electrical work for a photovoltaic system. Because of this, *Chapters 1-4* along with *Articles 690* and *705* of the *National Electrical Code (NEC)* will apply to the installation. *Article 705* is involved because the PV system and the systems provided by the electric utility can operate in parallel. It is vital that commercial electricians be aware of these requirements. Commercial photovoltaic installations tend to be larger and more complex than residential systems. Engineering and installation of commercial photovoltaic systems will often involve the electrical, structural, and roofing trades. This chapter will focus on electrical requirements of the *NEC* and related electrical installation techniques that are common in the industry.

Many of the *NEC* requirements for disconnecting means, grounding/bonding, labeling and photovoltaic system interconnection to the buildings electrical system exist because the photovoltaic system is a second supply of electricity for the building. A utility interactive solar photovoltaic system is not just another load for a building electrical system. The photovoltaic system is another source, much like an emergency generator. One big difference is that a photovoltaic supply cannot be turned off as easily as a generator. Much of the photovoltaic system will be energized anytime the sun is shining.

Proper installation is critical for building occupants and first responders in event of an emergency. To emphasize this point, a new *690.4(E)* was added to the 2011 *NEC*. It requires that all equipment and systems in *690.4(A)* through *(D)* and all associated wiring and interconnection be installed only by qualified persons. The term *Qualified Person* is defined in *NEC Article 100* as *One who has skills and knowledge related to the construction and operation of the electrical equipment and installations and has received safety training to recognize and avoid the hazards involved.* *

---

*Reprinted with permission from NFPA 70-2011.

THE PHOTOVOLTAIC EFFECT

When particles of light (photons) are absorbed by specialized combinations of semiconductors (diodes), electrical current is produced. This is known as the photovoltaic effect. Although first observed in the middle of the 1800s, it was not until the space race of the 1950s and 1960s that practical applications were realized. The most common material used today for production of photovoltaic cells is crystalline silicon. Crystalline silicon is fabricated into wafers and then doped (infused) with phosphorus gas to create the photovoltaic semiconductor. Further processing produces a functioning photovoltaic cell. Completed cells are combined to produce the modules used for photovoltaic systems. Over the years photovoltaic modules have become much more efficient and costs of manufacturing have dropped. The increased expense and environmental concerns associated with the use of fossil fuels have helped to make photovoltaic systems a cost effective source of electricity.

THE BASIC UTILITY INTERACTIVE PHOTOVOLTAIC SYSTEM

The photovoltaic effect can be put to practical use by supplying a commercial building with some or all of the electrical power needed to operate. System sizes range from 2 kW to 1 mW and larger. The photovoltaic modules in the array produce direct current electricity when exposed to sunlight. Direct current from the modules supplies an inverter that converts the dc into alternating current (ac) synchronized with utility electricity.

The electrical energy produced by a photovoltaic system can be stored through the use of batteries that then supply the building's electrical needs at night or on overcast days. A large enough system like this would not even need a utility supply of electricity. Such stand-alone systems are not as common or as practical as utility interactive photovoltaic systems. The normal utility electrical supply is always present in a utility interactive photovoltaic system. Electrical energy produced by the photovoltaic system reduces the amount of electricity supplied by the utility. If the photovoltaic system produces more electricity than is required for the building electrical load, the surplus electricity is sold to the utility. Bidirectional electrical metering equipment installed by the utility accurately measure the power produced by the PV system. The utility is required to pay for this privately produced electricity at predetermined rates. Many electrical utilities subsidize the installation of utility interactive photovoltaic systems. In addition, there are various state and federal programs that help to offset the cost of installation. A separate permit for installation of utility interactive PV systems is usually required by local jurisdictions. Figures 22-1 and 22-2 identify the basic photovoltaic components and system configurations.

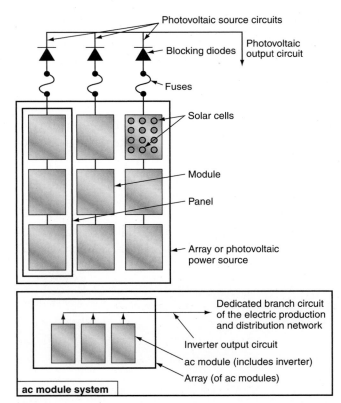

Notes:
1. These diagrams are intended to be a means of identification for photovoltaic system components, circuits, and connections.
2. Disconnecting means required by *Article 690, Part III*, are not shown.
3. System grounding and equipment grounding are not shown. See *Article 690, Part V.*

FIGURE 22-1 Identification of solar photovoltaic system components. (*Figure 690.1(A) reprinted with permission from NFPA-70, 2011.*)

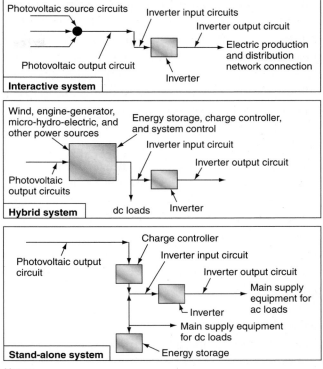

Notes:
1. These diagrams are intended to be a means of identification for photovoltaic system components, circuits, and connections.
2. Disconnecting means and overcurrent protection required by *Article 690* are not shown.
3. System grounding and equipment grounding are not shown. See *Article 690, Part V.*
4. Custom designs occur in each configuration, and some components are optional.

FIGURE 22-2 Identification of solar photovoltaic system components in common system configurations. (*Figure 690.1(B) reprinted with permission from NFPA 70-2011.*)

UTILITY INTERACTIVE PHOTOVOLTAIC SYSTEM COMPONENTS

Modules

Photovoltaic modules are the basic building block of a utility interactive photovoltaic system. Individual photovoltaic semiconductor cells are electrically connected together, sealed in a protective material, and mounted to a frame at the factory. Modules are produced using different technologies, which result in modules of differing efficiencies, sizes, and production costs. The amount of electrical power or watts produced varies as well. All modules are tested and listed to *UL Standard 1703*. Manufacturers' specification sheets will list all pertinent information and allow comparison of modules. Two parameters of interest to the designer and installer are module short circuit current (I_{sc}) and open circuit voltage (V_{oc}). The design and installation of a complete system will be dictated in part by these parameters. Many *NEC* rules of *Article 690* are based on these two factors. Electricity produced by photovoltaic modules is direct current (dc). Manufacturers will specify mounting and grounding points for modules in accordance with the listing standards. Photovoltaic modules are produced with wire leads for connection to other components of the photovoltaic system.

Array

An array is simply a collection of modules that form a complete photovoltaic generating unit. See Figure 22-3 for an example of a ground-mounted array of modules. The array will consist of a specific quantity of modules connected in series/parallel strings. This term also includes associated structural supports and mounting racks. Arrays may be fixed, adjustable, or of the tracking type. Fixed arrays are installed at a predetermined tilt angle to the sun exposure. Adjustable tilt arrays can be manually adjusted to maximize output as the sun changes position from summer to winter. Motor-controlled automatic sun-tracking arrays are the most efficient because modules will always attain maximum sun exposure. Automatic sun-tracking arrays are more expensive to install and maintain. The gain in output must offset this additional expense to make the system cost-effective.

Combiner Boxes

Series-connected strings of modules are connected in parallel to other series strings in the combiner box. Combiner boxes may be fused or non-fused and are produced with or without integral disconnects. Listed combiner boxes are required by *NEC 690.4(D)*. See Figure 22-4. Photovoltaic source circuits (module string conductors) are direct current and operate in the 400–600 volt range. All source circuit conductors and equipment (including combiner boxes) must be approved for use with up to 600 volts dc. Smaller commercial PV installations may not need to combine module string conductors. In this case all string conductors will be installed to the inverter. Exposed USE-2 or PV conductors will be spliced to the usual THWN type conductors installed in the conduit or raceway from the array to the inverter. Splices such as this take place in a junction box known as a transition box.

Disconnects

Disconnecting means are required in various locations throughout a photovoltaic system. *Part III*

FIGURE 22-3 A ground-mounted array of modules. (*Delmar/Cengage Learning*)

FIGURE 22-4 Fused combiner box. (*Delmar/Cengage Learning*)

of *NEC Article 690* provides requirements for installation of disconnecting means in a photovoltaic system. *NEC 690.14(C)* requires a disconnecting means for the source (array) ungrounded conductors, located either on the exterior of the building or nearest the point of entrance of photovoltaic system conductors. The exception to this section permits these conductors to pass through the structure if contained in a metallic raceway or enclosure; see *NEC 690.31(E)*. The dc grounded conductor is not permitted to be interrupted by disconnect contacts and must pass straight through the disconnect enclosure. Any disconnect used for direct current must be suitable for use with dc and will be labeled as such by the manufacturer. Specific instructions that manufacturers provide for using a disconnect with dc must be followed. For example, some manufacturers will require that two poles of a knife blade disconnect are used to interrupt a single dc circuit conductor. Direct current is more difficult to interrupt than alternating current, which is why this arrangement may be required.

Disconnecting means for inverters must be provided as well. Some inverters will have integral disconnects; those that don't will require installation of separate disconnects. Inverters will have connections to both dc (from array) and ac (from utility) circuits. AC output conductors of an inverter will be energized even when the dc input is turned off. Why? Because the ac output of the inverter is connected to existing energized ac premises or service conductors. These conductors will still be energized even though the inverter is not producing electrical current.

NEC 690.15 requires that equipment such as an inverter is able to be disconnected from ungrounded conductors of all sources. Where that equipment is energized from two sources (such as inverters) the disconnecting means shall be grouped and identified. All equipment disconnecting means must be within 50 feet and in sight of the equipment. A disconnecting means is a function, not always a separate device. For example, if an inverter is located adjacent to an existing electrical service panel and supplies that service panel through a circuit breaker in the panel, then the circuit breaker could fulfill the inverter ac disconnect requirement. Local electrical utilities may require installation of a dedicated utility disconnect in addition to *NEC* required disconnects.

As with all electrical components of photovoltaic systems, disconnects must be appropriate for installation conditions (indoor or outdoor) and circuiting (ac, dc, voltage, current, and number of poles).

Inverters

The equipment that converts the direct current produced by modules into an alternating current (ac) output is known as the inverter. A utility interactive inverter will automatically synchronize the inverter output to the utility system. Proper phase and voltage relationships must be maintained between an inverter output and the ac voltage and phase as supplied by the utility. Inverter outputs range from 2 kW to 500 kW and more. See Figure 22-5 for an example of a medium size utility interactive inverter. Commercial photovoltaic systems with multiple inverters producing a megawatt or more are becoming more common.

The *NEC* requires all utility interactive inverters to be listed to *UL 1741*. PV system designers and installers should be aware of two key requirements of the standard. The first is known as anti-islanding. Utility interactive inverters are required to automatically turn off when utility power is lost. This prevents an inverter from feeding electricity back into the utility electrical grid, which would be extremely dangerous for utility workers. Remember

FIGURE 22-5   A utility interactive inverter.
(*Delmar/Cengage Learning*)

that the 240 volts produced by an inverter can produce 12,000 volts or more on the primary side of utility transformers if the inverter were allowed to back feed the utility system. The utility transformer would be operating as a step-up transformer in effect. Another important requirement of *UL 1741* is dc ground-fault protection (GFP). Ground faults in PV source circuits (array) do not result in the high fault currents produced by utility electrical services. Photovoltaic modules are a current limited source. Short-circuit current from a module is slightly higher than normal module output to the inverter, not enough to open an overcurrent device (fuse or circuit breaker). Ground-fault current will flow through bonded exposed metallic structures and raceways to the inverter and then back to the source (modules). Heating and arcing caused by fault current can cause fires and is a shock hazard as well. For these reasons, a ground-fault protection system is required. The system is required to perform four functions:

1. Detect a ground fault in the photovoltaic array.
2. Interrupt the ground-fault current.
3. Indicate that a ground fault occurred.
4. Disconnect the faulted circuit.

Ground-fault protection is very important for photovoltaic systems. A ground fault can continue as long as modules are producing current, which happens anytime the sun is shining. A single stage ground fault (positive or negative conductor to grounded surface) will be interrupted by the GFP system. The actual short circuit will still exist until repaired, however. A 2-stage ground fault with both positive and negative conductors shorted together through common metallic structures/raceways on the roof will continue regardless of inverter action. Remember that modules in the array are the source of electrical current. If there is a complete path of current for the positive and negative conductors on the roof, current will flow.

Either the positive or negative source (array) conductor will be grounded. Grounding of this current carrying conductor will occur in the inverter and is determined by inverter design. There are *NEC* rules for identification of dc system grounded conductors as there are with ac system grounded conductors; see *NEC 200.6*.

The 2011 *NEC* includes a new requirement for arc-fault protection when the dc source and/or output circuits on or in a building are operating at or above 80 volts. Arc-fault protection for the dc source/output circuits is required because damaging arc faults have occurred that the ground fault protection system cannot prevent. *NEC 690.11* requires a listed system that will perform four functions:

1. Detect and interrupt arc faults in the dc PV source and output circuits.
2. Disable or disconnect either the inverter, charge controller, or the system components that are a part of the arcing circuit.
3. Require a manual reset of the disabled or disconnected equipment.
4. Annunciation of the arc fault.

UTILITY INTERACTIVE PHOTOVOLTAIC PLANS

A permit and associated inspection(s) by the AHJ (Authority Having Jurisdiction) will be required for photovoltaic systems in most jurisdictions. Commercial PV projects will typically require engineered plans. The plans usually have to be submitted to the AHJ for review and comment prior to issuance of a permit. A utility interactive solar photovoltaic plan will consist of an electrical single line diagram, a plot or site plan, structural plan and roof penetration (flashing) details. Depending on the size and complexity of a project, there could be several pages of structural and electrical details. Manufacturers' specifications and installation instructions for key components should be provided because these provide the basis for the design.

Site or Plot Plan

The site or plot plan will indicate the geographic locations of principle PV components and existing electrical service or affected panelboards. Photovoltaic module quantity and locations will be indicated along with locations of all disconnects, combiner boxes, inverters, and junction boxes. Conduit or raceway types and routing should also be indicated.

Single-Line Diagram

A single-line diagram is an electrical schematic diagram for the utility interactive PV system. Relative physical location of components is not the purpose of this plan sheet. A simplified electrical connection diagram is the goal. Quantity of modules in the array along with the quantity of modules per string must be indicated. Electrical size and specifications for the dc and ac disconnects, combiner boxes, inverters, and ac combiner panels are necessary information. Existing electrical service size, main circuit breaker size, and the size of any existing ac panels that are part of the design will need to be indicated. All conductor types and sizes along with conduit/raceway types and sizes must be shown. The overcurrent device (fuses, circuit breakers) ratings are needed to verify correct sizing of these components.

Grounding and bonding is of extreme importance for the utility interactive photovoltaic system and must be accurately indicated on the single-line diagram. Types of existing or new grounding electrodes must be specified along with all grounding electrode conductor sizes and connection points. Individual PV modules are required to be grounded per the manufacturer's instructions so connection of the equipment grounding conductor to the module is of extreme importance and must be detailed.

Finally, it is critical that the method of utility interconnection be specified. The *NEC* permits two basic methods in *705.12*. The supply side method is permitted by *705.12(A)*. This method is basically a service tap ahead of the service disconnect, and many of the requirements of *NEC Article 230* (Services) will apply. The service tap ahead of the main disconnect must be made in a manner which does not violate the listing of the service equipment. It is not always easy to make this connection, and the service equipment manufacturer may have to be consulted. Once this connection is made, the tap conductors are considered service-entrance conductors and must be treated as such. Keeping these conductors separate from all other conductors and located either outside of the building or inside, nearest the point of entry into the building are requirements of *NEC Article 230*. These service conductors must terminate in another service disconnecting means located adjacent to the existing service. Overcurrent protection will have to be provided.

A service load side connection of the PV system is permitted by *NEC 705.12(D)*, but there are limitations. The dedicated circuit breaker must be suitable for back feed because current from the inverter will be flowing from the load side to the line side of the circuit breaker. Also, the sum of all overcurrent devices supplying current to panelboard busbars cannot exceed 120% of the busbar rating. Adding the ampere rating of the panelboard main circuit breaker to the ampere rating of the inverter connection circuit breaker would have to equal no more than 120% of the panelboard or service buss rating. For example a 600-amp-rated service (buss rating) with a 600-amp main circuit breaker would be limited to a 120-amp circuit breaker for the inverter load side connection. (600-A $\times$ 120% = 720 − 600 = 120 A) The method of connection must be indicated with enough detail so that applicable *NEC* requirements can be verified. A basic single-line diagram for a utility interactive system using a load side connection is illustrated in Figure 22-6.

Calculations

Calculations are needed to support the single-line diagram. The designer will use information from manufacturers' product specification sheets to make calculations to design the system. Sizing of module strings, arrays, inverters, disconnects, overcurrent devices, and the conductors/raceways that connect all of these components together will be based on calculations. Required calculations can be broken down into two broad categories: ac and dc.

First, we will look at dc calculations. At a minimum there will be two dc calculations required. Quantity of modules in a series string will be limited by the open circuit voltage (V_{oc}) of the modules. The V_{oc} (provided by the manufacturers specification sheet) of each module in a series string is added together to provide total string voltage. For example, a string of 10 modules, each with a V_{oc} of 50 volts, would have a total string V_{oc} of 500 volts (10 modules $\times$ 50 volts = 500 volts). There is one other factor that must be considered to comply with *NEC 690.7*: temperature. *NEC 690.7* defines the maximum PV system voltage as the sum of the

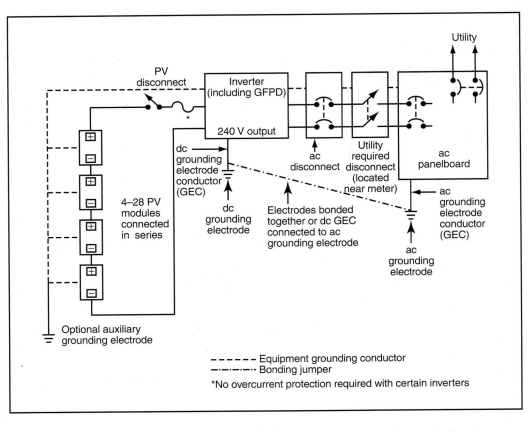

FIGURE 22-6   Single-line diagram for a utility interactive photovoltaic system.
(*Reprinted with permission from National Electrical Code Handbook Exhibit 690.6(C) NFPA 2008.*)

series-connected modules corrected to the lowest expected ambient temperature. Module V_{oc} goes up as ambient temperature goes down. A colder module will have a higher V_{oc} than a warmer module. The V_{oc} provided by the manufacturer results from a standard test condition (STC) of 77°F(25°C). Actual V_{oc} of the module/string will depend on ambient temperature where installed. If a manufacturer does not provide the coefficient of change (voltage per degree) for a module, then Table 22-1 (*NEC Table 690.7*) must be used. Table 22-1 (*NEC Table 690.7*) provides temperature correction factors for ambient temperatures from −40°F to 76°F (−40°C to 24°C). If the PV system were installed in a location where ambient temperature could be as low as 25°F, the temperature correction factor would be 1.12 from Table 22-1 (*NEC 690.7*). String V_{oc} is multiplied by this factor to provide the maximum system dc voltage. See Table 22-1. As long as the temperature-corrected V_{oc} remains less than the rating of system conductors and equipment

(600 volts), the system is in compliance. Quantity of modules in a series string and the lowest ambient temperature possible are the two variables that must be considered by the designer.

The second important dc calculation involves current. Maximum dc source current is defined by *NEC 690.8(A)* as the sum of parallel module rated short circuit currents multiplied by 125%. The module short circuit current (I_{sc}) is determined by a standard test condition (STC) irradiance of 1000W/ m^2. Actual solar noontime sunlight will often be more intense than the standard test condition, and this is why *NEC 690.8(A)* requires the calculation. *NEC 690.8(B)(1)* requires a second multiplication by 125% for sizing of circuit conductors and overcurrent devices. This is because the currents are continuous (3 hours or more). Conductor ampacity will be determined by *NEC 690.8(B)(2)*. Two options are permitted. The current value from *690.8(A)* can be multiplied by 125% without consideration for conditions of use (temperature and

TABLE 22-1

Table 690.7 Voltage Correction Factors for Crystalline and Multicrystalline Silicon Modules*

CORRECTION FACTORS FOR AMBIENT TEMPERATURES BELOW 77°F (25°C). (MULTIPLY THE RATED OPEN CIRCUIT VOLTAGE BY THE APPROPRIATE CORRECTION FACTOR SHOWN BELOW.)

Ambient Temperature (°C)	Factor	Ambient Temperature (°F)
24 to 20	1.02	76 to 68
19 to 15	1.04	67 to 59
14 to 10	1.06	58 to 50
9 to 5	1.08	49 to 41
4 to 0	1.10	40 to 32
−1 to −5	1.12	31 to 23
−6 to −10	1.14	22 to 14
−11 to −15	1.16	13 to 5
−16 to −20	1.18	4 to −4
−21 to −25	1.20	−5 to −13
−26 to −30	1.21	−14 to −22
−31 to −35	1.23	−23 to −31
−36 to −40	1.25	−32 to −40

*Reprinted with permission from NFPA 70-2011

TABLE 22-2

▶Table 310.15(B)(3)(c) Ambient Temperature Adjustment for Circular Raceways Exposed to Sunlight On or Above Rooftops*◀

Distance Above Roof to Bottom of Conduit	Temperature Adder °F	°C
0 to ½ (13 mm)	60	33
Above ½ in. (13 mm) to 3½ (90 mm)	40	22
Above 3½ in. (90 mm) to 12 in. (300 mm)	30	17
Above 12 in. (300 mm) to 36 in. (900 mm)	25	14

*Reprinted with permission from NFPA 70-2011

wire fill), or the current value from *690.8(A)* can be used without the additional 125% multiplier if the correction factors for conditions of use are applied. Multiplying the I_{sc} by 125% twice will provide the current for sizing all circuit conductors, disconnects, and overcurrent devices. Note that 125% × 125% = 156%. Multiplying module I_{sc} by 156% will provide the result required by *NEC 690.8(B)(2)(a)*. Remember that electrical current is the same in all parts of a series circuit. The I_{sc} of a module is the same amount of current that will flow in the entire series circuit. If a series string of modules has an I_{sc} of 5 amps, for example, then conductors and overcurrent devices must be sized for 156% of 5 amps (1.56 × 5), or 7.8 amps. When several strings are combined into a single output, the larger output conductors must be sized in the same way.

The ac side (inverter output) of a utility interactive photovoltaic system will require calculations to size conductors and components as well. *NEC 690.8(A)(3)* defines maximum inverter output current as the inverter continuous output rating. This information will be found in the manufacturer's specifications for the inverter. Dividing the inverter power rating (kW) by output voltage is another way to determine output current. A 10 kW (10,000 watt) inverter operating at 240 volts will have a rated output current of 42 amps (10,000 ÷ 240). *NEC 690.8(B)(1)* requires the designer to size inverter output conductors and overcurrent devices for 125% of inverter output current rating. Inverter output current is continuous (3 hours or more), and that is the reason for the requirement. With the 10 kW inverter example above, output conductors and overcurrent devices would have to be sized for 53 amps (125% × 42 amps).

Ampacity correction for ambient temperature must be considered for both the dc and ac system conductors. Have you ever left metallic tools or materials out in the sun on a hot day? You probably noticed that metallic objects in sunlight will attain temperatures well above ambient temperature. This phenomenon is recognized and addressed by Table 22-2 [*NEC Table 310.15(B)(3)(c)*]. Ambient temperature adders are provided by the table to determine temperature inside of conduits on a roof. See Table 22-2. This temperature is then added to the ambient temperature in Table 22-3 [*NEC Table 310.15(B)(2)(a)*] for application of the temperature correction factors of Table 22-3. Large commercial rooftop arrays often have long conduit installations exposed to sunlight, as shown in Figure 22-7. It is vital that conductors in the conduits are sized properly to avoid overheating. By applying *Table 310.15(B)(3)(c)* and the temperature correction factors of Table 4-4 [formerly located below *NEC Table 310.16*, now *NEC Table 310.15(B)*

TABLE 22-3

Table 310.15(B)(3)(a) Adjustment Factors for More Than Three Current-Carrying Conductors in a Raceway or Cable*

Number of Conductors[1]	Percent of Values in Tables 310.15(B)(16) through 310.15(B)(19) as Adjusted for Ambient Temperature if Necessary
4–6	80
7–9	70
10–20	50
21–30	45
31–40	40
41 and above	35

[1]Number of Conductors is the total number of conductors in the raceway or cable adjusted in accordance with 310.15(B)(5) and (6).
*Reprinted with permission from NFPA 70-2011.

FIGURE 22-7   Commercial buildings often have conduit installations that are installed across a roof and are exposed to the rays of the sun. (*Delmar/Cengage Learning*)

(2)(a)], the designer will ensure conductor insulation remains within its temperature rating. Another variable that will affect conductor ampacity is the quantity of current carrying conductors installed in a raceway or cable. *NEC Table 310.15 (B)(3)(a)* provides adjustment factors based on the number of current-carrying conductors in the raceway or cable. The correction factors provided by this table must be applied along with the ambient temperature correction factors to all conductors in the system.

UTILITY INTERACTIVE PHOTOVOLTAIC SYSTEM INSTALLATION

The first order of business when starting a new commercial PV installation is to review the approved project plans and specifications. It is critical to verify that the site matches what is indicated on the approved plan. Any discrepancies should be reported to the engineer of record. A utility interactive PV system will ultimately be integrated into the existing electrical system for the building. Does the existing electrical system match the design? Are locations of the existing service and panelboards accurate? Can the existing grounding electrode system be accessed? This is the time to ask these questions and inquire about any local requirements (AHJ and utility). Material, tool, and manpower needs based on the project schedule can be assessed at this point to keep the project running smoothly. Personnel protective equipment (PPE) requirements should be assessed so that this safety equipment will be onsite when needed. Logistics such as site access, employee parking, and working hours should be coordinated with the owner or manager of the building.

Mounting of Array

A photovoltaic module array for a commercial building may be either ground or roof mounted. Available space is often the determining factor. A ground-mounted array may be constructed so that it can be used as covered vehicle parking. Large commercial arrays take up a lot of space, and it is not uncommon for a large commercial building to have most of the roof covered with modules. This can be a concern for local fire fighters who could need access to a roof for fire fighting. The *NEC* does not address this issue, but some jurisdictions may have requirements that restrict the footprint of the array on a roof. Array size and footprint will be reviewed for compliance with local requirements during the plan review process.

Whether installed on the roof or in a parking lot, structural elements for support of the array will be the first priority. A ground-mounted array will have structural footings and framing designed by a structural engineer. A roof-mounted array may make use of existing supports, but many times additional structural framing is required. These additional elements will be detailed on the structural plan. The electrician is not normally directly responsible for the structural installation but may be involved in the layout of supports. Locations for conduit stub-ups for electrical components (whether underground or on the roof) must be coordinated with structural elements to access the final locations of combiner boxes, junction boxes, disconnects, inverters, panelboards, and services.

Once the structural layout and installation is started, the electrician can begin the electrical work. The electrical installation can be thought of as two separate systems: dc and ac. The dc system involves the conductors and components from the modules to the inverter. The ac system includes everything from the inverter output to the ultimate utility connection. Multiple ac panelboards, transformers, and disconnects may be required.

The order of the electrical work will depend on the scheduling of the other elements of the project. For example, array source conductors may have to be installed as modules are attached to the support racks. A large array of modules will be connected in a specific series-parallel combination per the single-line diagram. Source conductors that connect all modules must be placed and identified to insure the circuiting. Support of these conductors is of utmost importance. Array source conductors are permitted to be exposed, so good workmanship is the only protection against short circuits. Sharp edges must be avoided when routing the conductors. Only type *USE-2* or the newer type *PV* conductors are permitted to be installed as exposed source conductors. Cable ties used to secure the source conductors must be rated for ambient temperature and ultraviolet (UV) light exposure.

Module Grounding

Grounding of individual modules may have to be accomplished as the modules are installed as well. Module manufacturers will provide very specific grounding information in the installation instructions. The low currents associated with module ground faults make these connections extremely

FIGURE 22-8 Connection of equipment grounding conductor to module frames. (*Delmar/Cengage Learning*)

important. Ground-fault current must have a reliable low impedance path from the modules to the inverter for the life of the system. Module frames are anodized, clear-coated, or mil-finished aluminum. Aluminum oxidizes quickly, and copper conductors must be terminated properly to ensure a good connection. Using the grounding hardware and module frame grounding location specified in the module installation instructions is the only way to ensure connection integrity for the life of the system. The equipment grounding conductor attachment to module frames can be seen in Figure 22-8.

Installation of Combiner Boxes

Array source conductors will usually be routed to fused combiner boxes. Each source will be connected through a fuse in the combiner box to a larger output conductor. Larger systems may have 50 to 100 strings or more combined in the box to provide dc current for the inverter. A disconnect for the output of the combiner box is not required by the *NEC* but can be a very good idea. Without a combiner disconnect, the only way to interrupt the combined module current is to remove all source circuit fuses—not very practical in an emergency situation! Don't ever forget that modules are producing current anytime the sun is shining. Safe work practices must be observed. In essence modules are energized from the moment the installation is started.

With both roof- and ground-mounted arrays, the electrical installation is outdoors. Equipment and conductors rated for the conditions of use are required. NEMA 3R junction boxes and raintight conduit fittings will be required. Some installations with long exposed conduit runs may require expansion couplings to allow for movement caused by thermal expansion. See *NEC 300.7(B)*. All electrical equipment and raceways must be supported and secured.

Large Equipment

The inverter(s), transformer(s), and switchboard(s) in larger systems will usually be located in a central equipment yard. Large equipment may require the use of cranes or forklifts to position. Anchoring of all equipment will be required. Look to the equipment installation instructions and engineered drawings for specific requirements. Equipment that may require service while energized must meet the minimum working space requirements of *NEC 110.26*. This would always apply to inverters and switchboards. Some jurisdictions will apply the working space requirements to disconnects and transformers. Depth of the working space will vary per *NEC Table 110.26 (A)(1)*. Equipment voltage and surrounding conditions are the two variables. Minimum width for working space is 30 in. (762 mm) or the width of the equipment, whichever is greater. All equipment doors or hinged panels must be able to swing 90°. Working space is a primary concern with layout and installation of equipment. Requirements for working space around equipment over 600 volts are found in *Part III* of *NEC Article 110*.

Wiring

With the equipment and raceways in place, all conductors can be installed. As with all electrical installations, conductors must be installed without damaging insulation. Conductors must be color coded per *NEC* requirements. Grounded conductors, whether dc or ac, must be identified with white or gray insulation per *NEC 200.6*. Equipment grounding conductors must be green or bare in accordance with *NEC 250.119*. Ungrounded conductors should be identified per project specifications and all conductors tested to verify insulation integrity after installation.

Special rules are now in place in *690.31(E)* to control the wiring of dc PV source and output circuits inside a building. Included are requirements for wiring within 10 in. (25 cm) of the roof decking except where directly below the roof surface covered by PV modules and associated equipment. Additional requirements are in place for wiring installed in flexible metal conduit or Type MC cable.

The following wiring methods and enclosures that contain photovoltaic power source conductors are required by *690.31(E)(1)* to be marked with the wording "Photovoltaic Power Source" by means of permanently affixed labels or other approved permanent marking:

- Exposed raceways, cable trays, and other wiring methods.
- The covers or enclosures of pull boxes and junction boxes.
- Conduit bodies in which any of the available conduit openings are unused.

The labels or markings are required to be visible after installation. PV power circuit labels must appear on every section of the wiring system that is separated by enclosures, walls, partitions, ceilings, or floors. Spacing between labels and/or markings must not be more than 10 ft (3 m). Labels required by this section shall be suitable for the environment where they are installed.

Grounding Electrode System

A grounding electrode system for a utility interactive photovoltaic installation is required in accordance with *NEC 690.47*. The NEC permits various methods of installation for the grounding electrode system, but all methods include a bond to the existing grounding electrode system for the building. The designer will specify a grounding electrode system on the approved plan, and the system must be installed as designed. Pay attention to the grounding electrode conductor size and routing. Grounding electrode conductors are required to be continuous (not spliced unless by irreversible means) and bonded to metallic enclosures that they pass through, by *NEC 250.64*. Additional grounding electrodes may be required by the *NEC* or the designer. It is very important that all grounding electrodes for a utility interactive PV system and the building grounding electrode system are bonded together.

Utility Point of Connection

Larger commercial utility interactive PV systems will often make use of multiple inverters. The ac output of multiple inverters will be combined in a standard panelboard or switchboard which then supplies the building and/or utility through either a supply side or load side point of connection as permitted by *NEC 705.12*. *NEC 690.64* directs that the point of connection comply with *705.12* because that section covers the connection of interconnected electric power production sources. The method of connection will be selected by the system designer and detailed on the single-line diagram. Size of the utility interactive PV system relative to size of the existing electrical service will usually be the deciding factor. The load side connections permitted by *NEC 705.12(D)* are limited by the size of the existing service switchboard and main circuit breaker. *NEC 705.12(D)(7)* requires an inverter circuit breaker to be installed at the opposite end of the panelboard from the main circuit breaker to prevent possible overloading of the busbar(s). A load side connection is illustrated in Figure 22-9.

System size may require a designer to specify the supply-side connection permitted by *NEC 705.12(A)*. Supply-side connections are essentially a service tap, and the applicable provisions of *NEC Article 230* will apply. If the service tap is to be made in the existing service switchboard, then the tap must be accomplished in such a way that the listing of the switchboard is not compromised. A line side service tap box may be required to make the connection. See Figure 22-10 for a diagram of a supply-side connection. Service conductors are not subject to the same *NEC* overcurrent and grounding rules as the premises wiring system. Routing and physical protection requirements for service conductors found in *NEC Article 230* exist for this reason. A second service disconnect and

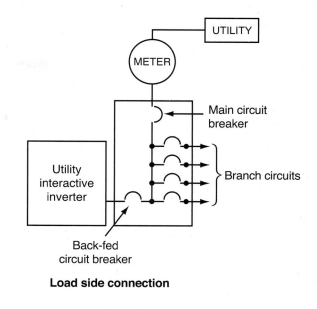

Load side connection

FIGURE 22-9 Solar PV system connected to the load side of the service disconnecting means. (*Delmar/Cengage Learning*)

related overcurrent protection along with specific labeling will be required for this type of connection. Requirements of *NEC Articles 230, 690,* and *705* as well as those of the utility will be considered by the system designer. It is critical that the electrician install the system as designed with either a supply or load side connection.

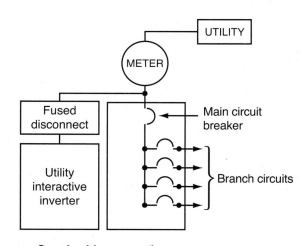

Supply side connection

FIGURE 22-10 Solar PV system connected to the line side of the service disconnecting means. (*Delmar/Cengage Learning*)

Labels, Signs, and Equipment Marking

A utility interactive PV system is a second supply of electricity for a building. Portions of the system are always energized. A second electrical service for a building may even be required. For these reasons the *NEC* has specific labeling requirements for the utility interactive PV system. Of course all standard *NEC* labeling requirements will still apply. All equipment and overcurrent devices must be permanently identified with the nomenclature specified in the approved plan. In addition, *NEC Article 690* has specific labeling requirements for various components of the system:

> *690.5(C)*—Inverter ground-fault warning label
> *690.7(E)*—Bipolar system warning label
> *690.17*—PV disconnect warning label
> *690.53*—DC system rating label
> *690.54*—AC system rating label
> *705.12(D)(7)*—Inverter output connection circuit-breaker position label

NEC 705.10 requires that a permanent plaque or directory that indicates all interconnected power sources for the premises be installed at the service and the location(s) of the power production source(s). All service disconnecting means are required to be permanently marked by *NEC 230.70(B)*.

⎓⟍⟋⎓ SYSTEM CHECKOUT AND COMMISSIONING

Once the system construction is complete, the process of initial start up or commissioning can begin. Before any system components are energized, a thorough inspection of the system must be performed. A final inspection by the AHJ may be required at this point as well. Start the inspection with the modules in the array.

- Are all modules installed and secured?
- Are source conductors of the correct type and size?
- What about securing and protection of the conductors? It is especially critical that the conductors are installed correctly because they

are usually operating in the 400–600 volt range, exposed to the elements, and cannot be easily de-energized.

- Grounding of modules per the manufacturer's installation instructions is critical and must be verified.
- Combiner and junction boxes must be securely installed with covers in place.
- Are direct current (dc) rated fuses of the correct amperage installed in fused combiner boxes and fused disconnects?
- Conduits must be supported in compliance with their controlling article.
- Conduit couplings and connectors must be made up tight and should be checked. The dc source conductors operate at over 250 volts. This means that the conduit bonding requirements of *NEC 250.97* will apply.
- DC disconnects should be looked at to verify correct ratings and proper termination of all conductors.
- All conductor (dc and ac) terminations throughout the entire system must be tightened to manufacturer-specified torque values.
- Temperature and conductor size ratings for all lugs and splicing devices should be checked.
- Is the grounding electrode system installed per the design? Verify that all grounding electrode conductors are of the correct size and type. They must be continuous and bonded to all metallic enclosures that they pass through according to *NEC 250.64*.
- Correct terminations of conductors in the inverter will have to be confirmed.
- Locations for the connection of the equipment grounding conductor(s) and grounding electrode conductor in the inverter are very important. The function of the ground fault detection/interruption system in the inverter will depend on correct termination of these conductors in the inverter.
- *NEC 110.7* requires all completed wiring installations to be free of short circuits and ground faults. Ohmmeters and high voltage

meg-ohm meters are used to verify the integrity of conductor insulation.

- Check all equipment for proper labeling and the working space for compliance with *NEC 110.26*.

The National Electrical Testing Association (NETA) has specific procedures for conductor insulation testing. Project specifications should be referenced for requirements in regard to any testing. Third-party testing may be required. All test results must be documented. Field evaluations by a Nationally Recognized Testing Laboratory (NRTL) for any nonlisted equipment may be required and must be completed prior to commissioning.

The inverter could have specific commissioning instructions. A manufacturer may require that a factory technician perform the startup to prevent possible damage and maintain the inverter warranty. An interactive inverter may need connections completed in a specific sequence so that synchronization with the utility system is realized. The general procedure for startup is to check voltage and current levels from the source (modules) to inverter closing disconnects as needed to accomplish this. Voltage and current levels should all be compared with expected values per the system design. Measuring module source dc voltage levels at the combiner boxes will confirm that all module series strings are properly connected.

Once the system is operable, some calibration may be required to attain maximum performance. Calibration of the system may involve adjustment of the array tilt angles if an adjustable array is installed. Motorized automatic sun tracking arrays will require initial adjustment as well.

SUMMARY

Although utility interactive photovoltaic systems are constructed with many of the same electrical materials used in the industry, this does not mean that there are not differences. Significant portions of the array will always be energized anytime the sun is up. These always "hot" conductors will usually be operating close to the 600-volt insulation rating and exposed to weather. An array short circuit will not produce enough fault current for an overcurrent device to be effective. Contrast this with a typical roof-mounted HVAC unit. The 480-volt branch circuit for the HVAC unit is installed in conduit, originates at an overcurrent device (circuit breaker) in a panelboard, and usually terminates in a fused disconnect at the unit. Turning off the building main service disconnect will de-energize the HVAC branch circuit but not the photovoltaic array conductors! Good workmanship in regard to module source string conductors and grounding/bonding of modules is very important. Inverter ac output to the building is a second supply of electrical power, and frequently a second electrical service connection is made for these systems. Location, disconnecting means, labeling, and grounding/bonding for this part of the system are vital.

Utility interactive photovoltaic systems are becoming more efficient by the day. Continued improvement of the systems along with a never-ending need for increased electrical power means that electricians who are able to install and maintain photovoltaic systems will be in demand for many years to come.

REVIEW

Refer to the *National Electrical Code* when necessary. Where applicable, responses should be written in complete sentences.

1. Is the electrical current produced by photovoltaic modules ac or dc?

2. To size module circuit (source) conductors and components, module string short circuit current (I_{sc}) must be multiplied by _____.

3. As module temperature decreases, module open circuit voltage (V_{oc}) _____.

4. Photovoltaic modules must be grounded per the _____ instructions.

5. A group of modules that form a complete photovoltaic power source is known as a(n) _____.

6. All conductors and components of array circuiting must be rated for _____.

7. Which types of conductors are permitted to be installed as exposed array circuiting?

8. Part _____ of *NEC Article 690* provides requirements for installation of disconnects in a photovoltaic system.

9. How is the grounded dc conductor of a photovoltaic system required to be identified?

    _____

    _____

10. Describe the function of a combiner box.

    _____

    _____

11. Describe the function of a utility interactive inverter.

    _____

    _____

12. What are two key requirements of *UL 1741* for utility interactive inverters?

    _____

    _____

13. List the four functions that the ground fault protection system for a utility interactive inverter must perform.

    _____

    _____

14. Which *NEC* section provides the requirements for grounding electrode systems in a photovoltaic installation? _____

15. What are the two types of utility connection permitted by the *NEC*?

    _____

    _____

16. Which *NEC* section provides the utility connection requirements for utility interactive photovoltaic systems? _____

17. The electrical schematic for a utility interactive photovoltaic system is known as the

    _____.

18. Which plan sheet will indicate the geographic locations of the photovoltaic system components? _____

19. Which *NEC* section provides working space requirements? _____

20. The process of initial startup for a utility interactive photovoltaic system is known as

    _____.

Appendix A

ELECTRICAL SPECIFICATIONS

General Clauses and Conditions

The General Clauses and Conditions and the Supplementary General Conditions are hereby made a part of the Electrical Specifications.

Scope

The electrical contractor shall furnish all labor and material to install the electrical equipment shown on the drawings herein specified, or both. The contractor shall secure all necessary permits and licenses required and in accepting the contract agrees to have all equipment in working order at the completion of the project.

Materials

All materials used shall be new, shall bear the label of a Nationally Recognized Testing Laboratory (NRTL), and shall meet the requirements of the drawings and specifications.

Workmanship

All electrical work shall be in accordance with the requirements of the *National Electrical Code*, shall be executed in a workmanlike manner, and shall present a neat and symmetrical appearance when completed.

Show Drawings

The contractor shall submit for approval descriptive literature for all equipment installed as part of this contract.

Motors

All motors shall be installed by the contractor furnishing the motor. All wiring to the motors and the connection of motors will be completed by the electrical contractor. All control wiring will be the responsibility of the contractor furnishing the equipment to be controlled.

Wiring

The wiring shall meet the following requirements:

a. The branch-circuit conductors will be copper not smaller than 12 AWG.

b. Branch-circuit conductors will be installed in electrical metallic tubing, electrical nonmetallic tubing or Type MC cable.

c. Except for the Owner's panelboard, three spare trade size ¾ EMT shall be installed from each flush-mounted panel into the ceiling joist space above the panel.

d. Feeders shall be installed in either rigid metal conduit, electrical metallic tubing, rigid nonmetallic conduit, or Type MC cable.

e. Conductors shall have an insulation with a 90°C rating, Types THHN/THWN, or XHHW. All conductors shall be copper.

Switches

Switches shall be ac general-use, snap switches, specification grade, 20 amperes, 120–277 volts, either single-pole, 3-way, or 4-way, as shown on the plans.

Time Clock

Furnish and install a 120 volt, two circuit, time clock to control the front and rear entry luminaires and the outdoor luminaires. The clock shall have astronomic control and integral battery backup.

Receptacles

The electrical contractor shall furnish and install, as indicated in blueprints E–1, E–2, and E–3, receptacles meeting NEMA standards and listed by Underwriters Laboratories, Inc., or equivalent.

Panelboards

The electrical contractor shall furnish panelboards, as shown on the plans and detailed in the panelboard schedules. Panelboards shall be listed by a Nationally Recognized Testing Laboratory. All panelboard interiors will have an ampere rating not less than as shown in the panelboard schedules and be rated at 75°C for supply conductors. Cabinets will be of painted galvanized sheet steel and will provide wiring gutters, as required by the *National Electrical Code*. Covers will be suitable for either flush or surface installation and shall be equipped with a keyed lock and a directory cardholder. All panelboards shall have an equipment grounding bus bonded to the cabinet. See the panelboard schedules for a listing of the required poles and overcurrent devices.

Molded-Case Circuit Breakers

Molded-case circuit breakers shall be installed in branch-circuit panelboards as indicated on the panelboard schedules following the Specifications. Each breaker shall provide inverse time delay under overload conditions and magnetic tripping for short circuits. The breaker operating mechanism shall be trip free, and multiple units will be internal common trip. Breakers shall have sufficient interrupting rating to interrupt the short-circuit current that is available at the line terminals of the panelboard or disconnect switch, but not less than 10,000 RMS symmetrical amperes.

Motor-Generator

The electrical contractor will furnish and install a motor-generator assembly capable of delivering 20 kilovolt-amperes at 208Y/120 volts, 3 phase. The prime mover engine shall be for use with one of the following fuels in the order of preference:

1. Utility supplied natural gas

2. On-site propane tank

3. On-site diesel fuel

The unit shall be liquid cooled, complete with 12-volt engine-starting batteries and battery charger, mounted on antivibration mounts with all necessary accessories, including mufflers, exhaust piping, fuel tanks, fuel lines, and remote derangement annunciator. The unit shall be listed as an assembly by a qualified Nationally Recognized Electrical Testing Laboratory.

Transfer Switch

The electrical contractor will furnish and install a complete listed automatic load transfer switch rated not less than 150 amperes at 208Y/120 volts, 3 phase. The switch control shall sense a loss of power on any phase and signal the motor-generator to start. When generator power reaches 80% of voltage and frequency, the switch shall automatically transfer the load to the generator source. When the normal power has been restored for a minimum of five minutes, the switch will reconnect the load to the utility power and shut off the motor-generator. Switch shall be in a NEMA 1 enclosure.

Luminaires

The electrical contractor will furnish and install luminaires, as described in the schedule shown in Chapter 15 of *Electrical Wiring Commercial* Edition 14 as published by Delmar, Cengage Learning. Luminaires shall be complete with diffusers and lamps where indicated. Ballast, when required, shall be Class P and have a sound rating of A. Each ballast is to be individually fused. The fuse size and type shall be as recommended by the ballast manufacturer. All rapid-start ballasts shall be Class E. Fluorescent lamps are to be T8 high-efficiency energy-saving type unless otherwise indicated.

Service Entrance and Equipment

The electrical contractor will furnish and install a service entrance as shown on the Electrical Power Distribution Riser Diagram, which is found in the electrical plans. Transformers and all primary service will be installed by the power company. The electrical contractor will provide a concrete pad and conduit rough-in for a pad-mounted transformer as required by the power company.

The electrical contractor will fully comply with *NEC Article 250* by installing a grounding electrode system. The grounding electrode system shall consist of not less than a connection to the main water pipe and to the concrete encased grounding electrode.

The electrical contractor will furnish and install PVC conduit from the electric utility to the service equipment at the building and pull the size and number of conductors shown on the electrical drawings from the transformer to the service equipment. The electrical contractor shall properly terminate the underground service conductors at the service equipment. The underground service conductors at the transformer will be terminated by the electric utility. All service conductors shall be marked at each end with corresponding colors according the the following schedule:

> Phase A: Black
>
> Phase B: Red
>
> Phase C: Blue
>
> Neutral: White

The electrical contractor will furnish and install service-entrance equipment, as shown in the plans and detailed herein. The equipment will consist of six heavy-duty switches, and the metering equipment as shown on the riser diagram. All equipment will be in NEMA Type 3R enclosures. The service equipment shall be rated not less than 35,000 RMS symmetrical amperes shown on service schedule.

The metering will be located as shown on the riser diagram and shall consist of 6 meters. Five of these meters shall be for the occupants of the building and one meter shall serve the owner's equipment. One meter shall be of the current transformer type as shown on the riser diagram. Coordinate all metering equipment with the electric utility.

Switches shall be rated to interrupt 200,000 symmetrical RMS amperes when used with fuses having an equal rating. They shall have rejection type R fuse clips for Class R fuses.

Fuses

a. Fuses 601 amperes and larger shall have an interrupting rating of 200,000 symmetrical RMS amperes. They shall provide time-delay of not less than 4 seconds at 500% of their ampere rating. They shall be current limiting and of silver-sand Class L construction.

b. Fuses 600 amperes and less shall have an interrupting rating of 200,000 symmetrical RMS amperes. They shall be of the rejection Class RK1 type, having a time delay of not less than 10 seconds at 500% of their ampere rating.

c. Plug fuses shall be dual-element, time-delay type with Type S base.

d. All fuses shall be so selected to ensure positive selective coordination as indicated on the schedule.

e. All lighting ballasts (fluorescent, mercury vapor, or others) shall be protected on the supply side with an appropriate fuse in a suitable approved fuseholder mounted within or on the fixture. Fuses shall be sized as recommended by the ballast manufacturer.

f. Spare fuses shall be provided in the amount of 20% of each size and type installed, but in no case shall less than three spares of a specific size and type be supplied. These spare fuses shall be delivered to the owner at the time of acceptance of the project.

g. Fuse identification labels, showing the size and type of fuses installed, shall be placed inside the cover of each switch.

Low-Voltage Remote-Control Switching

A low-voltage, remote-control switch system shall be installed in the drugstore as shown on the plans and detailed herein. All components shall be specification grade and constructed to operate on 24-volt control power. The transformer shall be a 120/24-volt, energy-limiting type for use on a Class II signal system.

Cabinet. A metal cabinet matching the panelboard cabinets shall be installed for the installation of relays and other components. A barrier will separate the control section from the power wiring.

Relay. A 24-volt ac, split-coil design relay rated to control 20 amperes of tungsten or fluorescent lamp loads shall be provided.

Switches. The switches shall be complete with wall plate and mounting bracket. They shall be normally open, single-pole, double-throw, momentary contact switches with on–off identification.

Rectifiers. A heavy-duty silicon rectifier with 7.5-ampere continuous duty rating shall be provided.

Wire. The wiring shall be in 2- or 3-conductor, color-coded, 20 AWG or larger wire.

Rubber Grommet. An adapter will be installed on all relays to isolate the relay from the metal cabinet to reduce noise.

Switching Schedule. Connections will be made to accomplish the lighting control as shown in the switching schedule, Table A-1.

TABLE A-1

Drugstore low-voltage, remote-control switching schedule.

Switch	Relay	Area Served	Branch Circuit	Master Control Switch
RCa	A	Main area lighting "a"	1	5
RCb	B	Main area lighting "b"	1	5
RCc	C	Main area lighting "c"	1	5
RCd	D	Cash register area	2	1
RCe	E	Storage area lights	2	2
RCf	F	Toilet lights and fan	3	2
RCg	G	Pharmacy "a"	4	3
RCh	H	Pharmacy "b"	3	3
RCi	I	Pharmacy "c"	3	3
RCj	J	Outdoor sign	12	6
RCk	K	Storage area fan	3	
RCl	L	Copier area light	3	8
RCm	M	Spare		

HEATING AND AIR-CONDITIONING SPECIFICATIONS

Only those sections that pertain to the electrical work are listed here.

Heating and Air-Conditioning Equipment

Packaged and split system heating and air-conditioning equipment will be furnished and installed by the HVAC contractor. The five HVAC units in the in five rental areas shall be furnished and installed by the HVAC contractor, as shown in Table A-2:

TABLE A-2

HVAC equipment.

HVAC AND HEATING EQUIPMENT

	Bakery	Drugstore		Beauty Salon	Insurance Office	Real Estate Office
KW Unit Heater (Storage)	10.0 KW	7.5 KW		5.0 KW	5.0 KW	N/A
Minimum Circuit Ampacity	34.7	26		17.4	17.4	N/A

HVAC Equipment

		Bakery	Drugstore		Beauty Salon	Insurance Office	Real Estate Office
Voltage		208/230, 3-Ph.	208/230, 3-Ph.		208/230, 3-Ph.	208/230, 3-Ph.	208/230, 3-Ph.
Cooling Capacity/ Compressor RLA		7.5 ton/25	6 ton/22.4		5 ton/17.6	7.5 ton/25	4 ton/14.6
Compressor LRA	Outdoor Unit	164	149	Package Unit	123	164	91
Condenser Motor		1 Ph. 3.1 fla	1 Ph 3.1 fla		1 ph 0.4 hp 2.5 fla	1 ph 0.7 hp 3.3 fla	1 ph 0.3 hp 2.0 fla
Minimum Branch Circuit Selection Current		34.4	31.1		37.5/42.4	51.4/58.6	37.5/42.4
Maximum Fuse amps		60/60	60/60		45/45	60/60	40/45
Maximum HACR Type Circuit Breaker		60/60	60/60		45/45	60/60	40/45
Evaporator Motor	Indoor Unit	3 ph 1 hp, 3.5 fla	3 ph 1.0 hp 3.5 fla		3 ph 1.0 hp 5.0 fla	3 ph 1.0 hp 3.6 fla	3 ph 1.0 hp 5.0 fla
KW Heat Strip		13.5/18.0	13.5/18.0		9.0/12.0	13.5/18.0	9.0//0
Minimum Branch Circuit Selection Current		53.1/58.6	51.3/66.9				
Maximum Fuse amps		60/60	60/70				
Maximum HACR Type Circuit Breaker		60/60	60/70				

Toilet Room Heating

	Heater Rating (Watts)	Continuous Load Factor	Adjusted Rating	Minimum Circuit Ampacity	Conductors (2 in EMT)	Overcurrent Device	12 AWG 90°C Ampacity	Conductors (4 in EMT)
Men's Toilet Room	1500	125%	1875	9.0	12 AWG	20 Amperes	30	24
Women's Toilet Room	2000	125%	2500	12.0	12 AWG	20 Amperes	30	24

Note: Installing 4, 12 AWG current-carrying conductors in one raceway or Type MC cable is permitted as 80 percent of the 90°C allowable ampacity of 30 amperes exceeds the minimum circuit ampacity and rating of the overcurrent protective device.

All wiring and electrical connections shall be completed by the electrical contractor. The space heaters in the storage areas in the basement and the wall heaters in the public toilet rooms on the second floor shall be furnished and installed by the electrical contractor.

Heating Control

In all tenant areas, the heating and cooling will be controlled by a combination heating/cooling thermostat located in the proper area, as shown on the electrical plans.

A wall thermostat shall be located for each unit heater in the basement storage areas, as shown on the electrical plans.

The electrical heaters installed in the toilet rooms on the second floor shall be controlled by a wall thermostat that is tamper resistant.

Sump Pump

The plumbing contractor will furnish and install an electric motor-driven, fully automatic sump pump. Motor will be ½ horsepower, 120 volts, single phase. Overload protection will be provided by a manual motor starter.

Electric Water Heaters

Electric water heaters will be furnished and installed by the plumbing contractor according to the Table A-3. The electrical contractor shall install wiring for the water heaters including disconnecting means where required and making all electrical connections.

TABLE A-3

Electric water heaters in the Commercial Building.

Tenant	Size	kW	Volts	Phase	Amperes	125% Factor	Branch Circuit Rating	Size/AWG THHN/ THWN	How Connected
Women's toilet, 2nd floor	2.5 gal	1.5 kW	120	1	12.5	15.6	20	12	Cord-and-plug.
Men's toilet, 2nd floor	2.5 gal	1.5 kW	120	1	12.5	15.6	20	12	Cord-and-plug.
Insurance Office	6 gal	1.65 kW	120	1	13.75	17.2	20	12	Hard wired. 20 A disconnect.
Real Estate Office	6 gal	1.65 kW	120	1	13.75	17.2	20	12	Hard wired. 20 A disconnect.
Bakery	80 gal	30 kW	208	3	83.3	104.1	110	2	Hard wired. 200 A disconnect.
Beauty Salon	120 gal	24 kW	208	3	66.6	83.3	90	4	Hard wired. 100 A disconnect.
Drugstore	80 gal	18 kW	208	3	50	62.5	70	4	Hard wired. 100 A disconnect.
Owner (in Basement)	50 gal	6 kW	208	3	16.7	20.9	25	10	Hard wired. Circuit breaker in panelboard serves as disconnect.

A disconnect switch or circuit breaker must be located within sight of the appliance or, if out of sight, be capable of being locked in the open position. *NEC 422.31(B)*.

Electrical values for the electric water heaters installed in the Commercial Building.

Refer to *NEC 422.10, 422.11,* and *422.13*.

FIRE ALARM SYSTEM

The electrical contractor shall furnish and install a fully functional intelligent, microprocessor-based fire alarm system in compliance with NFPA 72 *Fire Alarm Code* and NFPA 70 *National Electrical Code*. It shall include, but not be limited to, alarm initiating devices, alarm notification appliances, Fire Alarm Control Panel (FACP), auxiliary control devices, annunciator, and all wiring as shown on the drawings and specified herein.

The fire alarm system shall comply with requirements of NFPA 72 for Local Protected Premises Signaling Systems except as modified and supplemented by this specification. The system field wiring and all initiating and annunciating devices shall be supervised either electrically or by software-directed polling of field devices.

The Secondary Power Source of the fire alarm control panel will be capable of providing at least 24 hours of backup power with the ability to sustain 5 minutes in alarm at the end of the backup period.

Alarm, trouble, and supervisory signals from all intelligent reporting devices shall be encoded on NFPA Style 4 (Class B), Style 6 (Class A), or Style 7 (Class A) Signaling Line Circuits (SLC).

Initiation Device Circuits (IDC) shall be wired Class B (NFPA Style B) or Class A (NFPA Style D) as part of an addressable device connected by the SLC Circuit.

Notification Appliance Circuits (NAC) shall be wired Class B (NFPA Style Y) or Class A (NFPA Style Z) as part of an addressable device connected by the SLC Circuit.

All circuits shall be power-limited, per UL 864 9th edition requirements.

A single ground fault or open circuit on the system Signaling Line Circuit shall not cause system malfunction, loss of operating power, or the ability to report an alarm when wired NFPA Style 6/7.

Alarm signals arriving at the main FACP shall not be lost following a primary power failure or outage of any kind until the alarm signal is processed and recorded.

The CPU shall receive information from all intelligent detectors to be processed to determine whether normal, alarm, or trouble conditions exist for each detector. The software shall automatically compensate for the accumulation of dust in each detector up to allowable limits. The information shall also be used for automatic detector testing and for the determination of detector maintenance conditions.

The fire alarm control panel shall be capable of T-tapping Class B (NFPA Style 4) Signaling Line Circuits (SLCs). Systems that do not allow or have restrictions in, for example, the amount of T-taps, length of T-taps, and so on, are not acceptable.

Wiring

Wiring shall be in accordance with local, state, and national codes (e.g., *NEC Article 760*) and as recommended by the manufacturer of the fire alarm system. Number and size of conductors shall be as recommended by the fire alarm system manufacturer, but not smaller than 18 AWG for initiating device circuits and signaling line circuits, and 14 AWG (1.63 mm) for notification appliance circuits.

All wire and cable shall be listed and/or approved by a recognized testing agency for use with a protective signaling system. Wire and cable not installed in conduit shall have a fire-resistance rating suitable for the installation as indicated in *NEC 760* (e.g., FPLR).

All field wiring shall be electrically supervised for open circuit and ground fault.

Listing of Equipment

The system shall have proper listing and/or approval from the following nationally recognized agencies:

UL	Underwriters Laboratories, Inc.
ULC	Underwriters Laboratories Canada
FM	Factory Mutual
MEA	Material Equipment Acceptance (NYC)
CSFM	California State Fire Marshal

Fire Alarm Control Panel

The FACP shall be a NOTIFIER NFW-50 (FireWarden-50), or approved equal, and shall contain a microprocessor-based central processing unit (CPU). The CPU shall communicate with and control the following types of equipment used to make up the system: intelligent addressable smoke and thermal (heat) detectors, addressable modules, printer, annunciators, and other system-controlled devices.

The fire alarm control panel shall be connected to a separate dedicated branch circuit, maximum 20 amperes. This circuit shall be labeled at the main power distribution panel as "FIRE ALARM." Fire alarm control panel primary power wiring shall be 12 AWG. The control panel cabinet shall be grounded securely to either a cold water pipe or grounding rod. The control panel enclosure shall feature a quick removal chassis to facilitate rapid replacement of the FACP electronics.

Elevator Recall

Smoke detectors will be installed in the elevator hoist shaft. An alarm from such devices will signal the elevator to initiate emergency procedures. All lift call buttons, door buttons, and signals will become inoperative in the lift bank serving the machine room. Lifts will immediately be sent to the main floor of egress (ground level), where they will be decommissioned until the alarm condition has been cleared or manually taken over by Fire Department personnel.

Smoke detectors will be installed in each elevator lobby. These detectors will function to signal the elevator to recall to the primary floor of egress (ground level) in the event of an alarm. Detectors on the first floor will signal the elevator to recall to the secondary floor of egress.

PANELBOARD WORKSHEETS AND LOAD CALCULATION FORMS

Worksheets are provided for every tenant space and the Owner's space. How the worksheets are used to determine the loads to be supplied by each panelboard is explained in detail in Chapter 4.

In addition, a load calculation form is included for each tenant space and the owner's space. The method of using the load calculation form for the Drugstore is explained in detail in Chapter 3.

All panelboard schedules and the one line diagram for the service are included on sheet E–4 of the blueprints.

Finally, the load calculation for the service equipment is provided.

TABLE A-4(A)

Beauty Salon panelboard worksheet.

A Load/Area Served	B Quantity	C Load Each VA[1]	D Calculated or connected VA	E Load Type	F Load Modifier	G Ambient Temp. (°F/°C)	H Correction Factor[2]	I Current-Carrying Conductors	J Adjustment Factor[3]	K Minimum Allowable Ampacity	L Conductor Size (AWG)	M Allowable Ampacity[4]	N Overcurrent Protection	O Circuit Number
Luminaires:														
Style E	15	14	210	C	1.25	86/30	1	2	1	2.2	12	20	20	1
Style F	4	64	256	C	1.25	86/30	1	2	1	2.7	12	20	20	1
Style Q	1	24	24	N	1	86/30	1	2	1	0.2	12	20	20	1
B storage	1	64	64	N	1	86/30	1	4	0.8	0.7	12	20	20	2
B-EM storage	2	70	140	N	1	86/30	1	4	0.8	1.5	12	20	20	2
Receptacles station 1	2	180	360	R	1	86/30	1	6	0.8	3.8	12	20	20	4
Receptacles station 2	2	180	360	R	1	86/30	1	6	0.8	3.8	12	20	20	5
Receptacles station 3	2	180	360	R	1	86/30	1	6	0.8	3.8	12	20	20	6
Receptacle dryer 1	1	1440	1440	R	1	86/30	1	6	0.8	15.0	12	20	20	7
Receptacle dryer 2	1	1440	1440	R	1	86/30	1	6	0.8	15.0	12	20	20	8
Receptacle dryer 3	1	1440	1440	R	1	86/30	1	6	0.8	15.0	12	20	20	9
Receptacles east & west walls	2	180	360	R	1	86/30	1	2	1	3.0	12	20	20	10
Receptacles basement storage	4	180	720	R	1	3	1	4	0.8	7.5	12	20	20	3
Washer/dryer	1	4000	4000	N	1	86/30	1	3	1	19.2	10	30	30	11–13
Water heater	1	24,000	24,000	C	1.25	86/30	1	3	1	83.3	4	85	90	12–14–16
HVAC equipment[5]	1	13,500	13,500	H	1	86/30	1	3	1	37.5	8	50	45	15–17–19
Receptacle for HVAC servicing	1	1500	1500	R	1	86/31	1	3	1	12.5	8	50	45	17
Heat for basement	1	5000	5000	H	1.25	86/31	1	3	1	17.4	12	20	20	18–20–22

[1]Loads of fluorescent luminaires are figured at lamp wattage due to very high power factor ballasts.
[2]Correction factor is based upon temperature where the conductors are actually installed.
[3]The adjustment factor is determined by the number of current-carrying conductors in the same raceway of cable.
[4]Assume 75°C terminals.
[5]Minimum circuit ampacity includes 125% of larger heating load.

TABLE A-4(B)

Beauty Salon load calculation form.

	Count[1]	VA/Unit[2]	NEC Load[3]	Connected Load[4]	Continuous Loads[5]	Noncon-tinuous[6]	Neutral Load[7]
General Lighting:							
NEC 220.12	281.5	3	845				
Connected loads from worksheet				490			
Totals:			845	490	845		845
Storage Area:							
NEC 220.12	264	0.25	66				
Connected loads from worksheet				204			
Totals:			66	204		204	204
Other Loads:							
Receptacle outlets	12	180	2160			2160	2160
Hair dryer receptacles	3	1440		4320		4320	4320
Receptacle for HACR	1	1500		1500		1500	1500
Motors & Appliances:	Amperes	VA/Unit					
Washer/dryer	1	4000				4000	832
Water heater	1	24,000		24,000	24,000		
AC outdoor package unit	1	13,500		10,800	10,800		
Unit heater (storage)	1	5000		5000		5000	
			Total Loads	35,644.5	17,184	9860.5	
Minimum feeder load 125% of continuous loads plus 100% of noncontinuous:						61,740	9860.5
				Minimum ampacity:		171.5	27.4

Minimum 2/0 AWG ungrounded and 10 AWG neutral for load. Minimum neutral 6 AWG per *215.2(A)(1)*.

3/0 AWG ungrounded conductors installed for OCP and 2 AWG neutral installed.

Minimum size EMT = $(0.2679 \times 3) + 0.1158 = 0.9195$ sq. in. = Trade size 2

[1]Count: Square footage of given area, the number of a given type load, or the ampere rating of the given type load.
[2]VA/Unit: Specific unit load values required by *NEC Article 220*, or actual load values for the given type load.
[3]*NEC* Load: count × *NEC* VA/Unit. Also the "calculated load."
[4]Connected Load: Actual load connected to the electrical system.
[5]Continuous Loads: The maximum current is expected to continue for three hours or more. Column also used to apply 125% to the largest motor load.
[6]Noncontinuous: Loads that are not continuous.
[7]Neutral Load: The maximum unbalanced load calculated according to *NEC 220.61*.

TABLE A-5(A)

Bakery panelboard worksheet.

	A	B	C	D	E	F	G	H	I	J	K	L	M	N	O
	Load/Area Served	Quantity	Load Each VA[1]	Calculated or connected VA	Load Type	Load Modifier	Ambient Temp. (°F/°C)	Correction Factor[2]	Current-Carrying Conductors	Adjust-ment Factor[3]	Minimum Allowable Ampacity	Conductor Size (AWG)	Allowable Ampacity[4]	Over-current Protec-tion	Circuit Number
Luminaires:															
	C luminaires	15	61	915	C	1.25	86/30	1	2	1	9.5	12	20	20	1
	C-EM luminaires	3	67	201	C	1.25	86/30	1	2	1	2.1	12	20	20	1
	E luminaires	3	14	42	C	1.25	86/30	1	2	1	0.4	12	20	20	1
	N luminaire	1	34	34	C	1.25	86/30	1	2	1	0.4	12	20	20	1
	P luminaire	4	34	136	C	1.25	86/30	1	2	1	1.4	12	20	20	1
	Q luminaire	1	24	24	N	1	86/30	1	2	1	0.2	12	20	20	1
	T luminaire	7	9	63	C	1.25	86/30	1	2	1	0.7	12	20	20	1
	Exit signs	3	4.6	13.8	C	1.25	86/30	1	2	1	0.1	12	20	20	1
	B (basement)	5	64	320	N	1	86/30	1	2	1	2.7	12	20	20	2
	B-EM (basement)	3	70	210	N	1	86/30	1	2	1	1.8	12	20	20	2
	Receptacles (basement)	7	180	1260	R	1	86/30	1	2	1	10.5	12	20	20	3
	Receptacles sales	8	180	1440	R	1	86/30	1	2	1	12.0	12	20	20	4
	Recept cash register	1	240	240	R	1	86/30	1	2	1	2.0	12	20	20	5
	Recept work area	7	180	1260	R	1	86/30	1	2	1	10.5	12	20	20	6
	Recept toilet room	1	180	180	R	1	86/30	1	2	1	1.5	12	20	20	7
	Exhaust fan toilet	1	120	120	N	1	86/30	1	2	1	1.0	12	20	20	7
	Sign	1	1200	1200	C	1.25	86/30	1	2	1	12.5	12	20	20	8
	Doughnut machine	1	2990	2990	M	1.25	86/30	1	3	1	10.4	12	20	20	9–13
	Disposal	1	7920	7920	M	1.25	86/30	1	3	1	27.5	10	30	50	10–14
	Dishwasher	1	8604	8604	N	1	86/30	1	3	1	23.9	10	30	30	15–19
	Multimixers and dough divider[5]	1	5584	5584	M	1	86/30	1	3	1	15.5	12	20	20	16–20
	Oven	1	16,000	16,000	H	1.25	86/30	1	3	1	55.6	6	65	60	21–25
	Exhaust fan, work room	1	348	348	M	1.25	86/30	1	2	1	3.6	12	20	20	7
	HVAC outdoor unit[5]	1	12,384	12,384	H	1	86/30	1	3	1	34.4	8	50	60	22–26
	HVAC indoor unit[5]	1	19,116	19,116	C	1	86/30	1	3	1	53.1	6	65	60	27–31
	Water heater	1	30,000	30,000	C	1.25	86/30	1	3	1	104.2	2	115	110	28–32
	Heater, basement	1	10,000	10,000	C	1.25	86/30	1	3	1	34.7	8	50	40	33–37

[1]Loads of fluorescent luminaires are figured at lamp wattage due to very high power factor ballasts.
[2]Correction factor is based upon temperature where the conductors are actually installed.
[3]The adjustment factor is determined by the number of current-carrying conductors in the same raceway or cable.
[4]Assume 75°C terminals.
[5]125% factor included in "Load Each."

TABLE A-5(B)

Bakery feeder load calculation.

	Count[1]	VA/Unit[2]	NEC Load[3]	Connected Load[4]	Continuous Loads[5]	Noncon-tinuous[6]	Neutral Load[7]
General Lighting:							
NEC 220.12	1070	3	3210				
Connected loads from worksheet				1424			
Totals:			3210	1424	3210		3210
Storage Area:							
NEC 220.12	603	0.25	151				
Connected load from worksheet				134			
Totals:				134		151	151
Other Loads:							
Receptacle outlets	24	180	4320			4320	4320
Recep cash register	1	240		240		240	240
Receptacle for HVAC	1	1500		1500		1500	1500
Sign outlet	1	1200	1200		1200		1200
Motors & Appliances:	Count[1]	VA/Unit					
Doughnut machine[9]	1	2990		2990		1944	
Disposal[9]	1	7920		7920		5148	
Dishwasher[9]	1	8604		8604		5593	
Multimixers & dough divider[9]	1	5584		5584		3630	
Oven[9]	1	16,000		16,000		10,400	
Exhaust fans	2	468		468		468	468
HVAC outdoor unit	1	10,116		10,116			
HVAC indoor unit[8]	1	14,760		14,760	14,760		
Water heater[9]	1	30,000		30,000		19,500	
Unit heater (storage)[8]	1	10,000		10,000		10,000	
				Total Loads	19,170	62,893	11,089
			Minimum feeder load 125% of continuous loads plus 100% of noncontinuous:			86,855	11,089
					Minimum ampacity:	241.3	30.8

Minimum 250 kcmil ungrounded and 8 AWG neutral for load. Minimum neutral 4 AWG per *215.2(A)(1)*.

300 kcmil ungrounded installed for OCP [*240.4(B)*] and 4 AWG neutral installed.

Minimum size EMT = (0.4608 × 3) + 0.0824 = 1.4648 sq. in. = 2–1/2 in.

[1]Count: Square footage of given area, the number of a given type load, or the ampere rating of the given type load.
[2]VA/Unit: Specific unit load values required by *NEC Article 220*, or actual load values for the given type load.
[3]*NEC* Load: Count x *NEC* VA/Unit. Also the "calculated load."
[4]Connected Load: Actual load connected to the electrical system.
[5]Continuous Loads: The maximum current is expected to continue for three hours or more. Column also used to apply 125% to the largest motor load
[6]Noncontinuous: Loads that are not continuous.
[7]Neutral Load: The maximum unbalanced load calculated according to *NEC 220.61*.
[8]Because the heating and cooling loads are noncoincident, only the larger load is included.
[9]A 65% factor is applied to commercial cooking equipment as allowed by *NEC 220.56*.

TABLE A-6(A)

Insurance Office panelboard worksheet.

A	B	C	D	E	F	G	H	I	J	K	L	M	N	O
Load/Area Served	Quantity	Load Each VA[1]	Calculated or connected VA	Load Type	Load Modifier	Ambient Temp. (°F/°C)	Correction Factor[2]	Current-Carrying Conductors	Adjustment Factor[3]	Minimum Allowable Ampacity	Conductor Size (AWG)	Allowable Ampacity[4]	Overcurrent Protection	Circuit Number
Luminaires:														
E luminaires	6	14	84	C	1.25	86/30	1	2	1	0.9	12	20	20	1
E-EM luminaires	2	20	40	C	1.25	86/30	1	2	1	0.4	12	20	20	1
G luminaires	14	80	1120	C	1.25	86/30	1	2	1	11.7	12	20	20	1
Q Luminaire	1	24	24	C	1	86/30	1	2	1	0.2	12	20	20	1
G-EM luminaires	1	86	86	C	1.25	86/30	1	2	1	0.9	12	20	20	1
B luminaires basement	2	64	86	N	1	86/30	1	2	1	0.7	12	20	20	2
B-EM luminaires basement	1	70	70	N	1	86/30	1	2	1	0.6	12	20	20	2
Receptacles basement	4	180	720	R	1	86/30	1	2	1	6.0	12	20	20	3
Multioutlet raceway	7	2	360	R	1	86/30	1	2	1	3.0	12	20	20	4
Multioutlet raceway	7	2	360	R	1	86/30	1	2	1	3.0	12	20	20	6
Multioutlet raceway	6	2	360	R	1	86/30	1	2	1	3.0	12	20	20	8
Multioutlet raceway	11	2	360	R	1	86/30	1	2	1	3.0	12	20	20	5
Multioutlet raceway	11	2	360	R	1	86/30	1	2	1	3.0	12	20	20	7
Computer recept	6	200	1200	R	1	86/30	1	2	1	10.0	12	20	20	9
Computer recept	6	200	1200	R	1	86/30	1	2	1	10.0	12	20	20	11
Computer recept	6	200	1200	R	1	86/30	1	2	1	10.0	12	20	20	13
Floor receptacles	4	180	720	R	1	86/30	1	2	1	6.0	12	20	20	12
Coffee recept	1	1200	1200	R	1	86/30	1	2	1	10.0	12	20	20	14
Recepts waiting	6	180	1080	R	1	86/30	1	2	1	9.0	12	20	20	15
Water heater	1	1650	1650	C	1.25	86/30	1	2	1	17.2	12	20	20	16
Space heater basement	1	7500	7500	C	1.25	86/30	1	3	1	26.0	10	30	30	17–19–21
Packaged rooftop HVAC unit[5]	1	11,936	11,936	H	1	86/30	1	3	1	33.2	6	65	60	18–20–22
Recept for HVAC	1	1500	1500	R	1	86/30	1	2	1	12.5	12	20	20	24

[1]Loads of fluorescent luminaires figured at lamp wattage due to very high power factor ballasts.
[2]Correction factor is based upon temperature where the conductors are actually installed.
[3]The adjustment factor is determined by the number of current-carrying conductors in the same raceway or cable.
[4]Assume 75°C terminals.
[5]125% of largest motor is included in the minimum circuit selection current.

TABLE A-6(B)

Insurance Office feeder load calculation.

	Count[1]	VA/Unit[2]	NEC Load[3]	Connected Load[4]	Continuous Loads[5]	Noncon-tinuous[6]	Neutral Load[7]
General Lighting:							
NEC 220.12	1302	3	3906				
Connected loads from worksheet				1663			
Totals:			3906	1663	3906		3906
Storage Area:							
NEC 220.12	230	0.25	58				
Connected load from worksheet				156			
Totals:			58	156		156	156
Other Loads:							
Receptacle outlets	22	180	3960			3960	3960
Computer recepts	18	200		3600	3600	0	3600
Recept for coffee	1	1200		1200		1200	1200
Receptacle for HVAC	1	1500		1500		1500	1500
Motors & Appliances:	Amperes	VA/Unit					
Water heater	1	1650		1650			1650
AC outdoor package unit	1	18,504		14,796	14,796		
Unit heater (storage)	1	7500		7500		7500	
				Total Loads	22,302	14,316	15,972
Minimum feeder load 125% of continuous loads plus 100% of noncontinuous:						42,194	15,972
					Minimum ampacity:	117.2	44.4

Minimum 1 AWG ungrounded and 8 AWG neutral for load. Minimum neutral 6 AWG per *215.2(A)(1)*.

1/0 AWG ungrounded conductors installed for OCP and 4 AWG neutral installed.

Minimum size EMT = $(0.1855 \times 3) + 0.0824 = 0.6389$ sq. in. = trade size 1–1/2.

[1]Count: Square footage of given area, the number of a given type load, or the ampere rating of the given type load.
[2]VA/Unit: Specific unit load values required by *NEC Article 220*, or actual load values for the given type load.
[3]*NEC* Load: Count × *NEC* VA/Unit. Also the "calculated load."
[4]Connected Load: Actual load connected to the electrical system.
[5]Continuous Loads: The maximum current is expected to continue for 3 hours or more. Column also used to apply 125% to the largest motor load.
[6]Noncontinuous: Loads that are not continuous.
[7]Neutral Load: The maximum unbalanced load calculated according to *NEC 220.61*.

TABLE A-7(A)

Drugstore panelboard worksheet.

A	B	C	D	E	F	G	H	I	J	K	L	M	N	O
Load/Area Served	Quantity	Load Each VA¹	Calculated or connected VA	Load Type	Load Modifier	Ambient Temp. (°F/°C)	Correction Factor²	Current-Carrying Conductors	Adjustment Factor³	Minimum Allowable Ampacity	Conductor Size (AWG)	Allowable Ampacity⁴	Overcurrent Protection	Circuit Number
Luminaires:														
Style I	27	61	1647	C	1.25	86/30	1	2	1	17.2	12	20	20	1
Style G	7	80	560	C	1.25	86/30	1	2	1	5.8	12	20	20	2
Style G-EM	2	86	172	C	1.25	86/30	1	2	1	1.8	12	20	20	2
Style E2 cash register	3	26	78	C	1.25	86/30	1	2	1	0.8	12	20	20	2
Style E	2	14	28	C	1.25	86/30	1	2	1	0.3	12	20	20	3
Style E-EM	1	20	20	N	1	86/30	1	2	1	0.2	12	20	20	3
Style N	2	34	68	N	1	86/30	1	2	1	0.6	12	20	20	3
Style Q	2	24	48	N	1	86/30	1	2	1	0.4	12	20	20	3
Exit signs	2	5	9	C	1.25	86/30	1	2	1	0.1	12	20	20	3
Style B storage	5	64	320	N	1	86/30	1	2	1	2.7	12	20	20	4
Style B-EM storage	4	70	280	N	1	86/30	1	2	1	2.3	12	20	20	4
Receptacles first floor, north	6	180	1080	R	1	86/30	1	2	1	9.0	12	20	20	5
Receptacles first floor, south	5	180	900	R	1	86/30	1	2	1	7.5	12	20	20	6
Receptacles, cash registers	2	230	460	R	1	86/30	1	2	1	3.8	12	20	20	7
Receptacles, floor, toilet, & janitor areas	4	180	720	R	1	86/30	1	2	1	6.0	12	20	20	7
Receptacle for AC	1	1500	1500	R	1	86/30	1	2	1	12.5	12	20	20	8
Receptacles storage	8	180	1440	R	1	86/30	1	2	1	12.0	12	20	20	9
Receptacle, copier	1	1250	1250	R	1	86/30	1	2	1	10.4	12	20	20	10
Exhaust fan, toilet, & work rooms	2	120	240	C	1	86/30	1	2	1	2.0	12	20	20	3
Electric doors	2	360	720	N	1	86/30	1	2	1	6.0	12	20	20	11
Sign	1	1200	1200	C	1.25	86/30	1	2	1	12.5	12	20	20	12
Water heater	1	4500	4500	C	1.25	86/30	1	2	1	15.6	12	20	20	13-15-17
Air conditioner, outdoor unit⁵	1	10,725	10,725	H	1	100/38	0.88	3	1	33.9	8	45	60	14-16-18
Air conditioner, indoor unit	1	14,760	14,760	H	1.25	86/30	1	3	1	51.3	6	65	60	19-21-23
Space heater storage	1	7500	7500	H	1.25	86/30	1	3	1	26.0	10	30	30	20-22-24

¹Loads of fluorescent luminaires figured at lamp wattage due to very high power factor ballasts.
²Correction factor is based upon temperature where the conductors are actually installed.
³The adjustment factor is determined by the number of current-carrying conductors in the same raceway or cable.
⁴Assume 75°C terminals.
⁵The "Load Each VA" includes 125% of largest motor.

TABLE A-7(B)

Drugstore feeder load calculation.

	Count[1]	VA/Unit[2]	NEC Load[3]	Connected Load[4]	Continuous Loads[5]	Noncontinuous[6]	Neutral Load[7]
General Lighting:							
NEC 220.12	1395	3	4185				
Connected loads from worksheet				2630			
Totals:			4185	2630	4185		4185
Storage Area:							
NEC 220.12	858	0.25	215				
Connected load from worksheet				600			
Totals:			215	600		600	600
Other Loads:							
Receptacle outlets	23	180	4140			4140	4140
Copy machine	1	1250		1250		1250	1250
Receptacle for HACR	1	1500		1500		1500	1500
Sign outlet	1	1200	1200		1200		1200
Cash registers	2	240		480		480	480
Electric doors	2	360		720		720	720
Motors & Appliances:	Amperes	VA/Unit					
Water heater	1	4500		4500	4500		
Unit heater (storage)	1	7500		7500	7500		
AC outdoor unit	1	9645		9645			
AC indoor unit[8]	1	14,760		14,760	14,760		
Exhaust fans	2	120		240		240	240
Total Loads				32,145		8930	14,315
Minimum feeder load 125% of continuous loads plus 100% of noncontinuous:						49,111	14,315
Minimum ampacity:						136.4	39.8

Minimum 1 AWG ungrounded and 8 AWG neutral for load. Minimum neutral 6 AWG per *215.2(A)(1)*.

1/0 AWG ungrounded conductors installed for OCP and 4 AWG neutral installed.

Minimum size EMT = $(0.1855 \times 3) + 0.0824 = 0.6389$ sq. in. = trade size 1–1/2.

[1]Count: Square footage of given area, the number of a given type load, or the ampere rating of the given type load.
[2]VA/Unit: Specific unit load values required by *NEC Article 220*, or actual load values for the given type load.
[3]*NEC* Load: Count × *NEC* VA/Unit. Also the "calculated load."
[4]Connected Load: Actual load connected to the electrical system.
[5]Continuous Loads: The maximum current is expected to continue for 3 hours or more. Column also used to apply 125% to the largest motor load.
[6]Noncontinuous: Loads that are not continuous.
[7]Neutral Load: The maximum unbalanced load calculated according to *NEC 220.61*.
[8]Because the heating and cooling loads are noncoincident, only the larger load is included.

TABLE A-8(A)

Real Estate Office panelboard worksheet.

A	B	C	D	E	F	G	H	I	J	K	L	M	N	O
Load/Area Served	Quantity	Load Each VA[1]	Calculated or connected VA	Load Type	Load Modifier	Ambient Temp. (°F/°C)	Correction Factor[2]	Current-Carrying Conductors	Adjustment Factor[3]	Minimum Allowable Ampacity	Conductor Size (AWG)	Allowable Ampacity[4]	Overcurrent Protection	Circuit Number
Luminaires:														
Style E2	8	26	208	C	1.25	86/30	1	2	1	2.2	12	20	20	1
Style E2-EM	3	32	96	C	1.25	86/30	1	2	1	1.0	12	20	20	1
Style N	1	34	34	C	1.25	86/30	1	2	1	0.4	12	20	20	1
Receptacles	9	180	1620	R	1	86/30	1	6	0.8	16.9	12	20	20	2
Receptacles, coffee	1	1200	1200	R	1	86/30	1	6	0.8	12.5	12	20	20	3
Water heater	1	1650	1650	C	1.25	86/30	1	3	1	17.2	4	85	90	4
HVAC package equipment[5]	1	13,500	13,500	H	1	86/30	1	3	1	37.5	8	50	45	5–7–9
Receptacle for HVAC servicing	1	1500	1500	R	1	86/31	1	3	1	12.5	8	50	45	6

[1]Loads of fluorescent luminaires figured at lamp wattage due to very high power factor ballasts.
[2]Correction factor is based upon temperature where the conductors are actually installed.
[3]The adjustment factor is determined by the number of current-carrying conductors in the same raceway or cable.
[4]Assume 75°C terminals.
[5]Minimum circuit ampacity includes 125% of larger heating load.

TABLE A-8(B)

Real Estate Office feeder load calculation.

	Count[1]	VA/Unit[2]	NEC Load[3]	Connected Load[4]	Continuous Loads[5]	Noncontinuous[6]	Neutral Load[7]
General Lighting:							
NEC 220.12	420.5	3	1262				
Connected loads from worksheet				338			
Totals:			1261.5	338	1262		1262
Storage Area:							
NEC 220.12	0	0.25	0				
Connected load from worksheet				0			
Totals:			0	0		0	0
Other Loads:							
Receptacle outlets	9	180	1620			1620	1620
Coffee receptacle	1	1200		1200		1200	1200
Receptacle for HACR	1	1500		1500		1500	1500
Motors & Appliances:	Amperes	VA/Unit					
Water heater	1	1650		1650	1650		1650
AC outdoor package unit	1	13,500		10,800	10,800		
Total Loads				13,712		4320	7232
Minimum feeder load 125% of continuous loads plus 100% of noncontinuous:						21,459	7232
Minimum ampacity:						59.6	20.1

Minimum 6 AWG ungrounded and 12 AWG neutral for load. Minimum neutral 10 AWG per *215.2(A)(1)*.

3 AWG ungrounded conductors installed for OCP and 8 AWG neutral installed.

Minimum size EMT = $(0.0973 \times 3) + 0.0366 = 0.3285$ sq. in. = Trade size 1

---

[1]Count: Square footage of given area, the number of a given type load, or the ampere rating of the given type load.
[2]VA/Unit: Specific unit load values required by *NEC Article 220*, or actual load values for the given type load.
[3]*NEC* Load: Count × *NEC* VA/Unit. Also the "calculated load."
[4]Connected Load: Actual load connected to the electrical system.
[5]Continuous Loads: The maximum current is expected to continue for three hours or more. Column also used to apply 125% to the largest motor load.
[6]Noncontinuous: Loads that are not continuous.
[7]Neutral Load: The maximum unbalanced load calculated according to *NEC 220.61*.

TABLE A-9(A)

Owner panelboard worksheet.

	A	B	C	D	E	F	G	H	I	J	K	L	M	N	O
	Load/Area Served	Quantity	Load Each VA[1]	Calculated or connected VA	Load Type	Load Modifier	Ambient Temp. (°F/°C)	Correction Factor[2]	Current-Carrying Conductors	Adjustment Factor[3]	Minimum Allowable Ampacity	Conductor Size (AWG)	Allowable Ampacity[4]	Overcurrent Protection	Circuit Number
Luminaires:															
	A basement	2	64	128	N	1	86/30	1	2	1	1.1	12	20	20	1
	A-EM basement	2	70	140	N	1	86/30	1	2	1	1.2	12	20	20	1
	B basement	1	64	64	N	1	86/30	1	2	1	0.5	12	20	20	1
	B-EM basement	2	70	140	N	1	86/30	1	2	1	1.2	12	20	20	1
	Basement exit sign	1	4.6	4.6	C	1.25	86/30	1	2	1	1.5	12	20	20	1
	E-EM Basement	1	20	20	N	1	86/30	1	2	1	0.2	12	20	20	1
	E first floor	4	14	56	C	1.25	86/30	1	2	1	0.6	12	20	20	2
	E-EM first floor	2	20	40	C	1.25	86/30	1	2	1	0.4	12	20	20	2
	E second floor	3	14	42	N	1	86/30	1	2	1	0.4	12	20	20	2
	E-EM second floor	2	20	40	N	1	86/30	1	2	1	0.3	12	20	20	2
	F second floor	2	64	128	C	1.25	86/30	1	2	1	1.3	12	20	20	2
	F-EM second floor	3	70	210	C	1.25	86/30	1	2	1	2.2	12	20	20	2
	N second floor	2	34	68	N	1	86/30	1	2	1	0.6	12	20	20	2
	1st & 2nd floor exit signs	3	4.6	13.8	C	1.25	86/30	1	2	1	0.14	12	20	20	2
	K Outdoor	8	29	232	C	1.25	86/30	1	2	1	2.42	12	20	20	1
	Exhaust fans, toilets	2	120	240	N	1	86/30	1	2	1	2.0	12	20	20	2
	Receptacles basement	4	180	720	R	1	86/30	1	2	1	6.0	12	20	20	3
	Receptacles, 1st and 2nd floors	5	180	900	R	1	86/30	1	2	1	7.5	12	20	20	18
	Receptacle, telephone	1	600	600	R	1	86/30	1	2	1	5.0	12	20	20	4
	Water heater, 2nd toilet	1	1500	1500	C	1.25	86/30	1	2	1	15.6	12	20	20	5
	Water heater, 2nd toilet	1	1500	1500	C	1.25	86/30	1	2	1	15.6	12	20	20	6
	Space heater, women	1	2000	2000	C	1.25	86/30	1	2	1	12.0	12	20	20	7–9
	Space heater, men	1	1500	1500	C	1.25	86/30	1	2	1	9.0	12	20	20	8–10
	Sump pump	1	1127	1127	C	1.25	86/31	1	2	1	12	12	20	20	11
	Water heater (basement)	1	6000	6000	C	1.25	86/30	1	3	1	20.8	10	30	30	12–14–16
	Elevator	1	9000	9000	M	See text	86/30	1	3	1	24.5	10	30	40	13–15–17
	Elevator equipment feeder	1	9600	9600	N	1	86/30	1	3	1	26.7	10	30	20	19–21–23
	Fire alarm system	1	600	600	C	1.25	86/30	1	2	1	6.3	12	20	20	22

[1]Loads of fluorescent luminaires figured at lamp wattage due to very high power factor ballasts.
[2]Correction factor is based upon temperature where the conductors are actually installed.
[3]The adjustment factor is determined by the number of current-carrying conductors in the same raceway or cable.
[4]Assume 75°C terminals.

TABLE A-9(B)

Owner feeder load calculation.

	Count[1]	VA/Unit[2]	NEC Load[3]	Connected Load[4]	Continuous Loads[5]	Noncontinuous[6]	Neutral Load[7]
General Lighting *NEC 220.12*:							
Halls, corridors, and stairways	534	0.5	267.0		57.5	209.5	
Owner's room and toilet rooms	237	3.5	829.5			829.5	
Connected load from Worksheet:				1056.0	434.0	622.0	
Totals:			1097	1056.0	57.5	1039.0	1097
Other Loads:							
Receptacle outlets	9	180	1620			1620	1620
Receptacle phone	1	600	600			600	600
Water heaters, toilet rooms	2	1500		3000	3000		3000
Water heater, Owner's room	1	6000		6000	6000		
Space heater, women's toilet	1	2000			2000		
Space heater, men's toilet	1	1500			1500		
Fire alarm system	1	600			600		600
Motors & Appliances:	Amperes	VA/Unit					
Elevator	1	9000			9000		
Elevator equipment	1	9600				9600	
Sump pump	1	1127				1127	1127
Total Loads					22,649	15,647	8044
Minimum feeder load 125% of continuous loads plus 100% of noncontinuous:						43,958	8044
Minimum ampacity:						122.1	22.3

Minimum 1 AWG ungrounded and 8 AWG neutral for load. Minimum neutral 6 AWG per *215.2(A)(1)*.

1/0 AWG ungrounded conductors installed for OCP and 4 AWG neutral installed.

Minimum size EMT = $(0.1855 \times 3) + 0.0824 = 0.6389$ sq. in. = trade size 1−½.

[1]Count: Square footage of given area, the number of a given type load, or the ampere rating of the given type load.
[2]VA/Unit: Specific unit load values required by *NEC Article 220*, or actual load values for the given type load.
[3]*NEC* Load: Count × *NEC* VA/Unit. Also the "calculated load."
[4]Connected Load: Actual load connected to the electrical system.
[5]Continuous Loads: The maximum current is expected to continue for 3 hours or more. Column also used to apply 125% to the largest motor load.
[6]Noncontinuous: Loads that are not continuous.
[7]Neutral Load: The maximum unbalanced load calculated according to *NEC 220.61*.

TABLE A-10

Service feeder load calculation form.

	Count[1]	VA/Unit[2]	NEC Load[3]	Connected Load[4]	Continuous Loads[5]	Noncon-tinuous[6]	Neutral Load[7]
General Lighting:							
NEC 220.12	5160	3	15,480				
Connected loads from worksheets				4887			
Totals:			15,480	4887	15,480	0	15,480
Storage Area:							
NEC 220.12	2580	0.25	645				
Connected load from worksheet				311.6			
Totals:			645	3116	0	3116	3116
Other Loads:							
Receptacle outlets	96	180	17,280	0	0	10,000	10,000
7280 VA @ 50%	0	0	3640	0	0	3640	3640
Other receptacle outlets	23	9780	2400	21,900	6000	18,300	24,300
Motors & Appliances:	Count[1]	VA/Unit					
Kitchen equipment[9]	6	71,098	0	71,098	36,992	9222	0
Exhaust fans	4	708	0	708	0	708	708
HACR equipment[8]	12	118,640	0	109,532	54,656	44,760	0
Washer/dryer (BS)	1	4000	0	0	0	4000	832
Water heaters	7	39,300	0	40,800	40,800	0	6300
Elevator	1	9000	0	9000	0	9000	0
Elevator equipment	1	9600	0	9600	0	9600	0
Sump pump	1	1127	0	1127	0	1127	1127
				Total Loads	153,928	113,473	65,503
		Minimum feeder load 125% of continuous loads plus 100% of noncontinuous:				305,882	65,503
					Minimum ampacity:	850	182

Minimum (3) 350 kcmil ungrounded and (3) 1/0 AWG Neutral in parallel.

Minimum size Schedule 40 PVC = $(0.5242 \times 3) + 0.1825 = 1.7551$ sq. in. = $2\frac{1}{2}$ in.

[1]Count: Square footage of given area, the number of a given type load, or the ampere rating of the given type load.
[2]VA/Unit: Specific unit load values required by *NEC* Article 220, or actual load values for the given type load.
[3]*NEC* Load: Count × *NEC* VA/Unit. Also the "calculated load."
[4]Connected Load: Actual load connected to the electrical system.
[5]Continuous Loads: The maximum current is expected to continue for three hours or more. Column also used to apply 125% to the largest motor load.
[6]Noncontinuous: Loads that are not continuous.
[7]Neutral Load: The maximum unbalanced load calculated according to *NEC 220.61*.
[8]Because the heating and cooling loads are noncoincident, only the larger load is included.
[9]A 65% factor is applied to commercial cooking equipment as allowed by *220.56*.

TABLE A-11

Luminaire/lamp/ballast schedule for the basement of the Commercial Building.

LUMINAIRE/LAMP/BALLAST SCHEDULE

BASEMENT

Style	Catalog Number	Lamp Type & Color	No. of Lamps	Ballast Type	Ballast Voltage	Mounting	Description	Qty.	Input Watts Per Luminaire/Total Input Watts
A	WSC-232-UNV-ER81	T8 Bipin 32 watt 2800 lumens per lamp. Rapid start 48" 3500°K	2	Note 1	Note 2	Surface	Decorative wrap. Snap fit nonglare acrylic lens. Ballast cover removable without tools. Suitable for damp locations.	2	64/128
A-EM	WSC-232-UNV-EL4-ER81	T8 Bipin 32 watt 2800 lumens per lamp. Rapid start 48" 3500°K	2	Note 1	Note 2	Surface	Decorative wrap. Snap fit nonglare acrylic lens. Ballast cover removable without tools. Suitable for damp locations. See Note 3.	2	70/140
B	IAF-232-UNV-ER81	T8 Bipin 32 watt 2800 lumens per lamp. Rapid start 48" 3500°K	2	Note 1	Note 2	Surface or chain	Industrial. White enamel finish. 8% uplight. Ballast cover removable without tools. Suitable for damp locations.	14	64/896
B-EM	IAF-232-UNV-EL4-ER81	T8 Bipin 32 watt 2800 lumens per lamp. Rapid start 48" 3500°K	2	Note 1	Note 2	Surface or chain	Industrial. White enamel finish. 8% uplight. Ballast cover removable without tools. Suitable for damp locations. See Note 3.	12	70/840
E-EM	ML709835 ICAT120D-EM-494P01	LED module 3500°K	LED module	Universal LED driver 120–277 volts	Note 2	Recessed	Recessed downlight with emergency battery backup. Dimmable. May be installed in insulated (IC) or noninsulated (non-IC) ceilings. See Note 3.	1	14/14
EXIT	EEX7	LED	LED module	LED driver	Note 2	Surface	Suitable for ceiling, wall, or corner end mounting	1	4.6/4.6

Area = 2378 Sq. Ft. (See Note 4)   Total Watts = 2023   Lighting Power Density = 0.851 Watts/Sq. Ft.

Notes: 1. High-efficiency Class P electronic. Power factor: 98%. Ballast factor: 0.88.
2. Universal input voltage: 120–277 V ±10%.
3. Luminaire has an emergency nickel cadmium battery pack. In the event of a power outage, electronic circuitry instantly senses the loss of power and keeps the luminaire partially illuminated for a minimum of 90 minutes. When normal power is restored, battery pack automatically goes into charge mode. Battery charges while luminaire is energized. Full recharge time: 24 hours.
4. Square footage based on outside building dimensions minus one foot for each of the outside walls.

TABLE A-12

Luminaire/lamp/ballast schedule for the first floor of the Commercial Building.

LUMINAIRE/LAMP/BALLAST SCHEDULE

FIRST FLOOR

Style	Catalog Number	Lamp Type & Color	No. of Lamps	Ballast Type	Ballast Voltage	Mounting	Description	Qty.	Input Watts Per Luminaire/Total Input Watts
C	VT3-232DR UNV-EL4-EB81-WL-8L-M4-GT4	T8 Bipin-32 watt 2800 lumens per lamp. Instant start 48" 3500°K	2	Note 1	Note 2	Surface or chain	Vapor tight. Fiberglass housing. Acrylic lens. Excellent choice for indoor or outdoor applications where weathering, high temps, humidity, dust, or corrosive fumes are present.	15	61/915
C-EM	VT3-232DR-UNV-EL4-EB81-WL-8L-M4-GT4	T8 Bipin 32 watt 2800 lumens per lamp. Instant start 48" 3500°K	2	Note 1	Note 2	Surface or chain	Vapor tight. Fiberglass housing. Acrylic lens. Excellent choice for indoor or outdoor applications where weathering, high temps, humidity, dust, or corrosive fumes are present. See Note 3.	3	67/201
E	ML709835 ICAT120D-494P06	LED module 3500°K	LED module	Universal LED driver 120–277 volts	Note 2	Recessed	Recessed downlight. Dimmable. May be installed in insulated (IC) or noninsulated (non-IC) ceilings. Lumens: 813	10	14/140
E-EM	ML709835 CAT120D-EM-494P06	LED module 3500°K	LED module	Universal LED driver 120–277 volts	Note 2	Recessed	Recessed downlight. with emergency battrery backup. Dimmable. May be installed in insulated (IC) or noninsulated (non-IC) ceilings Lumens: 813 See Note 3.	4	20/80
E2	H750TD010-ML712835-TUNVD010-494P06	LED module 3500°K	LED module	Universal driver 120–277 volts	Note 2	Recessed	Recessed downlight. Dimmable. For installation in noninsulated (non-IC) ceilings Lumens: 1356	3	26/78

(continued)

TABLE A-12

Luminaire/lamp/ballast schedule for the first floor of the Commercial Building. (*continued*)

Style	Catalog Number	Lamp Type & Color	No. of Lamps	Ballast Type	Ballast Voltage	Mounting	Description	Qty.	Input Watts Per Luminaire/ Total Input Watts
G	2GR8-332A-UNV-EB82	T8 Bipin 32 watt 2800 lumens per lamp. Instant start 48" 3500°K	3	Note 1	Note 2	Recessed Troffer for dropped ceiling grid lay-in 3-3/4" depth	Acrylic prismatic lens.	7	80/560
G-EM	2GR8-332A-UNV-EL4-EB82	T8 Bipin 32 watt 2800 lumens per lamp. Instant start 48" 3500°K	3	Note 1	Note 2	Recessed Troffer for dropped ceiling grid lay-in 3-3/4" depth	Acrylic prismatic lens. See Note 3.	2	86/172
I	RCG-SS-232-UNV-EB82	T8 Bipin 32 watt 2800 lumens per lamp. Instant start 48" 3500°K	2	Note 1	Note 2	Recessed Strip light for dropped ceiling grid lay-in 5-1/8" deep	Recessed strip light. Recommended for mass merchandise and retail sales areas. Reflector reduces glare. Suitable for damp locations. See Note 3.	27	61/1647
K	LDWP-FC-2A-ED	Light-emitting diodes (LED)	1	Electronic LED driver	Note 2	Surface	Lumens: 2400 Exterior LED wall light. Cast aluminum. Wall pack. Suitable for wet locations. Full light cut-off door.	8	29/232
N	BC-217A-UNV-ER81	T8 Bipin 17 watt 1400 lumens per lamp. Rapid start 24" 3500°K	2	Note 1	Note 2	Surface	Wall light. Opal acrylic reflector. All-purpose wall bracket.	3	34/102
P	2RDI-217FLS-UNV-EB81	T8 Bipin 17 watt 1400 Lumens per lamp. Rapid start 24" 3500°K	2	Note 1	Note 2	Recessed	Direct/indirect luminaire. Suitable for damp locations. Indirect frosted white acrylic lamp shields. Ballast removeable from above or below.	4	34/136

TABLE A-12

Luminaire/lamp/ballast schedule for the first floor of the Commercial Building. (continued)

Q	CEL7035SD	75 watt MR16 Halogen	2	No ballast.	Dual-voltage 120/277 VAC	Recessed	3	24/72	Lumens: 1137 Emergency lighting. Solid state charger for nickel cadmium battery. Gives appearance of a ceiling. When power is lost, doors open automatically. When power is restored, doors close automatically. Maintains emergency lighting for 90 minutes. Self-tests every 28 days.
T	L805SML FL830P	LED module 3000°K	Three 3-watt White LEDs	LED driver	120	Flexible track mounting	7	9/63	Lumens: 312 Track light. Adjustable heads. White housing. Flood, narrow flood, spot lighting available.
EXIT	EEX7	LED	LED module	LED driver	Note 2	Surface	6	4.6/28	Suitable for ceiling, wall, or corner end mounting

Area = 2378 Sq. Ft. (See Note 4)   **Total Watts = 4426**   **Lighting Power Density = 1.861 Watts/Sq. Ft.**

Notes: 1. High-efficiency Class P electronic. Power Factor 98%. Ballast Factor 0.88.
2. Universal input voltage: 120–277 V ± 10%.
3. Luminaire has an emergency nickel cadmium battery pack. In the event of a power outage, electronic circuitry instantly senses the loss of power and keeps the luminaire partially illuminated for a minimum of 90 minutes. When normal power is restored, battery pack automatically goes into charge mode. Battery charges while luminaire is energized. Full recharge time: 24 hours.
4. Square footage based on outside building dimensions minus one foot for each of the outside walls.

TABLE A-13

Luminaire/lamp/ballast schedule for the second floor of the Commercial Building.

LUMINAIRE/LAMP/BALLAST SCHEDULE

SECOND FLOOR

Style	Catalog Number	Lamp type & Color	No. of Lamps	Ballast Type	Ballast Voltage	Mounting	Description	Qty.	Input Watts Per Luminaire/Total Input Watts
E	ML709835 ICAT120D-494P06	LED module 3500°K	LED module	Universal driver 120–277 volts	120	Recessed	Lumens: 813 Dimmable. May be installed in insulated (IC) or noninsulated (non-IC) ceilings.	24	14/336
E-EM	ML709835 ICAT120D-EM-494P06	LED module 3500°K	LED module	Universal driver 120–277 volts	120	Recessed	Lumens: 813 Recessed downlight with emergency battery backup. Dimmable May be installed in insulated (IC) or noninsulated (non-IC) ceilings. See Note 3.	4	14/56
E2	H750TD010-ML712835-TUNVD010-494P06	LED module 3500°K	LED module	Universal driver 120–277 volts	120	Recessed	Lumens: 813 Recessed downlight. Dimmable. For noninsulated (non-IC) ceilings.	8	26/208
E2-EM	H750TD010-ML12835-TUND010-EM-494P06	LED module 3500°K	LED module	Universal driver 120–277 volts	120	Recessed	Lumens: 813 Recessed downlight with emergency battery backup. Dimmable. For noninsulated (non-IC) ceilings. See Note 3.	3	32/96
F	GR8-232A-UNV-EB81	T8 Bipin 32 watt 2800 lumens per lamp. Instant start 48" 3500°K	2	Note 1	Note 2	Recessed Troffer for dropped ceiling grid lay-in 3-3/4" depth	Acrylic lens.	6	61/366
F-EM	GR8-232A-UNV-EL4-EB81	T8 Bipin 32 watt 2800 lumens per lamp. Instant start 48" 3500°K	2	Note 1	Note 2	Recessed Troffer for dropped ceiling grid lay-in 3-3/4" depth	Acrylic lens. See Note 3.	3	58/174

TABLE A-13

Luminaire/lamp/ballast schedule for the second floor of the Commercial Building. (continued)

Symbol	Catalog Number	Lamps	Lamp Description	Ballast	Voltage	Mounting	Description	Quantity	Watts
G	2GR8-332A-UNV-EB82	3	T8 Bipin 32 watt 2800 lumens per lamp. Instant start 48" 3500°K	Note 1	Note 2	Troffer for dropped ceiling grid lay-in 3-3/4" depth	Lumens: 2800 Acrylic lens. See Note 3.	14	80/1120
G-EM	2GR8-332A-UNV-EL4-EB82	3	T8 Bipin 32 watt 2800 lumens per lamp. Instant start. 48" 3500°K	Note 1	Note 2	Troffer for dropped ceiling grid lay-in 3-3/4" depth	Lumens: 2800 Acrylic lens. See Note 3.	1	86/85
N	BC-217A-UNV-ER81	2	T8 Bipin 17 watt 1400 lumens per lamp. Rapid start. 24" 3500°K	Note 1	Note 2	Surface	Wall light. Opal acrylic reflector. All-purpose wall bracket.	3	34/102
Q	CEL7035SD	2	75 watt MR16 halogen	No ballast.	Dual-voltage 120/277 VAC	Recessed	Lumens: 1137 Emergency lighting. Solid-state charger for nickel cadmium battery. Gives appearance of a ceiling. When power is lost, doors open automatically. When power is restored, doors close automatically. Maintains emergency lighting for 90 minutes. Self-tests every 28 days.	2	75/150
EXIT	EEX7	LED module	LED	LED driver	Note 2	Surface	Suitable for ceiling, wall, or corner end mounting.	2	4.6/9.2

Area = 2378 Sq. Ft.   (See Note 4)   **Total Watts = 2702**   **Lighting Power Density = 1.136 Watts/Sq. Ft.**

Notes:
1. High-efficiency Class P electronic. Power Factor: 98%. Ballast Factor: 0.88.
2. Universal input voltage: 120–277 V ±10%.
3. Luminaire has an emergency nickel cadmium battery pack. In the event of a power outage, electronic circuitry instantly senses the loss of power and keeps the luminaire partially illuminated for a minimum of 90 minutes. When normal power is restored, battery pack automatically goes into charge mode. Battery charges while luminaire is energized. Full recharge time: 24 hours.
4. Square footage based on outside building dimensions minus 1 foot for each of the outside walls.

Appendix B: Useful Formulas

HOW TO USE TABLES B–1 AND B–2 TO CALCULATE VOLTAGE LOSS

1. Multiply length of circuit in feet (one way) by the current (in amperes) by the number shown in the table for the kind of current (3-phase, single-phase, power factor) and the size of conductor.

2. Put a decimal in front of the last 6 digits. This is the approximate voltage drop to be expected in the circuit.

3. If you have conductors in parallel, one extra step is needed: for two conductors in parallel per phase, divide answer in Step 2 by 2; for three conductors in parallel per phase, divide answer in Step 2 by 3; for four conductors in parallel per phase, divide answer in Step 2 by 4.

EXAMPLE

A 208Y/120-volt building service is made up of three 500-kcmil copper conductors per phase. The length of the service conductors between the switchboard and the transformer secondary is approximately 25 feet. The wiring method is steel conduit. Find the approximate voltage drop when the load is 1000 amperes, 90 percent power factor.

1. $25 \times 1000 \times 81 = 2{,}025{,}000$
2. 2.025000
3. $\dfrac{2.025000}{3} = 0.675$ volts loss

HOW TO USE TABLES B–1 AND B–2 TO SELECT CONDUCTOR SIZE

1. Multiply length of circuit in feet (one way) by the current (in amperes).

2. Multiply the permitted voltage loss by 1,000,000.

3. Divide Step 2 by Step 1.

4. Look in the column that shows the type of conductor, current, and power factor for the number found in Step 3 that is equal to or less than that number.

EXAMPLE

Find the size copper conductors needed to limit the voltage loss on a 180-foot run. The wiring method is steel conduit. The system is 3-phase. The load is 40 amperes, 80 percent power factor.

1. $180 \times 40 = 7200$

2. $5.5 \times 1,000,000 = 5,500,000$

3. $\dfrac{5,500,000}{7200} = 764$

4. In the table, in the proper column, find the number 764 or one smaller. The smaller number is 745. Go to the left column and find the minimum conductor size to be 6 AWG.

TABLE B-1

Aluminum Conductors—Ratings and Volt Loss**

Conduit	Wire Size	Ampacity Type T, TW, UF (60° C Wire)	Type RHW, THHW (wet), THW, THWN, XHHW (wet), USE (75° C Wire)	Type RHH, THHW, THHN (dry), XHHW (dry) (90°C Wire) RHW-2, THW-2 (90° C wet or dry)	Direct Current	Volt Loss Three Phase (60 Cycle, Lagging Power Factor.) 100%	90%	80%	70%	60%	Single Phase (60 Cycle, Lagging Power Factor.) 100%	90%	80%	70%	60%
Steel Conduit	12	20*	20*	25*	6360	5542	5039	4504	3963	3419	6400	5819	5201	4577	3948
	10	25	30*	35*	4000	3464	3165	2836	2502	2165	4000	3654	3275	2889	2500
	8	30	40	45	2520	2251	2075	1868	1656	1441	2600	2396	2158	1912	1663
	6	40	50	60	1616	1402	1310	1188	1061	930	1620	1513	1372	1225	1074
	4	55	65	75	1016	883	840	769	692	613	1020	970	888	799	708
	3	65	75	85	796	692	668	615	557	497	800	771	710	644	574
	2	75	90	110	638	554	541	502	458	411	640	625	580	529	475
	1	85	100	115	506	433	432	405	373	338	500	499	468	431	391
	0	100	120	135	402	346	353	334	310	284	400	407	386	358	328
	00	115	135	150	318	277	290	277	260	241	320	335	320	301	278
	000	130	155	175	252	225	241	234	221	207	260	279	270	256	239
	0000	155	180	205	200	173	194	191	184	174	200	224	221	212	201
	250	170	205	230	169	148	173	173	168	161	172	200	200	194	186
	300	190	230	255	141	124	150	152	150	145	144	174	176	173	168
	350	210	250	280	121	109	135	139	138	134	126	156	160	159	155
	400	225	270	305	106	95	122	127	127	125	110	141	146	146	144
	500	260	310	350	85	77	106	112	113	113	90	122	129	131	130
	600	285	340	385	71	65	95	102	105	106	76	110	118	121	122
	750	320	385	435	56	53	84	92	96	98	62	97	107	111	114
	1000	375	445	500	42	43	73	82	87	89	50	85	95	100	103
Non-magnetic Conduit (Lead Covered Cables Or Installation In Fiber Or Other Non-magnetic Conduit. Etc.	12	20*	20*	25*	6360	5542	5029	4490	3946	3400	6400	5807	5184	4557	3926
	10	25	30*	35*	4000	3464	3155	2823	2486	2147	4000	3643	3260	2871	2480
	8	30	40	45	2520	2251	2065	1855	1640	1423	2600	2385	2142	1894	1643
	6	40	50	60	1616	1402	1301	1175	1045	912	1620	1502	1357	1206	1053
	4	55	65	75	1016	883	831	756	677	596	1020	959	873	782	688
	3	65	75	85	796	692	659	603	543	480	800	760	696	627	555
	2	75	90	100	638	554	532	490	443	394	640	615	566	512	456
	1	85	100	115	506	433	424	394	360	323	500	490	455	415	373
	0	100	120	135	402	346	344	322	296	268	400	398	372	342	310
	00	115	135	150	318	277	281	266	247	225	320	325	307	285	260
	000	130	155	175	252	225	234	223	209	193	260	270	258	241	223
	0000	155	180	205	200	173	186	181	171	160	200	215	209	198	185
	250	170	205	230	169	147	163	160	153	145	170	188	185	177	167
	300	190	230	255	141	122	141	140	136	130	142	163	162	157	150
	350	210	250	280	121	105	125	125	123	118	122	144	145	142	137

TABLE B-1

(continued)

Conduit	Wire Size	Type T, TW, UF (60° C Wire)	Type RHW, THHW (wet), THW, THWN, XHHW (wet), USE (75° C Wire)	Type RHH, THHW, THHN (dry), XHHW (dry) (90°C Wire) RHW-2, THW-2 (90° C wet or dry)	Direct Current	Three Phase (60 Cycle, Lagging Power Factor.)					Single Phase (60 Cycle, Lagging Power Factor.)				
		Ampacity	Ampacity	Ampacity		Volt Loss 100%	90%	80%	70%	60%	100%	90%	80%	70%	60%
	400	225	270	305	106	93	114	116	114	111	108	132	134	132	128
	500	260	310	350	85	74	96	100	100	98	86	111	115	115	114
	600	285	340	385	71	62	85	90	91	91	72	98	104	106	105
	750	320	385	435	56	50	73	79	82	82	58	85	92	94	95
	1000	375	445	500	42	39	63	70	73	75	46	73	81	85	86

*The overcurrent protection for conductor types marked with an (*) shall not exceed 15 amperes for 12 AWG, or 25 amperes for 10 AWG after correction factors for ambient temperature and adjustment factors for the number of conductor have been applied.

**Figures are L–L for both single-phase and 3-phase. Three-phase figures are average for the three phases.
(*Courtesy Cooper-Bussmann*)

TABLE B-2

Copper Conductors—Ratings and Volt Loss**

Conduit	Wire Size	Type TW, UF (60° C Wire)	Type RHW, THHW (wet), THW, THWN, XHHW (wet), USE (75° C Wire)	Type RHH, THHW, THHN (dry), XHHW (dry) (90°C Wire) RHW-2, THW-2 (90° C wet or dry)	Direct Current	Three Phase (60 Cycle, Lagging Power Factor.)					Single Phase (60 Cycle, Lagging Power Factor.)				
		Ampacity	Ampacity	Ampacity		Volt Loss 100%	90%	80%	70%	60%	100%	90%	80%	70%	60%
Steel Conduit	14	20*	20*	25*	6140	5369	4887	4371	3848	3322	6200	5643	5047	4444	3836
	12	25*	25*	30*	3860	3464	3169	2841	2508	2172	4000	3659	3281	2897	2508
	10	30	35*	40*	2420	2078	1918	1728	1532	1334	2400	2214	1995	1769	1540
	8	40	50	55	1528	1350	1264	1148	1026	900	1560	1460	1326	1184	1040
	6	55	65	75	982	848	812	745	673	597	980	937	860	777	690
	4	70	85	95	616	536	528	491	450	405	620	610	568	519	468
	3	85	100	110	490	433	434	407	376	341	500	501	470	434	394
	2	95	115	130	388	346	354	336	312	286	400	409	388	361	331
	1	110	130	150	308	277	292	280	264	245	320	337	324	305	283
	0	125	150	170	244	207	228	223	213	200	240	263	258	246	232
	00	145	175	195	193	173	196	194	188	178	200	227	224	217	206
	000	165	200	225	153	136	162	163	160	154	158	187	188	184	178
	0000	195	230	260	122	109	136	140	139	136	126	157	162	161	157
	250	215	255	290	103	93	123	128	129	128	108	142	148	149	148
	300	240	285	320	86	77	108	115	117	117	90	125	133	135	135
	350	260	310	350	73	67	98	106	109	109	78	113	122	126	126
	400	280	335	380	64	60	91	99	103	104	70	105	114	118	120

(continued)

TABLE B-2

(*continued*)

Conduit	Wire Size	Ampacity Type T, TW, UF (60° C Wire)	Type RHW, THHW (wet), THW, THWN, XHHW (wet), USE (75° C Wire)	Type RHH, THHW, THHN (dry), XHHW (dry) (90°C Wire) RHW-2, THW-2 (90° C wet or dry)	Direct Current	Volt Loss Three Phase (60 Cycle, Lagging Power Factor.) 100%	90%	80%	70%	60%	Single Phase (60 Cycle, Lagging Power Factor.) 100%	90%	80%	70%	60%
	500	320	380	430	52	50	81	90	94	96	58	94	104	109	111
	600	355	420	475	43	43	75	84	89	92	50	86	97	103	106
	750	400	475	535	34	36	68	78	84	88	42	79	91	97	102
	1000	455	545	615	26	31	62	72	78	82	36	72	84	90	95
Non-magnetic Conduit (Lead Covered Cables Or Installation In Fibre Or Other Non-magnetic Conduit. Etc.	14	20*	20*	25*	6140	5369	4876	4355	3830	3301	6200	5630	5029	4422	3812
	12	25*	25*	30*	3860	3464	3158	2827	2491	2153	4000	3647	3264	2877	2486
	10	30	35*	40*	2420	2078	1908	1714	1516	1316	2400	2203	1980	1751	1520
	8	40	50	55	1528	1350	1255	1134	1010	882	1560	1449	1310	1166	1019
	6	55	65	75	982	848	802	731	657	579	980	926	845	758	669
	4	70	85	95	616	536	519	479	435	388	620	599	553	502	448
	3	85	100	110	470	433	425	395	361	324	500	490	456	417	375
	2	95	115	130	388	329	330	310	286	259	380	381	358	330	300
	1	110	130	150	308	259	268	255	238	219	300	310	295	275	253
	0	125	150	170	244	207	220	212	199	185	240	254	244	230	214
	00	145	175	195	193	173	188	183	174	163	200	217	211	201	188
	000	165	200	225	153	133	151	150	145	138	154	175	173	167	159
	0000	195	230	260	122	107	127	128	125	121	124	147	148	145	140
	250	215	255	290	103	90	112	114	113	110	104	129	132	131	128
	300	240	285	320	86	76	99	103	104	102	88	114	119	120	118
	350	260	310	350	73	65	89	94	95	94	76	103	108	110	109
	400	280	335	380	64	57	81	87	89	89	66	94	100	103	103
	500	320	380	430	52	46	71	77	80	82	54	82	90	93	94
	600	355	420	475	43	39	65	72	76	77	46	75	83	87	90
	750	400	475	535	34	32	58	65	70	72	38	67	76	80	83
	1000	455	545	615	26	25	51	59	63	66	30	59	68	73	77

*The overcurrent protection for conductor types marked with an (*) shall not exceed 15 amperes for 14 AWG, or 20 amperes for 12 AWG, and 30 amperes for 10 AWG after correction factors for ambient temperature and adjustment factors for the number of conductor have been applied.

**Figures are L–L for both single-phase and 3-phase. Three-phase figures are average for the three phases.
(*Courtesy Cooper-Bussmann*)

Watts Wheel: The formulas on the outer ring can be used to determine the result in the inner ring. For example, if the voltage and resistance are known and it is desired to determine the watts, divide the voltage squared by the resistance.

A heater operates at 240 volts and the resistance has been found to be 4.8 ohms. What is the wattage of the heater?

$$\text{Watts} = \frac{\text{Voltage}^2}{\text{Resistance}} = \frac{240^2}{4.8} = \frac{57,600}{4.8} = 12,000 \text{ Watts}$$

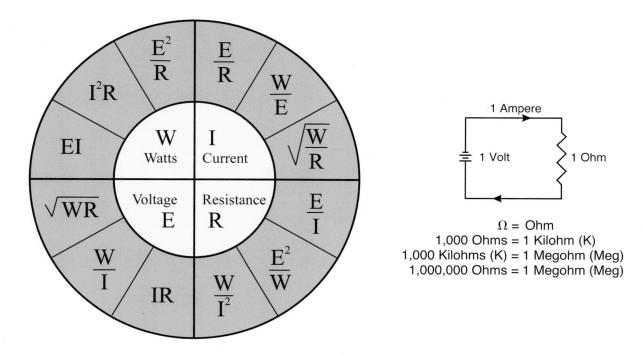

FIGURE B-1   Watts Wheel

TABLE B-3			
Useful Formulas			
To Find	**Single Phase**	**Three-Phase**	**Direct Current**
AMPERES when kilovolt-amperes are known	$\dfrac{kVA \times 1000}{E}$	$\dfrac{kVA \times 1000}{E \times 1.732}$	Not applicable
AMPERES when horsepower is known	$\dfrac{hp \times 1000}{E \times \% \text{ eff.} \times pf}$	$\dfrac{hp \times 746}{E \times 1.732 \times \% \text{ eff.} \times pf}$	$\dfrac{hp \times 746}{E \times \% \text{ eff.}}$
AMPERES when kilowatts are known	$\dfrac{kw \times 1000}{E \times pf}$	$\dfrac{kW \times 1000}{E \times 1.732 \times pf}$	$\dfrac{kW \times 1000}{E}$
WATTS	$E \times I \times pf$	$E \times I \times 1.732 \times pf$	$E \times I$
KILOWATTS	$\dfrac{I \times E \times pf}{1000}$	$\dfrac{I \times E \times 1.732 \times pf}{1000}$	$\dfrac{I \times E}{1000}$
VOLT-AMPERES	$I \times E$	$I \times E \times 1.732$	$I \times E$
KILOVOLT-AMPERES	$\dfrac{I \times E}{1000}$	$\dfrac{I \times E \times 1.732}{1000}$	Not applicable
HORSEPOWER	$\dfrac{I \times E \times \% \text{ eff.} \times pf}{746}$	$\dfrac{I \times E \times 1.732 \times \% \text{ eff.} \times pf}{746}$	$\dfrac{I \times E \times \% \text{ eff.}}{746}$

I = amperes   E = volts   kW = kilowatts   kVA = kilovolt-amperes
hp = horsepower   % eff. = percent efficiency   pf = power factor

Appendix C: NEMA Enclosure Types

TABLE C-1

NEMA Enclosure Types

In *Non-Hazardous Locations*, the specific enclosure Types, their applications, and the environmental conditions they are designed to protect against, **when completely and properly installed**, are as follows:

Type 1 Enclosures constructed for indoor use to provide a degree of protection to personnel against access to hazardous parts and to provide a degree of protection of the equipment inside the enclosure against ingress of solid foreign objects (falling dirt).

Type 2 Enclosures constructed for indoor use to provide a degree of protection to personnel against access to hazardous parts; to provide a degree of protection of the equipment inside the enclosure against ingress of solid foreign objects (falling dirt); and to provide a degree of protection with respect to harmful effects on the equipment due to the ingress of water (dripping and light splashing).

Type 3 Enclosures constructed for either indoor or outdoor use to provide a degree of protection to personnel against access to hazardous parts; to provide a degree of protection of the equipment inside the enclosure against ingress of solid foreign objects (falling dirt and windblown dust); to provide a degree of protection with respect to harmful effects on the equipment due to the ingress of water (rain, sleet, snow); and that will be undamaged by the external formation of ice on the enclosure.

Type 3R Enclosures constructed for either indoor or outdoor use to provide a degree of protection to personnel against access to hazardous parts; to provide a degree of protection of the equipment inside the enclosure against ingress of solid foreign objects (falling dirt); to provide a degree of protection with respect to harmful effects on the equipment due to the ingress of water (rain, sleet, snow); and that will be undamaged by the external formation of ice on the enclosure.

Type 3S Enclosures constructed for either indoor or outdoor use to provide a degree of protection to personnel against access to hazardous parts; to provide a degree of protection of the equipment inside the enclosure against ingress of solid foreign objects (falling dirt and windblown dust); to provide a degree of protection with respect to harmful effects on the equipment due to the ingress of water (rain, sleet, snow); and for which the external mechanism(s) remain operable when ice laden.

Type 3X Enclosures constructed for either indoor or outdoor use to provide a degree of protection to personnel against access to hazardous parts; to provide a degree of protection of the equipment inside the enclosure against ingress of solid foreign objects (falling dirt and windblown dust); to provide a degree of protection with respect to harmful effects on the equipment due to the ingress of water (rain, sleet, snow); that provide an additional level of protection against corrosion and that will be undamaged by the external formation of ice on the enclosure.

Type 3RX Enclosures constructed for either indoor or outdoor use to provide a degree of protection to personnel against access to hazardous parts; to provide a degree of protection of the equipment inside the enclosure against ingress of solid foreign objects (falling dirt); to provide a degree of protection with respect to harmful effects on the equipment due to the ingress of water (rain, sleet, snow); that will be undamaged by the external formation of ice on the enclosure that provides an additional level of protection against corrosion; and that will be undamaged by the external formation of ice on the enclosure.

(continued)

TABLE C-1

(*continued*)

Type 3SX	Enclosures constructed for either indoor or outdoor use to provide a degree of protection to personnel against access to hazardous parts; to provide a degree of protection of the equipment inside the enclosure against ingress of solid foreign objects (falling dirt and windblown dust); to provide a degree of protection with respect to harmful effects on the equipment due to the ingress of water (rain, sleet, snow); that provides an additional level of protection against corrosion; and for which the external mechanism(s) remain operable when ice laden.
Type 4	Enclosures constructed for either indoor or outdoor use to provide a degree of protection to personnel against access to hazardous parts; to provide a degree of protection of the equipment inside the enclosure against ingress of solid foreign objects (falling dirt and windblown dust); to provide a degree of protection with respect to harmful effects on the equipment due to the ingress of water (rain, sleet, snow, splashing water, and hose directed water); and that will be undamaged by the external formation of ice on the enclosure.
Type 4X	Enclosures constructed for either indoor or outdoor use to provide a degree of protection to personnel against access to hazardous parts; to provide a degree of protection of the equipment inside the enclosure against ingress of solid foreign objects (windblown dust); to provide a degree of protection with respect to harmful effects on the equipment due to the ingress of water (rain, sleet, snow, splashing water, and hose directed water); that provide an additional level of protection against corrosion; and that will be undamaged by the external formation of ice on the enclosure.
Type 5	Enclosures constructed for indoor use to provide a degree of protection to personnel against access to hazardous parts; to provide a degree of protection of the equipment inside the enclosure against ingress of solid foreign objects (falling dirt and settling airborne dust, lint, fibers, and flyings); and to provide a degree of protection with respect to harmful effects on the equipment due to the ingress of water (dripping and light splashing).
Type 6	Enclosures constructed for either indoor or outdoor use to provide a degree of protection to personnel against access to hazardous parts; to provide a degree of protection of the equipment inside the enclosure against ingress of solid foreign objects (falling dirt); to provide a degree of protection with respect to harmful effects on the equipment due to the ingress of water (hose directed water and the entry of water during occasional temporary submersion at a limited depth); and that will be undamaged by the external formation of ice on the enclosure.
Type 6P	Enclosures constructed for either indoor or outdoor use to provide a degree of protection to personnel against access to hazardous parts; to provide a degree of protection of the equipment inside the enclosure against ingress of solid foreign objects (falling dirt); to provide a degree of protection with respect to harmful effects on the equipment due to the ingress of water (hose directed water and the entry of water during prolonged submersion at a limited depth); that provide an additional level of protection against corrosion and that will be undamaged by the external formation of ice on the enclosure.
Type 12	Enclosures constructed (without knockouts) for indoor use to provide a degree of protection to personnel against access to hazardous parts; to provide a degree of protection of the equipment inside the enclosure against ingress of solid foreign objects (falling dirt and circulating dust, lint, fibers, and flyings); and to provide a degree of protection with respect to harmful effects on the equipment due to the ingress of water (dripping and light splashing).
Type 12K	Enclosures constructed (with knockouts) for indoor use to provide a degree of protection to personnel against access to hazardous parts; to provide a degree of protection of the equipment inside the enclosure against ingress of solid foreign objects (falling dirt and circulating dust, lint, fibers, and flyings); and to provide a degree of protection with respect to harmful effects on the equipment due to the ingress of water (dripping and light splashing).
Type 13	Enclosures constructed for indoor use to provide a degree of protection to personnel against access to hazardous parts; to provide a degree of protection of the equipment inside the enclosure against ingress of solid foreign objects (falling dirt and circulating dust, lint, fibers, and flyings); to provide a degree of protection with respect to harmful effects on the equipment due to the ingress of water (dripping and light splashing); and to provide a degree of protection against the spraying, splashing, and seepage of oil and non-corrosive coolants.

For more detailed and complete information, see NEMA Standards Publication 250–2003, *Enclosures for Electrical Equipment (1000 Volts Maximum)*.

Reprinted with permission from the National Electrical Manufacturers Association.

▶The *NEC 110.28* and *Table 110.28* contain similar information.◀

Appendix D: Outside Air Temperatures for Selected U.S. Cities

TABLE D-1

Outdoor Air Temperatures for Selected U.S. Cities

NORTHERN CITIES				SUNBELT CITIES			
City	Cooling Months	2% Design Temp., °F	Highest Temp. on Record, °F	City	Cooling Months	2% Design Temp., °F	Highest Temp. on Record, °F
Wichita, KS	June–Aug.	100	113	Phoenix, AZ	May–Sept.	107	122
Salt Lake City, UT	June–Aug.	96	107	Las Vegas, NV	May–Sept.	104	116
Omaha, NE	June–Sept.	94	107	Fort Worth, TX	May–Sept.	97	113
Kansas City, MO	June–Aug.	94	109	Sacramento, CA	May–Sept.	97	115
Denver, CO	June–Aug.	93	104	Houston, TX	June–Sept.	96	109
St. Louis, MO	May–Sept.	92	107	Little Rock, AR	June–Sept.	94	112
Philadelphia, PA	June–Aug.	92	104	Memphis, TN	May–Sept.	93	108
New York, NY	June–Aug.	91	106	Norfolk, VA	June–Sept.	93	104
Chicago, IL	June–Aug.	91	104	Albuquerque, NM	May–Sept.	92	107
Indianapolis, IN	June–Aug.	90	104	Louisville, KY	June–Sept.	92	106
Columbus, OH	June–Aug.	90	104	New Orleans, LA	May–Sept.	92	102
Minneapolis/ St. Paul, MN	June–Aug.	90	105	Tampa, FL	May–Sept.	91	99
Boston, MA	June–Aug.	90	102	Atlanta, GA	May–Sept.	91	105
Detroit, MI	June–Aug.	89	104	Wilmington, NC	May–Sept.	91	104
Portland, OR	June–Aug.	88	107	Baltimore, MD	May–Sept.	90	105
Pittsburgh, PA	June–Aug.	88	103	Miami, FL	May–Sept.	90	98
Seattle, WA	June–Aug.	83	100	Honolulu, HI	May–Sept.	88	95
Fairbanks, AK	June–Aug.	80	96	San Diego, CA	June–Oct.	81	111
San Francisco, CA	May–Sept.	78	103	Los Angeles, CA	June–Oct.	80	112

Sources: 2% design temperatures derived from ASHRAE 2% design data for the months indicated. Highest temperatures from National Climatic Data Center, NESDIS National Environmental Satellite, Data and Information Services.

Appendix E: Metric System of Measurement

The following table provides useful conversions of English customary terms to SI terms, and SI terms to English customary terms. The metric system is known as the International System of Units (SI), taken from the French *Le Système International d'Unités*. Whenever the slant line / is found, say it as "per." Be practical when using the values in the following table. Present-day use of calculators and computers provides many "places" beyond the decimal point. You must decide how accurate your results must be when performing a calculation. When converting centimeters to feet, for example, one could be reasonably accurate by rounding the value of 0.03281 to 0.033. When a manufacturer describes a product's measurement using metrics but does not physically change the product, it is referred to as a "soft conversion." When a manufacturer actually changes the physical measurements of the product to standard metric size or a rational whole number of metric units, it is referred to as a "hard conversion." When rounding off numbers, be sure to round off in such a manner that the final result does not violate a "maximum" or "minimum" value, such as might be required by the *National Electrical Code*. The rule for discarding digits is that if the digits to be discarded begin with a 5 or higher, increase the last digit retained by one; for example: 6.3745, if rounded to three digits, would become 6.37. If 6.3745 were rounded to four digits, it would become 6.375. The following information has been developed from the latest U.S. Department of Commerce and the National Institute of Standards and Technology publications covering the metric system.

TABLE E-1

Multiply This Unit(s)	By This Factor	To Obtain This Unit(s)
acre	4 046.9	square meters (m²)
acre	43 560	square feet (ft²)
ampere hour	3 600	coulombs (C)
angstrom	0.1	nanometers (nm)
atmosphere	101.325	kilopascals (kPa)
atmosphere	33.9	feet of water (at 4°C)
atmosphere	29.92	inches of mercury (at 0°C)
atmosphere	0.76	meters of mercury (at 0°C)
atmosphere	0.007 348	tons per square inch
atmosphere	1.058	tons per square foot
atmosphere	1.0333	kilograms per square centimeter
atmosphere	10 333	kilograms per square meter
atmosphere	14.7	pounds per square inch

(continued)

TABLE E-1

(continued)

Multiply This Unit(s)	By This Factor	To Obtain This Unit(s)
bar	100	kilopascals (kPa)
barrel (oil, 42 U.S. gallons)	0.158 987 3	cubic meters (m^3)
barrel (oil, 42 U.S. gallons)	158.987 3	liters (L)
board foot	0.002 359 737	cubic meters (m^3)
bushel	0.035 239 07	cubic meters (m^3)
Btu	778.16	feet-pounds
Btu	252	grams-calories
Btu	0.000 393 1	horsepower-hours
Btu	1 054.8	joules (J)
Btu	1.055 056	kilojoules (kJ)
Btu	0.000 293 1	kilowatt-hours (kWH)
Btu per hour	0.000 393 1	horsepower (hp)
Btu per hour	0.293 071 1	watts (W)
Btu per degree Fahrenheit	1.899 108	kilojoules per kelvin (kJ/K)
Btu per pound	2.326	kilojoules per kilogram (kJ/kg)
Btu per second	1.055 056	kilowatts (kW)
calorie	4.184	joules (J)
calorie, gram	0.003 968 3	Btus
candela per foot squared (cd/ft^2)	10.763 9	candelas per meter squared (cd/m^2)

Note: The former term *candlepower* has been replaced with the term *candela*.

Multiply This Unit(s)	By This Factor	To Obtain This Unit(s)
candela per meter squared (cd/m^2)	0.092 903	candelas per foot squared (cd/ft^2)
candela per meter squared (cd/m^2)	0.291 864	footlambert*
candela per square inch (cd/in^2)	1550.003	candelas per square meter (cd/m^2)

Celsius = (Fahrenheit − 3^2) × ⁵⁄₉

Celsius = (Fahrenheit − 3^2) × 0.555555

Celsius = (0.556 × Fahrenheit) − 17.8

Note: The term *centigrade* was officially discontinued in 1948 and was replaced by the term *Celsius*. The term *centigrade* may still be found in some publications.

Multiply This Unit(s)	By This Factor	To Obtain This Unit(s)
centimeter (cm)	0.032 81	feet (ft)
centimeter (cm)	0.393 7	inches (in.)
centimeter (cm)	0.01	meters (m)
centimeter (cm)	10	millimeters (mm)
centimeter (cm)	393.7	mils
centimeter (cm)	0.010 94	yards
circular mil	0.000 005 067	square centimeters (cm^2)
circular mil	0.785 4	square mils (mil^2)
circular mil	0.000 000 785 4	square inches (in.^{2})
cubic centimeter (cm^3)	0.061 02	cubic inches (in.^{3})
cubic foot per second	0.028 316 85	cubic meters per second (m^3/s)
cubic foot per minute	0.000 471 947	cubic meters per second (m^3/s)
cubic foot per minute	0.471 947	liters per second (L/s)
cubic inch (in.^{3})	16.39	cubic centimeters (cm^3)
cubic inch (in.^{3})	0.000 578 7	cubic feet (ft^3)

TABLE E-1

(continued)

Multiply This Unit(s)	By This Factor	To Obtain This Unit(s)
cubic meter (m³)	35.31	cubic feet (ft³)
cubic yard per minute	12.742 58	liters per second (L/s)
cup (c)	0.236 56	liters (L)
decimeter	0.1	meters (m)
dekameter	10	meters (m)
Fahrenheit = (⁹⁄₅ Celsius) + 32		
Fahrenheit = (Celsius × 1.8) + 32		
fathom	1.828 804	meters (m)
fathom	6.0	feet
foot	30.48	centimeters (cm)
foot	12	inches
foot	0.000 304 8	kilometers (km)
foot	0.304 8	meters (m)
foot	0.000 189 4	miles (statute)
foot	304.8	millimeters (mm)
foot	12 000	mils
foot	0.333 33	yard
foot, cubic (ft³)	0.028 316 85	cubic meters (m³)
foot, cubic (ft³)	28.316 85	liters (L)
foot, board	0.002 359 737	cubic meters (m³)
cubic feet per second (ft³/s)	0.028 316 85	cubic meters per second (m³/s)
cubic feet per minute (ft³/min)	0.000 471 947	cubic meters per second (m³/s)
cubic feet per minute (ft³/min)	0.471 947	liters per second (L/s)
foot, square (ft²)	0.092 903	square meters (m²)
footcandle	10.763 91	lux (lx)
footlambert*	3.426 259	candelas per square meter (cd/m²)
foot of water	2.988 98	kilopascals (kPa)
foot pound	0.001 286	Btus
foot pound-force	1 055.06	joules (J)
foot pound-force per second	1.355 818	joules (J)
foot pound-force per second	1.355 818	watts (W)
foot per second	0.304 8	meters per second (m/s)
foot per second squared	0.304 8	meters per second squared (m/s²)
gallon (U.S. liquid)	3.785 412	liters (L)
gallons per day	3.785 412	liters per day (L/d)
gallons per hour	1.051 50	milliliters per second (mL/s)
gallons per minute	0.063 090 2	liters per second (L/s)
gauss	6.452	lines per square inch
gauss	0.1	millitesla (mT)
gauss	0.000 000 064 52	webers per square inch
grain	64.798 91	milligrams (mg)
gram (g) (a little more than the weight of a paper clip)	0.035 274	ounce (avoirdupois)

(continued)

TABLE E-1

(continued)

Multiply This Unit(s)	By This Factor	To Obtain This Unit(s)
gram (g)	0.002 204 6	pound (avoirdupois)
gram per meter (g/m)	3.547 99	pounds per mile (lb/mile)
grams per square meter (g/m²)	0.003 277 06	ounces per square foot (oz/ft²)
grams per square meter (g/m²)	0.029 494	ounces per square yard (oz/yd²)
gravity (standard acceleration)	9.806 65	meters per second squared (m/s²)
quart (U.S. liquid)	0.946 352 9	liters (L)
horsepower (550 ft · lbf/s)	0.745 7	kilowatts (kW)
horsepower	745.7	watts (W)
horsepower hours	2.684 520	megajoules (mJ)
inch per second squared (in/s²)	0.025 4	meters per second squared (m/s²)
inch	2.54	centimeters (cm)
inch	0.254	decimeters (dm)
inch	0.025 4	meters (m)
inch	25.4	millimeters (mm)
inch	1 000	mils
inch	0.027 78	yards
inch, cubic (in.³)	16 387.1	cubic millimeters (mm³)
inch, cubic (in.³)	16.387 06	cubic centimeters (cm³)
inch, cubic (in.³)	645.16	square millimeters (mm²)
inches of mercury	3.386 38	kilopascals (kPa)
inches of mercury	0.033 42	atmospheres
inches of mercury	1.133	feet of water
inches of water	0.248 84	kilopascals (kPa)
inches of water	0.073 55	inches of mercury
joule (J)	0.737 562	foot pound-force (ft · lbf)
kilocandela per meter squared (kcd/m²)	0.314 159	lambert*
kilogram (kg)	2.204 62	pounds (avoirdupois)
kilogram (kg)	35.274	ounces (avoirdupois)
kilogram per meter (kg/m)	0.671 969	pounds per foot (lb/ft)
kilogram per square meter (kg/m²)	0.204 816	pounds per square foot (lb/ft²)
kilogram meter squared (kg · m²)	23.730 4	pounds foot squared (lb · ft²)
kilogram meter squared (kg · m²)	3 417.17	pounds inch squared (lb · in.²)
kilogram per cubic meter (kg/m³)	0.062 428	pounds per cubic foot (lb/ft³)
kilogram per cubic meter (kg/m³)	1.685 56	pounds per cubic yard (lb/yd³)
kilogram per second (kg/s)	2.204 62	pounds per second (lb/s)
kilojoule (kJ)	0.947 817	Btus
kilometer (km)	1 000	meters (m)
kilometer (km)	0.621 371	miles (statute)
kilometer (km)	1 000 000	millimeters (mm)
kilometer (km)	1 093.6	yards
kilometer per hour (kg/hr)	0.621 371	miles per hour (mph)
kilometer, square (km²)	0.386 101	square miles (mile²)
kilopound-force per square inch	6.894 757	megapascals (MPa)

TABLE E-1

(continued)

Multiply This Unit(s)	By This Factor	To Obtain This Unit(s)
kilowatts (kW)	56.921	Btus per minute
kilowatts (kW)	1.341 02	horsepower (hp)
kilowatts (kW)	1 000	watts (W)
kilowatt-hour (kWh)	3 413	Btus
kilowatt-hour (kWh)	3.6	megajoules (MJ)
knots	1.852	kilometers per hour (km/h)
lamberts*	3 183.099	candelas per square meter (cd/m²)
lamberts*	3.183 01	kilocandelas per square meter (kcd/m²)
liter (L)	0.035 314 7	cubic feet (ft³)
liter (L)	0.264 172	gallons (U.S. liquid)
liter (L)	2.113	pints (U.S. liquid)
liter (L)	1.056 69	quarts (U.S. liquid)
liter per second (L/s)	2.118 88	cubic feet per minute (ft³/min)
liter per second (L/s)	15.850 3	gallons per minute (gal/min)
liter per second (L/s)	951.022	gallons per hour (gal/hr)
lumen per square foot (lm/ft²)	10.763 9	lux (lx); plural: luces
lumen per square foot (lm/ft²)	1.0	footcandles
lumen per square foot (lm/ft²)	10.763	lumens per square meter (lm/m²)
lumen per square meter (lm/m²)	1.0	lux (lx)
lux (lx)	0.092 903	lumens per square foot (footcandle)
maxwell	10	nanowebers (nWb)
megajoule (MJ)	0.277 778	kilowatt hours (kWh)
meter (m)	100	centimeters (cm)
meter (m)	0.546 81	fathoms
meter (m)	3.280 9	feet
meter (m)	39.37	inches
meter (m)	0.001	kilometers (km)
meter (m)	0.000 621 4	miles (statute)
meter (m)	1 000	millimeters (mm)
meter (m)	1.093 61	yards
meter, cubic (m³)	1.307 95	cubic yards (yd³)
meter, cubic (m³)	35.314 7	cubic feet (ft³)
meter, cubic (m³)	423.776	board feet
meter per second (m/s)	3.280 84	feet per second (ft/s)
meter per second (m/s)	2.236 94	miles per hour (mph)
meter, square (m²)	1.195 99	square yards (yd²)
meter, square (m²)	10.763 9	square feet (ft²)
mho per centimeter (mho/cm)	100	siemens per meter (S/m)

Note: The older term "mho" has been replaced with "siemens." The term "mho" may still be found in some publications.

micro inch	0.025 4	micrometers (mm)
mil	25.4	micrometers (mm)
mil	0.025 4	millimeters (mm)

(continued)

TABLE E-1

(continued)

Multiply This Unit(s)	By This Factor	To Obtain This Unit(s)
mil	2.540×10^{-3}	centimeters (cm)
mil	8.333×10^{-5}	feet
mil	0.001	inches
mil	2.540×10^{-8}	kilometers
mil	2.778×10^{-5}	yards
miles per hour	1.609 344	kilometers per hour (km/h)
miles per hour	0.447 04	meters per second (m/s)
miles per gallon	0.425 143 7	kilometers per liter (km/L)
miles	1.609 344	kilometers (km)
miles	5 280	feet
miles	1 609	meters (m)
miles	1 760	yards
miles (nautical)	1.852	kilometers (km)
miles squared	2.590 000	kilometers squared (km²)
millibar	0.1	kilopascals (kPa)
milliliter (mL)	0.061 023 7	cubic inches (in.³)
milliliter (mL)	0.033 814	fluid ounces (U.S.)
millimeter (about the thickness of a dime)	0.1	centimeters (cm)
millimeter (mm)	0.003 280 8	feet
millimeter (mm)	0.039 370 1	inches
millimeter (mm)	0.001	meters (m)
millimeter (mm)	39.37	mils
millimeter (mm)	0.001 094	yards
millimeter, square (mm²)	0.001 550	square inches (in.²)
millimeter, cubic (mm³)	0.000 061 023 7	cubic inches (in.³)
millimeter of mercury	0.133 322 4	kilopascals (kPa)
ohm	0.000 001	megohms
ohm	1 000 000	micro ohms
ohm circular mil per foot	1.662 426	nano ohms meter (no · m)
oersted	79.577 47	amperes per meter (A/m)
ounce (avoirdupois)	28.349 52	grams (g)
ounce (avoirdupois)	0.062 5	pounds (avoirdupois)
ounce, fluid	29.573 53	milliliters (mL)
ounce (troy)	31.103 48	grams (g)
ounce per foot, square (oz/ft²)	305.152	grams per meter squared (g/m²)
ounces per gallon (U.S. liquid)	7.489 152	grams per liter (g/L)
ounce per yard, square (oz/yd²)	33.905 7	grams per meter squared (g/m²)
pica	4.217 5	millimeters (mm)
pint (U.S. liquid)	0.473 176 5	liters (L)
pint (U.S. liquid)	473.177	milliliters (mL)
pound (avoirdupois)	453.592	grams (g)
pound (avoirdupois)	0.453 592	kilograms (kg)
pound (avoirdupois)	16	ounces (avoirdupois)

TABLE E-1

(*continued*)

Multiply This Unit(s)	By This Factor	To Obtain This Unit(s)
poundal	0.138 255	newtons (N)
pound foot (lb · ft)	0.138 255	kilograms meter (kg · m)
pound foot per second	0.138 255	kilograms meter per second (kg · m/s)
pound foot, square (lb · ft²)	0.042 140 1	kilograms meter squared (kg · m²)
pound-force	4.448 222	newtons (N)
pound-force foot	1.355 818	newton meters (N · m)
pound-force inch	0.112 984 8	newton meters (N · m)
pound-force per square inch	6.894 757	kilopascals (kPa)
pound-force per square foot	0.047 880 26	kilopascals (kPa)
pound per cubic foot (lb/ft³)	16.018 46	kilograms per cubic meter (kg/m³)
pound per foot (lb/ft)	1.488 16	kilograms per meter (kg/m)
pound per foot, square (lb/ft²)	4.882 43	kilograms per meter squared (kg/m²)
pound per foot, cubic (lb/ft³)	16.018 5	kilograms per meter cubed (kg/m³)
pound per gallon (U.S. liquid)	119.826 4	grams per liter (g/L)
pound per second (lb/s)	0.453 592	kilograms per second (kg/s)
pound inch, square (lb · in.²)	292.640	kilograms millimeter squared (kg · mm²)
pound per mile	0.281 849	grams per meter (g/m)
pound square foot (lb · ft²)	0.042 140 11	kilograms square meter (kg/m²)
pound per cubic yard (lb/yd³)	0.593 276	kilograms per cubic meter (kg/m³)
quart (U.S. liquid)	946.353	milliliters (mL)
square centimeter (cm²)	197 300	circular mils
square centimeter (cm²)	0.001 076	square feet (ft²)
square centimeter (cm²)	0.155	square inches (in²)
square centimeter (cm²)	0.000 1	square meters (m²)
square centimeter (cm²)	0.000 119 6	square yards (yd²)
square foot (ft²)	144	square inches (in.²)
square foot (ft²)	0.092 903 04	square meters (m²)
square foot (ft²)	0.111 1	square yards (yd²)
square inch (in.²)	1 273 000	circular mils
square inch (in.²)	6.451 6	square centimeters (cm²)
square inch (in.²)	0.006 944	square feet (ft²)
square inch (in.²)	645.16	square millimeters (mm²)
square inch (in.²)	1 000 000	square mils
square meter (m²)	10.764	square feet (ft²)
square meter (m²)	1 550	square inches (in.²)
square meter (m²)	0.000 000 386 1	square miles
square meter (m²)	1.196	square yards (yd²)
square mil	1.273	circular mils
square mil	0.000 001	square inches (in.²)
square mile	2.589 988	square kilometers (km²)
square millimeter (mm²)	1 973	circular mils
square yard (yd²)	0.836 127 4	square meters (m²)

(*continued*)

TABLE E-1

(*continued*)

Multiply This Unit(s)	By This Factor	To Obtain This Unit(s)
tablespoon (tbsp)	14.786 75	milliliters (mL)
teaspoon (tsp)	4.928 916 7	milliliters (mL)
therm	105.480 4	megajoules (MJ)
ton (long) (2,240 lb)	1 016.047	kilograms (kg)
ton (long) (2,240 lb)	1.016 047	metric tons (t)
ton, metric	2 204.62	pounds (avoirdupois)
ton, metric	1.102 31	tons, short (2,000 lb)
ton, refrigeration	12 000	Btus per hour
ton, refrigeration	4.716 095 9	horsepower-hours
ton, refrigeration	3.516 85	kilowatts (kW)
ton (short) (2,000 lb)	907.185	kilograms (kg)
ton (short) (2,000 lb)	0.907 185	metric tons (t)
ton per cubic meter (t/m^3)	0.842 778	tons per cubic yard (ton/yd^3)
ton per cubic yard (ton/yd^3)	1.186 55	tons per cubic meter (ton/m^3)
torr	133.322 4	pascals (Pa)
watt (W)	3.412 14	Btus per hour (Btu/hr)
watt (W)	0.001 341	horsepower
watt (W)	0.001	kilowatts (kW)
watt-hour (Wh)	3.413	Btus
watt-hour (Wh)	0.001 341	horsepower-hours
watt-hour (Wh)	0.001	kilowatt-hours (kW/hr)
yard	91.44	centimeters (cm)
yard	3	feet
yard	36	inches
yard	0.000 914 4	kilometers (km)
yard	0.914 4	meters (m)
yard	914.4	millimeters
yard, cubic (yd^3)	0.764 55	cubic meters (m^3)
yard, square (yd^2)	0.836 127	square meter (m^2)

*These terms are no longer used, but may still be found in some publications.

Appendix F: Glossary

ELECTRICAL TERMS COMMONLY USED IN THE APPLICATION OF THE *NEC*

The following terms are used extensively in the application of the *National Electrical Code*. When added to the terms related to electric phenomena they constitute a "culture literacy" for the electrical construction industry. Many of these definitions are quotes or paraphrases of definitions given by the *NEC*; these have been marked with an *.

Accent Lighting: Directional lighting to emphasize a particular object or to draw attention to a part of the field in view.

Accessible (equipment):* Admitting close approach; not guarded by locked doors, elevation, or other effective means.

Accessible (wiring methods):* Capable of being removed or exposed without damaging the building structure or finish, or not permanently closed in by the structure or finish of the building.

Accessible, Readily:* Capable of being reached quickly for operation, renewal, or inspections, without requiring those to whom ready access is requisite to climb over or remove obstacles or to resort to portable ladders, and so forth.

Addendum: Modification (change) made to the construction documents (plans and specifications) during the bidding period.

Adjustment Factor: A multiplier that is used to reduce the ampacity of a conductor when there are more than three current-carrying conductors in a raceway or cable installed without maintaining spacing for a continuous length longer than 24 in. (600 mm). See *NEC 310.15(B)(3)(a)*.

AL/CU: Terminal marking on switches and receptacles rated 30 amperes and greater, suitable for use with aluminum, copper, and copper-clad aluminum conductors. If not marked, suitable for copper conductors only.

Allowable Ampacity: The values given in *NEC Table 310.15(B)(16), Table 310.15(B)(17), Table 310.15(B)(18),* and *Table 310.15(B)(19)*.

Alternate Bid: Amount stated in the bid to be added or deducted from the base bid amount proposed for alternate materials and/or methods of construction.

Ambient Lighting: Lighting throughout an area that produces general illumination.

American National Standards Institute (ANSI): An organization that identifies industrial and public requirements for national consensus standards and coordinates and manages their development, resolves national standards problems, and ensures effective participation in international standardization. ANSI does not itself develop standards. Rather, it facilitates development by establishing consensus among qualified groups. ANSI ensures that the guiding principles—consensus, due process, and openness—are followed.

Ampacity:* The maximum current in amperes that a conductor can carry continuously under the conditions of use without exceeding its temperature rating.

Ampere: The measurement of intensity of rate of flow of electrons in an electric circuit. An ampere is the amount of current through a resistance of 1 ohm under a pressure of 1 volt.

Appliance:* Utilization equipment, generally other than industrial, normally built in standardized sizes or types, that is installed or connected as a unit to perform one or more functions such as clothes washing, air conditioning, food mixing, deep frying, and so forth.

Approved:* Acceptable to the authority having jurisdiction.

Arc-Fault Circuit-Interrupter (AFCI):* A device intended to provide protection from the effects of arc faults by recognizing characteristics unique to arcing and by functioning to de-energize a circuit when an arcing fault is detected.

Architect: One who designs and supervises the construction of building or other structures.

As-Built Drawings: Contract drawings that reflect changes made during the construction process.

As-Built Plans: When required, a modified set of working drawings prepared for a construction project that includes all variances from the original working drawings that occurred during the project construction.

Authority Having Jurisdiction (AHJ): The organization, office, or individual responsible for approving equipment, materials, an installation, or a procedure. An AHJ is usually a governmental body that has legal jurisdiction over electrical installations. The AHJ has the responsibility for making interpretations of the rules, for deciding upon the approval of equipment and materials, and for granting the special permission where required. Where a specific interpretation or deviation from the *NEC* is given, get it in writing. See *90.4* and *Annex H* of the *NEC*.

Average Rated Life (lamp): How long it takes to burn out the lamp. For example, a 60-watt lamp is rated 1000 hours. The 1000-hour rating is based on the point in time when 50% of a test batch of lamps burns out and 50% are still burning.

Backlight: Illumination from behind a subject directed substantially parallel to a vertical plane through the optical axis of the camera.

Ballast: Used for energizing fluorescent lamps. It is constructed of a laminated core and coil windings or solid-state electronic components.

Ballast Efficacy: The total amount of light energy (lumens) emitted from a light source divided by the total lamp and ballast power (watts) input, expressed in lumens per watt.

Ballast Factor: A measure of the light output (lumens) of a ballast and lamp combination in comparison to an ANSI Standard "reference" ballast operated with the same lamp. It is a measure of how well the ballast performs when compared to the reference ballast.

Ballast Power Factor: A measure of how efficiently the lighting system utilizes electric current. It is the ratio of input watts to the input volt-amperes.

Bare Lamp: A light source with no shielding or guarding.

▶**Bathroom:** An area including a basin with one or more of the following: a toilet, a urinal, a tub, a shower, a bidet, or similar plumbing fixtures.◀

Bid: An offer or proposal of a price.

Bid Bond: A written form of security executed by the bidder as principal and by a surety for the purpose of guaranteeing that the bidder will sign the contract, if awarded the contract, for the stated bid amount.

Bid Price: The stipulated sum stated in the bidder's bid.

Bonded (Bonding): Connected to establish electrical continuity and conductivity.

▶**Bonding Conductor or Jumper:*** A conductor to ensure the required electrical conductivity between metal parts required to be electrically connected.◀

Bonding Jumper, Equipment:* The connection between two or more portions of the equipment grounding conductor.

Bonding Jumper, Main:* The connection between the grounded circuit conductor and the equipment grounding conductor at the service.

▶**Bonding Jumper, Supply-Side:*** A conductor installed on the supply side of a service or within a service equipment enclosure(s), or for a separately derived system, that ensures the required electrical conductivity between metal parts required to be electrically connected.◀ See *NEC 250.2*.

Bonding Jumper, System:* The connection between the grounded circuit conductor and the equipment grounding conductor at a separately derived system.

Branch Circuit:* The circuit conductors between the final overcurrent device protecting the circuit and the outlet(s).

Branch Circuit, Appliance:* A branch circuit supplying energy to one or more outlets to which appliances are to be connected; such circuits are to have no permanently connected luminaires not a part of an appliance.

Branch Circuit, General Purpose:* A branch circuit that supplies two or more receptacles or outlets for lighting and appliances.

Branch Circuit, Individual:* A branch circuit that supplies only one utilization equipment.

Branch Circuit, Multiwire:* A branch circuit consisting of two or more ungrounded conductors that have a voltage between them, and a grounded conductor that has equal voltage between it and each ungrounded conductor of the circuit and that is connected to the neutral or grounded conductor of the system.

Branch-Circuit Selection Current:* The value in amperes to be used instead of the rated-load current in determining the ratings of motor branch-circuit conductors; disconnecting means; controllers; and branch-circuit short-circuit and ground-fault protective devices wherever the running overload protective device permits a sustained current greater than the specified percentage of the rated-load current. The value of branch-circuit selection current will always be equal to or greater than the marked rated-load current; see *NEC 440-2.*

Brightness: In common usage, the term *brightness* usually refers to the strength of sensation that results from viewing surfaces or spaces from which light comes to the eye.

Building:* A structure that stands alone or is cut off from adjoining structures by fire walls, with all openings therein protected by approved fire doors.

Building Code: The legal requirements set up by the prevailing various governing agencies covering the minimum acceptable requirements for all types of construction.

Building Inspector/Official: A qualified government representative authorized to inspect construction for compliance with applicable building codes, regulations, and ordinances.

Building Permit: A written document issued by the appropriate governmental authority permitting construction to begin on a specific project in accordance with drawings and specifications approved by the governmental authority.

Candela: The international unit (SI) of luminous intensity. Formerly referred to as candle, as in candlepower.

Candlepower: A measure of intensity mathematically related to lumens.

Carrier: A wave having at least one characteristic that may be varied from a known reference value by modulation.

Carrier Current: The current association with a carrier wave.

CC: A marking on wire connectors and soldering lugs indicating they are suitable for use with copper-clad aluminum conductors only.

CC/CU: A marking on wire connectors and soldering lugs indicating they are suitable for use with copper or copper-clad aluminum conductors only.

Change Order: A written document between the owner and the contractor signed by the owner and the contractor authorizing a change in the work or an adjustment in the contract sum or the contract time. A contract sum and the contract time may be changed only by a change order.

Circuit Breaker:* A device designed to open and close a circuit by nonautomatic means and to open the circuit automatically on a predetermined overcurrent without damage to itself when properly applied within its rating.

CO/ALR: Terminal marking on switches and receptacles rated 15 and 20 amperes that are suitable for use with aluminum, copper, and copper-clad aluminum conductors. If not marked, they are suitable for copper or copper-clad conductors only.

Conductor:*

> **Bare:** A conductor having no covering or electrical insulation whatsoever. (See "Conductor, Covered.")
>
> **Covered:** A conductor encased within material of composition or thickness that is not recognized by this *Code* as electrical insulation. (See "Conductor, Bare.")
>
> **Insulated:** A conductor encased within material of composition and thickness that is recognized by this *Code* as electrical insulation.

Continuous Load:* A load where the maximum current is expected to continue for 3 hours or more.

Contract: An agreement between two or more parties, especially one that is written and enforceable by law.

Contract Modifications: After the agreement has been signed, any additions, deletions, or modifications of the work to be done are accomplished by change order, supplemental instruction, and field order. They can be issued at any time during the contract period.

Contractor: A properly licensed individual or company that agrees to furnish labor, materials, equipment, and associated services to perform the work as specified for a specified price.

Coordination (Selective):* Localization of an overcurrent condition to restrict outages to the equipment affected, accomplished by the choice of overcurrent-protective devices or settings.

Correction Factor: A multiplier that is used to modify the allowable ampacity of a conductor when the ambient temperature is less or greater than 86°F (30°C). These multipliers are found in *NEC Table 310.15(B) (2)(a).* Sometimes referred to as a derating factor.

CSA: Canadian Standards Association. CSA that develops safety and performance standards in Canada for electrical products similar to but not always identical to those of UL (Underwriters Laboratories) in the United States.

CU: A marking on wire connectors and soldering lugs indicating they are suitable for use with copper conductors only.

Current: The flow of electrons through an electrical circuit, measured in amperes.

Derating Factor: See "Adjustment Factor" and "Correction Factor."

Details: Plans, elevations, or sections that provide more specific information about a portion of a project component or element than smaller scale drawings.

Diagrams: Nonscaled views showing arrangements of special system components and connections not possible to clearly show in scaled views. A *schematic diagram* shows circuit components and their electrical connections without regard to actual physical locations. A *wiring diagram* shows circuit components and the actual electrical connections.

Dimmer: A switch with components that permits variable control of lighting intensity. Some dimmers have electronic components; others have core and coil (transformer) components.

Direct Glare: Glare resulting from high luminances or insufficiently shielded light sources in the field of view. Usually associated with bright areas, such as luminaires, ceilings, and windows that are outside the visual task of region being viewed.

Disconnecting Means:* A device or group of devices, or other means by which the conductors of a circuit can be disconnected from their source of supply.

Downlight: A small, direct lighting unit that guides the light downward. It can be recessed, surface mounted, or suspended.

Drawings: Graphic representations of the work to be done. They show the relationship of the materials to each other, including sizes, shapes, locations, and connections. The drawings may include schematic diagrams showing such things as mechanical and electrical systems. They may also include schedules of structural elements, equipment, finishes, and other similar items.

Dry Niche Luminaire:* A luminaire intended for installation in the wall of a pool or fountain in a niche that is sealed against the entry of pool water.

Dwelling Unit:* A single unit, providing complete and independent living facilities for one or more persons, including permanent provisions for living, sleeping, cooking, and sanitation.

Effective Ground-Fault Current Path: See "Ground-Fault Current Path, Effective."

Efficacy: The total amount of light energy (lumens) emitted from a light source divided by the total lamp and ballast power (watts) input, expressed in lumens per watt.

Elevations: Views of vertical planes, showing components in their vertical relationship, viewed perpendicularly from a selected vertical plane.

Equipment:* A general term including fittings, devices, appliances, luminaires, apparatus, machinery, and the like used as a part of, or in connection with, an electrical installation.

Feeder:* All circuit conductors between the service equipment or the source of a separately derived system and the final branch-circuit overcurrent device.

Fill Light: Supplementary illumination to reduce shadow or contract range.

Fire Rating: The classification indicating in time (hours) the ability of a structure or component to withstand fire conditions.

Flame Detector: A radiant energy-sensing fire detector that detects the radiant energy emitted by a flame.

Fluorescent Lamp: A lamp in which electric discharge of ultraviolet energy excites a fluorescing coating (phosphor) and transforms some of that energy to visible light.

Footcandle: The unit used to measure how much total light is reaching a surface. One lumen falling on 1 square foot of surface produces an illumination of 1 footcandle.

Fully Rated System: All overcurrent devices installed have an interrupting rating greater than or equal to the specified available fault current. (See "Series-Rated System.")

Fuse: An overcurrent protective device with a fusible link that operates and opens the circuit on an overcurrent condition.

General Conditions: A written portion of the contract documents set forth by the owner stipulating the contractor's minimum acceptable performance requirements, including the rights, responsibilities, and relationships of the parties involved in the performance of the contract. General conditions are usually included in the book of specifications but are sometimes found in the architectural drawings.

General Lighting: Lighting designed to provide a substantially uniform level of illumination throughout an area, exclusive of any provision for special lighting.

Glare: The sensation produced by luminance within the visual field that is sufficiently greater than the luminance to which the eyes are adapted to cause annoyance, discomfort, or loss in visual performance and visibility.

Ground:* The earth.

Ground Fault:* An unintentional, electrically conducting connection between a normally current-carrying conductor of an electrical circuit, and the normally non-current-carrying conductors, metallic enclosures, metallic raceways, metallic equipment, or earth.

Ground-Fault Circuit-Interrupter (GFCI):* A device intended for the protection of personnel that functions to de-energize a circuit or portion thereof within an established period of time when a current to ground exceeds the values established for a Class A device.

▶**Informational Note:** Class A ground-fault circuit-interrupters trip when the current to ground is 6 mA or higher and do not trip when the current to ground is less than 4 mA. For further information, see UL 943, *Standard for Ground-Fault Circuit Interrupters.*◀

Ground-Fault Current Path:* An electrically conductive path from the point of a ground fault on a wiring system through normally non-current-carrying conductors, equipment, or the earth to the electrical supply source. This definition is found in *NEC 250.2.*

Ground-Fault Current Path, Effective:* An intentionally constructed, low-impedance electrically conductive path designed and intended to carry current under ground-fault conditions from the point of a ground fault on a wiring system to the electrical supply source and that facilitates the operation of the overcurrent protective device or ground-fault detectors on high-impedance grounded systems; see *NEC 250.2* and *250.4(A)(5).*

Ground-Fault Protection of Equipment:* A system intended to provide protection of equipment from damaging line-to-ground fault currents by operating to cause a disconnecting means to open all ungrounded conductors of the faulted circuit. This protection is provided at current levels less than those required to protect conductors from damage through the operation of a supply circuit overcurrent device.

Grounded (Grounding):* Connected (connecting) to ground or to a conductive body that extends the ground connection.

Grounded Conductor:* A conductor used to connect equipment or the grounded circuit of a wiring system to a grounding electrode or electrodes.

Grounded, Solidly:* Connected to ground without inserting any resistor or impedance device.

▶**Grounding Conductor, Equipment (EGC):*** The conductive path(s) installed to connect normally non-current carrying metal parts of equipment together and to the system grounded conductor or to the grounding electrode conductor.◀

Grounding Electrode:* A conducting object through which a direct connection to earth is established.

Grounding Electrode Conductor:* A conductor used to connect the system grounded conductor or the equipment to a grounding electrode or to a point on the grounding electrode system.

HACR: A circuit breaker that has been tested and found suitable for use on heating, air-conditioning, and refrigeration equipment. The circuit breaker is marked with the letters *HACR.* Equipment that is suitable for use with HACR circuit breakers will be marked for use with HACR-type circuit breakers.

Halogen Lamp: An incandescent lamp containing a halogen gas that recycles tungsten back onto the tungsten filament surface. Without the halogen gas, the tungsten would normally be deposited onto the bulb wall.

Heat Alarm: A single- or multiple-station alarm responsive to heat.

Hermetic Refrigerant Motor-Compressor: A combination consisting of a compressor and motor, both of which are enclosed in the same housing, with no external shaft or shaft seals; the motor operates in the refrigerant. This definition is found in *Article 440.*

High-Intensity Discharge Lamp (HID): A general term for a mercury, metal halide, or high-pressure sodium lamp.

High-Pressure Sodium Lamp: A high-intensity discharge light source in which the light is primarily produced by the radiation from sodium vapor.

Horsepower (tools): Horsepower is a measurement of motor torque multiplied by speed. Horsepower also refers to the rate of work (power) an electric motor is capable of delivering. (See "Torque," "Horsepower, Rated," and "Horsepower, Maximum Developed.")

Horsepower, Maximum Developed: If a motor is required to work harder than its idle speed, the motor becomes overloaded and must develop extra horsepower. The most horsepower that can be drawn from a motor to handle this extra effort is referred to as its maximum developed horsepower. (See "Horsepower," "Torque," and "Horsepower, Rated.")

Horsepower, Rated: Rated horsepower is a motor's running torque at its rated running speed. The motor can be run continuously at its rated horsepower without overheating. If the motor is required to give an extra spurt of effort while running, the motor becomes overloaded and develops extra horsepower to compensate. The most horsepower that can be drawn from a motor to handle this extra effort is its maximum developed horsepower. (See "Torque," "Horsepower," and "Horsepower, Maximum Developed.")

Household Fire Alarm System: A system of devices that produces an alarm signal in the household for the purpose of notifying the occupants of the presence of a fire so that they will evacuate the premises.

Identified:* (as applied to equipment) Recognizable as suitable for the specific purpose, function, use, environment, application, and so on, where described in a particular *Code* requirement.

Identified* (conductor): The insulated grounded conductor. For sizes 6 AWG or smaller, the insulated grounded conductor shall be identified by a continuous white or gray outer finish or by three continuous white stripes on other than green insulation along its entire length. For sizes larger than 6 AWG, the insulated grounded conductor shall be identified either by a continuous white or gray outer finish or by three continuous white stripes on other than green insulation along its entire length, or at the time of installation by a distinctive white or gray marking at its terminations. This marking is required to encircle the insulation.

Identified* (terminal): The identification of terminals to which a grounded conductor is to be connected shall be substantially white in color. The identification of other terminals shall be of a readily distinguishable different color; see *NEC 200.9*.

IEC: The International Electrotechnical Commission is a worldwide standards organization. These standards may differ from those of Underwriters Laboratories. Some standards have been harmonized with the UL Standards. Some electrical equipment might conform to a specific IEC standard, but might not conform to the UL standard for the same item.

Illuminance: The amount of light energy (lumens) distributed over a specific area expressed as footcandles (lumens/square foot) or luce (lumens/square meter).

Illumination: The act of illuminating or the state of being illuminated.

Immersion Detection Circuit Interrupter (IDCI): A device integral with grooming appliances that will shut off the appliance when it is dropped in water.

In Sight From:* Where this *Code* specifies that one equipment shall be *in sight from*, *within sight from*, or *within sight*, and so on, of another equipment, the specified equipment is to be visible and not more than 50 ft (15 m) from the other.

Incandescent Filament Lamp: A lamp that provides light when a filament is heated to incandescence by an electric current. Incandescent lamps are the oldest form of electric lighting technology.

Indirect Lighting: Lighting by luminaire that distributes 90–100% of the emitted light upward.

Inductive Load: A load that is made up of coiled or wound wire that creates a magnetic field when energized. Transformers, core and coil ballasts, motors, and solenoids are examples of inductive loads.

▶**Informational Note:*** Explanatory material, such as references to other standards, references to related sections of the *NEC*, or information related to a *Code* rule, is included in this *Code* in the form of *Informational Notes*. Such notes are informational only and are not enforceable as requirements of this *Code*.◀ See *NEC 90.5(C)*.

Instant Start: A circuit used to start specially designed fluorescent lamps without the aid of a starter. To strike the arc instantly, the circuit utilizes higher open-circuit voltage than is required for the same length preheat lamp.

International Association of Electrical Inspectors (IAEI): A not-for-profit educational organization cooperating in the formulation and uniform application of standards for the safe installation and use of electricity, and collecting and disseminating information relative thereto. The IAEI is made up of electrical inspectors, electrical contractors, electrical apprentices, manufacturers, and governmental agencies.

Interrupting Rating:* The highest current at rated voltage that a device is identified to interrupt under standard test conditions.

►**Intersystem Bonding Termination:*** A device that provides a means for connecting bonding conductors for communications systems to the grounding electrode system.◄

Isolated Ground Receptacle: A grounding-type device in which the equipment ground contact and terminal are electrically isolated from the receptacle mounting means.

Joule: The work required to continuously produce one *watt* of *power* for one *second*.

Kilowatt (kW): One thousand watts equals 1 kilowatt.

Kilowatt-hour (kWh): One thousand watts of power in 1 hour. One 100-watt lamp burning for 10 hours constitutes 1 kilowatt-hour. Two 500-watt electric heaters operating for 1 hour constitutes 1 kilowatt-hour.

Kitchen: An area with a sink and permanent provisions for food preparation and cooking.

Labeled:* Equipment or materials to which has been attached a label, symbol, or other identifying mark of an organization that is acceptable to the authority having jurisdiction and concerned with product evaluation, which maintains periodic inspection of production of labeled equipment or materials and by whose labeling the manufacturer indicates compliance with appropriate standards or performance in a specified manner.

Lamp: A generic term for a man-made source of light.

Light: The term generally applied to the visible energy from a source. Light is usually measured in lumens (a unit of luminous flux in terms of candelas), candelas (a unit of luminous intensity), or candlepower (luminous intensity expressed in candelas). When light strikes a surface, it is either absorbed, reflected, or transmitted.

Lighting Outlet:* An outlet intended for the direct connection of a lampholder, a luminaire, or a pendant cord terminating in a lampholder.

Listed:* Equipment, materials, or services included in a list published by an organization that is acceptable to the authority having jurisdiction and concerned with evaluation of products or services, that maintains periodic inspection of production of listed equipment or materials, or periodic evaluation of services, and whose listing states that either the equipment, material, or services meets appropriate designated standards or has been tested and found suitable for a specified purpose.

Load: The electric power used by devices connected to an electrical system. Loads can be figured in amperes, volt-amperes, kilovolt-amperes, or kilowatts. Loads can be intermittent, continuous, periodic, short time, or varying. See the definition of "Duty" in the *NEC*.

Load Center: A common name for panelboards. A load center may not be as deep as a panelboard, and generally does not contain relays or other accessories as are available for panelboards. Circuit breakers are "plug-in" as opposed to "bolt-in" types used in panelboards. Manufacturers' catalogs will show both load centers and panelboards. The UL standards do not differentiate.

Location, Damp:* Locations protected from weather and not subject to saturation with water or other liquids but subject to moderate degrees of moisture. Examples of such locations include partially protected locations under canopies, marquees, roofed open porches, and like locations, and interior locations subject to moderate degrees of moisture, such as some basements, some barns, and some cold-storage warehouses.

Location, Dry:* A location not normally subject to dampness or wetness. A location classified as dry may be temporarily subject to dampness or wetness, as in the case of a building under construction.

Location, Wet:* Installations underground or in concrete slabs or masonry in direct contact with the earth; in locations subject to saturation with water or other liquids, such as vehicle washing areas; and in unprotected locations exposed to weather.

Locked-Rotor Current: The steady-state current taken from the line with the rotor locked and with rated voltage and frequency applied to the motor.

Low-Pressure Sodium Lamp: A discharge lamp in which light is produced from sodium gas operating at a partial pressure.

Lumens: The SI unit of luminous flux. The units of light energy emitted from the light source.

Luminaire:* A complete lighting unit consisting of a light source such as a lamp or lamps, together with the parts designed to position the light source and connect it to the power supply. It may also include parts to protect the light source, ballast, or distribute the light. A lampholder itself is not a luminaire. Prior to the *National Electrical Code* adopting the international definition of *luminaire*, the commonly used term in the United States was and in most instances still is *lighting fixture*.

Luminaire Efficiency: The total lumen output of the luminaire divided by the rated lumens of the lamps inside the luminaire.

Lux: The SI (International System) unit of illumination. One lumen uniformly distributed over an area of 1 square meter.

Maximum Overcurrent Protection (MOP): A term used with equipment that has a hermetic motor compressor(s). MOP is the maximum ampere rating for the equipment's branch-circuit overcurrent protective device. The nameplate will specify the ampere rating and type (fuses or circuit breaker) of overcurrent protective device to use.

Mercury Lamp: A high-intensity discharge light source in which radiation from the mercury vapor produces visible light.

Metal Halide Lamp: A high-intensity discharge light source in which the light is produced by the radiation from mercury together with halides of metals such as sodium and scandium.

Minimum Supply Circuit Conductor Ampacity (MCA): A term used with equipment that has a hermetic motor compressor(s). MCA is the minimum ampere rating value used to determine the proper size conductors, disconnecting means, and controllers. MCA is determined by the manufacturer, and is marked on the nameplate of the equipment.

National Electrical Code **(NEC):** The electrical code published by the National Fire Protection Association. This *Code* provides for practical safeguarding of persons and property from hazards arising from the use of electricity. It does not become law until adopted by Federal, state, and local laws and regulations. The *NEC* is not intended to be a design specification or an instruction manual for untrained persons.

National Electrical Manufacturers Association (NEMA): NEMA is a trade organization made up of many manufacturers of electrical equipment. They develop and promote standards for electrical equipment.

National Fire Protection Association (NFPA): Located in Quincy, MA. The NFPA is an international standards making organization dedicated to the protection of people from the ravages of fire and electric shock. The NFPA is responsible for administrating the process for developing the *National Electrical Code, Installation of Sprinkler Systems, Life Safety Code, National Fire Alarm Code*, and over 295 other codes, standards, and recommended practices.

The NFPA may be contacted at (800) 344-3555 or on their Web site at http://www.nfpa.org.

Nationally Recognized Testing Laboratory (NRTL): The term used to define a testing laboratory that has been recognized by OSHA; for example, Underwriters Laboratories.

Neutral Conductor:* The conductor connected to the neutral point of a system that is intended to carry current under normal conditions.

Neutral Point:* The common point on a wye connection in a polyphase system or midpoint on a single-phase, 3-wire system, midpoint of a single-phase portion of a 3-phase delta system, or midpoint of a 3-wire, direct-current system.

No Niche Luminaire:* A luminaire intended for installation above or below the water without a niche; see *NEC 680.2.*

Noncoincidental Loads: Loads that are not likely to be on at the same time. Heating and cooling loads would not operate at the same time; see *NEC 220.60.*

Notations: Words on working drawings, providing specific information.

Occupational Safety and Health Administration (OSHA): OSHA is a division of the U.S. Department of Labor. OSHA develops, administers, and enforces regulations relating to safety in the workplace. Electrical regulations are covered in *Part 1910, Subpart S*. The *NEC* must still be referred to, in conjunction with OSHA regulations.

Ohm: A unit of measure for electric resistance. An ohm is the amount of resistance that will allow 1 ampere to flow under a pressure of 1 volt.

Outlet:* A point on the wiring system at which current is taken to supply utilization equipment.

Overcurrent:* Any current in excess of the rated current of equipment or the ampacity of a conductor. It may result from overload, short circuit, or ground fault.

Overcurrent Device: Also referred to as an overcurrent protection device. A form of protection that operates when current exceeds a predetermined value. Common forms of overcurrent devices are circuit breakers, fuses, and thermal overload elements found in motor controllers.

Overload:* Operation of equipment in excess of normal, full-load rating, or of a conductor in excess of rated ampacity that, when it persists for a sufficient length of time, would cause damage or dangerous overheating. The current is contained in its normal path. A fault, such as a short circuit or ground fault, is not an overload.

Owner: An individual or corporation that owns a real property.

Panelboard:* A single panel or group of panel units designed for assembly in the form of a single panel; including buses, automatic overcurrent devices, and equipped with or without switches for the control of light, heat, or power circuits; designed to be placed in a cabinet or cutout box placed in or against a wall or partition and accessible only from the front.

Performance Bond: (1) A written form of security from a surety (bonding) company to the owner, on behalf of an acceptable prime or main contractor or subcontractor, guaranteeing payment to the owner in the event the contractor fails to perform all labor, materials, equipment, or services in accordance with the contract. (2) The surety companies generally reserve the right to have the original prime or main or subcontractor remedy any claims before paying on the bond or hiring other contractors.

Photometry: The pattern and amount of light that is emitted from a luminaire, normally represented as a cross-section through the fixture distribution pattern.

Plan Checker or Reviewer: A term sometimes used to describe a building department official who examines construction documents for compliance with applicable *Code* requirements.

Plans: See "Working Drawings."

Power Factor: A ratio of actual power (W or kW) being used to apparent power (VA or kVA) being drawn from the power source.

Power Supply: A source of electrical operating power including the circuits and terminations connecting it to the dependent system components.

Preheat Fluorescent Lamp Circuit: A circuit used on fluorescent lamps wherein the electrodes are heated or warmed to a glow stage by an auxiliary switch or starter before the lamps are lighted.

Preliminary Drawings: The drawings that precede the final approved drawings. Usually stamped "PRELIMINARY."

Prime Contractor: Any contractor having a contract directly with the owner. Usually the main (general) contractor for a specific project.

Project: A word used to represent the overall scope of work being performed to complete a specific construction job.

Proposal: A written offer from a bidder to the owner, preferably on a prescribed proposal form, to perform the work and to furnish all labor, materials, equipment, and/or service for the prices and terms quoted by the bidder.

Punch List (Inspection List): A list of items of work requiring immediate corrective or completion action by the contractor prepared by the owner or an authorized representative.

Qualified Person:* One who has skills and knowledge related to the construction and operation of the electrical equipment and installations and has received safety training to recognize and avoid the hazards involved.

Raceway:* An enclosed channel of metal or nonmetallic materials designed expressly for holding wires, cables, or busbars, with additional functions as permitted in this *Code*. Raceways include, but are not limited to, rigid metal conduit, rigid nonmetallic conduit, intermediate metal conduit, liquidtight flexible conduit, flexible metallic tubing, flexible metal conduit, electrical nonmetallic tubing, electrical metallic tubing, underfloor raceways, cellular concrete floor raceways, cellular metal floor raceways, surface raceways, wireways, and busways.

Rapid-Start Fluorescent Lamp Circuit: A circuit designed to start lamps by continuously heating or preheating the electrodes. Lamps must be designed for this type of circuit. This is the modern version of the "trigger start" system. In a rapid-start two-lamp circuit, one end of each lamp is connected to a separate starting winding.

Rated-Load Current: The rated-load current for a hermetic refrigerant motor-compressor is the current resulting when the motor-compressor is operated at the rated load, rated voltage, and rated frequency of the equipment it serves. This definition is found in *Article 440*.

Rate of Rise Detector: A device that responds when the temperature rises at a rate exceeding a predetermined value.

Recognized: Refers to a product or component that is incomplete in construction features or limited in performance capabilities. A "Recognized" product is intended to be used as a component part of equipment that has been "listed." A "Recognized" product must not be used by itself. A listed product may contain a number of components that are been "Recognized."

Receptacle:* A contact device installed at the outlet for the connection of a single contact device. A single receptacle is a single contact device with no other contact device on the same yoke. A multiple receptacle is two or more contact devices on the same yoke.

Receptacle Outlet:* An outlet where one or more receptacles are installed.

Resistive Load: An electric load that produces heat. Some examples of resistive loads are the elements in an electric range, ceiling heat cables, and electric baseboard heaters.

Root-Mean-Square (RMS): The square root of the average of the square of the instantaneous values of current or voltage. It is the effective or heating value of a sinusoidal waveform. For example, assume the RMS value of voltage line-to-neutral is 120 volts. During each electrical cycle, the voltage rises from zero to a peak value ($120 \times 1.4142 = 169.7$ volts), back through zero to a negative peak value, then back to zero. The RMS value is 0.707 of the peak value ($169.7 \times 0.707 = 120$ volts). The root-mean-square value is what an electrician reads on an ammeter or voltmeter.

Schedules: Tables or charts that include data about materials, products, and equipment.

Scope: A written range of view or action; outlook; hence, room for the exercise of faculties or function; capacity for achievement; all in connection with a designated project.

Sections: Views of vertical cuts through and perpendicular to components, showing their detailed arrangement.

▶**Separately Derived System:*** A premises wiring system whose power is derived from a source of electric energy or equipment other than a service. Such systems have no direct connection from circuit conductors of one system to circuit conductors of another system, other than connections through the earth, metal enclosures, metallic raceways, or equipment grounding conductors. ◀

Series-Rated System: Where a circuit breaker is used on a circuit having an available fault current higher than the marked interrupting rating by being connected on the load side of an acceptable overcurrent protective device having a higher rating, the circuit breakers are considered to be a series-rated system. A series-rated system can be either a tested combination of the main circuit breaker or fuse, and downstream circuit breaker in accordance to UL *Standard 489* or a system selected under engineering supervision in existing installations. (See "Fully Rated System.")

Service:* The conductors and equipment for delivering energy from the serving utility to the wiring system of the premises served.

Service Conductors:* The conductors from the service point to the service disconnecting means.

▶**Service Conductors, Overhead:*** The overhead conductors between the service point and the first point of connection to the service-entrance conductors at the building or other structure. ◀

▶**Service Conductors, Underground:*** The underground conductors between the service point and the first point of connection to the service-entrance conductors in a terminal box, meter or other enclosure, inside or outside the building wall.

Informational Note: Where there is no terminal box, meter, or other enclosure, the point of connection is considered to be the point of entrance of the service conductors into the building. ◀

▶**Service Drop:*** The overhead conductors between the utility distribution system and the service point. ◀

Service Equipment:* The necessary equipment, usually consisting of a circuit breaker or switch and fuses, and their accessories, located near the point of entrance of supply conductors to a building or other structure, or an otherwise defined area, and intended to constitute the main control and means of cutoff of the supply.

▶**Service Lateral:*** The underground conductors between the utility distribution system and the service point. ◀

Service Point:* A service point is the point of connection between the facilities of the serving utility and the premises wiring.

▶**Informational Note:** The service point can be described as the point of demarcation between where the serving utility ends and the premises wiring begins. The serving utility generally specifies the location of the service point based on their conditions of service. ◀

Shall: Indicates a mandatory requirement.

Short Circuit: A connection between any two or more conductors of an electrical system that allows current to travel along a path having essentially no (or a very low) electrical impedance. The current flow is outside of its intended path, thus the term *short circuit*. A short circuit is also referred to as a fault.

Should: Indicates a recommendation or that which is advised but not required.

Site: The place where a structure or group of structures was or is to be located (as in *construction site*).

Smoke Alarm: A single or multiple station alarm responsive to smoke.

Smoke Detector: A device that detects visible or invisible particles of combustion.

Sone: A sone is a unit of loudness equal to the loudness of a sound of 1 kilohertz at 40 decibels above the threshold of hearing of a given listener.

Specifications: Text setting forth details such as description, size, quality, performance, and workmanship, etc. Specifications that pertain to all of the construction trades involved might be subdivided into "General Conditions" and "Supplemental General Conditions." Further subdividing the specifications might be specific requirements for the various contractors such as electrical, plumbing, heating, masonry, and so forth. Typically, the electrical specifications are found in Division 16.

Split-Wired Receptacle: A receptacle that can be connected to two branch circuits. These receptacles may also be used so one receptacle is live at all times, and the other receptacle is controlled by a switch. The terminals on these receptacles usually have breakaway tabs so the receptacle can be used either as a split-wired receptacle or as a standard receptacle.

Standard Network Interface (SNI): A device usually installed by the telephone company at the demarcation point where its service leaves off and the customer's service takes over. This is similar to the service point for electrical systems.

Starter: A device used in conjunction with a ballast for the purpose of starting an electric discharge lamp. Sometimes used to refer to a motor controller.

Structure:* That which is built or constructed.

Subcontractor: A qualified contractor subordinate to the prime or main contractor.

Surface-Mounted Luminaire: A luminaire mounted directly on the ceiling.

Surge Arrester:* A protective device for limiting surge voltages by discharging or bypassing surge current; it also prevents continued flow of follow current while remaining capable of repeating these functions.

Surge-Protective Device (SPD):* A protective device for limiting transient voltages by diverting or limiting surge current; it also prevents continued flow of follow current while remaining capable of repeating these functions and is designated as follows:

Type 1: *Permanently connected SPDs intended for installation between the secondary of the service transformer and the line side of the service disconnect overcurrent device.*

Type 2: *Permanently connected SPDs intended for installation on the load side of the service disconnect overcurrent device, including SPDs located at the branch panel.*

Type 3: *Point of utilization SPDs.*

Type 4: *Component SPDs, including discrete components, as well as assemblies.*

Informational Note: *For further information on Type 1, Type 2, Type 3, and Type 4 SPDs, see UL 1449, Standard for Surge Protective Devices.*

Suspended (pendant) Luminaire: A luminaire hung from a ceiling by supports.

Switches:*

General-Use Switch: A switch intended for use in general distribution and branch circuits. It is rated in amperes, and it is capable of interrupting its rated current at its rated voltage.

Motor-Circuit Switch: A switch, rated in horsepower, capable of interrupting the maximum operating overload current of a motor of the same horsepower rating as the switch at the rated voltage.

Symbols: A graphic representation that stands for or represents another thing. A symbol is a simple way to show such things as lighting outlets, switches, and receptacles on an electrical working drawing. The American Institute of Architects has developed a comprehensive set of symbols that represents just about everything used by all building trades. The National Electrical Contractors Association has produced a *National Electrical Installation Standard NEIS 100, Symbols for Electrical Construction Drawings.* When an item cannot be shown using a symbol, then a more detailed explanation using a notation or inclusion in the specifications is necessary.

T & M: An abbreviation for a contracting method called Time and Material. A written agreement between the owner and the contractor wherein payment is based on the contractor's actual cost for labor, equipment, materials, and services plus a fixed add-on amount to cover the contractor's overhead and profit.

Task Lighting: Lighting directed to a specific surface or area that provides illumination for visual tasks.

Terminal: A screw or a quick-connect device where a conductor(s) is intended to be connected.

Thermocouple: A pair of dissimilar conductors so joined at two points that an electromotive force is developed by the thermoelectric effects when the junctions are at different temperatures.

Thermopile: More than one thermocouple connected together. The connections may be series, parallel, or both.

Torque: Torque is a measurement of rotation or turning force. Torque is measured in ounce inches (oz. in.), ounce feet (oz. ft.), and pound feet (lb. ft.). (See "Horsepower," "Horsepower, Rated," and "Horsepower, Maximum Developed.")

Troffer: A recessed lighting unit, usually long and installed with the opening flush with the ceiling. The term is derived from "trough" and "coffer."

UL: Underwriters Laboratories is an independent not-for-profit organization that develops standards, and tests electrical equipment to these standards.

Ungrounded:* Not connected to ground or a conductive body that extends the ground connection. The ungrounded conductor is generally referred to as the "hot" or "live" conductor.

Volt: The difference of electric potential between two points of a conductor carrying a constant current of 1 ampere, when the power dissipated between these points is equal to 1 watt. A voltage of 1 volt can push 1 ampere through a resistance of 1 ohm.

Volt-ampere: A unit of power determined by multiplying the voltage and current in a circuit. A 120-volt circuit carrying 1 ampere is 120 volt-amperes.

Voltage (nominal):* A nominal value assigned to a circuit or system for the purpose of conveniently designating its voltage class (e.g., 120/240 volts, 480Y/277 volts, 600 volts).

The actual voltage at which a circuit operates can vary from the nominal within a range that permits satisfactory operation of equipment.

Voltage (of a circuit):* The greatest root-mean-square (effective) difference of potential between any two conductors of the circuit concerned.

Voltage Drop: Also referred to as IR drop. Voltage drop is most commonly associated with conductors.

A conductor has resistance. When current is flowing through the conductor, a voltage drop will be experienced across the conductor. Voltage drop across a conductor can be calculated using Ohm's law: $E = I \times R$.

Voltage to Ground:* For grounded circuits, the voltage between the given conductor and that point or conductor of the circuit that is grounded; for ungrounded circuits, the greatest voltage between the given conductor and any other conductor of the circuit.

Watertight:* Constructed so that moisture will not enter the enclosure under specified test conditions.

Watt: A measure of true power. A watt is the power required to do work at the rate of 1 joule per second. Wattage is determined by multiplying voltage times amperes times the power factor of the circuit: $W = E \times I \times PF$.

Weatherproof:* Constructed or protected so that exposure to the weather will not interfere with successful operation. Rainproof, raintight, or watertight equipment can fulfill the requirements for weatherproof where varying weather conditions other than wetness, such as snow, ice, dust, or temperature extremes, are not a factor.

Wet Niche Luminaire:* A luminaire intended for installation in a forming shell mounted in a pool. This definition is found in *NEC Article 680*.

Working Drawings: Views of horizontal planes, showing components in their horizontal relationship. A set of construction drawings. Also called plans.

Work Order: A written order, signed by the owner or a representative, of a contractual status requiring performance by the contractor without negotiation of any sort.

Zoning: Restrictions of areas or regions of land within specific geographical areas based on permitted building size, character, and uses as established by governing urban authorities.

Appendix G: Web Sites

The following list of Internet World Wide Web sites has been included for your convenience. These are current as of time of printing. Web sites are a "moving target" with continual changes. We do our utmost to keep these Web sites current. If you are aware of Web sites that should be added, deleted, or that are different from those shown, please let us know. Most manufacturers will provide catalogs and technical literature upon request either electronically or by mail. You will be amazed at the amount of electrical and electronic equipment information that is available on these Web sites.

You can search the Web using a search engine such as http://www.google.com, http://www.dogpile.com, http://www.yahoo.com, or http://www.ask.com. Just type in a key word (for example: fuses) and you will be led to many Web sites relating to fuses. It's easy!

Web sites marked with an asterisk * are closely related to the "home automation" market.

ACCUBID
http://www.accubid.com

Acme Electric Corporation
http://www.acmepowerdist.com

Active Home*
http://www.X10-beta.com/activehome/

Active Power
http://www.activepower.com

Adalet
http://www.adalet.com

Advanced Cable Ties
http://www.actfs.com/

AEMC Instruments
http://www.aemc.com

AFC Cable Systems
http://www.afcweb.com

Air-Conditioning & Refrigeration Association
http://www.ahrinet.org/

Alcan Cable
http://www.cable.alcan.com

Allen-Bradley
http://www.ab.com

Allied Moulded Products
http://www.alliedmoulded.com

Allied Tube & Conduit
http://www.alliedtube.com

Alpha Wire
http://www.alphawire.com

The Aluminum Association, Inc.
http://www.aluminum.org

American Council for an Energy Efficient Economy
http://www.aceee.org

American Gas Association
http://www.aga.org

American Heart Association
http://www.americanheart.org

American Institute of Architects
http://www.aia.org

American Lighting Association
http://www.americanlightingassoc.com

American National Standards Institute
http://www.ansi.org

American Pipe & Plastics, Inc.
http://www.ampipe.com

American Power Conversion
http://www.apcc.com

American Public Power Association
http://www.appanet.org

American Society for Testing and Materials
http://www.astm.org

American Society of Heating, Refrigerating, and
Air-Conditioning Engineers
http://www.ashrae.org

American Society of Mechanical Engineers
http://www.asme.org

American Society of Safety Engineers
http://www.asse.org

Amprobe
http://www.amprobe.com

Anixter (data & telecommunications cabling)*
http://www.anixter.com

Appleton Electric Co.
http://www.appletonelec.com

ArcWear
http://www.arcwear.com

Arlington Industries, Inc.
http://www.aifittings.com

Arrow Fasteners
http://www.arrowfastener.com

Arrow Hart Wiring Devices
http://www.arrowhart.com

Assisted Living: The World of Assisted Technology
http://www.abledata.com

Associated Builders & Contractors
http://www.abc.org

Association of Cabling Professions
http://www.wireville.com

Association of Home Appliance Manufacturers
http://www.aham.org

Automated Home Technologies*
http://www.automatedhomenashville.com/

Automatic Switch Co. (ASCO)
http://www.asco.com

AVO Training Institute
http://www.avotraining.com

Baldor Motors & Drives
http://www.baldor.com

R. W. Beckett Corporation
(oil burner technical info.)
http://www.beckettcorp.com

Belden Wire & Cable Co.
http://www.belden.com

Bell, Hubbell Inc.
http://www.hubbell-bell.com

Bender
http://www.bender.org

Best Power
http://www.bestpower.com

BICC General
http://www.generalcable.com

BICSI (Building Industry Consulting Services
International, a telecommunications assoc.)*
http://www.bicsi.org

BidStreet
http://www.BidStreetUSA.com

Black and Decker
http://www.blackanddecker.com

The Blue Book (construction)
http://www.thebluebook.com

Bodine Company
http://www.bodine.com

boltswitch, inc.
http://www.boltswitch.com

Brady Worldwide, Inc.
http://www.whbrady.com

Bridgeport Fittings Inc.
http://www.bptfittings.com

Broan
http://www.broan.com

Bryant–Hubbell Inc.
http://www.hubbell-bryant.com

Burndy Electrical Products
http://www.fciconnect.com

Cable Telecommunications Engineers*
http://www.scte.org

Caddy
http://www.erico.com

Canadian Standards Association
http://www.csa-international.org

Cantex
http://www.cantexinc.com

Carlon, A Lamson & Sessions Co.
http://www.carlon.com

Carol Cable
http://www.generalcable.com

Casablanca Fan Company
http://www.casablancafanco.com

Caterpillar
http://www.cat.com

Caterpillar
http://www.cat-engines.com

CEBus Industry Council*
http://www.cebus.org

CEE News magazine
http://www.ceenews.com

Cengage Delmar Learning
http://www.delmarlearning.com

Cengage Delmar Learning (electrical)
http://www.delmarelectric.com

Center for Disease Control (CDC)
http://www.cdc.gov

Centralite*
http://www.centralite.com

CEPro*
http://www.ce-pro.com

Channellock Inc
http://www.channellock.com

Chromalox
http://www.chromalox.com

Circon Systems Corporation*
http://www.circon.com

Coleman Cable
http://www.colemancable.com

Columbia Lighting
http://www.columbia-ltg.com

CommScope* (coax cable mfg.)
http://www.commscope.com

Communications Systems, Inc.
http://www.commsystems.com

Computer and Telecommunications Page*
http://www.cmpcmm.com/cc/

ConnectHome*
http://www.connecthome.com

Construction Specifications Institute
http://www.csinet.org

Construction Terms/Glossary
http://www.constructionplace.com/glossary.asp

Consumer Electronic Association (CEA)*
http://www.ce.org

Consumer Product Safety Commission (CPSC)
http://www.cpsc.gov

Continental Automated Building Association (CABA)*
http://www.caba.org

Contrast Lighting (recessed)
http://www.contrastlighting.com

Cooper Bussmann, Inc.
http://www.cooperbussman.com

Cooper Industries
Here you will find links to all Cooper Divisions for electrical products, lighting, wiring devices, power and hand tools.
http://www.cooperindustries.com

Cooper Lighting
http://www.cooperlighting.com

Copper (technical information)
http://energy.copper.org

Copper Development Association
http://www.copper.org

Craftsman Tools
http://www.craftsman.com

Crouse-Hinds
http://www.crouse-hinds.com

Cummins Power Generation*
http://www.cumminspower.com

Custom Electronic Design & Installation Association*
http://www.cedia.net

Custom Electronic Professional*
http://www.ce-pro.com

Eaton
http://www.eaton.com/EatonCom/Markets/Electrical/SelectRegion/index.htm

DABMAR
http://www.dabmar.com

Danaher Power Solutions
http://www.danaherpowersolutions.com

Daniel Woodhead Company
http://www.danielwoodhead.com

Day-Brite Lighting
http://www.daybrite.com

Department of Energy
http://www.energy.gov

Dimension Express (to obtain roughing-in dimensions from all appliance manufacturers)
http://www.dexpress.com

F. W. Dodge
http://www.fwdodge.com

DOMOSYS Corporation*
http://www.domosys.com

Douglas Lighting Control
http://www.douglaslightingcontrol.com

Dremel
http://www.dremel.com

Dual-Lite
http://www.dual-lite.com

Dynacom Corporation
http://www.dynacomcorp.com

Eaton (links to all divisions)
http://www.eatonelectrical.com

EC Online
http://www.econline.com

EC & M magazine
http://www.ecmweb.com

Edison Electric Institute
http://www.eei.org

Edwards Signaling and Security Products
http://www.edwards-signals.com

ELAN Home Systems*
http://www.elanhomesystems.com

Electric-Find
http://www.electric-find.com

Electric Smarts
http://www.electricsmarts.com

Electri-flex
http://www.electriflex.com

Electrical Apparatus Service Association
http://www.easa.com

Electrical Contracting & Engineering News
http://www.ecpzone.com

Electrical Contractor magazine
http://www.ecmag.com

The Electrical Contractor Network
http://www.electrical-contractor.net

Electrical Designers Reference
http://www.edreference.com

Electrical Distributor magazine
http://www.tedmag.com

Electrical Reliability Services, Inc.
(formerly Electro-Test)
http://www.electro-test.com

Electrical Safety
http://www.electrical-safety.com

Electrical Safety Foundation International
http://www.esfi.org

Electrician.com
http://www.electrician.com

Electro-Test Inc.
http://www.electro-test.com

Electronic House*
http://www.electronichouse.com

Electronic Industry Alliance*
http://www.eia.org

Emerson
http://www.emersonfans.com

Encore Wire Limited
http://www.encorewire.com

Energy Efficiency & Renewable Energy Network
(U.S. Dept. of Energy)
http://www.eere.energy.gov

Energy Star
http://www.energystar.gov

Environmental Protection Agency
http://www.epa.gov

Erico
http://www.erico.com

Essex Group, Inc.*
http://www.essexgroup.com

Estimation Inc.
http://www.estimation.com

ETL Semko
http://www.intertek-etlsemko.com/

Exide Technologies
http://www.exide.com

Factory Mutual (FM Global)
http://www.fmglobal.com

Faraday
http://www.faradayfirealarms.com

FCI/Burndy (telecom connectors)
http://www.fciconnect.com

Federal Communications Commission*
http://www.fcc.gov

Federal Pacific
http://www.federalpacific.com

Federal Signal Corp.
http://www.federalsignal.com

Ferraz-Shawmut
http://www.ferrazshawmut.com

Fiber Optic Association, Inc.*
http://www.thefoa.org

Fibertray
http://www.ditel.net

First Alert
http://www.firstalert.com

Flex-Core
http://www.flex-core.com

Fluke Corporation
http://www.fluke.com

Fluke Networks*
http://www.flukenetworks.com

Fox Meter Inc.
http://www.arcfaulttester.com

Fox Meter Inc.
http://www.foxmeter.com

Gardner Bender
http://www.gardnerbender.com

GE Appliances
http://www.geappliances.com

GE Electrical Distribution & Control
http://www.ge.com/edc

GE Industrial Systems
http://www.geindustrial.com

GE Lighting
http://www.gelighting.com

General Cable
http://www.generalcable.com

Generators:

Coleman Powermate
http://www.powermate.com/

Cummins Power Generation
http://www.cumminspower.com

Generac Power Systems
http://www.generac.com

Gen-X
http://www.genxnow.com

Gillette Generators, Inc.
http://www.gillettegenerators.com

Kohler Power Systems
http://www.kohlergenerators.com

Onan Corporation
http://www.onan.com

Reliance Electric
http://www.reliance.com

Genesis Cable Systems*
http://www.genesiscable.com

Global Engineering Documents
http://www.global.ihs.com

Grayline, Inc. (tubing)
http://www.graylineinc.com

Greenlee Inc.
http://www.greenlee.com

Greyfox Systems*
http://www.legrand.us/OnQ.aspx?NewLookRedir=yes

Guth Lighting
http://www.guth.com

The Halex Company
http://www.halexco.com

Hand Tools Institute
http://www.hti.org.

Handicapped Suggestions
The World of Assisted Technology
http://www.abledata.com

Hatch Transformers, Inc
http://www.hatchtransformers.com

Heavy Duty
http://www.milwaukeetool.com

Heyco Products, Inc.
http://www.heyco.com

Hilti Corporation
http://www.hilti.com

Hinkley Lighting
http://www.hinkleylighting.com

HIOKI (measuring instruments)
http://www.hioki.co.jp/

Hitachi Cable Manchester, Inc.*
http://www.hcm.hitachi.com

Hoffmann
http://www.hoffmanonline.com

Holophane
http://www.holophane.com

Holt, Mike Enterprises
http://www.mikeholt.com

Home Automation and Networking Association*
http://www.homeautomation.org

Home Automation, Inc.*
http://www.homeauto.com

Home Automation Links*
http://www.asihome.com

Home Automation Systems*
http://www.smarthome.com

Home Controls, Inc.*
http://www.homecontrols.com

Home Director, Inc.*
http://www.homedirector.net

HomeStar*
http://www.wellhome.com/home

Homestore*
http://www.homestore.com

Home Touch*
http://www.home-touch.com

Home Toys*
http://www.hometoys.com

Honeywell
http://www.honeywell.com

Honeywell*
http://www.honeywell.com/yourhome

Housing & Urban Development
http://www.hud.gov

Houston Wire & Cable Company
http://www.houwire.com

Hubbell Incorporated (links to all divisions)
http://www.hubbell.com

Hubbell Incorporated
http://www.hubbell-premise.com

Hubbell Lighting Incorporated
http://www.hubbell-ltg.com

Hubbell Wiring Devices
http://www.hubbell-wiring.com

Hunt Control Systems, Inc.
http://www.huntdimming.com

Hunter Fans
http://www.hunterfan.com

Husky
http://www.mphusky.com

ICC
http://www.iccsafe.org/Pages/default.aspx

Ideal Industries, Inc.
http://www.idealindustries.com

Illuminating Engineering Society of North America
http://www.iesna.org

ILSCO
http://www.ilsco.com

Independent Electrical Contractors
http://www.ieci.org

Institute of Electrical and Electronic Engineers, Inc.
http://www.ieee.org

Instrumentation, Systems, and Automation Society
http://www.isa.org

Insulated Cable Engineers Association (ICEA)
http://www.icea.net

Insulated tools (certified insulated products)
http://www.insulatedtools.com

Intel's Home Network
http://www.intel.com/

Intermatic. Inc.
http://www.intermatic.com

International Association of Electrical
Inspectors (IAEI)
http://www.iaei.org

International Association of Plumbing &
Mechanical Officials
http://www.iapmo.org

International Brotherhood of Electrical Workers
http://www.ibew.org

International Code Council (ICC)
http://www.iccsafe.org

International Dark-Sky Association
http://www.darksky.org

International Electrotechnical Commission (IEC)
http://www.iec.ch/

International Organization for Standardization (ISO)
http://www.iso.org

iSqFt*
http://www.isqft.com/

Jacuzzi
http://www.jacuzzi.com

Jefferson Electric
http://www.jeffersonelectric.com

Johnson Controls, Inc.
http://www.johnsoncontrols.com

Joslyn
http://www.joslynmfg.com

Juno Lighting, Inc.
http://www.junolighting.com

Kenall Lighting
http://www.kenall.com

Kichler Lighting
http://www.kichler.com

Kidde (home safety)
http://www.kiddeus.com

King Safety Products
http://www.kingsafety.com

King Wire Inc.
http://www.kingwire.com

Klein Tools, Inc.
http://www.kleintools.com

Kohler Generators
http://www.kohler.com

Lamp recycling information (sponsored by NEMA)
http://www.lamprecycle.org

LEDs Magazine
http://www.ledsmagazine.com

Leviton Mfg. Co., Inc (links to all divisions)
http://www.leviton.com

Lew Electric
http://www.lewelectric.com

Liebert
http://www.liebert.com

Lighting
http://www.lightsearch.com

Lighting Research Center
http://www.lrc.rpi.edu

Lightning Protection Institute
http://www.lightning.org

Lightolier
http://www.lightolier.com

Lightolier Controls*
http://www.lolcontrols.com

Litetouch*
http://www.litetouch.com

Littelfuse, Inc.
http://www.littelfuse.com

LonWorks*
http://www.lonworks.echelon.com

Lucent Technologies*
http://www.lucent.com

Lumisistemas, S.A. de C.V
http://www.lumisistemas.com

Lutron Electronics Co., Inc.
http://www.lutron.com

Lutron Home Theater*
http://www.classic.lutron.com

MagneTek (links to all divisions)
http://www.magnetek.com

Makita Tools
http://www.makita.com

Manhattan/CDT
http://www.manhattancdt.com

McGill
http://www.mcgillelectrical.com

Megger
http://www.megger.com/us/index.asp

MET Laboratories, Inc.
http://www.metlabs.com

MI Cable Company
http://www.micable.com

Midwest Electric Products, Inc.
http://www.midwestelectric.com

Milbank Manufacturing Co.
http://www.milbankmfg.com

Milwaukee Electric Tool
http://www.milwaukeetool.com

Minerallac
http://www.minerallac.com

Molex Inc.
http://www.molex.com

Motor Control
http://www.motorcontrol.com

Motorola Lighting Inc.
http://www.motorola.com

MTU Onsite Energy
http://www.mtuonsiteenergy.com/mtuonline/

Multiplex Technology
http://www.multiplextechnology.com/

National Association of Electrical Distributors
http://www.naed.org

National Association of Home Builders
http://www.nahb.org

National Association of Manufacturers
http://www.nam.org

National Association of Radio and Telecommunications
Engineers, Inc.
http://www.narte.com

National Conference of States on Building
Codes and Standards
http://www.ncsbcs.org

National Electrical Contractors Association (NECA)
http://www.necanet.org

National Electrical Manufacturers Association (NEMA)
http://www.nema.org

National Fire Protection Association
http://www.nfpa.org

National Fire Protection Association
National Electrical Code® information
http://www.nfpa.org

National Institute for Occupational Safety & Health
http://www.cdc.gov/niosh

National Institute of Standards and Technology (NIST)
http://www.nist.gov

National Joint Apprenticeship and
Training Committee (NJATC)
http://www.njatc.org

National Lighting Bureau
http://www.nlb.org

National Resource for Global Standards
http://www.nssn.org

National Safety Council
http://www.nsc.org

National Spa and Pool Institute
http://www.nspi.org

National Systems Contractors Association*
http://www.nsca.org

NECDirect
http://www.necdirect.org

Newton's Electrician
http://www.electrician.com

Nora Lighting
http://www.noralighting.com

NuTone
http://www.nutone.com

Occupation Health & Safety
http://www.ohsonline.com

Occupational Safety & Health Administration
http://www.osha.gov

Okonite Co.
http://www.okonite.com

Olflex Cable
http://www.olflex.com

Onan Corporation
http://www.onan.com

ONEAC Corporation
http://www.oneac.com

OnQ Technologies*
http://www.onqtech.com

Optical Cable Corporation*
http://www.occfiber.com

Ortronics Inc.
http://www.ortronics.com

Osram Sylvania Products, Inc.
http://www.sylvania.com

Overhead door information: Door & Access
System Manufacturers Association
http://www.dasma.com

OZ/Gedney
http://www.o-zgedney.com

Panasonic Lighting
http://www.panasonic.com/MHCC/pl/technoi.htm

Panasonic Technologies
http://www.panasonic.com

Panduit Corporation
http://www.panduit.com

Pass and Seymour, Legrand
http://www.legrand.us/PassAndSeymour.
aspx?NewLookRedir=yes

Pelican Rope Works
http://www.pelicanrope.com

Penn-Union Corp.
http://www.penn-union.com

Phase-A-Matic, Inc.
http://www.phase-a-matic.com

Philips
http://www.philips.com

Philips Advance
http://www.advance.philips.com/default.aspx

Philips Lamps
http://www.ALTOlamp.com

Philips Lighting Company
http://www.philipslighting.com

Philtek Power Corporation
http://www.philtek.com

Plant Engineering On-Line
http://www.plantengineering.com

Popular Home Automation*
http://www.phoenixcontact.com

Porter Cable
http://www.porter-cable.com

Power & Tel.
http://www.ptsupply.com/enterprise.asp

Power Quality Monitoring
http://www.pqmonitoring.com

Power-Sonic Corporation
http://www.power-sonic.com

Premise Wiring Products
http://www.unicomlink.com

Prescolite
http://www.prescolite.com

Progress Lighting
http://www.progresslighting.com

RACO
http://www.hubbell-raco.com

Radix Wire Company
http://www.radix-wire.com

Rayovac Corporation
http://www.rayovac.com

RCA*
http://www.rca.com

Refrigeration Service Engineering Society
http://www.rses.org

Regal Fittings
http://www.regalfittings.com

Reliable Power Meters
http://www.reliablemeters.com

Remke Industries, Inc.
http://www.remke.com

Rhodes, M. H. (timers)
http://www.mhrhodes.com

Ridgid Tool Company
http://www.ridgid.com

Robertson Transformer Co.
http://www.robertsontransformer.com

Robroy Industries – Conduit Div.
http://www.robroy.com

Rockwell Automation
http://www.automation.rockwell.com

RotoZip Tool Corp.
http://www.rotozip.com

Russelectric
http://www.russelectric.com

Safety Technology International, Inc.
http://www.sti-usa.com

S&C Electric Company
http://www.sandc.com

SBCCI
http://www.sbcci.org

Sea Gull Lighting
http://www.seagulllighting.com

Seatek, Inc.
http://www.SeatekCo.com

    Digital Monitoring Products*
http://www.dmpnet.com

    Home Automation*
http://www.homeauto.com

Honeywell Security*
http://www.security.honeywell.com/sce

NAPCO Security Systems*
http://www.napcosecurity.com

Siemens (main site)
http://www.siemens.com

Siemens Energy & Automation
http://www.sea.siemens.com

The Siemon Company*
http://www.siemon.com

Sensor Switch Inc.
http://www.sensorswitchinc.com

Shat-R-Shield
http://www.shat-r-shield.com

Silent Knight
http://www.silentknight.com

Simplex Grinnell
http://www.simplexgrinnell.com

Simpson Electric Co.
http://www.simpsonelectric.com

Skil Tools
http://www.skil.com

Smart Home*
http://www.smarthome.com

Smart Home Pro*
http://www.smarthomepro.com

Smart House, Inc.*
http://www.smart-house.com

Smoke & Fire Extinguisher Signs
http://www.smokesign.com

Sola/Hevi Duty
http://www.sola-hevi-duty.com

Southwire
http://www.southwire.com

SP Products, Inc.
http://www.spproducts.com

A.W. Sperry Instruments, Inc.
http://www.awsperry.com

Stanley Tools
http://www.stanleyworks.com

Starfield Controls*
http://www.starfieldcontrols.com

Steel Tube Institute
http://www.steeltubeinstitute.org

Straight Wire*
http://www.straightwire.com

Square D, Groupe Schneider
http://www.squared.com

Square D (Homeline)
http://www.squared.com/retail

Square D, Power Logic
http://www.powerlogic.com

Steel Tube Institute, Conduit Committee
http://www.steelconduit.org

Sunbelt Transformers
http://www.sunbeltusa.com

Superior Electric
http://www.superiorelectric.com

Superior Essex Group
http://www.superioressex.com

Suttle
http://www.suttleonline.com

Sylvania
http://www.sylvania.com

Sylvania Lighting International
http://www.sli-lighting.com

Telecommunications Industry Association (TIA)*
http://www.tiaonline.org

Terms /Construction /Glossary
http://www.constructionplace.com/glossary.html

Thermostats, Communicating*
http://www.homeauto.com

This Old House
http://www.thisoldhouse.org

Thomas & Betts Corporation
http://www.tnb.com

Thomas Lighting
http://www.thomaslighting.com

Thomas Register
http://www.thomasregister.com

3M Electrical Products
http://www.3m.com/elpd

Tork
http://www.tork.com

Touch-Plate Lighting Controls
http://www.touchplate.com

Trade Service Corporation
http://www.tradeservice.com

Trade Service Systems
http://www.tradepower.com

Triplett Corporation
http://www.triplett.com

TYCO (links to all divisions)
http://www.tyco.com

Tyco Electronics
http://www.tycoelectronics.com/components/default.aspx

Tyton Hellermann
http://www.hellermanntyton.us

UEI Test Equipment
http://www.ueitest.com

Underwriters Laboratories
http://www.ul.com

UL Standards Department
http://ulstandardsinfonet.ul.com

Unistrut (Tyco)
http://www.unistrut.com

Unity Manufacturing
http://www.unitymfg.com

Universal Lighting Technologies
http://www.universalballast.com

USA Wire and Cable, Inc.
http://www.usawire-cable.com

U.S. Department of Energy (DOE)
http://www.doe.gov

U.S. Department of Labor
http://www.dol.gov

U.S. Department of Labor, Employment & Training Administration
http://www.doleta.gov

U.S. Environmental Protection Agency (EPA)
http://www.epa.gov

Ustec* (structured wiring)
http://www.legrand.us/OnQ.aspx?NewLookRedir=yes

Vantage Automation & Lighting Control*
http://www.vantagecontrols.com

Watt Stopper
http://www.wattstopper.com

Waukesha Electric Systems
http://www.waukeshaelectric.com

Werner Ladder Company
http://www.wernerladder.com

Westinghouse
http://www.westinghouselighting.com/

Westinghouse Lighting
http://www.westinghouselighting.com

Wheatland Tube Company
http://www.wheatland.com

Wireless Industry*
http://www.compnetworking.about.com/

Wireless Thermostats*
http://www.wirelessthermostats.com

Wiremold
http://www.wiremold.com

Woodhead Industries
http://www.danielwoodhead.com

X-10*
http://www.x10.com

X-10 Pro Automation Products*
http://www.x10pro.com

Zenith Controls, Inc.*
http://www.geindustrialsystems.com

Zircon
http://www.zircon.com

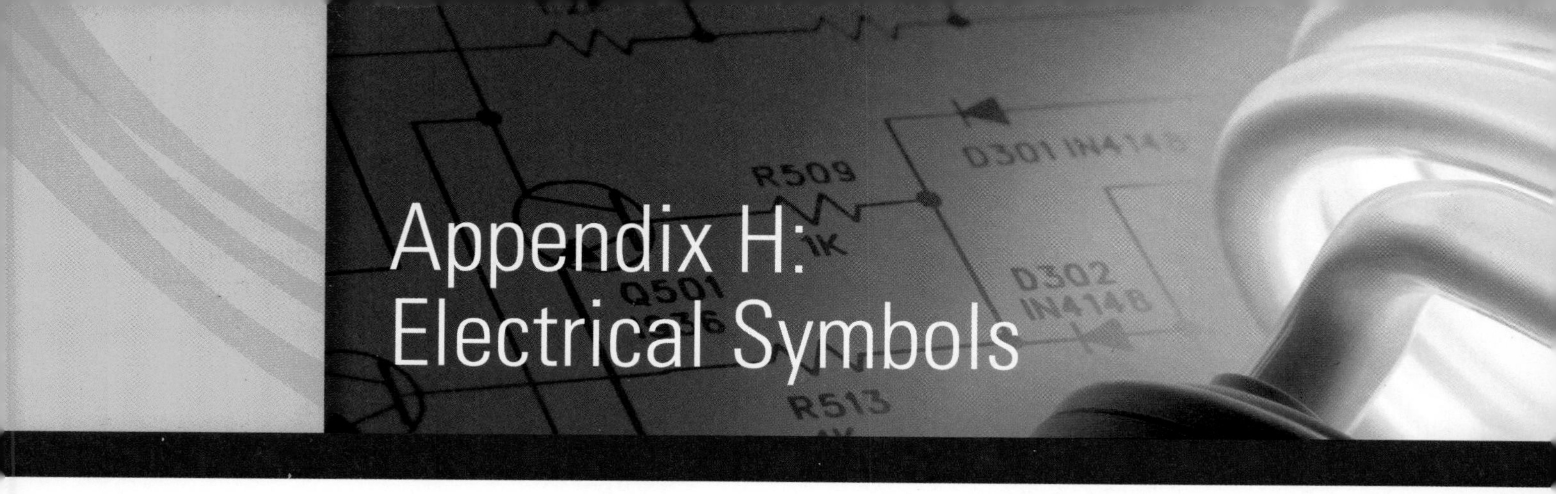

ELECTRICAL SYMBOLS

RACEWAYS—BOXES AND BUSWAYS	
PREFERRED SYMBOL	**DESCRIPTION**
J	Underfloor raceway system junction box, flush floor mounted.
P ▪ 2	Power pole with devices indicated in the specifications and on the drawing. "P" indicates type, "2" indicates circuit.
T ▪	Telecom pole with devices indicated in the specifications and on the drawings. "T" indicates type.
TP ▪ 2	Telecom/Power pole with devices indicated in the specifications and on the drawings. "TP" indicates type, "2" indicates power circuit.
PB OR □	Pull box—size as indicated or required.
TR TR TR	Cabletray size as indicated.
TR TR TR	Cabletray size as indicated, concealed.
BW BW BW CTB	Busway with cable tap box, rating, and type as indicated on drawings.
BW BW BW F	Busway with plug-in device as indicated, shown with fused disconnect.
⊠	Busway feeding up.
⊿	Busway feeding down.
BW	Busway expansion joint.
WW WW WW	Wireway, size as indicated or required.

FIGURE H-1 Raceways, boxes, and busways. (*Courtesy NECA NEIS 100-2006*)

SECURITY SYSTEM SYMBOLS	
PREFERRED SYMBOL	DESCRIPTION
C ◁ WP	CCTV camera. "WP" indicates weatherproof exterior camera.
CCTV	CCTV coaxial cable outlet and power outlet.
MTV	CCTV monitor outlet.
B ○	Doorbell.
B /	Door buzzer.
B ⊨	Door chime.
DR	Electric door opener.
ES	Electric door strike.
IC	Intercom unit—flush MTD.
MI	Master intercom and directory unit.
MD	Motion detector.
ML	Security door alarm magnetic lock.
CR WP	Security card reader. "WP" indicates weatherproof.
SCP	Security control panel.
DC	Security door contacts.
•	Security exit push button.

SECURITY SYSTEM SYMBOLS	
PREFERRED SYMBOL	DESCRIPTION
K	Security keypad.
ID	Infrared detector.
UD	Ultrasonic detector.
DA	Door alarm.
P	Panic bar.

FIGURE H-2 Security system symbols. (*Courtesy NECA NEIS 100-2006*)

FIRE ALARM COMMUNICATIONS AND PANEL SYMBOLS	
PREFERRED SYMBOL	**DESCRIPTION**
M	Fire alarm master box.
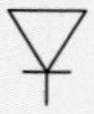 (firefighter phone symbol)	Firefighter phone.
FT	Coded transmitter.
DK	Drill key switch.
K	Key repository (knox box).
FAA	Annunciator panel.
FACP	Fire alarm control panel.
EVAC	Voice evacuation panel.
FATC	Fire alarm terminal cabinet.
BATT	Battery pack and charger.
ASFP	Air-sampling control/detector panel with associated air-sampling piping network.
TPR	Transponder.
IAM	Individual addressable module.

FIRE ALARM COMMUNICATIONS AND PANEL SYMBOLS	
PREFERRED SYMBOL	**DESCRIPTION**
ZAM	Zone adapter module.
CZAM	Control zone adapter module.
MZAM	Monitor zone adapter module.
SYN	Synchronized module.
KHS	Kitchen hood system.

FIGURE H-3 Fire alarm communications and panel symbols. (*Courtesy NECA NEIS 100-2006*)

FIIRE ALARM INDICATOR SYMBOLS	
PREFERRED SYMBOL	**DESCRIPTION**
 CR	Control relay.
DH	Door holder.
 F	Horn and strobe.
 F	Mini-horn and strobe.
 H	Horn unit only.
 S	Strobe unit only.
 F	Bell and strobe.
 F	Buzzer and strobe.
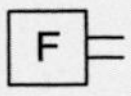 F	Chime and strobe.
 F S	Speaker and strobe.
	LED pilot light.
 F WP	Indicating beacon. "WP" indicates weatherproof.
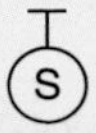 S	Speaker—ceiling mounted.

FIGURE H-4 Fire alarm indicator symbols. (*Courtesy NECA NEIS 100-2006*)

FIRE ALARM SENSOR SYMBOLS	
PREFERRED SYMBOL	DESCRIPTION
F	Manual pull station.
S	Smoke detector.
S_D	Duct smoke detector with two auxilliary contacts.
RTS	Remote station for duct-mounted smoke detectors.
S_A	Area-type smoke detector used at ductwork opening.
S_E	Elevator recall with auxilliary contacts.
S_SC	Self-contained smoke detector —single-station type.
S_V	Smoke detector—visual and audible signal.
B_R	Beam smoke detector "S"— indicates sending unit "R"—indicates receiver.
C	Carbon monoxide detector. Line voltage with battery backup.
F	Flame detector.
H	Automatic heat detector (135°F rate of rise).
H_F	Automatic heat detector. "F"—indicates fixed temperature 190°F.
H_2	Hydrogen detector.
D/FS	Motor-operated fire/smoke duct damper.

FIRE ALARM SENSOR SYMBOLS	
PREFERRED SYMBOL	DESCRIPTION
FS	Waterflow switch.
PS	Low-pressure switch.
TS	Tamper switch.
PIV	Post-indicator valve.
EOL	End-of-line resistor.

FIGURE H-5   Fire alarm sensor symbols. (*Courtesy NECA NEIS 100-2006*)

COMMUNICATIONS—TELEDATA SYMBOLS	
PREFERRED SYMBOL	DESCRIPTION
▽	Data outlet.
 ▽F	Data outlet floor type. "F" indicates flush mounted. "S" indicates surface mounted.
▼	Telephone/data outlet.
 ▼F	Telephone/data outlet floor type. "F" indicates flush mounted. "S" indicates surface mounted.
▼F	Telephone outlet.
▼W	Telephone outlet—wall mounted.
 ▼F	Telephone outlet floor type. "F" indicates flush mounted. "S" indicates surface mounted.

FIGURE H-6 Communications-teledata symbols.
(*Courtesy NECA NEIS 100-2006*)

If a number Indicates Number of wires Pulled To it

COMMUNICATIONS—EQUIPMENT	
PREFERRED SYMBOL	DESCRIPTION
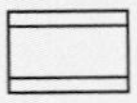 ▭	Equipment cabinet.
▬	Equipment rack—wall mounted.
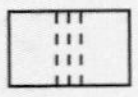 ▤	Equipment rack—free standing.
 TCC	Terminal cabinet with 3/4" plywood backing.
 ▱	Plywood backboard.

FIGURE H-8 Communications equipment.
(*Courtesy NECA NEIS 100-2006*)

COMMUNICATIONS—AUDIO/VISUAL	
PREFERRED SYMBOL	DESCRIPTION
 ⊡	Call-in switch.
 TV	Cable antenna system outlet. (CATV)
 TV M	Master antenna system outlet. (MATV)
 M	Microphone outlet—floor mounted.
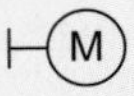 ⊣M	Microphone outlet—wall mounted.
S	Speaker—ceiling mounted.
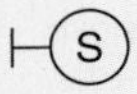 ⊣S	Speaker—wall mounted.
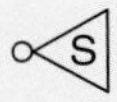 ⊲S	Speaker horn.
 ⊣P ↕	Speaker bidirectional paging—wall mounted.
 ◇P ↕	Speaker bidirectional paging—ceiling mounted.
IC	Intercom unit—flush mounted.
MI	Master intercom and directory unit.
VC	Volume control.

FIGURE H-7 Communications audio/visual.
(*Courtesy NECA NEIS 100-2006*)

Marine Grade

SCHEMATIC AND ONE-LINE DIAGRAM SYMBOLS	
PREFERRED SYMBOL	**DESCRIPTION**
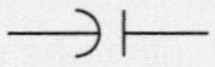	Capacitor.
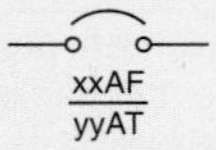 xxAF / yyAT	Circuit breaker (open). "xxAF" indicates frame size. "yyAT" indicates trip size.
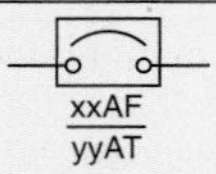 xxAF / yyAT	Circuit breaker (enclosed). "xxAF" indicates frame size. "yyAT" indicates trip size.
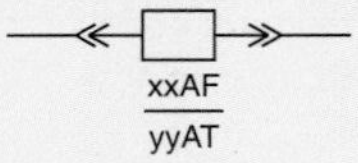 xxAF / yyAT	Primary draw out–type circuit breaker. "xxAF" indicates frame size. "yyAT" indicates trip size.
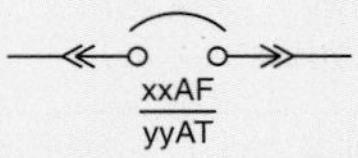 xxAF / yyAT	Low-voltage draw out–type circuit breaker. "xxAF" indicates frame size. "yyAT" indicates trip size.
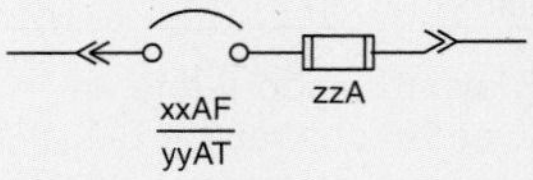 xxAF / yyAT zzA	Low-voltage draw out–type circuit breaker with current limiting fuses. "xxAF" indicates frame size. "yyAT" indicates trip size. "zzA" indicates fuse rating.
	Contact, normally open (NO) ("TC"—with timed closing).
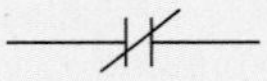	Contact, normally closed (NC) ("TO"—with timed opening).
CT	Current transformer cabinet.
 zzA	Fused cutouts. "zzA" indicates fuse rating.
	Disconnect switch unfused.
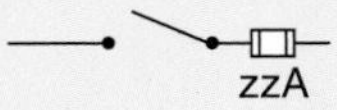 zzA	Disconnect switch air break with fuse. "zzA" indicates fused rating.
 zzA zzA	Fuse. "zzA" indicates fuse rating.
	Overload relay.
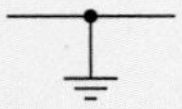	Grounding connection—system and or equipment.

SCHEMATIC AND ONE-LINE DIAGRAM SYMBOLS	
PREFERRED SYMBOL	**DESCRIPTION**
(K2)	Kirk key interlock system. "2"—indicates related kirk keys.
	Lightning arrester and grounding to protect all phases.
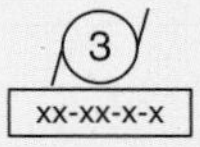 (3) xx-xx-x-x	Motor and label. "3" denotes horsepower.
(MO)	Motor operator for circuit breakers or switches.
	Network protector.
PANEL	Panelboard.
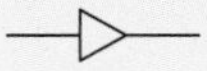	Pothead.
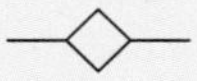	Stress cone.
	Resistor.
(ST)	Shunt trip.
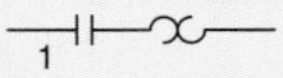 1	Magnetic starter with NEMA size indicated.
GFCI	Ground fault circuit interrupter, personnel protection.
 (G) xxx KW xxxV-xø GENERATOR	Generator.
 Tx xxx KVA xxxV xø xW PRI xxxY/xxxV xø xW SEC	Transformer, dry type, unless otherwise indicated. A delta connection is indicated on the primary and a grounded wye connection on the secondary.
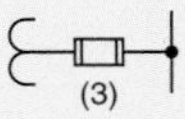 (3)	Potential transformer. "3"—indicates quantity.
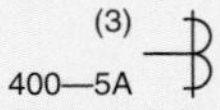 (3) 400—5A	Current transformer. "3"—indicates quantity, "400 5A" indicates ratio.

FIGURE H-9   Schematic and one-line diagram symbols. (*Courtesy NECA NEIS 100-2006*)

| SCHEMATIC AND ONE-LINE DIAGRAM SYMBOLS | |
| *(continued)* | |
PREFERRED SYMBOL	DESCRIPTION
△	3-phase, 3-wire delta connection.
 △ (grounded)	Corner-grounded delta.
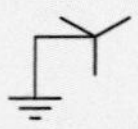 (wye symbol)	3-phase, 4-wire wye connection (grounded neutral).
AFD ③ xx-x-x	Adjustable frequency drive. "3" references detail number.
XX' xxxV BUSDUCT	Busduct or busway.
XX' xxxV WIREWAY	Wireway.

FIGURE H-9 Continued
(*Courtesy NECA NEIS 100-2006*)

| SCHEMATIC AND ONE-LINE DIAGRAM SYMBOLS—SWITCHES | |
PREFERRED SYMBOL	DESCRIPTION
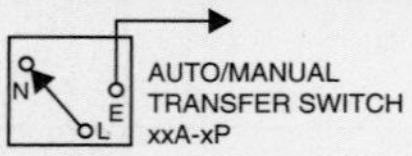 AUTO/MANUAL TRANSFER SWITCH xxA-xP	Transfer switch.
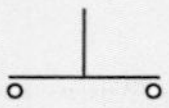 (push button symbol)	Push button (start).
 (push button symbol)	Push button (stop).
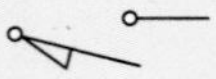 (limit switch symbol)	Limit switch.
 (flow switch symbol)	Flow switch.
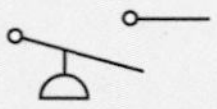 (pressure switch symbol)	Pressure switch.
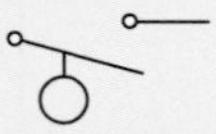 (float switch symbol)	Float switch.
 Ⓡ	Pilot light. Letter indicates color. Example: R = red.
 (solenoid symbol)	Solenoid.

FIGURE H-10 Schematic and one-line diagram symbols—switches. (*Courtesy NECA NEIS 100-2006*)

| MISCELLANEOUS | |
PREFERRED SYMBOL	DESCRIPTION
— G —	Ground bar. Length to be noted.
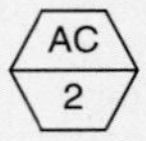 AC 2	Mechanical equipment tag number; refer to mechanical equipment schedule.
 K2	Equipment tag number; refer to equipment schedule. "K"—indicates kitchen "C"—indicates computer.
 B	Note symbol; refer to note as indicated.
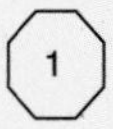 1	Feeder number; refer to "feeder schedule".
A	Typical/similar room or area layout symbol. "A"—indicates layout type.
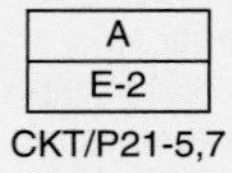 A E-2 CKT/P21-5,7	Typical layout symbol—refer to layout type "A" on drawing E-2; circuits to be used are as indicated.
 2 DESCRIPTION E-4 SCALE: N.T.S.	Detail header, indicating detail No. 2 on drawing E-4.
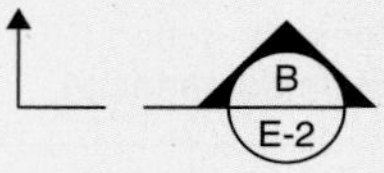 B E-2	Section identifier, indicating section "B" on drawing E-2. Left or right arrow.
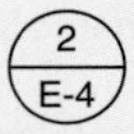 2 E-4	Detail identifier, indicating detail No. 2 on drawing E-4.

FIGURE H-11 Miscellaneous symbols.
(*Courtesy NECA NEIS 100-2006*)

ABBREVIATIONS			
ABBREVIATION	**DESCRIPTION**	**ABBREVIATION**	**DESCRIPTION**
1P	One pole	Δ	Delta
2P	Two pole	DET	Detector
3P	Three pole	DISC	Disconnect
4P	Four pole	DIST	Distribution
1P1W	One pole, one wire	DN	Down
1P2W	One pole, two wire	DWG	Drawing
2P2W	Two pole, two wire	DT	Dusttight(*)
2P3W	Two pole, three wire	E	Wired on emergency circuit
3P2W	Three pole, two wire	EA	Each
3P3W	Three pole, three wire	EC	Electrical contractor
3P4W	Three pole, four wire	EF	Exhaust fan
4P4W	Four pole, four wire	ELEC	Electric(al)
A	Ampere	EMER	Emergency
AC	Alternating current	EMT	Electric metallic tubing
AF	AMP frame	ENCL	Enclosure
AFCI	Arc-fault circuit interrupter	EOL	End of line
AFF	Above finished floor	EPO	Emergency power off
AFG	Above finished grade	EQUIP	Equipment
AHU	Air handling unit	EWC	Electric water cooler
AIC	Ampere interrupting capacity	EWH	Electric water heater
AL	Aluminum	EXIST.	Existing
AS	AMP switch	F	Flush
AT	AMP trip	FA	Fire alarm
ARCH	Architect	FBO	Furnished by others
ATS	Automatic transfer switch	FC	Fire protection contractor
AUD	Audiometer box connection	FCU	Fan coil unit
AUX	Auxiliary	FDN	Foundation
A/V	Audio visual	FIXT	Fixture
AWG	American wire gauge	FLA	Full load amps
BLDG	Building	FLEX	Flexible
C	Conduit (Generic term for raceway, Provide as specified.)	FLR	Floor
		FMC	Flexible metallic conduit
CAM	Camera	FRE	Fiberglass reinforced epoxy conduit
CAT	Catalog	FURN	Furniture
CATV	Cable television	GC	General contractor
CB	Circuit breaker	GEN	Generator
CKT	Circuit	GFCI	Ground fault circuit interrupter
COL	Column	GFPE	Ground fault protection equipment
C.T.	Current transformer	GND	Grounded
CU	Copper	GRC	Galvanized rigid conduit
CL	Centerline	HGT	Height
DC	Direct Current	HP	Horsepower

FIGURE H-12   Symbol abbreviations. (*Courtesy NECA NEIS 100-2006*)

	ABBREVIATIONS		*(continued)*
ABBREVIATION	**DESCRIPTION**	**ABBREVIATION**	**DESCRIPTION**
HV	High voltage	*NEC*	*National Electrical Code*
HVAC	Heating, ventilating, and air conditioning	NF	Nonfused
		NIC	Not in contract
HW	Hot water	NL	Night light
Hz	Hertz (cycle) per second	NM	Nonmetallic sheathed cable
IAM	Individual addressable module	NO	Normally open
IC	Intercommunication	NRTL	Nationally recognized testing lab
ID	Identification	#	Number
IG	Isolated ground	NTS	Not to scale
IMC	Intermediate metal conduit	O2	Oxygen
IPS	Interruptible power supply	OHD	Overhead door operator
IR	Passive infrared	P	Pole
JB	Junction box	PB	Pull box
KCMIL	Thousand circular mils	PC	Plumbing system contractor
K/O	Knockout	PE	Primary service
KVA	Kilovolt ampere	PH ∅	Phase
KVAR	Kilovolt ampere reactive	PNL	Panel(board)
KW	Kilowatt	PIV	Post indicating valve
LFMC	Liquidtight flexible metallic conduit	PP	Power panel
LFNC	Liquidtight flexible nonmetallic conduit	PR	Pair
LP	Lighting panelboard	PRI	Primary
LS	Limit switch	PT	Potential transformer
LTG	Lighting	PVC	Polyvinyl chloride conduit
LV	Low-voltage	PWR	Power
MAINT	Maintained	RE	Remove existing
MAU	Make-up air unit	REC	Recessed
MAX	Maximum	RECP	Receptacle
MC	Metal clad cable	REF	Roof exhaust fan
MCB	Main circuit breaker	RL	Relocate existing
MCC	Motor control center	RM	Room
MD	Motorized damper	RMC	Rigid metal conduit
MDP	Main distribution panel	RT	Raintight(*)
MISC	Miscellaneous	RTU	Rooftop unit
MFR	Manufacturer	RSC	Rigid steel conduit
MLO	Main lugs only	S	Surface mounted
MOD	Motor operated disconnect switch	SCH	Schedule
MTD	Mounted	SD	Smoke damper
MTG	Mounting	SE	Secondary electric service
MTS	Manual transfer switch	SEC	Secondary
N	North	SIG	Signal
N/A	Not applicable	SN	Solid neutral
NC	Normally closed	SP	Spare

FIGURE H-12 Continued (*Courtesy NECA NEIS 100-2006*)

ABBREVIATIONS			
ABBREVIATION	**DESCRIPTION**	**ABBREVIATION**	**DESCRIPTION**
SPKR	Speaker	UH	Unit heater
SPL	Splice	U.O.I.	Unless otherwise indicated
SS	Stainless steel	UPS	Uninterruptible power source
STP	Shielded twisted pair	UTIL	Utility
STL	Carbon steel	UTP	Unshielded twisted pair
SUSP	Suspended	V	Volt
SW	Switch	VT	Vaportight(*)
SWBD	Switchboard	Y	Wye
SWGR	Switchgear	W	Watt
TC	Telephone cabinet	W/	With
TCI	Telecommunications cabling installer	WH	Watthour
TCP	Temperature control panel	WP	Weatherproof
TEL/DATA	Telephone/data	WT	Watertight(*)
TEL	Telephone	XFMR	Transformer
TEMP	Temporary	XP	Explosion proof(*)
TERM	Terminal(s)	ZAM	Zone adapter module
TV	Television	+72	Mounting units to centerline above finished floor or grade
TYP	Typical		
UC	Under counter		
UG	Underground		

FIGURE H-12 Continued (*Courtesy NECA NEIS 100-2006*)

NURSE CALL SYSTEM	
PREFERRED SYMBOL	**DESCRIPTION**
NCA	Nurse call annunciator.
E	Emergency pull-cord station
◓	Dome light with tone.
NC	Nurse call—patient station. "A" = denotes connection to remote annunciator in emergency room. "PC" = denotes patient pull station. "SA" = denotes staff assist station.
DS	Duty station.
SS	Staff station.
NCC	Nurse call system central cabinet.

FIGURE H-13 Nurse call system. (*Courtesy NECA NEIS 100-2006*)

Appendix I: Bender Guide

FEATURES

Your IDEAL Bender has engineered features which include:

1. Arrow

   To be used with stub, offset and outer marks of saddle bends.

2. Rim Notch

   Locates the center of a saddle bend.

3. Star-Point

   Indicates the back of a 90° bend.

4. Degree Scale

   For offsets, saddles and those special situations.

5. A Choice

   High strength ductile iron or light weight aluminum.

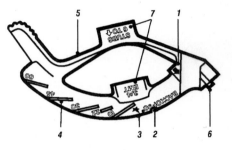

The above are features that lead to perfectly predictable and repeatable bends.

INSTRUCTIONS

Bend conduit with skill and professionalism. Take the guesswork out of bending.

Steps to Remember

Step 1. Measure your job.

Step 2. Mark you conduit using the recommended tables.

Step 3. Use your bender's engineered marks.

Note: Reference to the above Steps 1, 2, and 3 will be made throughout this booklet.

⟋⟍ DON'T FORGET

- When bending on the floor, pin the conduit to the floor. Use heavy foot pressure.
- When bending in the air, exert pressure as close to your body as possible.
- In case you overbend, use the back pusher or the expanded end of the bender handle to straighten your conduit to fit the job.

How to Bend a Stub

The stub is the most common bend. Note that your bender is marked with the "take-up" of the arc of the bender shoe.

💡 EXAMPLE

Consider making a 14" stub, using a 3/4" EMT conduit.

Step 1. The IDEAL bender indicates stubs 6" to ↑. Simply subtract the take-up, or 6", from the finished stub height. In this case 14" minus 6" = 8".
Step 2. Mark the conduit 8" from the end.
Step 3. Line up the Arrow on the bender with the mark on the conduit and bend to 90°.

Remember: Heavy Foot pressure is critical to keep the EMT in the bender groove and to prevent kinked conduit.

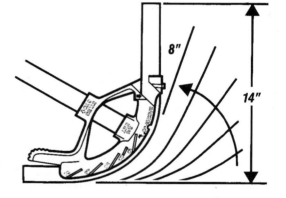

⟋⟍ HOW TO MAKE BACK-TO-BACK BENDS

A back-to-back bend produces a "U" shape in a single length of conduit. Use the same technique for a conduit run across the floor or ceiling which turns up or down a wall.

💡 EXAMPLE

Step 1. After the first 90° bend has been made, measure to the point where the back of the second bend is to be, "B".
Step 2. Measure and mark your conduit the same distance, mark "B".
Step 3. Align the mark on the conduit with the Star-Point on the bender and bend to 90°.

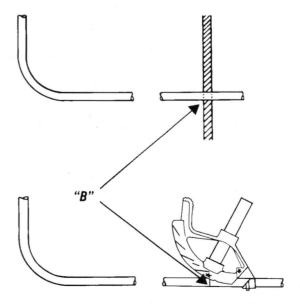

HOW TO MAKE AN OFFSET BEND

The offset bend is used when an obstruction requires a change in the conduit's plane.

Before making an offset bend, you must choose the most appropriate angles for the offset. Keep in mind that shallow bends make for easier wire pulling, steeper bends conserve space.

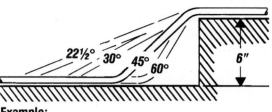

Example:

You must also consider that the conduit shrinks due to the detour. Remember to ignore the shrink when working away from the obstruction, but be sure to consider it when working into it.

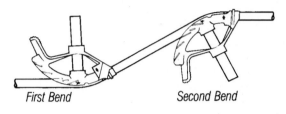

First Bend   *Second Bend*

EXAMPLE

Step 1. Measure the distance from the last coupling to the obstruction.

Step 2. Add the "shrink amount" from the Reference Table for Offset Bends to the measured distance and make your first mark. Your second mark will be placed at the "distance between bends." (Refer to the Reference Table for Offset Bends.)

Step 3. Align the Arrow with the first mark and using the Degree Scale bend to the chosen angle. Slide down the conduit and rotate conduit 180°, align the Arrow and bend as illustrated.

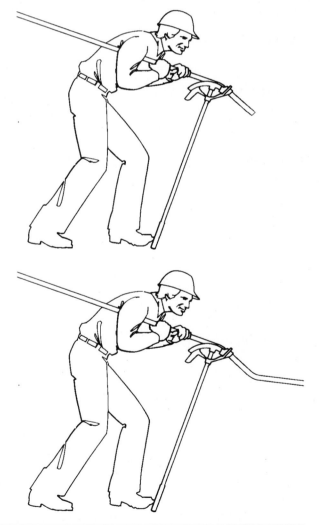

EXAMPLE

30° Bend with a 6" Offset Depth

Distance Between Bends ←12"→ Shrink Amount1-1/2"

REFERENCE TABLE FOR OFFSET BENDS

Offset Depth	22-1/2°		30°		45°		60°	
2"	5-1/4"	3/8"						
3"	7-3/4"	9/16"	6"	3/4"				
4"	10-1/2"	3/4"	8"	1"				
5"	13"	15/16"	10"	1-1/4"	7"	1-7/8"		
6"	15-1/2"	1-1/8"	12"	1-1/2"	8-1/2"	2-1/4"	7-1/4"	3"
7"	18-1/4"	1-5/16"	14"	1-3/4"	9-3/4"	2-5/8"	8-3/8"	3-1/2"
8"	20-3/4"	1-1/2"	16"	2"	11-1/4"	3"	9-5/8"	4"
9"	23-1/2"	1-3/4"	18"	2-1/4"	12-1/2"	3-3/8"	10-7/8"	4-1/2"
10"	26"	1-7/8"	20"	2-1/2"	14"	3-3/4"	12"	5"

HOW TO MAKE SADDLE BENDS

The saddle bend is similar to an offset bend, but in this case the same plane is resumed. It is used most often when another pipe is encountered.

Most common is a 45° center bend and two 22-1/2° outer bends, but you can use a 60° center bend and two 30° bends. Important: Use the same calculation for either set of angles.

 EXAMPLE:

Step 1. You encounter a 3" O.D. pipe 4 feet from the last coupling. The formula shown in the chart below indicates that for each inch of outside diameter of the obstruction, you must move your center mark ahead 3/16" per inch of obstruction height and make your outer marks 2-1/2" per inch of obstruction height from the center mark.

Step 2. The following table gives the actual mark spacings. In this example, the center mark is moved ahead 9/16" to 48-9/16". The outer marks are 7-1/2" from the center mark, or 41-1/16" and 56-1/16". Mark you conduit at these points.

If Obstruction Is	Move Your Center Mark Ahead	Make Outside Marks From Center Mark
1"	3/16"	2-1/2"
2"	3/8"	5"
3"	9/16"	7-1/2"
4"	3/4"	10"
5"	15/16"	12-1/2"
6"	1-1/8"	15"

Step 3.

(A) Align the center mark with the Rim Notch and bend to 45°.

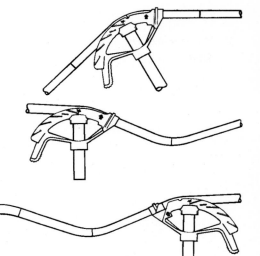

(B) Do not remove the conduit from the bender. Slide the bender down to the next mark and line up with the Arrow. Bend to 22-1/2° as indicated.

(C) Remove and reverse the conduit and locate the other remaining mark at the Arrow. Bend to 22-1/2° as indicated.

CAUTION: Be sure to line up all bends to be in the same plane.

HICKEYS

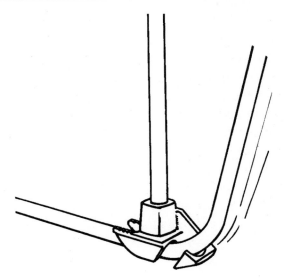

Hickeys require a different approach to bending. It is not a fixed radius device but rather one that requires several movements per bend. The hickey can give you the advantage of producing bends with a very tight radius.

Order Information

The IDEAL bender line gives you the engineering design, indicator marks and durability to bend conduit with ease and confidence.

IDEAL INDUSTRIES, INC.
Sycamore, IL 60178, U.S.A.
800-435-0705 Customer Assistance / Asistencia al cliente / Service après-vente
www.idealindustries.com

Courtesy of IDEAL INDUSTRIES, INC.

Code Index

Note: Page numbers in **bold type** reference non-text material, such as tables and figures.

Index

Note: Page numbers in **bold type** reference non-text material, such as tables and figures.